POWER GENERATION:

Air Pollution Monitoring and Control

POWER GENERATION:

Air Pollution Monitoring and Control

edited by

Kenneth E. Noll
Professor, Environmental Engineering
Illinois Institute of Technology, Chicago

Wayne T. Davis
Assistant Professor, Environmental Engineering
University of Tennessee, Knoxville

ANN ARBOR SCIENCE
PUBLISHERS INC
P.O. BOX 1425 • ANN ARBOR, MICH. 48106

230 Collingwood, P.O. Box 1425, Ann Arbor, Michigan 48106

Library of Congress Catalog Card Number 75-25330
ISBN 0-250-40118-5

Manufactured in the United States of America

Preface

This book provides a comprehensive treatment of air pollution control technology associated with power generation. Methods for the assessment of the environmental impact of power production, which involves atmospheric modeling and monitoring, and for the evaluation of the technology for power plant stack gas cleaning for sulfur and particles are presented. In order to provide a book that can be read selectively and be easily referenced the material is divided into six sections: (1) the present status of air pollution control, (2) prediction of pollutant dispersion, (3) monitoring of power plant effluents in the environment and stack, (4) operation of supplemental control systems, (5) removal of sulfur dioxide from stack gases, and (6) control of particles from power generation.

The section on atmospheric modeling provides methods for determining ground level concentrations including generalized calculation procedures, effects of downwash, calculation of background concentrations, estimation of general levels of pollution in urban areas, and air quality determination from emission data. Models specifically applicable to cooling towers are also provided.

Two comprehensive sections on air monitoring include chapters on design of air monitoring surveys, instrument calibration and operation, methods for plume monitoring with aircraft, and unique methods for continuous particle and sulfur monitoring in stacks as well as a detailed evaluation of monitoring methods for cooling towers. Supplemental control methods are evaluated, and a chapter is provided on TVA experiences in their use.

The section concerned with gaseous control includes detailed economic information on alternate processes for removal of sulfur dioxide from stack gases. Pilot plant experience, the results of process modification for various removal methods, and the effect of chemical additives on efficiency and operational characteristics are described. The section on particle removal discusses electrostatic precipitator application and provides information on the operational efficiency of electrostatic precipitators, which can be compared to present design equations and utilized for the development of new equations.

The effect of particle size and particle resistivity on collection efficiency is also discussed. Six chapters present a comprehensive evaluation of fabric filter application to the collection of fly ash from power plants. The results of pilot and full-scale tests are described, including the effect of various fabrics and different types of fly ash on collection efficiency and pressure drop. Methods for determining the costs associated with the application of fabric filters to control of fine particles from combustion sources and power generation are presented.

April, 1976

Kenneth Noll
Chicago, Illinois

Contents

Section I
Introduction

1

Status of Air Pollution Control

Arthur C. Stern
Department of Environmental Sciences and Engineering,
School of Public Health, University of North Carolina, Chapel Hill

The year 1975 marks the twentieth anniversary of federal air pollution legislation in the United States. Therefore, this may be a good time to see what we have done right and what we have done wrong in air pollution regulation in the United States during these two decades.

The first thing we did right was to remove the stricture that health was the only basis for public concern about air pollution. The expansion of the definition of concern in federal legislation to "public health and welfare, including injury to agricultural crops and livestock, damage to and the deterioration of property and hazards to air and ground transportation" was one of the most significant steps taken. The Clean Air Act includes the following definitive wording: "All language referring to effects on welfare includes, but is not limited to, effects on soils, water, crops, vegetation, man-made materials, animals, wildlife, weather, visibility and climate, damage to and deterioration of property, and hazards to transportation, as well as effects on economic values and on personal comfort and well-being." We take this broader concept of our area of concern for granted today, but twenty years ago to be released from the shackles of proof of injury to health was truly revolutionary.

Another thing we did right was to extend to those charged with the administration of the legislation the broadest possible choice of methods of accomplishing their research, investigative, training, technical and financial assistance responsibilities. Subject only to appropriation, allocation and impoundment restrictions imposed by higher authority, the federal operating agencies have had extreme freedom in deciding how to accomplish these responsibilities, leading to a wide variety of arrangements with a wide variety of institutions. Compared with the restrictions under which other programs have had to operate, this opportunity to fully develop both old and new ways to get things done has been a tremendous boon to air pollution control.

A third thing we did right was to channel much of the federal effort into building up state and local capability in air pollution control. Twenty years ago state activity and capability in air pollution control was almost nil. All the existing legislation did in the state with the greatest problem, California, was to authorize control activity by its counties. The concern and authority of our municipalities was limited to smoke, odor and fly ash. Even before federal financial assistance to state and local agencies was authorized, the authorized technical assistance was used from the start as a consciously manipulated means to cause states to develop their own air pollution control laws and agencies. Initially, this was not popular with the cities, where the then existing expertise dwelled. They felt that since their organizations were in being, the federal agency should work with them, rather than attempting to build up non-existent state authority. A somewhat converse situation occurred after the authorization of federal financial assistance to state and local control agencies, because the fledgling state agencies were not happy with the provision of federal funds directly to their cities and counties, rather than having these funds channeled through the state agency.

A fourth thing we did right was to separate the several forms of federal financial assistance to state and local agencies into various kinds of program grants rather than to establish formula grants that would have made funds available without concern for program content or progress. Other than stipulating the matching requirement, the federal legislation authorizing financial support of state and local air pollution control activities has been silent on the form of such support. I can well remember the soul-searching from which evolved the concept and definition of separate grants for planning, developing, establishing, improving and maintaining state and local programs. It was this concept and these definitions that forced state after state to pass state air pollution control legislation and to establish state air pollution control agencies, which built up the strength in the state agencies we have today. I am sure that few know that the organization of State and Territorial Air Pollution Agencies (STAPA), an attempt to give them increased strength through cooperative endeavor, evolved not from the state agencies themselves, but rather was conceived and initially organized by the federal air pollution agency.

A fifth thing we did right was to open up air pollution sources to inspection and monitoring nationally. Much longer than twenty years ago, there was no right of entry for federal and state agency representatives and only limited right of entry for the representatives of many municipal agencies. It has been the force of relatively recent federal and state air pollution legislation that has opened the gates of otherwise noncooperative polluters to what most of us will agree is proper inspection and monitoring.

A sixth thing we did right was to mandate federal jurisdiction over air pollution from the automobile and the airplane. Although we may well take issue with some of the details as to how this was done, the general idea of having one federal agency assume this responsibility, rather than 52 separate state and territorial agencies, has been for the better.

A seventh thing we did right was to allow citizen suits against actions taken under the federal air pollution act, or against failure to take actions that legally could and should have been taken under the act. The value of allowing such suits has been amply demonstrated by their effectiveness in bringing into sharp focus ambiguously and poorly drafted sections of the act, and administrative actions that deviate from the courts' interpretation of the law.

An eighth thing we did right was to establish a national system of new source performance standards for stationary sources. This has been tantamount to aligning the United States with other countries of the world that adhere to the "best practicable means" of control philosophy for new sources. Prior to the establishment of this system, we had a hodge-podge of a few jurisdictions each applying its own new source emission limit and vast areas of the nation where none was applied. The arguments as to whether there should have been more than one category of new source emission limits, based upon existing air quality, topography, meteorology and demography, or an invariant national emission standard, have been articulated previously and need not be repeated here. However, whether they be multivariant or invariant, the important thing is the "ounce of prevention" nationally applicable new performance standards provide.

Now let's look at some of the things we have done wrong. Over the years, the federal air pollution law, as it has been changed from amendment to amendment, has forced the federal air pollution agency from one misdirected enforcement path to another and to yet still another, each time with a largely wasted expenditure of time, manpower and money.

The first of these was the path of federal action against interstate pollution requiring numerous conferences, hearings, and reports, very few cases ever brought to court and minimal abatement of pollution. Although this path is still a part of the law, it is almost totally ignored today. However, there was a time when it totally absorbed the federal enforcement staff.

The second of these paths was that of the organization and implementation of interstate air quality control regions. This activity during its heyday also totally absorbed the federal enforcement staff and resulted in the delineation and promulgation of the boundaries of a large number of interstate regions, none of which ever functioned as their senatorial proponents envisaged and which remain as parts of the intrastate air quality control regions we have today. The effort expended in setting the boundaries of these regions was an educational process in that it forced us to develop and become accustomed to mathematical models; also, it taught the state and local agency personnel involved techniques and procedures in modeling and holding public hearings that they have had to use subsequently for other purposes. However, it was a very expensive educational process for the nation in the sense that it wasted time, manpower and money, and accomplished no pollution abatement.

The third of these paths is the air quality standard-state implementation plan path, on which we are now discontinuously engaged—discontinuously because of the zigs and zags required by court intervention resulting from citizen suits and litigation initiated by affected industries. The deadline dates, the detailed requirements of the law and these suits have made this path like a board game where you move four spaces forward only to land on a block that says "Go back six spaces and start over again." It is too early to say that this path, like the previous two, has also been a failure, but to date it certainly has not been a resounding success, despite the best efforts of the Environmental Protection Agency (EPA) to make it work. The faults are more with the way the law requires the path to go than with those piloting us along the path. The air quality management path is more appealing in concept than in practice. In other countries of the world, air quality standards are goals piously hoped to be eventually achieved but have a minimal role (except for new source stack height or emission limit computation) in the day-to-day abatement of pollution. Only in the United States and East Germany has the law required air quality standards to be the basis for a complete system of control of existing sources. We spend our time arguing details of the path, such as scrubbers versus intermittent control, sulfates versus sulfur dioxide, catalytic versus thermal afterburners and the exact numerical value of an air quality standard, rather than directing our attention to the validity of the pathway itself.

Another thing that we have done wrong is to have specified in the law the extent of reduction of automobile exhaust gases required to be accomplished, the dates they are to be accomplished, and the dates acceptable abatement of pollution from all sources is to be accomplished. A cardinal principle in drafting air pollution control legislation is to leave the numbers and dates out of the law, leaving them to be set by regulation by whatever procedure the law sets up to establish regulations. We are now paying dearly for the violation of this cardinal principle in our federal legislation.

A third thing we have done wrong is to have set up a system of dual air quality standards rather than a singular standard. Here again was something more attractive in concept than in practice. An air quality standard should be arrived at by a weighting of all societal values affected by it and that affect it. The protection of public health is only one of these values, albeit a tremendously important one. It should not be allowed to preempt the process, particularly since we find it so difficult to define what we mean by "attainment and maintenance of public health."

A fourth thing we have done wrong is to require that once the administrator of EPA has issued an air quality criteria document for a pollutant he *must* follow it sequentially by the issuance of a control technology document and the promulgation of an air quality standard, and *must* call upon the state to submit implementation plans. This has had a stultifying effect on the issuance of air quality criteria documents and the derivation therefrom of air quality standards. If the issuance of a control technology document and an air quality standard were discretionary rather than mandatory on the administrator, following issuance of an air quality criteria document, and if the requirement to submit implementation plans were also discretionary rather than mandatory, the administrator could issue air quality criteria documents and promulgate air quality standards for all pollutants for which he has accumulated sufficient information, instead of the present situation where the sequelae of such action serve as a powerful impediment to his doing so. It is important that the American people be informed of the levels of pollutants that are safe and unsafe by the timely issuance of air quality criteria documents and the promulgation of air quality standards. Impediments to this happening should be deleted from the Clean Air Act.

A fifth thing we have done wrong is to legislate and regulate in such a manner that in many situations land use planning has become subsidiary to air pollution control regulations rather than to have air pollution but one of several factors requiring consideration in determining the optimum use of a parcel of land. In the long run, it is harmful to the cause of air pollution control to have air pollution considerations overwhelm other considerations in situations where this should not be the case. The reason this situation was allowed to develop was our present dedication to the air quality standard-state implementation plan or, as it is more commonly known, "air quality management" path, noted earlier. Conflicts of this type are inherent in blind adherence to the air quality management approach, and this is one reason why we need to examine so carefully the validity of this pathway.

A sixth thing we have done wrong is to have ignored for so long in our air pollution laws and regulations the energy implications of our control requirements, and then when we finally ceased ignoring these implications to have overreacted in the opposite direction. Of all the things we have done wrong, this is perhaps the most understandable because certainly a decade ago fossil fuel usage and air pollution control were universally viewed as implacable opposites that never could be made mutually compatible. Yet we now know that there must be an accommodation between our national needs for energy and air pollution control.

In balance, we seem to have done more things right than wrong. If we are smart enough to keep what is right and correct what is wrong, the future course of air pollution control should be bright.

ACKNOWLEDGMENT

This chapter was first published in the *Journal of the Air Pollution Control Association*, July, 1975.

2

Environmental Impact Statements for Nuclear Power Facilities

William C. Yee
Environmental Engineer, Environmental Assessment Staff

J. Clement Burdick III
Chief, Environmental Assessment Staff

Peter A. Krenkel
Director, Division of Environmental Planning,
Tennessee Valley Authority

INTRODUCTION

With the passage of the National Environmental Policy Act of 1969 (NEPA),[1] federal agencies were directed to give full consideration to environmental factors along with the technical and economic factors in all of their activities. This paper discusses the implementation of the NEPA mandate by one federal agency, the Tennessee Valley Authority (TVA). Since the bulk of the power needs of TVA involve nuclear power plants, most of the following discussion is directed toward the development of environmental assessment procedures for nuclear plants. Attention is focused on projects that were underway at the time of passage of NEPA as well as those which commenced at a later time. Brief comments are also directed toward the assessment procedures employed for nonnuclear programs conducted by the agency.

THE TENNESSEE VALLEY AUTHORITY

The Tennessee Valley Authority is a federal corporation established by Congress in 1933, which is charged with development of the resources of the Tennessee Valley.[2] TVA is active in seven states surrounding the Tennessee Valley covering a watershed area of

41,000 square miles and having a power service area of some 80,000 square miles.

The specific missions authorized under the TVA Act[2] of the agency include the following activities:

1. Flood control in the Tennessee River and its tributaries,
2. Improvement of navigation in the Tennessee River,
3. Generation of electric power "primarily for the benefit of the people of the section as a whole . . . at the lowest possible rates . . . ,"
4. Provide agricultural and industrial development,
5. Encourage reforestation of all lands where practicable,
6. Improve the economic and social well-being of the people,
7. Participate in the interests of the national defense.

To fulfill these missions, TVA has historically employed personnel representing a broad range of educational disciplines. In addition, the overall role of resource development has entailed integral involvement with environmental concerns. To implement the program of integrated environmental assessment required by NEPA, it has been only necessary to restructure the internal organizational relationships between the diverse groups within the agency rather than to develop new professional capabilities.

ENVIRONMENTAL ASSESSMENT PROCESS

As a resource development agency, TVA is engaged in many types of activities which range in size from small research projects to large water control projects and the construction and operation of electric power generation facilities. To properly discharge its NEPA obligations, internal procedures have been developed so that potential environmental impacts and mitigation measures of each proposed activity are considered. These procedures are detailed in the document, TVA Code IX Environmental Quality.[3] If it is determined that a specific activity is a major federal action and will result in significant impacts on the human environment, then an environmental impact statement is prepared for the activity.

A flowsheet is provided in Figure 2.1 which shows the generalized procedure for environmental assessment of projects. The TVA organization in charge of initiating the project is responsible for preparing the Environmental Evaluation Record using inputs as appropriate from other agency organizations and outside consultants. This record and its conclusions are reviewed by the Division of Environmental Planning in consultation with the Division of Law. If it is determined that an environmental statement is not needed, this concludes the environmental review process relative to the need for an environmental statement. Depending upon the information de-

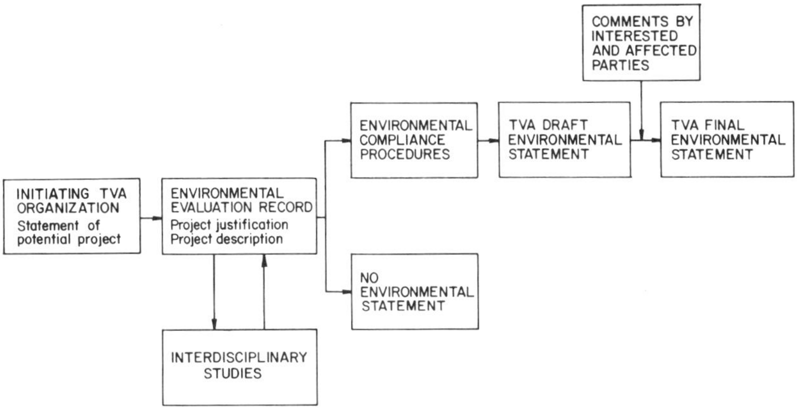

Figure 2.1. Flow diagram for preparation of environmental impact statements for TVA.

veloped, additional project design measures may be instituted at this point to improve the environmental aspects of the project.

If it is determined that a statement is needed, a time schedule for environmental compliance procedures to develop the statement is prepared and is added to the Environmental Evaluation Record. Various TVA organizations participate with the initiating organization in the development of a draft environmental statement. This statement is made available to various federal and state agencies, organizations, and the general public for comments. After evaluation of any comments received, responses are formulated as appropriate. A final environmental statement is then prepared which includes these comments and responses as well as any modifications which have been made in the project design. After internal review, the final environmental statement is filed with the Council on Environmental Quality. This fulfills the NEPA requirements of inclusion of environmental considerations in the decision-making process.

Examples of activities that normally require an environmental statement include: (a) large water resource development and water control projects, (b) new power generating facilities, (c) major transmission system additions, 500-kV and above, (d) addition of new navigation locks, (e) port facilities, (f) new town projects, and (g) system-wide vector control and watermilfoil control programs. Table 2.1 lists by category a number of projects on which TVA environmental statements have been issued.

Examples of activities that normally require only an environmental evaluation record include: (a) routine operation, maintenance, and minor upgrading of existing TVA facilities, (b) minor research, development, and joint demonstration projects, (c) technical and planning assistance to state and local organizations, (d) establishment

Table 2.1. TVA Final Environmental Statements Nonnuclear Programs.

Category	*Project*	*Date of Issuance*
Flood control	Briceville	May 29, 1973
	St. Paul	May 23, 1974
	Brainerd Area Flood Relief	July 12, 1974
Fossil plant modification	Widows Creek	January 15, 1973
Gas turbines	Allen Turbines 1-16	October 29, 1971
	Colbert Turbines 1-8	March 31, 1972
	Johnsonville Turbines 11-16	November 25, 1974
	Gallatin Turbines 1-4	December 20, 1974
Multi-purpose dam	Tellico	February 10, 1972
	Nolichucky	March 10, 1972
	Duck River	May 1, 1972 (Supp. June 18, 1974)
	Bear Creek	December 29, 1972
Reservoir property management	Watermilfoil Control	September 29, 1972
	Vector Control	April 10, 1974
Transmission lines	Shelby 500-kV Line	September 22, 1971
	Widows Creek-West Jefferson 500-kV Transmission Line	November 7, 1974
Waterway development	Yellow Creek Port	November 30, 1971
	Pickwick Lock	September 13, 1972

of environmental quality monitoring programs and field monitoring stations, (e) permits under section 26a of the TVA Act for minor structures, boat docks, and shoreline facilities, and (f) contracts for the sale, purchase, or interchange of electricity. Depending on circumstances, an environmental statement may or may not be required for such activities as major research and development projects and major modifications or additions to existing TVA activities or facilities.

NUCLEAR POWER PROGRAM ASSESSMENTS

TVA has a large commitment to nuclear power. Starting in 1967 with the Browns Ferry Nuclear Plant located in northern Alabama, more than 16,000 MWe of nuclear power at five plant sites are planned as shown in Figure 2.2 and Table 2.2. This is almost equal to TVA's present fossil fuel generating capacity. In addition, two new proposed nuclear plants were recently announced which will add an additional 5000 MWe capacity to the system. Thus, by 1985 it is expected that total nuclear generating capacity will be equal to all of the present combined hydro- and fossil-fuel electric-generating capacity. Figure 2.3 indicates the location of the five light water nuclear power plants of TVA and the proposed location of the Clinch River Breeder Reactor Demonstration Plant near Knoxville.

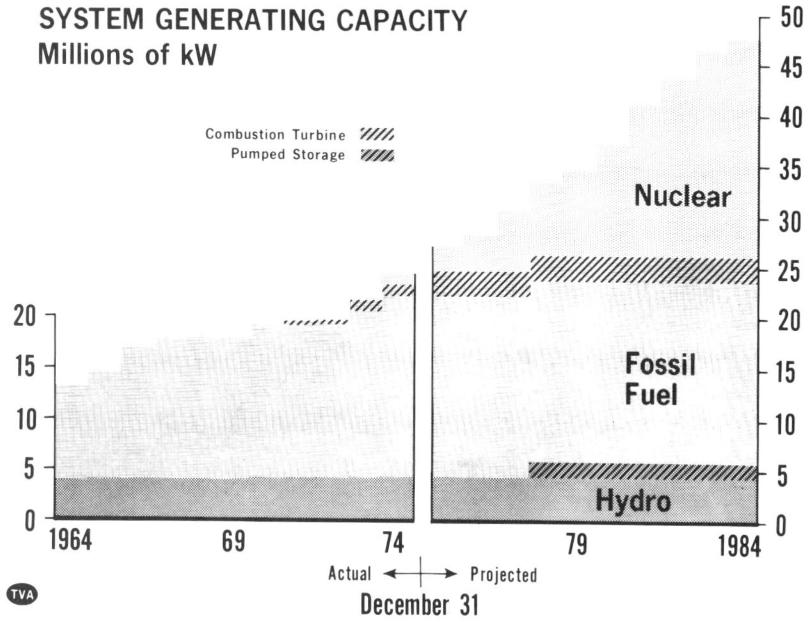

Figure 2.2. Growth of TVA System generating capacity.

The preparation of environmental impact statements and environmental reports is a major activity in connection with planning, scheduling, and licensing the construction and operation of nuclear power plant projects. Environmental review of TVA nuclear projects is unique in that the Atomic Energy Commission (now the Nuclear Regulatory Commission) as the federal licensing agency has NEPA obligations in regard to issuing construction permits and operating licenses, while TVA, also a federal agency, has NEPA obligations in regard to committing the government to building a nuclear plant. To avoid duplication of effort, a "lead agency" arrangement was made with the AEC for TVA's Browns Ferry, Sequoyah, and Watts Bar plants whereby TVA prepared the environmental statement in consultation with AEC. AEC initially limited its review of the statement to radiological matters, but later expanded the review to the complete TVA statement. For the Bellefonte Nuclear Plant, AEC determined that it would prepare its own impact statement for the project. For Bellefonte, the TVA draft environmental statement was submitted to AEC in lieu of the environmental report, pursuant to the provisions of Appendix D to 10 C.F.R. Part 50 which permitted such arrangements where the applicant is a federal agency.[4,5] Due to the voluminous nature of material required for an environmental report which is not essential to an environmental statement, for the pro-

Table 2.2. TVA Nuclear Plant Schedule.

Power Plant Project	*Reactor Type*	*Unit No.*	*Name Plate Power Rating (MW)*	*Total Plant Power (MW)*	*Anticipated Commercial Operation*	*Estimated[a] Cost (Millions)*
Browns Ferry	Boiling water	1	1152	3456	Aug. 1974	$ 815
		2	1152			
		3	1152			
Sequoyah	Pressurized water	1	1221	2442	Sept. 1977	675
		2	1221		May 1978	
Watts Bar	Pressurized water	1	1270	2540	Nov. 1978	805
		2	1270		Aug. 1979	
Bellefonte	Pressurized water	1	1332	2664	June 1980	1,000
		2	1332		March 1981	
Hartsville (proposed)	Boiling water	1	1287	5148	Dec. 1981	2,500
		2	1287		Dec. 1982	
		3	1287		June 1982	
		4	1287		June 1983	
X21 and X22 (considered for eastern area)	Boiling water	1	1287	2574	March 1985	1,600
		2	1287			
X24 and X25 (considered for western area)	Pressurized water	1	1407	2814	June 1985	1,900
		2	1407			

[a]As submitted to Congress January 1975.

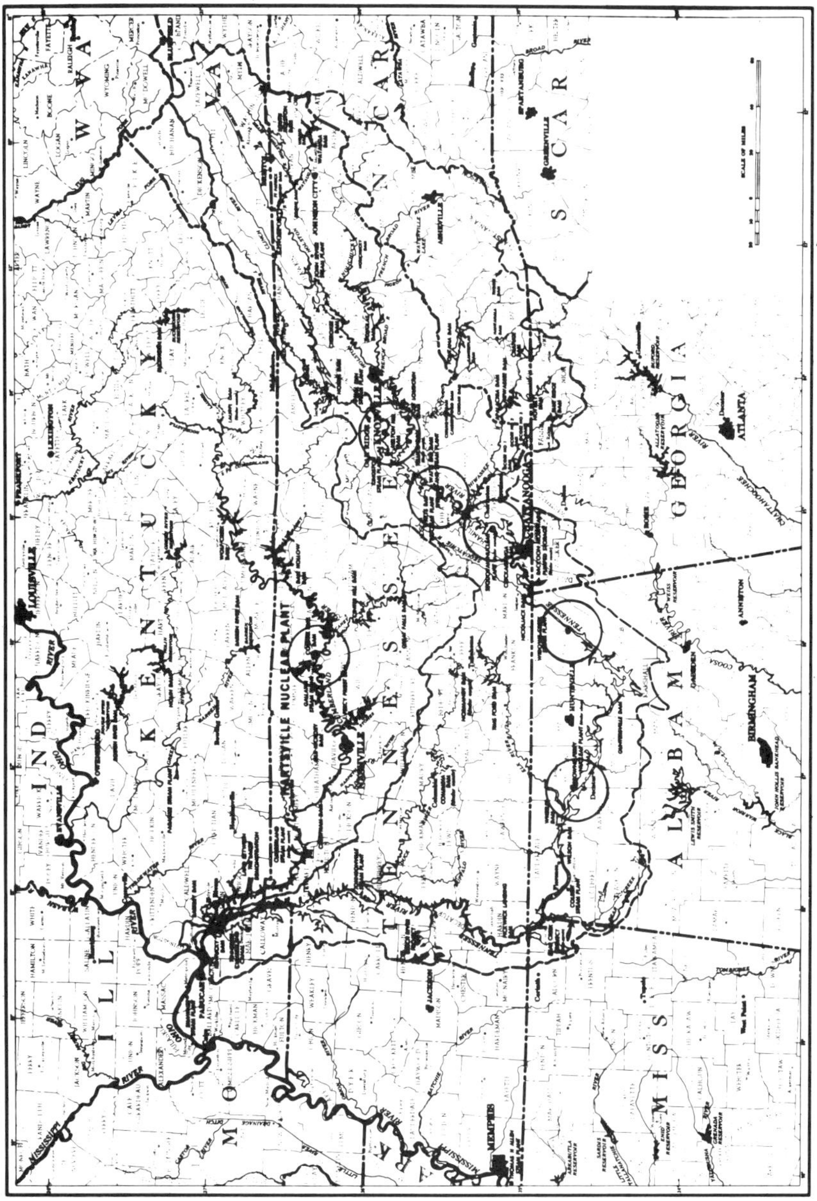

Figure 2.3. Nuclear power plants in the Tennessee Valley.

posed Hartsville Nuclear Plants TVA submitted an environmental report to AEC and also prepared a TVA environmental statement for the project.[6,7]

The Energy Reorganization Act of 1974 (42 U.S.C. 5801 *et seq.)* abolished the Atomic Energy Commission and gave the responsibility for nuclear plants to the newly established Nuclear Regulatory Commission (NRC). Currently, the relationship of NRC with TVA is the same as with the former AEC as explained in the previous paragraph. Figure 2.4 summarizes the presently existing licensing procedures employed by TVA/NRC.

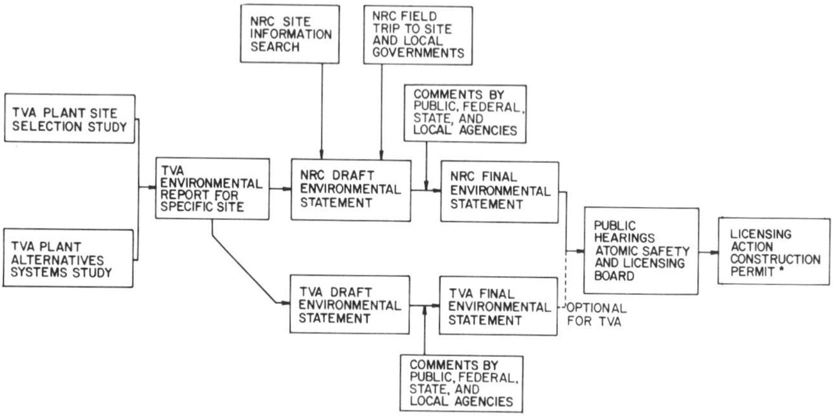

Figure 2.4. Flowsheet for NRC licensing action on nuclear power plants. Note: Environmental Review for operating licenses are conducted by NRC and TVA only to the extent that conditions change from those discussed in the NRC Final Statement at the construction permit stage.

EARLY NUCLEAR PLANT ENVIRONMENTAL STATEMENTS

Early nuclear plant environmental statements of TVA were developed in the transition period associated with the implementation of NEPA. Commitments to build had been made and construction activities were already underway on two plants, Browns Ferry and Sequoyah, when NEPA became effective. As mentioned previously, these two plants and the Watts Bar Plant received environmental analysis under a lead agency agreement with AEC. For these two plants, since construction impacts were largely completed when NEPA was enacted, attention in the statements was directed largely toward operational impacts. Of the many issues considered two will be discussed subsequently: —the effects of radioactivity on man, and the effects of thermal effluents on the biota of receiving waters.

Problems in development of these three early statements also occurred due to the rapid changes taking place in state and federal guidelines for both radwaste and thermal discharges, which became progressively more stringent in a comparatively short period of time. Other legislation enacted during this time frame produced new standards, criteria, or guidelines in other areas of environmental concern, *i.e.*, air, water, pesticides, noise, and solid wastes. Modifications in plant designs became necessary, even for those plants already under construction, in order to remain in compliance.

Browns Ferry Project

The case of TVA's first nuclear plant, Browns Ferry, located in northern Alabama, is an example of design modification due to changing environmental criteria.[8,9] Detailed designs were completed for gaseous and liquid radwaste systems and once-through method of cooling was chosen for dissipating the plant waste heat. Environmental analyses were complete and construction activities were underway. Specifically, the multi-part diffusers for heated water discharge were being installed. Then, due to the promulgation of lower permissible discharge concentrations of radioactivity and more stringent standards for thermal effluents, it was necessary to backfit the system design and repeat the impact analysis.

With regard to potential radioactivity releases, the AEC amended its regulations in 1970 to require that all radioactive emissions would be "as low as practicable."[10,11] In June 1971, AEC initiated a rulemaking proceeding to establish numerical design objectives for meeting the "as low as practicable" regulation for light-water-cooled nuclear power reactors.[12,13]

At Browns Ferry this resulted in modifications in the gaseous and liquid radwaste treatment systems to incorporate additional decontamination equipment. Hydrogen recombiners were added to reduce radiolytic gas volumes generated by the boiling water-type reactor operation. Charcoal beds were added to sorb and store nonnoble radioactive gases like iodine to allow their activity to decay. Additional treatment of liquid radwaste included evaporation for concentration into a low-volume package, which is more amenable for transporting to a long-term radioactive storage facility. These modifications resulted in a reduction in the release of gaseous and liquid radioactivity by a factor of 100. With this expanded treatment, the exposure of a man continuously located at the site boundary due to plant operation was calculated to be less than two percent of the dose due to natural background radiation, an exposure that is indistinguishable from background activity. Much effort during this time period was involved in the evaluation of the design, the redesign, reevaluation of radiological impact, and documentation of the methodology used in the derivation of the dose models employed.

This approach now appears to be overly conservative as demonstrated by the definition of the "as low as practicable" in the regulation as adopted.[4]

A considerable effort was also expended in the area of thermal discharges from the Browns Ferry plant into Wheeler Reservoir (Tennessee River). In compliance with the Water Quality Act of 1965,[14] the state of Alabama proposed[15] that cooling water discharges should not increase the ambient temperature of receiving water more than 10° F. nor should the temperature of the receiving water exceed 93° F. Prior to this time TVA had determined from experience at its Paradise coal-fired steam plant that once-through cooling would be an acceptable alternate at Browns Ferry. Further studies were conducted using physical and mathematical models to simulate the behavior of the heated discharge in the reservoir. Based upon these studies, it was found that a diffuser system could be designed to achieve sufficient mixing of the plant discharge with the river to remain in compliance with the proposed Alabama temperature standards. Installation of this system was therefore undertaken.

The Environmental Protection Agency proposed a more restrictive set of temperature standards for the state of Alabama which was published in the *Federal Register* of March 11, 1972.[16] In effect they reduced the temperature rise above ambient to no more than 5° F. and the maximum temperature to 86° F. Under these circumstances, it was determined that a diffuser system alone would not be adequate to ensure year-round compliance with the new standards and that supplemental cooling with mechanical draft cooling towers would be needed during periods of low river flow or high seasonal temperatures, such as in late summer. Figure 2.5 is an artist's conception of the Browns Ferry plant showing the cooling towers employed. This resulted in changes and additions to the environmental analysis which was incorporated in the final environmental statement to describe and evaluate new impacts such as new land requirements for the mechanical draft towers, modifications to the intake channel, and new assessments on fogging, drift, and liquid blowdown discharges related to the operation of the towers. It also required the analysis of impacts due to three modes of dissipating waste heat: once-through cooling alone, once-through with mechanical draft towers operating in a helper mode, and closed-cycle cooling with mechanical draft towers alone.

Sequoyah Project

Radioactive releases and thermal discharges were also of interest in the development of the environmental statement for the Sequoyah nuclear plant.[17] In this case, the concern was for gaseous releases of

Figure 2.5. Artist's conception of Browns Ferry Nuclear Plant with supplemental cooling towers.

radioactivity from a pressurized water reactor system (PWR) and the resultant potential annual dose to man. The alternative chosen to meet the "as low as practicable" criteria was to increase the storage time for gaseous radwaste thus extending the decay time from 45 to 60 days before releasing to the atmosphere. Auxiliary cooling capacity was added to the once-through cooling system to meet EPA approved, state of Tennessee temperature standards.[18] The more costly natural draft towers were selected in preference to mechanical draft towers because taller tower stacks improve plume dispersion and minimize potential impacts of ground fog considering meteorological conditions at the Sequoyah site.

Watts Bar Project

At the Watts Bar Plant[19] it was found that waterflow past the plant site (located two miles below Watts Bar Dam on the Tennessee River) was not adequate the year round for once-through cooling. In fact, no flow conditions existed about 10% of the time, on an annual basis, due to the required operation of the multi-purpose reservoir system. Thus, closed cycle rather than open cycle cooling was required, and natural draft cooling towers were selected for this application.

LATER NUCLEAR PLANT ENVIRONMENTAL STATEMENTS

For the Bellefonte and Hartsville projects, the TVA environmental impact statements were developed using the AEC Regulatory Guide 4.2 (now superceded by NRC Regulatory Guide 4.2, Revision 1, "Preparation of Environmental Reports for Nuclear Power Stations"). This guide is used as a reference work by all electrical utilities who submit environmental reports to AEC(NRC) as part of the nuclear plant licensing process. Figure 2.4 describes the process in effect for these two projects.

According to the guide, construction impacts (including socioeconomic considerations) should be given full treatment along with impacts of plant operations. Environmental impact analyses should be conducted based on at least one year of baseline data collected in an integrated environmental monitoring program at the specific site. In addition, discussion should be included on comparison of alternative sites to the nuclear plant site chosen; alternative feasible sources to the electric power that would be supplied by a specific nuclear plant; and alternative systems for treating plant effluents. Uniform benefit-cost considerations should also be established to quantify the various alternatives where possible so that a more objective balancing of costs and the benefits of a project can be prepared.

Bellefonte Project

The draft environmental statement for Bellefonte[20] prepared by TVA was accepted by the AEC as the equivalent of an environmental report from an electrical utility. As part of the independent review and assessment, AEC staff also visited the site and contacted state and local government representatives. Final environmental statements were issued on this project by two federal agencies, one by TVA and one by AEC. Both were utilized in the public hearing before the Atomic Safety and Licensing Board which resulted in the issuance of a construction permit to build the nuclear plant in December 1974.

Some points of interest relative to this case were of a local nature and included (a) mortality of fish eggs and larvae ("entrainment" mortality) in river water that is utilized in the plant coolant system, (b) access routes for roads, the railroad, and transmission lines, and (c) treatment of vegetation along transmission line rights-of-way.

To determine entrainment mortality in relation to the fish population in Guntersville Reservoir (Tennessee River), a monitoring program has been developed in which data on selected fish species will be gathered by age class in the region around the plant site. Access routes which minimize terrestrial impacts were selected by TVA and approved by AEC. Experimental right of way clearing programs are being used to establish another view of costs, impacts and public acceptance of these techniques.

Hartsville Project

The Hartsville nuclear plant has been called the "world's largest power plant." It is the largest single project proposed by TVA at one site totaling almost 5,000 MWe for the four nuclear units. Figure 2.6 provides an artist's conception of the Hartsville plant.

Experience gained in mitigation of socioeconomic impacts to local communities on previous nuclear projects undertaken by TVA has been employed to develop mitigation strategies for the Hartsville project. It is estimated that the number of construction workers at the building site will increase rapidly after the start of the project and peak at about 5000 people. Of this peak number, it is expected that 2700 workers will move into the vicinity of the construction site. These people together with their families will increase the population by about 6100 in the five-county area surrounding the site. This situation could tax certain community services such as public school education to the limit of their capability during the construction period. After construction activities subside, it is expected that the permanent operating personnel might number about 350–400.

In its final statement on the Hartsville project, TVA has identified the probable socioeconomic impacts due to construction activities

Figure 2.6. Artist's conception of Hartsville Nuclear Plant.

and has outlined what steps will be taken to mitigate these impacts. The areas of impacts and mitigation commitments discussed in the final statement included the following six broad categories: (a) housing, (b) education, (c) water and sewer facilities, (d) local government budgets, (e) health and medical services, and (f) transportation facilities.

ENVIRONMENTAL ASSESSMENT AND MONITORING PROGRAMS FOR NUCLEAR PLANTS

Because of its role as a resource development agency, TVA has historically developed capabilities for data acquisition to support its regional development programs. In addition, environmental improvement programs have been part of the activity of a number of organizations within TVA. In connection with the preparation of environmental impact statements, resources of the agency have been expanded to serve additional needs. This includes the development of an in-house capability (a) to assemble essential base-line data for proposed power projects, (b) to make assessments of environmental impacts that may result from the construction and operation of a specific power plant, and (c) to verify the projected environmental impacts with an integrated monitoring program in the region of the plant site.

Development of Integrated Monitoring Programs

Prior to the preparation of the Bellefonte environmental statement, specific monitoring programs were devised to verify specific assessments of operational impacts, such as radiological effects to man, and effects of thermal discharges to the aquatic biota. Since the Bellefonte plant was the first completely post-NEPA project, it provided the opportunity to prepare an integrated monitoring program that verified major anticipated impacts both during construction and during operation of the proposed plant. Development of this program required an interdisciplinary approach involving many of the organizational groups within TVA including, for example, the Divisions of Environmental Planning; Forestry, Fisheries, and Wildlife Development; Water Management; Agricultural Development; and Navigational Development and Regional Studies; and the Offices of Power; and Engineering Design and Construction.

The Bellefonte project was an interorganizational monitoring effort which allowed a rationale to be developed for a nonradiological environmental monitoring program[20] for both construction effects and effects of the operation of the plant. A major effort was devoted to a monitoring program for construction effects because the TVA application for a license to AEC was for a construction permit. Since this was an attempt to develop a monitoring program with limited field data at the Bellefonte site, it was agreed by TVA and AEC that the program would be periodically reviewed and various elements would be added or deleted by common agreement.

Details of the plan were very specific to the Bellefonte site. Sampling points, frequency of sampling, and analysis and interpretation were coordinated among the different TVA organizations.

The monitoring program was divided into monitoring for terrestrial impacts and for aquatic impacts. Terrestrial monitoring was designed to evaluate the following hypotheses:

1. Land use changes associated with construction and operation will result in changes or losses of agricultural crops and livestock, wildlife, wildlife habitat, forested areas, hunting opportunity, and nonconsumptive recreational activities.
2. Transmission line construction, operation, and maintenance will result in changes of losses of agricultural crops and livestock, wildlife, wildlife habitat, forested areas, hunting opportunity, and nonconsumptive recreational activities.
3. Operation of the facility may result in accumulations of toxic materials in plant and animal tissues or soils.
4. Operation of the facility may alter moisture regimes of natural ecosystems.

Construction impacts were treated as primary and secondary effects. An example of primary impact is modification or alteration of habitat by blasting, drilling, excavating, and clearing. Related secondary impacts might include increases in housing developments and trailer courts, use of recreation facilities, and use of available transportation facilities, such as railroads, barges, and automobiles. Vegetation and vertebrate parameters used for both primary and secondary impacts are to be monitored by observing changes in the nature of selected species of plant and animal life of the area. Aerial photography is to document changes in land use on the site and the surroundings.

Operational impacts are related mainly to elevated discharges from cooling towers. Possible effects include (a) changes in air and soil moisture, (b) changes in plant growth, (c) damage to foliage because of icing or accumulation of salts, (d) changes in incidence of disease or insect infestation, and (e) changes in the structural diversity of plant or vertebrate communities, including agricultural crops and livestock.

Aquatic monitoring for construction impacts included documentation of the long-term loss of habitat due to the construction of intake and discharge structures, channels, etc. Short-term effects were documented for the effects of turbidity and local siltation on phytoplankton and macrophytes, and benthos. No fish monitoring was undertaken because of the dynamic nature of fish distributions and the difficulties of separating construction effects from the natural variation in fish populations.

Operational monitoring of aquatic impacts included the following areas of activity: (a) heated water discharges, (b) impingement of organisms on trash-collecting screens, (c) entrainment of biota, and (d) discharge of chemicals.

After several years of field experience at industrial sites several changes in the philosophy of environmental monitoring have emerged. Instead of attempting to comprehensively monitor all changes in terrestrial impact, a greater emphasis has been placed on monitoring endangered or threatened species and on monitoring sensitive terrestrial ecosystems or habitats. Similarly, some procedures have been modified for monitoring aquatic impact. For example, it was originally proposed to conduct siltation monitoring. Further considerations indicated that siltation measurements using presently available techniques cannot be correlated from one set of data to another. In addition, siltation effects on biological systems are difficult to evaluate. Consequently, for the present, simple turbidity measurements will be used in conjunction with data on selected aquatic species to evaluate the short-term turbidity effect on aquatic species due to construction activities.

Cooling Tower Heat Dissipation Systems

Cooling towers have become an accepted alternative to once-through cooling for dissipating waste heat from power plants. They offer the advantage of utilizing less river water flow and thereby diminish the potential for entrainment mortality of aquatic species that is commonly encountered with once-through cooling.

In a typical operation, river water is used to transport waste heat from the power plant condenser to the cooling tower where heat is discharged to the atmosphere by evaporation. Remaining water along with new river water is recycled through this heat removal loop. Dissolved solids of the water in the loop are higher than that in water withdrawn from the river because of the evaporation process. But dissolved solids are maintained at a constant level by discharging a fraction of the total flow as "blowdown" and adding river water as "makeup" water to compensate for volume losses to blowdown and evaporation.

The potential environmental problems of cooling towers are linked to meteorological conditions and terrain in the vicinity of the plant site. Local ground fogging and icing can become worse when the air temperature and wind speed are low and the relative humidity is high. Under these conditions, the moisture-ladened plume from the cooling towers tends to settle rather than rise and dissipate in the atmosphere. This can cause prolonged hazardous driving conditions on nearby highways and greater potential frost damage to plant equipment such as electrical transmission lines and to surrounding vegetation.

Another area of concern is cooling tower drift. This consists of dissolved solids that are entrained with the plume of water vapor emitted from the tower. Evaporation during atmospheric transport can result in the generation of solid particulates which fall back to earth, and can have an impact on the terrestrial habitat in the region of the power plant. Some of the more significant parameters that affect the degree of impact include the following: composition of dissolved solids in the river water utilized; degree of concentration of the river water in the power plant condenser/cooling tower loop; height of cooling tower; and meteorological conditions.

Water chemistry is an important consideration in the assessment of drift impacts. Drift particulates resulting from the use of relatively clean river water tend to have less terrestrial impacts than if the river water used for cooling contains industrial, municipal, and agricultural wastes (particularly heavy metals) that had been discharged upstream of the power plant. Water in the condenser/cooling tower loop should not be recycled to the point where the dissolved solids have been concentrated to their solubility limits. The season of the year and amount of rainfall can also significantly influence the composition of the river water utilized and the drift particulates that are deposited.

Meteorological conditions in the region of the plant site and the nature of the surrounding terrain can influence the choice of cooling tower. A natural draft cooling tower is characterized by a tall stack (500–600 feet high) which retains larger droplets in the tower, permits the release of smaller droplet sizes at a greater height, and results in the dispersion of particulates over a greater area. A mechanical tower is considerably lower in height ($\sim$ 100 feet high) which results in the release of a greater mass flux of drift particulates including the larger droplets and higher deposition rates in the proximity of the tower. Where meteorological conditions are favorable, drift concentrations from mechanical draft towers at the site boundary and beyond can be environmentally acceptable even though such concentrations are higher than that for natural draft towers. Under these circumstances, mechanical draft towers may be selected on the basis of lower capital costs.

One of the many environmental factors in the choice between mechanical and natural draft cooling towers is an assessment of drift impacts on the terrestrial habitat. This is a two-step process consisting of prediction of the distribution of drift over the site and its surroundings, and an evaluation of the effect of the drift constituents on crops and livestock (if the area is predominantly agricultural land) and on foliage and wildlife (if the area is in its natural state). A number of predictive models of varying degrees of sophistication are available in published form which predict drift distribution. One of these is described by P. R. Slawson and T. M. C. Wigley[22] relative to their drift analysis of a TVA project. Input required for this model includes local meteorological conditions and technical specifications of cooling tower operation. One useful output of a mathematical model can be drift isopleth concentrations as a function of distance from the tower. All readily available models have yet to be confirmed by actual field data.

A procedure for drift assessment is the following: (a) define the land use pattern around the plant site boundary, (b) select chemical elements of interest (*i.e.*, heavy metals such as cadmium, or micronutrients such as copper) and determine the ratio,

$$\frac{\text{specific element}}{\text{dissolved solids} + \text{suspended solids}}$$

using at least one year of baseline data on the chemical composition of the river water that will be utilized in the cooling tower system, (c) choose a maximum value of drift concentration at the site boundary for a time period when conditions of environmental stress might be high (*i.e.*, during the autumn season just before harvesting crops on agricultural lands, when, typically, there is little rainfall to dilute effects of deposited drift particulates on vegetation), (d) estimate the concentration of a specific element in the deposited drift particulates by assuming it to be proportional to the ratio in (b), and (e) compare this concentration to any field data on drift studies that have been reported in the literature.

To date, very little field data are available on the effects of drift particulates on terrestrial plants and animals. Extensive studies are planned for large cooling tower installations when they are completed and are placed in operation (*e.g.*, natural draft towers at Potomac Electric Power Company's plant at Chalk Point, Maryland, in fall 1975[23] and mechanical draft towers at the TVA nuclear plant at Browns Ferry, Alabama commencing in 1976). In the interim period, publications in similar fields of research may be used to estimate potential impacts. Such information includes limiting values of certain elements in foliar sprays and normal concentrations and limiting values of selected chemical elements in soil such as those shown in Table 2.3.[21,24,25]

Recently, modeling techniques were employed by TVA to compare environmental effects of drift from natural draft and mechanical draft towers using the technique outlined above. The drift isopleth concentrations were a useful aid in identifying the tower system of

Table 2.3. Possible Limiting Application Rates of Selected Elements to Avoid Biological Effects Due to Cooling Tower Drift Based on Literature from Related Fields.

	Suggested Limits		
		Trace Element in Soil	
Chemical Element	*Foliar Sprays Recommended Concentration (lb/ac. yr)*	*Limiting Value (lb/ac. yr)*	*Normal Concentration (lb/ac. yr)*
Micronutrients			
Boron	0.1-0.5[24]	—	14-260[21]
Copper	0.3-0.5[24]	—	20-160[21]
Manganese	1-2[24]	—	40-6000[21]
Zinc	0.3-0.7[24]	1000[25]	20-600[21]
Trace elements			
Cadmium	—	100[25]	0.02-0.4[25]
Chromium	—	—	—
Lead	---	4000[25]	0.2-400[25]

choice so that drift effects offsite are minimal. The entire assessment process discussed above is therefore a useful tool for making preliminary judgments on the type of cooling tower to use. Monitoring programs described in the previous section are being developed to verify the predicted drift distribution and its effects on the terrestrial habitat.

ACKNOWLEDGMENT

Input and assistance on this paper by Samuel H. Calhoun, Ronald B. Maxwell, and Thomas A. Wojtalik of the Environmental Assessment Staff of the Division of Environmental Planning of the Tennessee Valley Authority is most gratefully acknowledged.

REFERENCES

1. National Environmental Policy Act of 1969, Pub. L. 91-190, 91st Congress, Sess. (1970), 83 Stat. 852.
2. Tennessee Valley Authority Act of 1933 [48 Stat. 58, as amended, 16 U.S.C. paragraphs 831-831dd (1970; Supp. III, 1973)].
3. Tennessee Valley Authority Code IX, Environmental Quality Management, *Federal Register* **39**, No. 32, 5671 (February 14, 1974).
4. "Tennessee Valley Authority Final Environmental Statement, Bellefonte Nuclear Plant Units 1 and 2" (May 24, 1974).
5. "Final Environmental Statement, Bellefonte Nuclear Plant Units 1 and 2, Tennessee Valley Authority," Docket Nos. 50-438 and 50-439, U.S. Atomic Energy Commission, Directorate of Licensing (June, 1974).
6. "Tennessee Valley Authority Final Environmental Statement, Hartsville Nuclear Plants" (May 23, 1975).

7. "Final Environmental Statement, Hartsville Nuclear Plants, Tennessee Valley Authority," Docket Nos. STN 50-518, STN 50-519, STN 50-520, and STN 50-521, U.S. Nuclear Regulatory Commission, Office of Nuclear Reactor Regulation (June, 1975).
8. "Tennessee Valley Authority, Draft Environmental Statement, Browns Ferry Nuclear Plant Units 1, 2, and 3" (July 14, 1971).
9. "Tennessee Valley Authority, Final Environmental Statement, Browns Ferry Nuclear Plant Units 1, 2, and 3" (September 1, 1972).
10. Appendix D to 10 CFR Part 50, September 9, 1971, Supp. 71-74, "Annual Report to Congress for 1970."
11. Amendments to 10 CFR Part 50 (35 FR 18385), December 3, 1970. See pp. 89 and 90 "Annual Report to Congress for 1970."
12. Proposed amendment to Appendix D of 10 CFR Part 50, published for comment (December 1, 1971) (36 FR 22851).
13. Appendix I 10 CFR part 50, *Federal Register* **40**, No. 87, 19439 (May 5, 1975).
14. Water Quality Act of 1965, Public Law 89-234, 89th Cong. 1st Sess. 79 Stat. 903 (1965).
15. Water Quality criteria adopted by Alabama Water Improvement Commission (May 5, 1967). Addendum (June 19, 1967).
16. Environmental Protection Agency (40 CFR, Part 120). "Interstate Waters of State of Alabama, Proposed Water Quality Standards," *Federal Register* **37**, No. 49, 5260 (March 11, 1972).
17. "Tennessee Valley Authority Final Environmental Statement, Sequoyah Nuclear Plant Units 1 and 2" (February 13, 1974).
18. "General Water Quality Criteria for the Definition and Control of Pollution in the Waters of Tennessee," Tennessee Water Quality Control Board (December 17, 1971), p. 5.
19. "Tennessee Valley Authority Final Environmental Statement, Watts Bar Nuclear Plant Units 1 and 2" (November 9, 1972).
20. "Tennessee Valley Authority Draft Environmental Statement, Bellefonte Nuclear Plant Units 1 and 2" (March 6, 1973).
21. Krauskopf, K. B. "Geochemistry of Micronutrients," In *Micronutrients in Agriculture,* (Madison, Wisconsin: Soil Science Society of America, 1972).
22. Wigley, T. M. L., and P. R. Slawson. "Drift Analysis for the Hartsville Nuclear Power Plant," report supplied to the Division of Environmental Planning, Tennessee Valley Authority, Chattanooga, Tennessee (September, 1974).
23. Pell, J. "Chalk Point Cooling Tower Project," *Cooling Tower Environment 1974*, ERDA CONF-740302 (May, 1975).
24. Wittwer, S. H. "Foliar Application of Nutrients," *Plant Food Rev.* No. 2 (1967).
25. Kloke, A. "Lead-Zinc-Cadmium Enrichment in Soils and Plants," *Staub* **34**, 20-24 (1974).

3

Eliminate Source Emission Codes for Coal-Refuse Fired Power Plants

Frederick A. Brunner
Sverdrup & Parcel and Associates, Inc., St. Louis, Missouri

Converting coal-fired, steam-electric power plants to operate on both municipal refuse and coal has become an item for serious national consideration. As large stationary sources, most power plants were early subjects for air quality standards and performance surveillance. In a relatively short period of time, the following operational changes have been suggested: (a) convert from coal to natural gas fuel (and back again, to a limited degree); (b) increase the stack height; (c) install more efficient electrostatic precipitators and other control devices; (d) obtain low sulfur-content fuel; and (e) remove gaseous sulfur emissions. Yet increased generating capacity is necessary to meet public demand. Over the last four years, the concept of dual-fuel (coal and refuse) conversion of power plants has been increasingly heralded as not only salvation from mounting solid waste problems but also as a means of using a fuel 15% lower in sulfur content than coal.

The enthusiastic endorsement showered on such conversion is understandable, but somewhat naive and premature. The heart of the endorsement is public support for any program that recycles waste resources—such as steel, aluminum, or glass—and uses the combustible waste portion for power generation.[1] It is mainly supported as a means of making a more manageable waste product by reducing the volume by 95%. Radio, television and news articles develop a large measure of the support. Yet economic justification does not always fully exist. Despite the fact that sanitary landfills are not always properly managed, many large metropolitan area landfills are still the most economic means of solid waste disposal. Wastepaper should, in fact, be added to the list of recoverable items if resource recovery is a goal worthy of support, since 34% of the total energy used to produce it is recovered but less than that amount of energy is required to recycle it. Since wastepaper prices fluctuate, the worth of paper recycling versus fuel utilization, therefore, is subject to change.

The difference which *is* worthy of public support is that in using the refuse as a fuel, priority is given to producing heat, and mass and volume reductions become secondary. Public support will be necessary for developing a fuel economy program while also achieving waste volume reduction through a process for which there is presently limited performance data. Existing power plant capabilities and state and federal air quality codes presently limit the potentially realizable benefits.

To properly assess the ramifications of these constraints, the design and operational risks and the uncertainties of undertaking to convert a power plant will now be reviewed. While conversion to dual-fuel has been under investigation for some time,[2-4] the first operational data was reported in 1975.[5] The ramifications, therefore, will be expressed in terms of calculated values (certainties), unverified values (uncertainties), and anticipated risks.

A number of design and operational studies have shown it is practical to have air-classified solid waste firing rates equal to 15–30% of the boiler's total heat-input requirements. Twenty percent replacement is generally accepted as an ideal target level, which is equivalent to about 38% replacement by weight. Stated another way, 2–3 lb of refuse are the equivalent energy replacement for 1 lb of coal. Thus for every ton of coal not mined, shipped and burned, 2–3 tons of municipal refuse are burned for energy input and volume reduction. These are obviously desirable benefits since refuse has only a 0.30% sulfur content—a considerable reduction in overall sulfur emissions.

Some of the unpublicized design data, however, are not quite as positive. Volumetrically, the refuse before burning is about twice as bulky as the equivalent weight of coal. Therefore, it requires greater storage area, which causes operational problems. In brief, since 2–3 times as many pounds of refuse are required to replace each pound of coal, and refuse is twice as bulky as coal, about five times as much refuse storage space is required than for the replaced coal. The leachate from both stored refuse and coal must be controlled and treated. Table 3.1 is a waste characteristic comparison between one type of coal and a particular municipal refuse.[5] Since refuse is subject to putrefaction, it cannot be stored for days and must be a "first-in, first-out" arrangement.

Table 3.1. Waste Characteristics of Coal and Refuse.[5]

Waste Parameter	*Coal*	*Refuse*
BOD	0 mg/l	-250 mg/l
COD	0	1000–1800
Dissolved oxygen	10–15	10–13
Dissolved solids	350–450	400–600
Suspended solids	0–40	100–300
Settleable solids	0	2–3
Oil and grease	0–5	50–75
Chloride	20–25	25
Nitrate	0–3	2–8
Iron	1	6–8
Copper	0–0.01	0.1–0.3
Lead	0	0.5
pH	8–8.5	8–9

The ash content of coal and refuse is considerably different. A typical low-sulfur content coal will have an ash content of about 9.4%, 80% of which will be fly ash. The remaining 20% is bottom ash. A "typical" classified municipal refuse fuel has an ash content of about 18%, 50% of which is fly ash and 50% is bottom ash. This coal replacement may appear to represent a reduction in the quantity of generated fly ash, but because of the 2:1 refuse-for-coal ratio and the higher refuse ash content, there is an approximate 40% increase in fly ash.[6] Based on a refuse replacement of 20% in the energy input, the *overall* fly ash load will increase by about 33%, and the overall bottom ash will increase by about 300%.

Both the benefits and the problems are "certainties" associated with the dual-fuel conversion system. There are methods of handling the associated waste from the conversion.

Because of the greatly increased quantity of bottom ash, the ash sluicing operation will triple. Since there is some uncertainty whether total combustion takes place, the bottom ash may have to be disposed of in a sanitary landfill and the sluicing water treated in a biological or physical/chemical manner. These facilities are not a part of most existing power plants. The fly ash collection system is, however. Generally, present-day plant systems consist of an electrostatic precipitator (EP). These were designed (more empirically than theoretically) for a certain type of coal and a certain plant operation. (This chapter will not explore the results of reduced fuel sulfur content since that has been more adequately handled by others.[7,8]) Increased collection efficiency, better electrical control, increased collection plate area, reduced migration velocity, and design redundancy are the results of improved EP design over the last 20 years.[9] An existing system, however, has all of these somewhat fixed when it is installed. In many cases, the electrical control system has been updated and improved, but space limitations prevent most other modifications. Even 20-year-old EP systems are efficient enough to satisfy various state and federal A.Q. regulations, however.

Now these older plants are seeing a new life. They can continue to provide peak power while using refuse fuel. Yet, the existing dust-collection system will now be called upon to handle an increase in fly-ash quantity, moisture content, temperature, gas volume, face velocity, particle size, and electrical resistivity, none of which were anticipated when they were designed. The expected result is a loss of collection efficiency. Then the only way to meet the emission codes will be to install a new system designed to deal with the new fuel and EP equipment improvements, which will conflict with the initial benefit of using a *slightly* modified existing plant. Most EP systems have a 30-year life, and the bulk of the EP units that will be affected by the dual-fuel conversion are now 20 years old, or younger. The remaining 10 years of useful life is not long enough to amortize the new equipment costs.

Some states have sought to establish a separate emission regulation for dual-fuel power plants based on combining the incinerator regulation for the refuse portion with the power plant regulation for the coal portion. A review of the incinerator and coal-fuel regulations shows they are similar. For the projects reviewed at a 20% energy-input replacement, the state combined-emission regulations require 96.7% or better collection efficiency. These same projects, however, seem to be operationally capable of achieving a collection efficiency of only 85–92%. These results are somewhat borne out by an operational study.[5] A figure depicting increased EP plate area according to the design year was reworked by inverting and superimposing it on a figure depicting decreasing EP migration velocity according to the design year.[8] The expected loss of efficiency due to fuel conversion was added to this composite figure (Figure 3.1). Thus it appears that the changed fuel and operational conditions are equivalent to a time and design efficiency regression.

Therefore, while it is desirable to use existing power plants for dual-fuel operation, plant systems should not be totally replaced but used to the maximum extent possible. The risk of halting operation due to exceeding state emission standards should be eliminated and recognition given to existing equipment limitations by testing each anticipated fuel-converted plant solely according to the limits of the ambient air standards in each specific area plant setting.

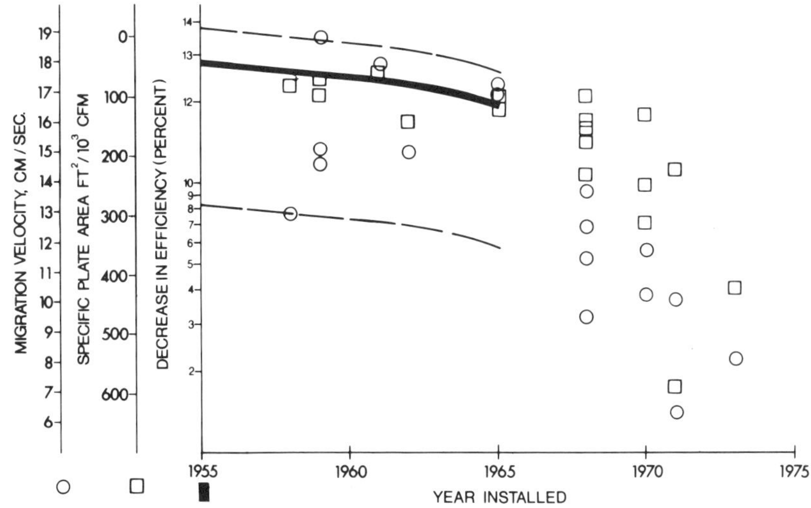

Figure 3.1. Regressed collection efficiency from dual-fuel conversion.

REFERENCES

1. Kasper, William C. "Power from Trash," *Environment* **16**, No. 2, 55 (March, 1974).
2. Roberts, R. M. and E. M. Wilson. "The Value of Refuse Energy and the Cost of its Recovery," ASME Paper No. 71-WA/Inc-3, Presented at the ASME Winter Annual Meeting, Washington, D.C. (November 28–December 2, 1971).
3. Sommerlad, R. E. and R. W. Bryers. "Steam Generator Aspects," ASME Paper No. 71-WA/Inc-2, Presented at the ASME Winter Annual Meeting, Washington, D.C. (November 28–December 2, 1971).
4. Konopka, A. P. "Air Pollution Aspects," ASME Paper No. 71-WA/Inc-1, Presented at the ASME Winter Annual Meeting, Washington, D.C. (November 28–December 2, 1971).
5. Shannon, L. J., M. P. Schrag, F. I. Honea and D. Bendersky. *St. Louis/Union Electric Refuse Firing Demonstration Air Pollution Test Report* (Washington, D.C. U.S. Environmental Protection Agency, 1974) p. 108.
6. Dille, Earl K. and David L. Klumb. "Recycling Solid Waste for Utility Fuel and Recovery of Other Resources," Presented to 1973 Frontiers of Power Technology, Oklahoma University.
7. Oglesby, Sabert, Jr. "Design Considerations for Electrostatic Precipitators," Presented at Specialty Conference, APCA St. Louis Section, St. Louis, Missouri (March, 1973).
8. Greco, Joseph. "Electrostatic Precipitators—An Operator's View," Presented at Specialty Conference, APCA St. Louis Section, St. Louis, Missouri (March, 1973).
9. Coy, D. W. and N. W. Frisch. "Specifying Precipitators for High Reliability" Presented at EPA/SR1 Symposium, Pensacola Beach, Florida (October, 1974).

4

Air Pollution Grandfather Clauses Are Better Than Land Use Zoning Limitations

Barry W. Muller
State of Rhode Island and Providence Plantations,
Department of Health, Division of Air Pollution Control,
Providence, Rhode Island

Benjamin Linsky
Professor, West Virginia University, Morgantown

INTRODUCTION

Although local and regional area land use planning and air pollution control agencies began to work together in the United States at least as far back as the early 1950s,[1,2] the accelerating interest of more agencies and specialists in both fields at local, state, and national levels has resulted in considerable confusion about some aspects of the potential of these two approaches to better air quality.

The major objective of this chapter is to explore and clarify one major element of confusion for a wide range of specialists in air pollution control and in land use planning and zoning. This element is the regulation of existing air pollution sources. Names such as grandfather, existing, or preexisting have been used to describe these conditions.

A preexisting land use exists as the result of the adoption of a new or changed zoning law to cover that piece of land. When a zoning classification is altered, some new activities or uses of differing characteristics will no longer be permitted to be constructed in the particular zoning district. However, an activity that lawfully predated the change in zoning is almost always allowed to continue as a "preexisting nonconforming" use.

The impact of preexisting nonconforming use upon land use law is paralleled in air pollution control by the influence of existing pollutant sources on air pollution control statutes. A particular source of air pollution emissions which exists precedent to enactment of an air pollution control law may not have been required to be quite as clean as that same type of source constructed after the enactment of the air pollution statute. Reasons for the preferential treatment of existing sources of air pollution include the technical limtiations of the control devices that must be added, the often higher costs of such control measure, and an unwillingness to penalize those owners who have made considerable investment and good faith efforts in reliance on old regulations.

Grandfather clauses in air pollution control statutes serve to relieve preexisting air pollution emissions sources from the full impact of tighter control limits which will be applied to new sources of emissions. Commonly, the grandfather exemption appears in emission limits, but it may be used in design standards, fuel regulations, and variance procedures as well.

The most prevalent use of the grandfather clause appears in the emission limits of air pollution control statutes. For standards of visible emissions, particulate emissions, sulfur compound emissions and nitrogen oxide emissions, some relaxation of standards for existing sources are dominant.

VISIBLE EMISSIONS

Visible emission limitations utilizing the Ringelmann shade and opacity scale are long-standing and familiar. The Ringelmann scale is a measure by which the relative density of emissions of grey shade may be compared and consists of five gradations in which 0 is equivalent to all white and 5 is equivalent to all black.[3] The equivalent opacity of other colors, including white, cream and orange is also widely used. Industry criticizes the use of such standards as being too subjective, but the readings taken by trained observers are reasonably consistent, precise and accurate. The visual standard is an indispensible enforcement tool; moreover, the appearance of an emission relates directly to aesthetic concerns. For these reasons, as well as the encouragement of citizen participation in initiating air pollution cases, many laws and regulations use the visible emission standards.

In its air pollution ordinance of 1969, the city of Cleveland[4] distinguished between all new and existing process and fuel-burning installations as applied to visible emissions. For all new sources, constructed after October 30, 1969, visible emissions were restricted to No. 1 Ringelmann or its equivalent opacity, except that emissions of a density not greater than No. 2 Ringelmann or its equivalent opacity were permitted for an aggregate period of 3 minutes in any hour not to exceed 9 minutes in any 8-hour period. Existing sources were permitted visible emissions not to exceed No. 2 Ringelmann or its equivalent opacity. Further, emissions of a density not greater than No. 3 Ringelmann or its equivalent opacity were to be allowed for an aggregate period of 5 minutes in any hour not to exceed 15 minutes in any 8-hour period. A West Virginia regulation[5] provides that existing coking operations were to meet a compliance schedule in terms of visible emissions. By January 1, 1971 existing coke works were limited to a visible emission of No. 3 Ringelmann or its equivalent opacity for a period of time aggregating 3 minutes per charge of each coke oven and 1 minute per push on each oven. In three succeeding six-month time frames, visible emissions were to be reduced on existing coke works to the point that by July 1, 1972, allowable emissions were to be limited to No. 2 Ringelmann or its equivalent opacity for a period aggregating no more than 2 minutes per charge on each oven and 1 minute per push on each oven. By comparison, all new coking facilities constructed after July 1, 1970 were limited to visible emissions not greater than No. 2 Ringelmann or its equivalent opacity but the duration of emissions was reduced to a period aggregating no more than 1.5 minutes per charge per oven and 0.5 minutes per push per oven.

In keeping with the federal New Source Performance Standards,[6] all states must, at a minimum, prescribe such standards that EPA promulgates to the applicable sources. The state of Illinois incorporated such visible emission standards for two types of new sources. For new fuel combustion emissions sources with actual heat input greater than 250 million Btu per hour, allowable opacity is restricted to 20% (No. 1 Ringelmann) with an exception up to 40% (No. 2 Ringelmann) for a period or periods aggregating 3 minutes in any 60-minute period limited to 3 times in any 24-hour period. For new Portland cement plants, visible emissions are limited to an opacity of 10% (No. 0.5 Ringelmann). Existing fuel combustion facilities with actual heat input greater than 250 million Btu per hour and existing Portland cement plants are required to meet the less stringent statewide visible emission standard:

> No person shall cause or allow the emission of smoke or other particulate matter from any other emission source into the atmosphere of an opacity greater than 30%.

An additional provision stipulates that emissions in existing sources up to 60% opacity (No. 3 Ringelmann) would be allowed for a period or periods aggregating 8 minutes in any 60-minute period limited to 3 times in a 24-hour period.[7]

In the state of Maryland, the 1972 regulations governing the control of visible emissions provide that for new sources, *e.g.*, "any installation for which the major proportion of the fuel-burning, incineration, processing or manufacturing equipment in the installation is erected on or after the effective date of the regulations (January 17, 1972)," no visible emissions are permitted. For existing and modified sources (those altered, changed or added to on or after the effective date of the regulations)[8] emissions no darker than No. 1 Ringelmann or its equivalent opacity are permitted.

PARTICULATE EMISSIONS

Particulate emissions limitations measured by stack sampling rather than by plume opacity are also subject to differences by the grandfather clause exemption. Older regulations and ordinances dealt primarily with visible emissions of particulate matter, but some sought to regulate particulate emissions by providing an emission limit, sometimes also incorporating a collection efficiency limit.

Other limits with broad or specific design criteria are also used. For example, the 1956 Charlotte, North Carolina Air Pollution Ordinance[9] required that all new installation of boilers, furnaces, or fuel-burning equipment for the burning of coal be equipped with "approved and effective" fly ash removers. This grandfather clause exempted existing facilities from this requirement.

The Bridgeport, Connecticut Air Pollution Control Ordinance of 1957 limited dust emissions to 0.85 lb/1000 lb of emission gases for the products of combustion for all gases emitted from any fuel-burning equipment or process furnace. Excepted from this limit was any equipment constructed or installed prior to the effective date of the ordinance, provided that the preexisting equipment employ dust collectors that would achieve a standard of 90% collection efficiency.[10]

The Seattle, Washington (1961), Wheeling, West Virginia (1958), and Tacoma, Washington (1955) air pollution ordinances had the same dust limitation as Bridgeport. In these ordinances, however, an exception to the dust emission limitation would be granted provided that dust collection equipment be employed. For sources constructed after the effective date of the control ordinance using dust collectors, an 85% collection efficiency would be required. Any source constructed prior to the effective date of the ordinance utilizing dust collection equipment was required to meet a collection efficiency of 75%. In most statutes, the collection efficiency standard has been eliminated as being too restrictive against the use of alternative means of air pollution control such as materials and process equipment changes, which may be employed to mitigate emissions.[11]

Present emission standards for particulate matter in some air pollution regulations allow existing sources to pollute more than new ones. In developing the particulate emissions standards for such sources, the state of Illinois utilizes two distinct process weight rate tables as a basis for a variety of industrial processes (Table 4.1). By comparison of the two it may be noted that particularly in the middle size ranges, emissions limitations for existing sources are considerably more lax than those for new sources. The agency adopted the more stringent table for both new sources and existing sources that had not taken steps toward compliance with earlier, more lenient regulations.[11] These more stringent restrictions were deemed appropriate as technological evidence indicated that the standards be tightened and to maintain compliance with the air quality standards in the Chicago area and other major metropolitan areas of the

Table 4.1. Particulate Standards for Process Emission Sources.[7]

Process Weight Rate (lb/hr)	*Process Weight Rate (ton/hr)*	*Allowable Emissions (lb/hr)*	
		New	*Existing*
100	0.05	0.55	0.55
200	0.10	0.77	0.87
400	0.20	1.10	1.40
600	0.30	1.35	1.83
800	0.40	1.58	2.22
1,000	0.50	1.75	2.58
1,500	0.75	2.40	3.38
2,000	1.00	2.60	4.10
4,000	2.00	3.70	6.52
6,000	3.00	4.60	8.56
8,000	4.00	5.35	10.40
10,000	5.00	6.00	12.00
20,000	10.00	8.70	19.20
30,000	15.00	10.80	25.20
40,000	20.00	12.50	30.50
50,000	25.00	14.00	35.40
60,000	30.00	15.60	40.00
70,000	35.00	17.00	41.30
80,000	40.00	18.20	42.50
90,000	45.00	19.20	43.60
100,000	50.00	20.50	44.60
200,000	100.00	29.50	51.20
300,000	150.00	37.00	55.40
400,000	200.00	43.00	58.60
500,000	250.00	48.50	61.00
600,000	300.00	53.00	63.10
700,000	350.00	58.00	64.90
800,000	400.00	62.00	66.20
900,000	450.00	66.00	67.70
1,000,000	500.00	67.00	69.00

For new sources:
Interpolated and extrapolated (up to process weight rates of 450 tons/hr) values of the data in the table shall be determined by using the equation:

$$E = 2.54\,(P)^{0.534}$$

Interpolated and extrapolated values of the data of the table for process weight greater or equal to 450 tons/hr shall be determined using the equation:

$$E = 24.8 (P)^{0.16}$$

For existing sources:

Interpolated and extrapolated values of the data in the table for process weight rates up to 30 tons/hr shall be determined by using the equation:

$$E = 4.10 (P)^{0.67}$$

and interpolated and extrapolated values of the data for process weight rates in excess of 30 tons/hr shall be determined by using the equation:

$$E = 55.0 (P)^{0.11} - 40.0$$

where in all cases:

E = allowable emission rate in lb/hr

and

P = process weight rate in tons/hr

state in the face of continued growth. In other parts of the state the more stringent regulations were necessary to reduce local problems, avoid unnecessary degradation of air that can be clean, and to leave the maximum room for future development.[11]

On the other hand, for those existing sources which made significant expenditures and good faith efforts to reduce emissions the more lenient particulate standards were retained (Table 4.1). The opinion of the Illinois Pollution Control Board explains this decision:

> [To disallow the more lenient standards in favor of the more stringent ones] might in many cases require scrapping of expensive new equipment to achieve marginal improvement in emissions that we do not think would be worth the cost. [The more lenient standards] apply if the source is now in compliance with its provisions; or if the owner had entered into contracts pursuant to a variance embodying a program that will achieve such compliance; or if he has commenced construction of facilities to achieve such compliance and obtains a variance to that effect in the immediate future.

The allowable concentration of particulate matter from new and existing incinerators burning more than 2000 lb of refuse per hour is limited to 0.08 grains per standard cubic foot (gr/scf) corrected to 12% carbon dioxide (CO_2) and for those burning more than 60,000 lb of refuse per hour the limitation is 0.05 gr/scf corrected to 12% carbon dioxide. From all other existing incinerators the permissible particulate loading is 0.2 gr/scf corrected to 12% CO_2 while comparable new incinerators must meet a standard of 0.1 gr/scf corrected to 12% CO_2.

Unique particulate emission limitations from large (greater than 250 million Btu per hour actual heat input) fossil fuel generating plants using solid fuel exclusively have been established since statewide emission inventories showed that such sources constitute a very substantial percentage of total particulate emitted.[11] The

Illinois Pollution Control Board adopted the federal standard for large new sources of 0.1 lb of particulate per million Btu heat input. However, exception to this rule was granted to existing sources of this nature permitting an emission rate up to 0.2 lb of particulates per million Btu heat input. Stipulations for allowing this rate were based on the efforts and monies these facilities had expended to comply with the former standard. A degradation clause was written into this exception providing that if such a facility was controlling emissions in the region between 0.1 and 0.2 lb of particulates per million Btu heat input, the "emission control of such source is not allowed to degrade more than 0.05 lb of particulate per million Btu heat input from the original design standard or equipment performance test conditions, whichever is stricter."

In the state of West Virginia, Regulation V[12] allows certain exemptions for thermal coal driers. Until September 1, 1971, existing coal thermal drying operations, constructed prior to March 1, 1970, were allowed to emit 0.15 gr/scf of exhaust gas. Thereafter, any such existing drier was limited to emissions prescribed in Table 4.2a. New thermal driers, constructed after March 1, 1970 are limited to the stricter emissions prescribed in Table 4.2b.

Table 4.2a. Permissible Emissions of Particulates from Coal Thermal Drying Operations Installed before March 1, 1970.[12]

Total Plant Volumetric Flow Rate (scf/min)	*Maximum Allowable Particulate Loading per Drier (gr/scf)*
120,000 or less	0.12
172,000	0.11
245,000	0.10
351,000	0.09
500,000 and above	0.08

Table 4.2b. Permissible Emissions of Particulates from Coal Thermal Drying Operations Installed after March 1, 1970.[12]

Total Plant Volumetric Flow Rate (scf/min)	*Maximum Allowable Particulate Loading per Drier (gr/scf)*
95,000 or less	0.10
111,000	0.09
163,000	0.08
240,000 and above	0.07

For the volumetric flow rate between any two consecutive volumetric flow rates stated above, limitations shall be determined by linear interpretation. For the purpose hereof, the total volumetric flow rate shall be the total standard cubic feet of dry gas passed through all thermal driers at one plant location.

SULFUR COMPOUND EMISSIONS

Federal New Source Performance Standards[6] prescribe two sulfur dioxide (SO_2) emissions limitations for new fossil fuel steam generating plants of greater than 250 million Btu per hour actual heat input: when liquid fuels are burned, allowable emissions are 0.8 lb of SO_2 per million Btu heat input; when solid fuels are burned, allowable emissions are 1.2 lb of SO_2 per million Btu actual heat input.

In the state of Illinois, existing fossil fuel generating plants of heat input greater than 250 million Btu per hour are permitted a more lenient emission rate, depending upon their location. For plants burning solid fuels in the Chicago, St. Louis (Illinois) and Peoria metropolitan areas, as well as in those major metropolitan areas throughout the state which have an average annual ambient SO_2 concentration greater than 60 $\mu g/m^3$, the allowable emission rate is 1.8 lb of SO_2 per million Btu heat input. Sources located outside of any major metropolitan area may emit up to 6.0 lb of SO_2 per million Btu actual heat input. Those sources throughout the state burning residual fuel oil may emit 1.0 lb of SO_2 per million Btu heat input.[7]

In the state of Florida, allowable SO_2 emissions from existing power plants greater than 250 million Btu per hour actual heat input are 1.5 lb of SO_2 per million Btu actual heat input and from those burning liquid fuel, 1.1 lb of SO_2 per million Btu actual heat input.[13]

Some regulations may prescribe a sulfur limitation in fuel. Cleveland[4] prohibits the use of fuel containing 2% sulfur or more by weight in existing sources while requiring new sources to use fuel with only a 1% sulfur content by weight.

NITROGEN OXIDE EMISSIONS

The state of Illinois prescribes emissions standards for both new and existing fuel combustion emission sources with actual heat input greater than 250 million Btu per hour. For new sources using gaseous, liquid, or solid fossil fuels, the allowable emissions respectively are 0.2, 0.3, and 0.7 lb of nitrogen oxides (NO_x) per million Btu actual heat input, which correspond to the federal New Source Performance Standards.[6] For existing sources the comparative allowable emission rates are 0.3, 0.3, and 0.9 lb of NO_x per million Btu actual heat input. The intent of the Illinois NO_x emission requirements on electric power stations is to control emissions by burner adjustments or alterations in firing procedures but not require nitrogen oxide removal systems on stacks.[7] The Illinois Pollution Control Board stated that:

> The technology for oil and gas firing control is well supported for both new and existing sources using such fuels. On the other hand, the application of such techniques to coal-fired furnaces is less clear cut. The regulation encourages the development of designs which will reduce NO_x

emissions from all new large fuel combustion units but specifies somewhat more lenient standards for existing coal-fired equipment.

While adopting the federal New Source Performance Standards for large power plants using gas and oil, Maryland applied a more stringent standard to new solid fuel-fired power plants: 0.5 lb of NO_x per million Btu actual heat input. However, for existing coal-fired power plants built before May 12, 1972, the state did not prescribe any nitrogen oxide emission standards. Instead, in the metropolitan areas of Baltimore and Washington, D.C. and surrounding counties, regulations dictate that the discharge of nitrogen oxides will not be "in excess of the rates attainable by the use of reasonably applicable control technology." Rather than setting an emission limitation for existing sources of nitrogen oxides a nebulous design standard was utilized.

CONCLUSIONS

"Preexisting nonconforming use" exemptions in most local land use zoning laws are delineated chiefly in provisions which allow such uses that continue relatively unchanged to be excepted from the newly established regulations. In air pollution control law, grandfather provisions permit less stringent limitations to be imposed on existing activities.

Although land use zoning and stationary source air pollution control laws share the common goal of protecting land uses, the basis of each type of law is distinctive. The fundamental concern of zoning has been the protection of the property rights and the preservation of the land. Air pollution control law is based on the rights of the individual to be free from unpleasant exposure and effects. Controls are placed on the land use activity as to emissions from equipment, design of equipment, input materials and fuels so that the individual will be given as much protection and freedom from displeasure as the community or state allows.

Zoning law, being predisposed towards the land, reflects lenient attitudes with respect to termination of a preexisting nonconforming use. The most significant attempt at termination of the preexisting nonconforming use, amortization, has met very limited acceptance and has proved to be a satisfactory control measure only in a few cases consisting primarily of nuisance and/or little monetary value.

Under air pollution control law, the termination of a polluting activity may be accomplished through a variety of devices, in addition to the fines and, on rare occasion, jail penalties imposed by a judge for violations of laws and legal regulations. A mechanism common to older laws was the sealing clause. Such restrictions limited the number of violations (usually three) a single source was permitted in a year's time. If this number was exceeded, the equipment could be sealed until compliance was assured.[10,14,15] In addition to emission violations, other ordinances provided that if work on an emissions source, for which an installation permit was required, was commenced without that permit, the equipment would be sealed.[14,16] While many present air pollution control regulations do not utilize the "sealing" clause, some have retained it.[17]

State air pollution control statutes also incorporate cease and desist provisions under which, if it is determined that a source is violating provisions of the law, rules, or regulations, the agency may compel the activity to enjoin operation until it cleans up.[18,19] Also by increasing the statutory or regulatory standards of design, emission or effects to a degree that is so restrictive that a preexisting nonconforming use cannot meet them, termination may result. This is true of open burning dumps and many types of incinerators in many areas of the United States.[20]

The power of the administrative variance board in air pollution proceedings is minimized by the temporary nature of the variance. Unlike its zoning counterpart which may invoke permanent changes on the use of the land, the impermanent nature of the air pollution control variance removes the arbitrary power of the variance board to promote nonconformity.

Air pollution control regulations may help reinforce the preexisting nonconforming land use restrictions. For example, to declare abandonment of a preexisting nonconforming use under zoning law requires proof of intent by the municipality or county that the owner had in mind to cease operations. Under one type of air pollution restriction, an abandoned emissions source may be subject to more stringent emissions standards if resumed.[13]

An intensification of a nonconforming land use operation without any identifiable element of physical vertical or horizontal change or extension cannot be enjoined by zoning law. However, if such intensification results in increased air pollution, new source standards may be applied.

While local land use zoning law is relatively static insofar as control of preexisting nonconforming uses are concerned, air pollution provisions can cope with the "rising expectations" of people along with improvements in the state-of-the-art of control, and increasing understanding of the mechanisms of pollutant dispersal and effects, as well as better recognition of undesirable effects.

REFERENCES

1. Linsky, Benjamin. "The Relationship between Air Pollution Planning and Zoning," presented at the 46th Annual Meeting of the Air Pollution Control Association, Baltimore, Maryland (May 26, 1953).
2. Linsky, Benjamin. "Air Pollution in a Community as Related to Zoning," presented at Marine City Chamber of Commerce Meeting (October 15, 1952).

3. State of Pennsylvania, County of Allegheny, Allegheny County Health Department. *Rules and Regulations, Article XVIII, Air Pollution Control* (Effective Date June 15, 1972).
4. City of Cleveland,Ohio. *Air Pollution Control Code* (Effective Date October 30, 1969).
5. State of West Virginia, West Virginia Air Pollution Control Commission. *Regulation VII, To Prevent and Control Particulate Air Pollution from Manufacturing Processes Operations* (Effective Date July 1, 1970).
6. Federal Register. *Standards of Performance for New Stationary Sources* **36**, No. 247, Part II, Washington, D.C. (Thursday, December 23, 1971).
7. State of Illinois. *Air Pollution Control Regulations,* Part II, Rules 201-208 (Effective Date April 14, 1972).
8. State of Maryland, State Department of Health and Mental Hygiene. *Rules and Regulations Governing the Control of Air Pollution,* Sections 10.03.36-41 (Effective Date February 12, 1974).
9. City of Charlotte, North Carolina. *Smoke Abatement Ordinance,* No. 323 (Effective Date April 25, 1956).
10. City of Bridgeport, Connecticut. *Air Pollution Control Ordinance* (Effective Date June 15, 1957).
11. Illinois Pollution Control Board. "In the Matter of Emission Standards," PCB R 71-23 (April 13, 1972).
12. State of West Virginia, West Virginia Air Pollution Control Commission. *Regulation V, to Prevent and Control Air Pollution from the Operation of Coal Preparation Plants and Coal Handling Operations* (Effective March 1, 1970).
13. State of Florida. *Rules of the State of Florida, Department of Pollution Control, Chapter 17-2 Air Pollution* (1971).
14. City of Chicago, Illinois. *Air Pollution Control Ordinance,* Chapter 17 (Effective Date May 1, 1959).
15. City of Tampa, Florida. *Smoke Ordinance,* Chapter 32 (Effective Date May 1, 1948).
16. City of New York, New York. *Rules and Regulations of the Department of Air Pollution Control,* Chapter 47 (Effective Date May 15, 1954).
17. State of Ohio, County of Montgomery, Montgomery County Combined General Health District. *Air Pollution Control Regulations* (Effective Date July 5, 1972).
18. State of Alabama. *Air Pollution Control Act of 1971.*
19. State of West Virginia. *Air Pollution Control Law of 1970,* Title 16 West Virginia Statutes.
20. Linsky, Benjamin. Private communication (September 21, 1974).

Section II
Predicting Dispersion of Emissions

5

A Simplified Technique Used to Evaluate Atmospheric Dispersion of Emissions from Large Power Plants

T. L. Montgomery, W. B. Norris,
F. W. Thomas, and **S. B. Carpenter**
Tennessee Valley Authority, Air Quality Branch,
Muscle Shoals, Alabama

INTRODUCTION

This chapter provides a simplified approach for estimating plume centerline surface concentrations identified with principal atmospheric plume dispersion conditions. It also presents procedures for calculating maximum concentrations with respect to downwind distance and wind speed. Although this procedure is designed for single source-type emissions, its applicability to multiple sources can be achieved through use of appropriate correction factors which are included. It is intended to be used for installations with stacks 2½ times or more the height of the power plant building, with effluent exit velocities equal to or greater than the mean wind speed and with furnace heat input equal to or greater than 10^9 Btu/hr connected to a single stack.

Investigators have developed numerous mathematical equations for predicting surface concentrations for various plants and stack heights.[1-7] Several full-scale field experiments and ambient monitoring programs have been conducted to evaluate and test the reliability of existing atmospheric dispersion and plume rise models.[4-6]

The calculation procedures have been compiled into a simple step-by-step method involving the use of three nomograms similar to those used previously by Gifford.[8] The nomogram technique does not require a detailed understanding of atmospheric dispersion processes. However, for completeness, a description of the atmospheric conditions identified with various dispersion processes is included.

The procedures presented are used to compute 1-hour average surface concentrations from fossil-fuel power plant emissions (although provision is made for modifying the results to 3-hour and 24-hour concentrations). Specifically, maximum ground-level concentrations are calculated using the coning, inversion breakup, and trapping models. In addition, the variation of surface concentration with distance is calculated in the case of the coning plume. While the technique presented here gives results for single-stack emissions, they may be adapted to multiple-stack sources by means of appropriate empirical correction factors, which are included in the description of the procedures.

NOMOGRAM PARAMETERS

An initial requirement in calculating surface concentrations of stack effluents is to determine the height to which the plume rises. Plumes from power plant stacks are usually emitted with high exit velocities and heat content and may rise to considerable heights because of the vertical momentum and buoyancy. Essential parameters for calculating plume rise, $\triangle h$, with this method are effluent exit velocity, V_s; inside radius of stack top, r; density of ambient air at stack top, ρ_a; density of stack gas at stack top, ρ_s; and mean wind speed between stack top and plume top, $\bar{u}$. For a plume with a given rise, $\triangle h$, the essential parameters needed for calculating surface concentrations are: distance downwind from source, x; height of stack, h_s; mean wind speed between stack top and plume top, $\bar{u}$; and emission rate, Q.

PROCEDURES FOR USING NOMOGRAMS

The following steps outline the procedures for estimating (a) maximum surface concentrations for coning, inversion breakup, and trapping dispersion models, and (b) surface concentrations for a given downwind distance and wind speed for the coning model only.

Before using the nomograms it is necessary to have the following data and to make one computation:

h_s = stack height (m)
Q = effluent emission rate (kg/sec)
r = inside radius of stack top (m)
$\bar{u}$ = mean wind speed between stack top and plume top (m/sec)*
V_s = stack gas exit velocity (m/sec)
x = downwind distance from source to where surface concentration is estimated (m)*

*It is not necessary to have these data if the maximum plume centerline surface concentration is to be computed.

ρ_a = density of ambient air at stack top (g/m^3)†
ρ_s = density of stack gas at stack top (g/m^3)‡

Compute the nomogram parameter $G = V_s r^2 \left[\frac{\rho_a - \rho_s}{\rho_a}\right]$ (m^3/sec)

Procedural steps for using the nomograms are:

Procedure 1

A. *Coning Dispersion Model — Maximum Plume Centerline Surface Concentration*

Step 1 — Project a line from the calculated G value on the left-hand G scale of Figure 5.1 to a value equal to h_s on the right-hand neutral Δh scale. Read the $\bar{u}$ value from the center scale. This is the $\bar{u}$ associated with the maximum plume centerline surface concentration, χ_{max}. (Chi, χ, is a symbol denoting 1-hour average surface concentrations.)

Step 2 — Calculate effective stack height, $h_e = 2h_s$ (m)

Step 3 — Locate this h_e value on the coning h_e scale of Figure 5.2. Project a line from this point tangent to the curved coning x scale. The point of tangency is the downwind distance where the maximum plume centerline surface concentration is expected to occur. Mark the point where the above projected line intersects the right-hand scale.

From this point project a line through the $\bar{u}$ value, center scale, obtained previously from Figure 5.1, to the coning h_e scale. Mark this point of intersection.

From this point project a line through the selected Q, center scale, to the right-hand scale. Note: If Q < 1 kg/sec, use Q = 1 kg/sec and the appropriate correction factor described later. The intersect value on this right-hand scale is the coning maximum 1-hour average plume centerline surface concentration, χ_{max}, in μg/m^3.

Step 4 — Multiply the result by any applicable correction factors described later.

B. *Inversion Breakup Model — Maximum Plume Centerline Surface Concentration*

---

†Ambient air density, ρ_a, is given by ρ_a=352,330/T_a at one standard atmosphere of pressure.

‡Stack gas density, ρ_s, may be found by subjecting the fuel or the stack gas to chemical analysis. For coal-fired plants, the stack gas is about 2% denser than dry air at the same temperature.

Step 1 — Project a line from the calculated G value on the left-hand G scale of Figure 5.1 through $\bar{u} = 3$ m/sec to the right-hand stable Δh scale. Read Δh in meters from this stable Δh scale. Calculate $h_d = (\Delta h + 100)$ (m).

Step 2 — Locate h_s on the left-hand h_s scale of Figure 5.3. Locate the h_d value obtained from step 1 on the center h_d scale. Connecting these two points, project a line to the right-

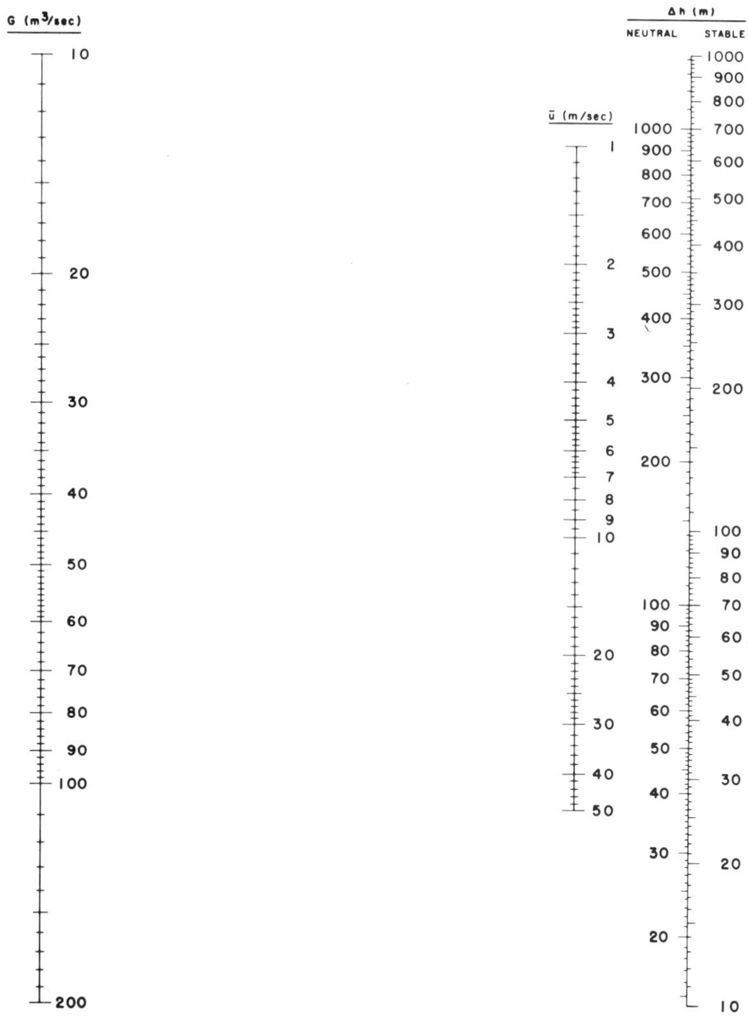

Figure 5.1. Buoyant plume rise, Δh, versus G and mean wind speed, $\bar{u}$, for neutral and stable atmospheric stability.

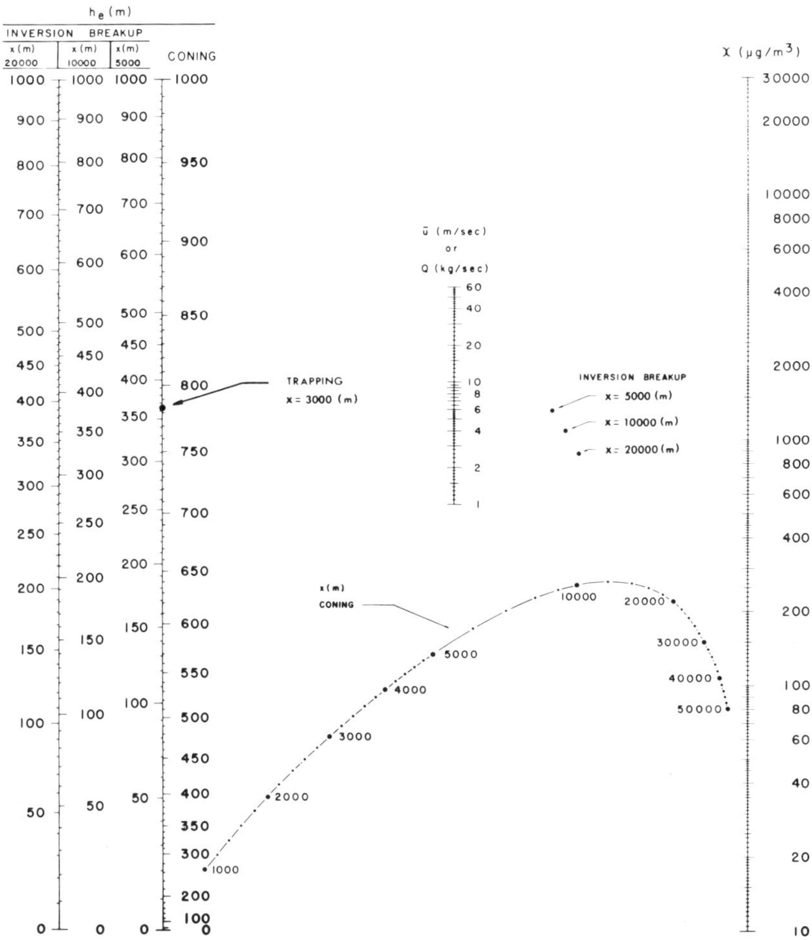

Figure 5.2. Plume centerline 1-hour average surface concentration, χ, versus downwind distance, x, stack height, h_e effluent emission rate, Q, and mean wind speed, ū.

hand x_{max} scale. The intersect point is the distance downwind, x_{max}, where the maximum inversion breakup surface concentration is expected to occur.

Step 3 — Calculate effective stack height, $h_e = (h_s + \Delta h)$ **(m)**.

Step 4 — Locate this h_e value on the inversion breakup h_e scale of Figure 5.2 that represents the x value (5,000, 10,000, or 20,000 m) that is most nearly equal to x_{max}. If the x = 10,000 or 20,000 m—h_e scale is utilized, project a horizon-

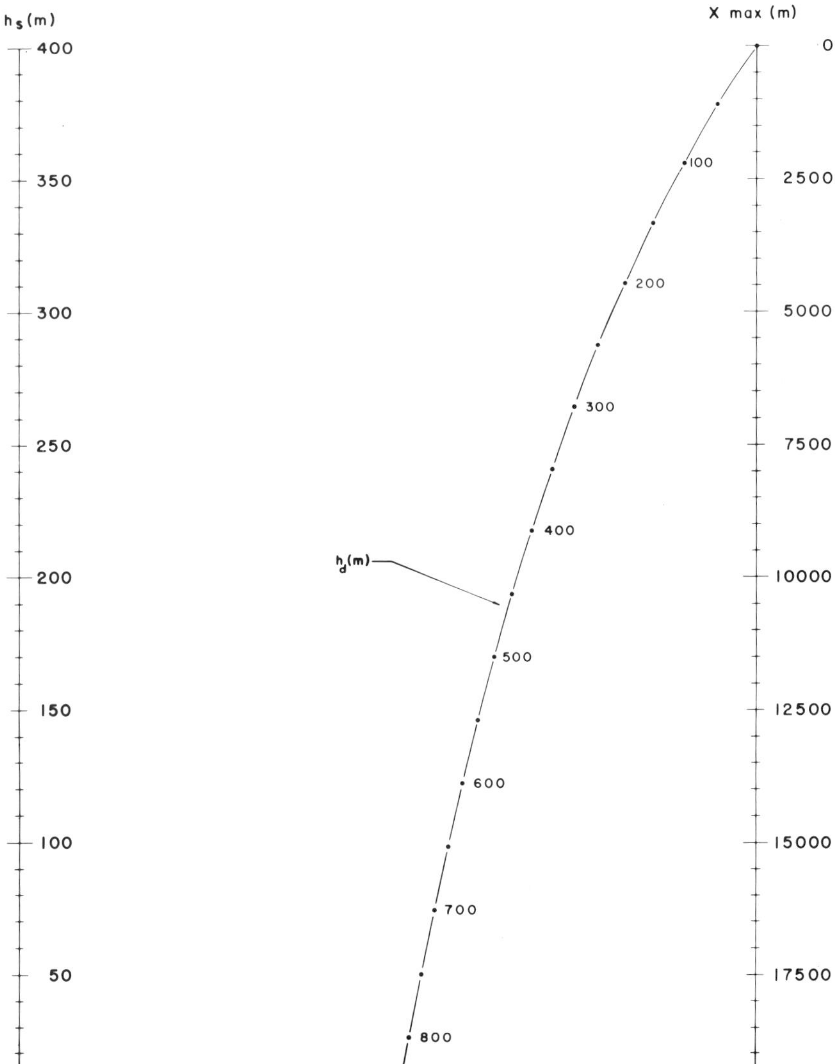

Figure 5.3. Distance, x_{max}, versus stack height, h_s, and h_d, for inversion break-up dispersion.

tal line to the $x = 5{,}000$ m—h_e scale and mark the point of intersection on the $x = 5{,}000$ m—h_e scale. If x_{max} is the most nearly equal to $x = 5{,}000$ m, mark the h_e value obtained from step 3 on that scale. From the resulting point on the inversion breakup $x = 5{,}000$ m—h_e scale project a line through the inversion breakup point, representing the x that is almost nearly equal to x_{max}, to the right-hand scale. Mark this point of intersection.

From this point project a line through $\bar{u} = 3$ m/sec to the coning h_e scale. Mark this point of intersection.

From this point project a line through the selected Q, center scale, to the right-hand scale. *Note:* If $Q < 1$ kg/sec, use $Q = 1$ kg/sec and the appropriate correction factor described later. The intersect value on this right-hand scale is the inversion breakup maximum 1-hour average plume centerline surface concentration, χ_{max}, in $\mu g/m^3$. *Note:* the χ_{max} is actually for $x = 5{,}000$, 10,000, or 20,000 m, an x value which may be significantly different from the previously calculated x_{max}. The error in χ_{max} will probably not be significant, however. But, a more accurate procedure would be to calculate two χ_{max} values, one for $x < x_{max}$ and one for $x > x_{max}$ and then calculate χ_{max} by linear interpolation.

Step 5 — Multiply the result by any applicable correction factor described later.

C. *Trapping Model — Maximum Plume Centerline Surface Concentration*

Step 1 — Project a line from the trapping, $x = 3{,}000$ m dot, located on the coning h_e scale of Figure 5.2 through the appropriate Q, center scale, to the right-hand scale. Note: If $Q < 1$ kg/sec, use $Q = 1$ kg/sec and the appropriate correction factor described later. The intersect value on this right-hand scale is the trapping maximum 1-hour average plume centerline surface concentration, χ_{max}, in $\mu g/m^3$.

Step 2 — Multiply the result by any applicable correction factor described later.

Procedure 2 — Coning Dispersion Model Only—Centerline Surface Concentration With Given x and ū

Step 1 — Locate the calculated G on the left-hand scale of Figure 5.1. Locate the selected $\bar{u}$ on the center scale.

Project a line from G through $\bar{u}$ to the right-hand neutral Δh scale. The intersect point on the right-hand neutral Δh scale is Δh in meters.

Step 2 — Calculate effective stack height, $h_e = (h_s + \Delta h)$ (m).

Step 3 — Locate this h_e value on the coning h_e scale of Figure 5.2. Locate the desired downwind distance, x, on the center curved coning x scale.

Project a line from the calculated coning h_e value through the desired coning x value to the right-hand scale. Mark this point of intersection. From this point project a line through the selected $\bar{u}$ (center scale) to the coning h_e scale. Mark this point of intersection. From this point project a line through the desired Q (center scale) to the right-hand scale. *Note:* If $Q < 1$ kg/sec, use $Q = 1$ kg/sec and the appropriate correction factor described later. The intersect value on this right-hand scale is the coning 1-hour average plume centerline surface concentration, χ, in $\mu g/m^3$ for the selected downwind distance and wind speed.

Step 4 — Multiply the result by any applicable correction factor described later.

CORRECTION FACTORS

The following correction factors should be applied, if required, when using the nomograms.

A. *Emission Rate Less Than 1 kg/sec.* For emission rates less than 1 kg/sec, multiply the result from Figure 5.2 by actual emission rate, Q (kg/sec).

B. *1-Hour to 3-Hour and 24-Hour Averaging Time.* Multiply the result from Figure 5.2 by 2/3 to obtain an estimate of the 3-hour average and by 1/5 to obtain an estimate of the 24-hour average surface concentration, χ, ($\mu g/m^3$).

C. *Multiple-Stack Source.* Multiply the result from Figure 5.2 by $N^{4/5}$, where N is the number of stacks of a given height.

N	$N^{4/5}$
2	1.7
3	2.4
4	3.0
5	3.6
6	4.2
7	4.7
8	5.3
9	5.8
10	6.3

If a multiple-stack source consists of stacks of approximately one height, and also stacks of a considerably different height, these should be treated as distinct groups of sources and the resulting concentrations from each group added to obtain the total concentration.

WORKED EXAMPLE—CONING DISPERSION MODEL

Problem

To illustrate the use of these nomograms, the expected 1-hour average plume centerline surface sulfur dioxide concentration, χ, at a distance, x, 8 km from the source, is determined for coning dispersion conditions. A large coal-fired power plant with four units (each connected to a single stack) has been selected for this example. The ambient air temperature, T_a, near stack top is 289° K. The mean wind speed, $\bar{u}$, between stack top and plume top is 6 m/sec. Design and operational characteristics of the plant are as follows:

$$V_s = 13 \text{ m/sec}$$
$$r = 2.5 \text{ m}$$
$$T_s = 464^\circ \text{ K}$$
$$h_s = 92 \text{ m}$$
$$Q = 1 \text{ kg/(sec-unit) } (SO_2)$$

Solution

The initial estimate will be for one unit. Required data, Q, V_s, r, $\bar{u}$, x, and h_s are given above. Obtain the density parameters, ρ_a and ρ_s, from the equation $\rho_a = 352, 330/T_a$, utilizing T_a and T_s together with the assumption that coal-fired power plant stack gases are 2% denser than dry air at the same temperature. The respective densities (g/m³) of ambient air and stack gas are:

$$\rho_a = 1219.1 \text{ gm/m}^3$$
$$\rho_s = 774.5 \text{ gm/m}^3$$

1. Solve for $G = V_s r^2 \left(\frac{\rho_a - \rho_s}{\rho_a}\right)$ where G is the input parameter on the left-hand scale of the nomogram in Figure 5.1.

   $$G = (13)\ (2.5)\ (2.5)\ \left(\frac{1219.1 - 774.5}{1219.1}\right)$$

   $G = 29.6 \text{ m}^3\text{/sec}$
2. Solve for plume rise, Δh, from Figure 5.1 (Step 1, Procedure 2, for Using Nomograms); $\Delta h = 200$ m.
   The effective stack height, h_e, is equal to stack height, h_s, plus plume rise, Δh (Step 2, Procedure 2, for Using Nomograms), or: $h_e = h_s + \Delta h = 92 + 200 = 292$ m.
3. Solve for plume centerline surface concentration, χ, (Step 3, Procedure 2, for Using Nomograms) from Figure 5.2; $\chi \simeq 155$ μg/m³.
4. Apply correction factors, if required (Step 4, Procedure 2, for Using Nomograms).
   a. This factor does not apply since the emission rate is equal to 1 kg/sec.

b. The factors do not apply since 1-hour, not 3-hour or 24-hour, averages are desired.
c. Multiply 155 $\mu g/m^3$ (the result from 3, characteristic of 1 stack), by $N^{4/5}$, where N = 4 and $N^{4/5} = 3.0$.
155 x 3.0 = 465 $\mu g/m^3$, approximately 0.17 ppm SO_2 at 60° F and 1 atmosphere (this is the plume centerline surface concentration resulting from all four units.)

METEOROLOGICAL CONSIDERATIONS

The following information is given for a better understanding of the meteorological basis for the models used in the technique presented above.

Atmospheric Stability

Atmospheric stability is a measure of the degree to which the atmosphere resists or enhances vertical motion. When a buoyant plume is emitted into a stable atmosphere there is minimal vertical dispersion and often considerable horizontal spreading, or meandering, particularly in light winds. When a plume is emitted into an unstable atmosphere, the plume in most cases is rapidly dispersed in the vertical. In neutral atmospheric stability conditions the plume is dispersed about equally in the vertical and horizontal directions with the rate of mixing usually between those identified with stable and unstable conditions.

Atmospheric stability can be determined from measurement of temperature change with height (vertical temperature gradient). Rates of temperature change with height are often expressed as vertical potential temperature gradient. The vertical potential temperature gradient (°K/100 m) = actual gradient (°K/100 m) + 1° K/100 m. In terms of vertical potential temperature gradient, atmospheric stability is classified as:

1. *Neutral* when potential temperature does not change with increasing height (zero vertical gradient).
2. *Stable* when potential temperature increases with increasing height (positive vertical gradient).
3. *Unstable* when potential temperature decreases with increasing height (negative vertical gradient).
4. *Inversion,* or extremely stable, when vertical potential temperature increases with increasing height at a rate greater than 1° K/100 m.

Neutral conditions usually occur with moderate to strong wind, greater than about 4 m/sec, with cloudy skies in the day and clear or cloudy skies at night; stable and inversion conditions with weak wind, < 1–2 m/sec and clear to partly cloudy skies at night through early morning; and unstable conditions with weak wind, < 1–2 m/sec, and strong daytime solar radiation.

Plume Configuration

The shape or configuration of a plume is largely dependent upon the temperature and wind structure throughout the plume environment. A résumé of the relevant plume configurations follows:

1. *Coning* — Identified with neutral conditions, the plume has a cone-shaped profile and an elliptical cross-section, the horizontal axis being the longer. Such a plume displays moderate spread both horizontally and vertically as it is carried downwind. It has been found that with the older and smaller-to-intermediate size power plant units, the coning plume is identified with maximum surface concentrations. Equation (2) gives the plume centerline surface concentration.
2. *Fanning* — Identified with stable (including inversion) conditions, the plume persists as a compact ribbon when transported downwind with minimal dispersion in the vertical, some horizontal spread and meandering, depending upon the wind direction persistency.
3. *Inversion Breakup* — Identified with the transition from a stable to a neutral atmosphere, the plume is initially embedded in an inversion or stable layer (fanning plume), usually below 300–400 m in the Tennessee Valley region. With the onset of daytime solar radiation and subsequent thermally induced mixing, neutral stability progressively builds upward from the surface through the plume environment. As the mixing penetrates the plume, the effluent is subsequently mixed uniformly to the ground. Relatively high concentrations of short duration, 30–45 minutes, occur in midmorning as the mixing penetrates the stable plume. Maximum concentrations occur from 5–50 km from the source. In the Tennessee Valley region inversion breakup may occur on 200 or 300 days per year; however, the magnitude of concentrations in many instances may not be significant because of the height of the stable plume and the degree of stability of the ambient layer. Also, inversion breakup is usually limited to a narrow path beneath the stable plume and the resultant concentrations do not normally occur at any one location more than a few times a year. An exception would be in an area where local channeling or interference of the plant emission from topographical features was common. Equation (4) gives the plume centerline surface concentration.

4. *Trapping* — Identified with a plume emitted into an atmosphere where vertical mixing is restricted by a stable layer above the plume, the effluent some time after release is uniformly mixed between the ground and the stable layer. The height of the stable layers associated with the highest surface concentrations has been observed to be between 750 and 1200 m above the ground. Maximum surface concentrations may persist 2–5 hours from midmorning through midafternoon and occur rélatively near the power plant source (3–10 km). The trapping dispersion condition, as applicable to large power plants, has been found to occur on about 30–40 days per year in the Tennessee Valley region. It is always identifiied with a well-developed surface high-pressure system marked with very stable and light-to-moderate wind speed through the lower 2000 m. Equation (8) gives the plume centerline surface concentration.

EQUATIONS

The following equations were used in constructing the nomograms, Figures 5.1-5.3. The derivation of the equations can be obtained from References 1-7. Assumptions and symbols are defined later.

Plume Rise

The plume rise equation is:

$$\Delta h = \frac{C(114m^{2/3})}{\bar{u}} \left[gV_s r^2 \left(\frac{\rho_a - \rho_s}{\rho_a} \right) \right]^{1/3} \qquad (1)$$

where C is a coefficient dependent upon atmospheric stability. The value $114m^{2/3}$ is obtained from $(1219m)^{2/3}$, which represents the 2/3 power of the mean downwind distance from the source where a buoyant plume from a tall stack ceases or nearly ceases to rise.

Dispersion Models, χ, and x_{max}

In the following equations, surface concentration, χ, is a function of x; and x_{max} is the specific x where χ reaches a valid maximum value. The χ at x_{max} is referred to elsewhere in this presentation as χ_{max}.

In each of the dispersion models presented below, σ_y and/or σ_z appear either explicitly or implicitly. They are given by:

$$\sigma_y = ax^p \text{ and } \sigma_z = bx^q,$$

where a, b, p, and q depend on atmospheric stability in the layer from stack top to plume top.

A. *Coning*

The equations for χ and x_{max} for coning dispersion are:

$$\chi = \frac{(2)\ (10^9)\ Q}{(2)\ 2\pi\ \sigma_y\ \sigma_z\ \bar{u}} \quad \exp\left[-1/2\left(\frac{h_e^2}{\sigma_z^2}\right)\right] \tag{2}$$

$$x_{max} = \left(\frac{h_e^2}{2b^2}\right)^{\frac{1}{2q}} \tag{3}$$

B. *Inversion Breakup*

$$\chi = \frac{(10^9)\ Q}{(2)\ (1.1)\ \sqrt{2\pi}\ \sigma_{yf}\ H_f\ \bar{u}} \tag{4}$$

$$x_{max} = \frac{\rho_a\ \sigma\ \bar{u}}{4\kappa}\left(H_f^2 - h_s^2\right) \tag{5}$$

$$\sigma_{yf} = \sigma_y + 0.47\ h_e\ (m) \tag{6}$$

$$H_f = 2.15\ \sigma_z + h_e\ (m) \tag{7}$$

C. *Trapping*

$$\chi = \frac{(10^9)\ Q}{(2)\ \sqrt{2\pi}\ \sigma_{yt}\ H_t\ \bar{u}} \tag{8}$$

$$x_{max} = 3{,}000\ m \tag{9}$$

$$\sigma_{yt} = \sigma_y + (H_t/1.1 - 2.15\ \sigma_z)\ (.047)\ (m) \tag{10}$$

H_t = height of trapping dispersion mixing layer (m)

An h_e for neutral atmospheric conditions should be used with the coning equations, and a stable h_e should be used with the inversion breakup equations. In Equation (2) the 2 in the numerator is a ground reflection factor, applicable to the coning dispersion model only. In Equations (2), (4), and (8), the 2 in the denominator converts instantaneous concentrations to which σ_y, σ_{yf}, σ_{yt}, σ_z, and H_f are related to 1-hour averages; the 10^9 factor is necessary since Q is expressed in terms of kilograms and χ is expressed in terms of micrograms. The 1.1 in the denominator of Equation (4) accounts for an approximate 10% minimum increase in the plume top height that occurs during the inversion breakup process.

ASSUMPTIONS

The equations used in developing the nomograms were adapted to the following assumptions.

1. Effluent is a conservative gas or consists of suspended particulates so small that diffusion is essentially gaseous.
2. Effluent is dispersed over relatively flat terrain with no significant interference with topographical features.
3. Effluent is emitted continuously.
4. Effluent is emitted from a single stack.
5. Ground is considered a perfect reflecting surface.
6. Mean wind direction is constant.

7. The following numerical values were inherent in the forms of the equations that were utilized to develop the nomograms, Figures 5.1-5.3.

A. *Plume Rise*

$$\frac{\Delta\theta}{\Delta z} = 0^\circ \text{ K/100 m for neutral plume rise}$$

$$\frac{\Delta\theta}{\Delta z} = 1.15^\circ \text{ K/100 m for stable plume rise}$$

B. *Coning* χ, χ_{max}, *and* x_{max}

$$\frac{\Delta\theta}{\Delta z} = 0^\circ \text{ K/100 m}$$

$a = 0.42 \text{ (m)}^{.25}$

$b = 0.39 \text{ (m)}^{.25}$

$p = 0.75$

$q = 0.75$

C. *Inversion Breakup* χ_{max} *and* x_{max}

$$\frac{\Delta\theta}{\Delta z} = 1.15^\circ \text{ K/100 m (before plume transition)}$$

$\bar{u} = 3$ m/sec (reasonable minimum $\bar{u}$)

$\rho_a = 1{,}220 \text{ g/m}^3$

$\sigma = 0.24 \text{ cal/(g °K)}$

$\kappa = 8 \times 10^3 \text{ cal/(m °K sec)}$

$a = 1.32 \text{ (m}^{.45})$

$b = 6.71 \text{ (m}^{.79})$

$p = 0.55$

$q = 0.21$

D. *Trapping* χ_{max} *and* x_{max}

$$\frac{\Delta\theta}{\Delta z} = 1.15^\circ \text{ K/100 m (before plume transition)}$$

$\bar{u} = 3$ m/sec (reasonable minimum $\bar{u}$)

$H_t = 762$ m (reasonable minimum mixing height—based on field experience)

$x_{max} = 3{,}000$ m (reasonable minimum x for trapping dispersion—based on field experience)

$a = 1.32 \text{ (m}^{.45})$

$b = 6.71 \text{ (m}^{.79})$

$p = 0.55$

$q = 0.21$

ACKNOWLEDGMENTS

The financial support and technical assistance furnished by the U.S. Environmental Protection Agency to TVA during its full-scale field studies of dispersion and plume rise are gratefully acknowledged. Also, the assistance of Dr. Frank Gifford and the staff of the Atmospheric Turbulence and Diffusion Laboratory, U.S. Department of Commerce, Oak Ridge, Tennessee, in preparation of the rapid-estimate technique presented above is likewise acknowledged with gratitude.

This article is a U.S. Government publication and therefore not subject to copyright. It was originally published as a paper in the May 1973 issue of the Journal of the Air Pollution Control Association and is reprinted with the concurrence of the Air Pollution Control Association.

NOMENCLATURE

a = crosswind dispersion coefficient corresponding to given atmospheric stability (m^{1-p})

b = vertical dispersion coefficient corresponding to given atmospheric stability (m^{1-q})

C = atmospheric stability coefficient for buoyant plume rise corresponding to given atmospheric stability (dimensionless)

Δh = plume rise (m)

$\frac{\Delta\theta}{\Delta z}$ = potential temperature gradient (°K/100 m)

g = 9.807, acceleration due to gravity (m/sec^2)

$G = V_s r^2 \left(\frac{\rho_a - \rho_s}{\rho_a}\right)$ (m^3/sec)

H_t = height of trapping dispersion mixing layer (m)

H_f = height of plume top prior to inversion breakup $= 2.15\,\sigma_z + h_e$ (m)

$h_d = H_f - h_s$ (m)

h_e = effective stack height $= h_s + \Delta h$ (m)

h_s = stack height (m)

N = number of power plant stacks of a given height (dimensionless)

p = crosswind dispersion exponent corresponding to a given atmospheric stability (dimensionless)

q = vertical dispersion exponent corresponding to a given atmospheric stability (dimensionless)

κ = eddy conductivity of the atmosphere [cal/(m °K sec)]

π = 3.1416 (dimensionless)

Q = effluent emission rate (kg/sec)

r = inside radius of stack top (m)

ρ_a = density of ambient air at stack top (g/m^3)

ρ_s = density of stack gas at stack top (g/m^3)

σ_y = standard deviation of crosswind normally distributed concentration of plume effluent for a specific downwind distance from source and specific atmospheric stability ($\sigma_y = ax^p$) (m)

σ_{yf} = standard deviation of crosswind normally distributed concentration of plume effluent for inversion breakup for a specific downwind distance from source ($\sigma_{yf} = \sigma_y + 0.47\ h_e$) (m)
σ_{yt} = standard deviation of crosswind normally distributed concentration of plume effluent for trapping dispersion for a specific downwind distance from the source [$\sigma_{yt} = \sigma_y + (H_t/1.1 - 2.15\ \sigma_z) * 0.47$] (m)
σ_z = standard deviation of vertical normally distributed concentration of plume effluent for a specific downwind distance from source and specific atmospheric stability ($\sigma_z = bx^q$) (m)
σ = specific heat of air at constant pressure [cal/(g °K)]
T_a = ambient air temperature at stack top (°K)
T_s = stack gas temperature at stack top (°K)
$\bar{u}$ = mean wind speed between stack top and plume top (m/sec)
V_s = stack gas exit velocity (m/sec)
x = distance downwind from source to where plume centerline surface concentration is estimated (m)
x_{max} = distance downwind from source where the maximum plume centerline surface concentration occurs (m)
χ = plume centerline 1-hour average surface concentration of effluent ($\mu g/m^3$)
χ_{max} = maximum plume centerline 1-hour average surface concentration of effluent for any downwind distance and wind speed ($\mu g/m^3$)

REFERENCES

1. Smith, Maynard, *et al. Recommended Guide for the Prediction of the Dispersion of Airborne Effluents* (New York: American Society of Mechanical Engineers, 1968).
2. Turner, D. Bruce. "Workbook of Atmospheric Dispersion Estimates," (Cincinnati, Ohio: U.S. Department of Health, Education, and Welfare, Public Health Service, National Center for Air Pollution Control, 1967).
3. U.S. Department of Atomic Energy. *Meteorology and Atomic Energy,* USAEC Division of Technical Information (1968).
4. Tennessee Valley Authority. "Full-Scale Study of Dispersion of Stack Gases — A Summary Report," Chattanooga, Tennessee (1964).
5. Tennessee Valley Authority. "Full-Scale Study of Plume Rise at Large Electric Generating Stations," Muscle Shoals, Alabama (1968).
6. Carpenter, S. B., *et al.* "Principal Plume Dispersion Models—TVA Power Plants," Paper No. 70-149, 63rd Annual Meeting of the Air Pollution Control Association, St. Louis, Missouri (1970).
7. Briggs, Gary A. "Plume Rise," T1D-25075 (Oak Ridge, Tennessee: U.S. Atomic Energy Commission, 1969).
8. Gifford, Frank A., Jr. "An Alignment Chart for Atmospheric Diffusion Calculations," *Bulletin of the American Meteorological Society,* **34** No. 3, 101-105; **No. 5, 216 (1953).**

6

Estimation of Downwash Effects

Gary A. Briggs
Air Resources Atmospheric Turbulence and Diffusion Laboratory,
NOAA, Oak Ridge, Tennessee

INTRODUCTION

This is a simplified approach to the calculation of ground-level concentrations of effluents from small industrial and fuel-burning installations. It is intended to serve as a first approximation to a very complex process. Because each stack, building, and terrain configuration is different, actual ground concentration may frequently differ from the values calculated here by a factor of two.[1]

The procedures given here were designed especially for source heights of less than 100 m; some of the simplifications made are not valid for large emissions. It is important to note that all *lengths are in meters* (m) and *velocities are in meters per second* (m/sec) in these formulas; this avoids needless reiteration of the formulas for different units. Table 6.1 provides all necessary conversion factors.

This chapter gives a method for predicting the occurrence of downwash. This is a common occurrence with small emissions, and greatly increases ground concentrations in the immediate vicinity downwind of the source.

ELEVATED SOURCE OR GROUND SOURCE?

The answer to this question can mean either a zero concentration or a very high concentration of effluent at the ground in the neighborhood of an emission. Does the plume keep its distance from the ground, and if so, what is its effective height, or is the plume brought to the ground very near the source? The latter can happen if the

Table 6.1. Conversion Factors.

1 m = 3.28 ft	1 cal/sec = 4.185 watt
10^3m = 0.621 mi	1 cal/g-°C = 1.00 Btu/lb-°F
1 sec = min/60 = hr/3600	1 watt = 1 joule/sec = 1 kg-m^2/sec^3
= day/86,400 = 30 days/259	1 ft = 0.305 m
10^6 = yr/3.14 • 10^7	1 mi = 1.61 • 10^3m
1 kg = 10^3 g = 10^6 mg = $10^9 \mu$g	1 lb = 0.454 kg
1 kg = 2.2 lb	1°R = 0.555 °K
1 °K = 1.8°R	°R = °F + 460
°K = °C + 273	1 Btu = 252 cal
1 cal = 1 g-cal = 10^{-3} kg-cal	1 ft/sec = 0.305 m/sec
1 cal = 0.00397 Btu	1 mph = 0.447 m/sec
1 m/sec = 3.28 ft/sec	1 ton/hr = 0.252 kg/sec
1 m/sec = 2.24 mph	1 ppm = $10^{-6}(m_o/24)$kg/m^3
	= $(m_o/24)$mg/m^3
1 kg/sec = 3.96 ton/hr	1 lb/hr = 0.126 g/sec
1 kg/m^3 = $10^6(24/m_o)$ppm	1 ton/mi^2 = 0.35 g/m^2
1 g/sec = 7.93 lb/hr	1 watt = 0.239 cal/sec
1 g/m^2= 2.85 ton/mi^2	1 Btu/lb-°F = 1.00 cal/g-°C

efflux velocity is too low, the stack is too short, or the emission is denser than air. Downwash of the plume due to terrain is also possible, particularly if there is an escarpment upwind of the source, but this case is relatively rare.

The answer to the above question can depend on the wind speed. It also can depend on the location of the stack relative to buildings and the wind direction. The great variety of possible building geometries gives ample reason for not expecting great accuracy from the following "rules-of-thumb."

Stack Aerodynamic Effect

An effluent emitted vertically from a stack can rise due to its momentum or can be brought downward by the low pressure in the wake of the stack. Which occurs depends on the ratio of the efflux velocity, v_s, to the crosswind velocity, u. Make the following computation, where D is the inside stack diameter and h_s is the source height above the ground:

$$h' = h_s + 2(v_s/u - 1.5)D \tag{1}$$

It is suggested that this be done for the following values of u: 1, 2.5, 4.5, 7, and 10 m/sec. The efflux velocity can be determined from direct measurement, from the amount of forced draft, from the rate of the process and relative proportions of its gaseous product (thermal expansion should be taken into account), or from visual or cinematographic estimates (if there are visible tracers in the effluent.) Building and stack measurements can be made directly, taken from drawings, or scaled from photographs.

If the effluent is emitted from a nonvertical stack or vent, set $h' = h_s$.

Equation (1) is based partly on wind tunnel observations of Sherlock and Stalker[2] who showed that downwash (negative rise) occurs when v_s/u is less than about 1.5 and that the plume downwashes about one stack diameter at $v_s/u = 1$. For high values of v_s/u, it is a conservative form of Equation (5.2) recommended by Briggs[3] for momentum rise; $2(v_s/u)D$ approximates the plume rise at the point where it is equal to 1.7 times the downwind distance, so essentially represents the very close-in plume rise. Buoyancy is neglected in this stage, since it does not cause a doubling of the plume rise until a distance $x = 10\ u\ v_s/(-g\ \Delta)$.

Building Effect

If the effluent is emitted from a stack or vent on or near a building, it may be brought downward by the flow of air over and around the building. Let l_b equal the lesser of the building height, h_b, or the building width perpendicular to the wind direction, w_b. If h' is less than $(h_b + 1.5\ l_b)$ and the point of emission is on the roof, anywhere within $l_b/4$ of the building, or within $3l_b$ directly downwind of the building, the plume can be considered to be within the regional of building influence. If the plume is within the region of building influence, there are several possibilities:

A. If h' in Equation (1) is less than $(h_b + 0.5l_b)$, part or all of the effluent is likely to circulate within the aerodynamic "cavity" that forms in the lee of the building. This cavity usually begins at the upwind edge of a flat roof or at the crest of a pitched roof (unless the crest is parallel to the wind). It grows to a height of about $(h_b + 0.5\ l_b)$ and a width a little greater than w_b, and extends over all lee sides of the building and downwind 2 to 3.5 l_b. Thus, effluents in the cavity region may affect persons on the ground and in the building. One must especially consider the placement of intake vents providing ventilation within the building. Following are some rough guidelines for estimating the concentration (χ) experienced in the cavity region. Let $\chi = KQ/(u\ l_b^2)$, where Q is source strength. If $H' > 0.35$, K is generally 1 or less throughout the cavity. If $H' < 0.35$, K is typically 1.5 and at most is 3.0, except on the side of the building where the effluent is emitted (for instance, the roof). Here, K can range up to 100 $(H' = (h' - b_b)/l_b)$, The concentration along the axis of the plume can be roughly approximated by $\chi = 4Q/(us^2)$, where s is the distance from the source measured along the axis. The airflow near buildings is complicated, and it is difficult to predict the trajectory of the plume axis. For example, in the cavity within $l_b/4$ of the roof, the flow is usually upwind.

B. If $h' > h_b$, compute $h'' = 2\,h' - (h_b + 1.5\,l_b)$ (2)
   If $h' < h_b$, compute $h'' = h' - 1.5\,l_b$ (3)

C. If h'' is greater than $l_b/2$, the plume remains an elevated source. If h'' is less than $l_b/2$, treat the plume as a ground source with an initial cross-sectional area $A = l_b^2$.

The above rules reduce to a simpler form in the case of a squat building, *i.e.*, when $h_b < w_b$: If $h' \geqslant 2.5\,h_b$, the plume escapes the region of building influence and $h'' = h'$; if $h' \leqslant 1.5\,h_b$, the plume downwashes into the building cavity [see (1) above] and also becomes a ground source with $A = h_b^2$ [see (3) above]; for in-between values of h', the plume remains elevated and $h'' = 2\,h' - 2.5h_b$ [see (2) above].

The method suggested here for accounting for the building effect is an interpolation of several rules-of-thumb respecting air flow around buildings. It is generally accepted that a building has very little effect on the airflow at 2½ building heights above the ground and above. On the other hand, the aerodynamic cavity downwind of a sharp-edged building develops to roughly 1½ building heights. It develops higher over a very wide (*i.e.*, squat) building, but the plume also has more distance in which to rise out of the cavity in this case. This method does allow some close-in plume rise to be considered with respect to escaping the cavity; however, it should be conservative since it does not allow for the lower wind speed near the building, which promotes greater rise. The cavity height may be less than 1½ building heights in the case of pitched or rounded roofs.

For a squat building, this method assumes that if $h' < 1.5\,h_b$, the plume behaves as if it were a ground source of initial area $A = h_b^2$. This gives concentrations in approximate agreement with those measured by Yang and Meroney[4] near the end of the cavity. The values of χ within the cavity adjacent to the building were estimated from measurements around cubes and rectangles by Halitsky.[5] Equation (2) is a linear interpolation formula giving $h'' = h'$ when $h' = 2.5\,h_b$ and $h'' = 0.5\,h_b$ when $h' = 1.5\,h_b$, thus giving a χ of the same order as that given for a ground source ($h' < 1.5\,h_b$, $A = h_b^2$).

For tall buildings, w_b replaces h_b as the characteristic cavity width and height above h_b. It is assumed that a roof level plume is not pulled all the way down to the ground within the cavity if $h_b > 2\,w_b$; hence, h'' is no more than $1.5\,w_b$ below h'.

Yang and Meroney[4] found that atmospheric stability had only a slight effect on concentrations immediately downwind of a building, so this effect is neglected here.

SUMMARY OF EQUATIONS AND WORKED EXAMPLE

Summary of Equations

Stack aerodynamic effect:

$$h' = h_s + 2D(v_s/u - 1.5). \tag{1}$$

Building effect: applies only if stack is on or near building or is within $3\, l_b$ downwind and $h' < h_b = 1.5\, l_b$, where l_b = lesser of h_b or w_b; if not, $h'' = h'$.

If $h' < h_b + 0.5\, l_b$, high concentrations may occur in building "cavity;"

If $h' > h_b$, compute $h'' = 2h' - (h_b + 1.5\, l_b)$ (2)

If $h' < h_b$, compute $h'' = h' - 1.5\, l_b$ (3)

If $h'' < l_b/2$, treat plume as ground source with $A = l_b^2$

Worked Example

Suppose we have a plant with a stack located on the roof of a long, flat building with the following specifications:

$h_b = l_b = 66$ ft $= 20$ m
$h_s = 99$ ft $= 30$ m
$r = 1.64$ ft $= 0.5$ m
$V_s = 16$ ft/sec $= 5$ m/sec
Q (SO_2) = 0.035 lb/sec = 0.016 kg/sec
Effluent temperature = 290 °F = 416 °K
Ambient temperature = 70 °F = 294 °K
Distance to nearest property line = 330 ft = 100 m

Solution

Stack aerodynamic effect: $h' = h_s + 4\,(V_s/u - 1.5)\,r$
$u = 1$ m/sec: $h' = 30 + 4\,(5/1 - 1.5)\,(0.5) = 37$ m
$u = 2.5$ m/sec: $h' = 30 + 4\,(5/2.5 - 1.5)\,(0.5) = 31$ m
$u = 5$ m/sec: $h' = 30 + 4\,(5/5 - 1.5)\,(0.5) = 29$ m
$u = 10$ m/sec: $h' = 30 + 4\,(5/10 - 1.5)\,(0.5) = 28$ m

Building effect: $h'' = 2h' - (h_b + 1.5\, l_b)$ if 30 m $< h' <$ 50 m
$h'' = h_b - l_b = 0$ if $h' <$ 30 m (downwash)
$u = 1$ m/sec: $h'' = 2\,(37) - 50 = 24$ m
$u = 2.5$ m/sec: $h'' = 2\,(31) - 50 = 12$ m
$u = 5$ m/sec: $h'' = 0$
$u = 10$ m/sec: $h'' = 0$
(plume becomes ground source when $h' = 30$ m, $h' = h_s + 4\,(V_s/u - 1.5)r$
Solving for u, $30 = 30 + 4\,(5/u - 1.5)\,(0.5)$, $u = 3.33$ m/sec

SYMBOLS AND DEFINITIONS

Symbol	*Definition*	*Units*
A	Initial cross-sectional area of a ground plume	m^2
D	Inside stack diameter	m
h	Effective source height (after stack aerodynamic, building, and buoyancy effects have been accounted for)	m
h′	Plume height after stack aerodynamic effect is accounted for	m
h″	Plume height after building effect is accounted for	m
h_b	Building height	m
h_s	Source height above the ground	m
K	Dimensionless concentration coefficient in cavity region	—
l_b	The lesser of h_b or w_b	m
Q	Source strength of a component of the effluent	kg/sec
u	Wind speed at source height or at an open location	m/sec
v_s	Average efflux velocity (volume flow rate ÷ area)	m/sec
w_b	Building width perpendicular to the wind direction	m
x	Distance downwind of the source	m
χ	Concentration of a component of the effluent	kg/m^3

REFERENCES

1. Smith, Maynard (Ed.) *Recommended Guide for the Prediction of the Dispersion of Airborne Effluents.* (New York: American Society of Mechanical Engineers, 1968).
2. Sherlock, R. H. and E. A. Stalker. "A Study of Flow Phenomena in the Wake of Smoke Stacks," *Engineering Research Bulletin No. 29* (Ann Arbor, Michigan: University of Michigan, 1941).
3. Briggs, G. A. *Plume Rise,* AEC Critical Review Series TID-25075 (1969).
4. Yang, B. T. and R. N. Meroney. "Gaseous Dispersion into Stratified Building Wakes," Fluid Dynamics and Diffusion Laboratory, Colorado State University, Fort Collins, CER70-71BTY-RNM-8 (1970).
5. Halitsky, J. "Gas Diffusion near Buildings," In *Meteorology and Atomic Energy,* USAEC Report TID-24190 (1968), pp. 221-255.

7

Dispersion of Sulfur Dioxide Emissions from Area Sources

F. A. Gifford and S. R. Hanna
Atmospheric Turbulence and Diffusion Laboratory,
National Oceanic and Atmospheric Administration,
Oak Ridge, Tennessee

INTRODUCTION

This chapter can be used in evaluating surface concentrations of sulfur dioxide emitted generally over an area. The most common example of an area source is provided by urban domestic and industrial heating, for which it is necessary to consider the sum of a large number of individual sources over an area of some convenient size. Given an area-source inventory, this chapter provides a simple, systematic procedure for calculating the surface concentration at any point, averaged over an hour, a day, or a year. The basic physical model of atmospheric diffusion employed is the same as for Chapter 5 on tall-stack emissions.

Area-source concentration calculations are a recent development compared with calculations of concentrations from isolated sources. There are discussions of some of the methods now available, but these all require the use of a high-speed electronic computer.[1] The method to be described, although very simple, has been shown[2] to provide results of a precision comparable with that of the more elaborate computer calculations and can be used with confidence for area-source calculations in urban air pollution and regional air quality studies. An even simpler area-source calculation[3] suitable for many air pollution control purposes is also described. The model is described here in such a way that all the calculations can be done by hand. For the purpose of making many applications of the model, it has also been programmed in Fortran.[4]

THE AREA-SOURCE GRID

The area source strength data are assumed to be given in the form of a regular "checkerboard" grid pattern. This area-source grid pattern is the usual one in which source information is provided. If the available source strength data are given in some other pattern, for instance by counties or districts, they should be modified to approximate a regular checkerboard pattern by choosing the largest grid-square size that will give an adequate representation of the given data. Squares of 5 x 5 km are fairly standard.

CALCULATION OF AREA-SOURCE SURFACE CONCENTRATION: ONE-HOUR PERIOD

It is assumed that the concentration is required at the center of a grid square, which is designated square "O." The procedure when the receptor is not centered on a grid square is covered under Correction Factors, below. The calculation requires evaluation of the following simple formula for the area-source concentration, χ_A:

$$u\chi_A = c_oQ_o + c_1Q_1 + c_2Q_2 + c_3Q_3 + c_4Q_4 + c_5Q_5 \quad (1)$$

The Q's are the given area source strength emissions per unit area per second in a line of source-grid squares centered on the square containing the receptor, *i.e.*, square "O" whose strength equals Q_o, and extending *upwind*. The c's are multipliers that depend on the meteorological conditions and the grid size. For a 5 x 5 km grid the multipliers are given in Table 7.1.

Table 7.1. Grid Multipliers for Meteorological Dispersion Conditions

	Neutral	*Stable*	*Unstable*
c_o:	153	331	137
c_1:	48	124	23
c_2:	28	73	12
c_3:	20	54	8.3
c_4:	16	44	6.7
c_5:	14	38	5.3

The area-source calculation, Equation (1), requires combination of these multipliers with upwind source strengths. Since this may have to be done for a number of receptor points and wind directions, it is convenient to introduce the multipliers by means of a multiplier grid, Figures 7.1–7.3, prepared from Table 7.1.

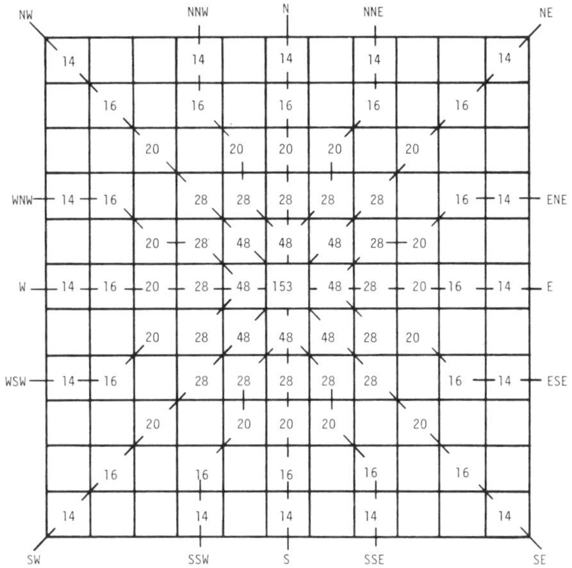

Figure 7.1. Multiplier grid for neutral meteorological conditions.

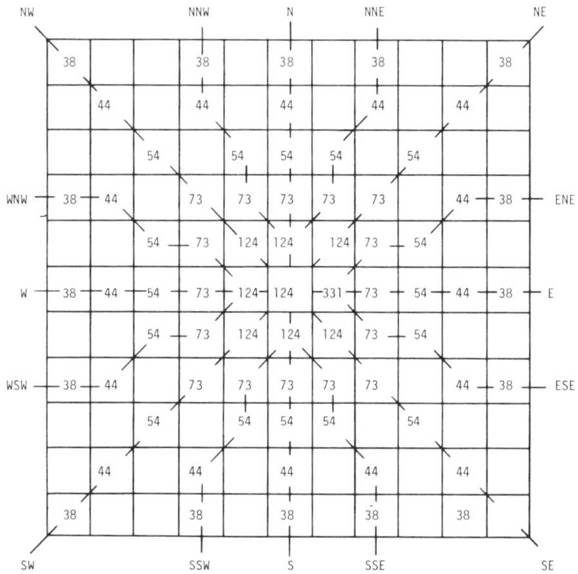

Figure 7.2. Multiplier grid for stable meteorological conditions.

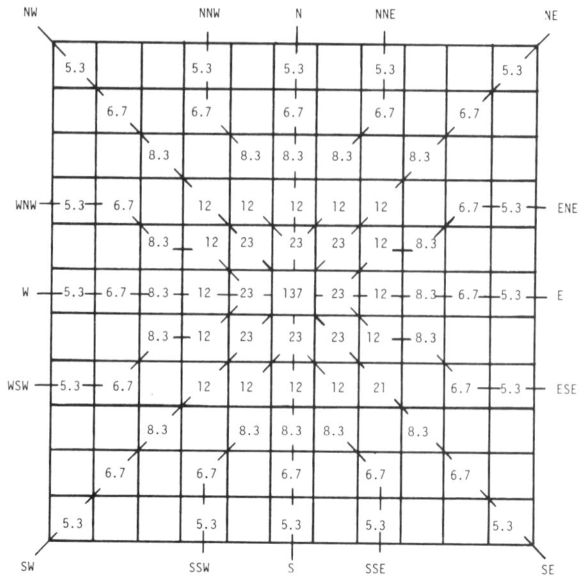

Figure 7.3. Multiplier grid for unstable meteorological conditions.

Step 1 — Plot the area-source strength data at the centers of 1/2-inch squares. This is the "source grid." It is convenient for the following steps to do this on tracing paper or, better, to make a transparency.

Step 2 — Select the multiplier grid, Figures 7.1, 7.2, or 7.3, appropriate to the given meteorological conditions. Center the source grid over the multiplier grid so that square "O" corresponds to the source grid square that contains the receptor point.

Step 3 — Six numbers on the multiplier grid, including the central number at square "O," correspond to each wind direction. Multiply these by the six numbers in the corresponding source-grid squares and sum these products.

Step 4 — Divide the result of Step 3 by the given mean wind speed, u. The result is the required hourly average area-source concentration value.

Step 5 — Relocate the multiplier grid to a new receptor square and repeat Steps 1 through 5.

EQUATIONS FOR AREA SOURCE

The equation for surface concentrations from an area source:

$$\chi_A = \left(\frac{2}{\pi}\right)^{1/2} \frac{(\Delta x/2)^{1-q}}{b(1-q)} \left[Q_o + \sum_{i=1}^{5} Q_i \left[(2i+1)^{1-q} - (2i-1)^{1-q}\right] \right] \quad (2)$$

is based directly, on Equation (2) of Chapter 5; Δx is the size of the source-strength grid, and the rest of the terms are defined in the nomenclature. Equation (2) must be multiplied by the frequency f_d of wind from the d direction, to obtain the annual average.

Comparing Equations (1) and (2), it can be seen that the dimensionless multipliers c_i, i= 0, 1, . . . are given by:

$$c_o = \left(\frac{2}{\pi}\right)^{1/2} \frac{(\Delta x/2)^{1-q}}{b(1-q)} \quad (3)$$

$$c_1 = \left(\frac{2}{\pi}\right)^{1/2} \frac{(\Delta x/2)^{1-q}\ (3^{1-q}-1)}{b(1-q)} \quad (4)$$

$$c_2 = \left(\frac{2}{\pi}\right)^{1/2} \frac{(\Delta x/2)^{1-q}\ (5^{1-q}-3^{1-q})}{b(1-q)} \quad (5)$$

and so on. The values appearing in Table 7.1 were calculated from these equations.

CALCULATION OF ANNUAL AVERAGE AREA-SOURCE CONCENTRATION

The procedure is the same as the previous one. The exception is that the annual average wind direction frequency must be included. It is assumed that a standard, 16-point wind direction frequency distribution (wind rose) is available, giving the fraction, f_d, of the time the wind blows *from* each direction, d. The average annual wind speed is used and neutral meteorological conditions are assumed to apply. Because the wind rose varies from place to place, it is necessary to calculate a new multiplier grid for each application (each city). The complete procedure is as follows.

Step 1 — Form an annual average multiplier grid by multiplying each of the values in the hourly average multiplier grid for *neutral meteorological conditions,* Figure 7.1, by the corresponding numbers in the wind direction frequency grid, Figure 7.4, after first summing the f_d's as indicated in Figure 7.4; f_1 is the relative frequency (fraction) of winds from the NNE, f_2 that from NE, . . ., f_{16} that from N.

f_{14}			f_{15}		f_{16}		f_1			f_2
	f_{14}		f_{15}		f_{16}		f_1		f_2	
		f_{14}		f_{15}	f_{16}	f_1		f_2		
f_{13}	f_{13}		f_{14}	f_{15}	f_{16}	f_1	f_2		f_3	f_3
		f_{13}	f_{13}	f_{14}	$f_{15}+f_{16}+f_1$	f_2	f_3	f_3		
f_{12}	f_{12}	f_{12}	f_{12}	$f_{11}+f_{12}+f_{13}$	1	$f_3+f_4+f_5$	f_4	f_4	f_4	f_4
		f_{11}	f_{11}	f_{10}	$f_7+f_8+f_9$	f_6	f_5	f_5		
f_{11}	f_{11}		f_{10}	f_9	f_8	f_7	f_6		f_5	f_5
		f_{10}		f_9	f_8	f_7		f_6		
	f_{10}		f_9		f_9		f_7		f_6	
f_{10}			f_9		f_8		f_7			f_6

Figure 7.4. Wind direction frequency grid.

Step 2 — Center the source-grid square containing the receptor point over the multiplier grid. It is convenient for this purpose to use a transparent source grid. (If the annual average source strength differs from the short period average value, then of course the annual source strengths must be used.)

Step 3 — Multiply *all* the number pairs of these two grids, and sum. (Where a square of the multiplier overlay falls outside the source grid area, the product is zero.)

Step 4 — Divide the result of Step 3 by the annual average wind speed to get χ_0.

Step 5 — Relocate the multiplier grid to a new receptor square and repeat Steps 1 through 5.

Correction Factors

The following procedures should be applied, if required, to the values determined by the foregoing procedure.

One-Hour to 24-Hour or Other Averaging Time

To calculate average area-source concentrations for 24-hour or other (*e.g.*, monthly) periods, follow the same procedures as for the annual average except that wind direction frequencies appropriate to the period in question must be used.

Grid Size Other Than 5 x 5 km

Multiply the value obtained at Step 4 by the appropriate correction factor shown in Table 7.2.

Table 7.2. Grid Size Correction Factors–Meteorological Conditions.

Grid Size (km)	*Neutral*	*Stable*	*Unstable*
1	.94	.97	.93
2	.97	.98	.96
3	.98	.99	.98
4	.99	1.00	.99
5	1.00	1.00	1.00
10	1.03	1.01	1.03
15	1.04	1.02	1.05
20	1.05	1.02	1.07
25	1.06	1.03	1.08

Receptor Point Not at Center of a Source-Grid Square

Calculate χ_0 for the center of the source-grid square containing the receptor point, and for the next nearest square beyond the receptor point. Interpolate linearly between these values.

Worked Example

Problem

The source inventory for sulfur dioxide assuming no decay for a (real) city is given in Figure 7.5. The grid spacing is 5 km. (1) Calculate the sulfur dioxide concentration for a receptor located at the center of source square with the heavy outline, for the case of a WNW wind at 10 mph (3.4 m/sec) under neutral meteorological conditions. (2) Given the following annual wind frequency distribution, calculate the annual average concentration for a receptor located at the center of the same source square.

.15	.16	.27	.18	.14	.07
.39	.45	.58	.27	.20	.07
.22	.64	1.42	.36	.24	.14
.10	.34	.50	.14	.05	.05
.08	.31	.13	.05	**.05**	.05
.05	.12	.26	.08	.10	.11

Figure 7.5. Source inventory grid, annual average particulate emissions ($\mu g/m^2$ sec).

Direction	f_d	*Direction*	f_d
NNE	.02	SSW	.05
NE	.04	SW	.05
ENE	.08	WSW	.05
E	.08	W	.05
ESE	.08	WNW	.12
SE	.05	NW	.14
SSE	.04	NNW	.07
S	.04	N	.02

Solution

Procedure for Hourly Average Concentration. Select the multiplier grid for neutral conditions, Figure 7.1. Center over this the square of the source inventory overlay (Figure 7.5) containing the receptor point. Corresponding to WNW wind, it is seen that Equation (1) gives:

$$\chi_0 = \frac{1}{u}(153 \times .05 + 48 \times .05 + 28 \times .50 + 20 \times .34 + 16 \times .22 + 14 \times 0) = 34.4/3.4 = 10. \mu g/m^3$$

Procedure for Annual Average Concentration. Form an annual average multiplier grid by multiplying the corresponding numbers in Figure 7.6 and Figure 7.1. The result is Figure 7.7. Figure 7.6 was obtained by substituting the actual wind direction frequencies in the above table into the annual wind frequency grid, Figure 7.4. The receptor point of Figure 7.5 is now centered over the multiplier grid, Figure 7.7, and all corresponding numbers are multiplied and added; the result equals 24.03. The annual average concentration is:

$$24.03/3.4 = 7.1\ \mu g/m^3$$

.14			.07		.02		.02			.04
	.14		.07		.02		.02		.04	
		.14		.07	.02	.02		.04		
.12	.12		.14	.07	.02	.02	.04		.08	.08
		.12	.12	.14	.11	.04	.08	.08		
.05	.05	.05	.05	.22	1.0	.24	.08	.08	.08	.08
		.05	.05	.05	.13	.05	.08	.08		
.05	.05		.05	.05	.04	.04	.05		.08	.08
		.05		.05	.04	.04		.05		
	.05		.05		.04		.04		.05	
.05			.05		.04		.04			.05

Figure 7.6. Wind direction frequency grid for example.

A Simple Approximation to the Area-Source Calculation. Area-source strength values do not in practice vary much from one grid square to another. Furthermore, the grid multipliers of Table 7.1 show that the contribution to χ_a from the zeroth square (strength =

Q_o) is weighted most heavily. This suggests the following simple approximation to Equation (1):

$$u\chi_a = CQ_o \tag{6}$$

where the coefficient C is obtained by summing the values given in Table 7.1, *i.e.*,

$$C = c_o + c_1 + c_2 + c_3 + c_4 + c_5 \tag{7}$$

thus C equals 279 for neutral conditions, 664 for stable conditions, and 192 for unstable conditions.

Corresponding to the conditions of the worked example, Equation (6) provides the result, $\chi_a = 4.1\ \mu g/m^3$. This is in reasonably good agreement with the results using Equation (1), especially the annual average value. Hanna[3] has made a number of comparisons and finds Equation (6) to be a valid, useful area-source formula.

2.52			.98		.28		.28			.56
	2.24		1.12		.32		.32		.64	
		2.80		1.40	.40	.40		.80		
1.68	1.92		3.92	1.96	.56	.56	1.12		1.28	1.12
		2.40	3.36	6.72	5.28	1.92	2.24	1.66		
.70	.80	1.00	1.40	10.56	153	11.52	2.24	1.60	1.28	1.12
		1.00	1.40	2.40	6.24	2.40	2.24	1.60		
.70	.80		1.40	1.40	1.12	1.12	1.40		1.28	1.12
		1.00		1.00	.80	.80		1.00		
	.80		.80		.64		.64		.80	
.70			.70		.56		.56			.70

Figure 7.7. Annual average multiplier grid for example.

NOMENCLATURE

b = vertical dispersion coefficient corresponding to given atmospheric stability (m^{1-q})
c = multiplier that depends on the meteorological conditions and the grid size
Δx = size of the source-strength grid (m)
f_d = fraction of the time the wind blows from each direction (dimensionless)
p = crosswind dispersion exponent corresponding to a given atmospheric stability (dimensionless)
q = vertical dispersion exponent corresponding to a given atmospheric stability (dimensionless)
Q_N = area source strengths of grid square N ($\mu g/m^2$ sec)
u = mean wind speed in plume region (m/sec)
χ_A = area source 1-hour average surface sulfur dioxide concentrations for the center of the source-grid square A ($\mu g/m^3$)

REFERENCES

1. Stern, A. C., (ed.) "Proceedings of Symposium on Multiple-Source Urban Diffusion Models," USEPA (1970).
2. Gifford, F. A. and S. R. Hanna. "Urban Air Pollution Modeling," *Proceedings of The Second Clean Air Conference*, H. M. England and W. T. Beery, eds. (New York: Academic Press, 1971), pp. 1146-1151.
3. Hanna, S. R. "A Simple Method of Calculating Dispersion from Urban Area Sources," *J. Air Poll. Control Assoc.*, 21, 774-777 (1971).
4. Hanna, S. R. "Description of ATDL Computer Model for Dispersion from Multiple Sources," *Industrial Air Pollution Control* (Ann Arbor: Ann Arbor Science, 1973), pp. 23-32.

8

Calculation of Cooling Tower Drift Deposition

P. R. Slawson and **T. M. L. Wigley**
Envirodyne Limited, Waterloo, Ontario, Canada

INTRODUCTION

The increasing demand for electricity along with the lack of sufficient quantities of water for once-through cooling and the concern for the effects of the rejection of waste heat on the aquatic environment has led to a greater dependence on mechanical and natural draft cooling towers. The warm moist plumes from these cooling towers contain water vapor and liquid water, the latter as both condensed and drift or carry-over droplets. Drift droplets, along with their dissolved and suspended solids that originate from the cooling water source and from additives used in the treatment of the condenser water, are dispersed and may subsequently be deposited on the surrounding terrain and vegetation. The environmental impact of this drift deposition and of the excess moisture in the cooling tower plumes, must be assessed as part of the overall environmental impact assessment for a proposed power plant. In the case of drift, the drift deposition rate (mass of deposited solids per unit impact area per unit time) can be calculated. The effect of excess moisture can be assessed by calculating the frequency of occurrence of visible (condensed) plume lengths, fog, icing, and increased relative humidity due to the cooling tower plume. These items are collectively referred to as "fogging potential." Both drift and "fogging" calculations require the details of representative climatic data for a given site.

There are several models in existence that calculate drift rate and fogging potential. Some of these will be referenced later in this chapter. This chapter will first describe a drift model (which, in principle, is similar to many others) and the application of this model to a specific plant site. The description of the model emphasizes the basic physical aspects and is general enough to give the reader an understanding of the problem from which less sophisticated and more easily used models may be developed. A thorough knowledge of vapor plume dynamics is essential to the development of a comprehensive drift model.

Over the past few years a considerable effort has been made towards increasing our understanding of the dynamics of vapor plumes. Much of this work has been based on the application, by Morton,[1] of the Morton, Taylor and Turner entrainment theory[2] to vapor plumes. The recent revival of interest in vapor plumes was begun by Csanady[3] and subsequently extended by Wigley and Slawson,[4,5] Hanna,[6] Slawson,[7] Richards,[8] Weil,[9] Wigley,[10] and others. The particular problem of condensed plume length prediction, and the possible approaches to this problem, have been summarized by Slawson *et al.*[11] and Wigley and Slawson.[12] The theoretical results developed by these authors are based on physical principles that have been well-proved by dry plume observations. Limited testing against full-scale condensed plume data has, to date, succeeded in confirming many predictions (*see*, for example, Slawson *et al.*[11]).

THEORETICAL CONSIDERATIONS

Due to the great complexity of the drift deposition problem the theoretical models are still relatively crude. They must synthesize state-of-the-art knowledge of droplet dynamics, plume behavior and atmospheric dispersion and must be based in part on theoretical results which have yet to be verified, or even tested, in the field. It is quite possible that predicted deposition rates using the best possible drift models are accurate to little better than an order of magnitude.

The drift problem is basically one of determining the life cycle (size and trajectory) of a "test" drift droplet, of arbitrary initial size, from the drift eliminators to its eventual deposition on the ground at some point downwind from a cooling tower. For large droplets this is a deterministic calculation, similar to determining the trajectory of a projectile (of variable size). For smaller droplets the calculation must be a statistical one based on diffusion theory. All droplets, during their life cycles, spend various times in four distinctly different environments (Figure 8.1): (a) inside the cooling tower shell, (b) inside the condensed (*i.e.*, visible) portion of the cooling tower plume, (c) inside the "dry" invisible portion of the plume, and (d) in the free atmosphere. The drift calculation must determine what happens to a droplet in each of these zones and how long a droplet remains in each zone. Calculations must also include the influence of downwash and multiple sources on plume trajectory, the effect of droplet washout by rain, and the influence of terrain height variations.

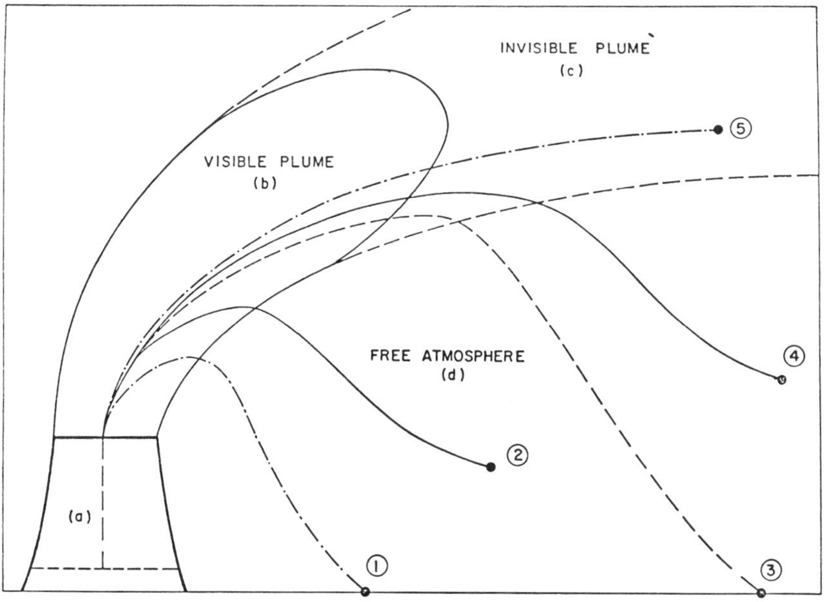

Figure 8.1. Different zones in a drift droplet's life cycle and a classification of the possible types of droplet trajectories. Types 1 and 3 reach ground-level. Types 2 and 4 fall out of the plume but evaporate before reaching ground level. Type 5 evaporates within the invisible part of the plume and so never leaves the plume.

Once the behavior of a test droplet is known, similar results for test droplets of various sizes must be summed to cover the initial distribution of droplet sizes. In principle this provides an estimate of the deposition rate for any given set of atmospheric and initial plume conditions. Climatic data can then be used to sum these deposition rates to find, for example, monthly or annual mean deposition rates, or they can be used to identify extreme conditions (and the frequency of such conditions) which lead to the highest deposition rates.

The following equations must be used to determine drift deposition rates: (a) the equations for cooling tower plume dynamics, (b) the droplet growth (or decay) question, (c) the trajectory equations for a droplet, and (d) an appropriate diffusion equation for a droplet cloud.

The boundary conditions which affect the solutions to these equations are: (1) atmospheric conditions (temperature, humidity and wind speed profiles), (2) initial plume conditions (temperature, moisture content and efflux velocity), (3) the cooling tower (or towers) design specifications (tower-top radius, tower height, number and relative position of towers), (4) initial drift droplet size distribution, chemical composition and drift rate, and (5) the surrounding terrain profile.

a. *Equations for cooling tower plume dynamics*

$$\frac{d}{dt}(VR^2) = \frac{2v_e}{R}VR^2 \tag{1}$$

$$\frac{d}{dt}(VR^2(\Delta q + \sigma)) = -GVR^2w \tag{2}$$

$$\frac{d}{dt}\left\{VR^2g\left(\frac{\Delta T^*}{T_a^*} - \frac{L\sigma}{C_pT_a^*}\right)\right\} = -N^2VR^2w \tag{3}$$

$$\frac{d}{dt}(VR^2w) = gVR^2\left(\frac{\Delta T^*}{T_a^*} - \sigma\right) \tag{4}$$

$$\frac{d}{dt}(VR^2v_x) = U\frac{d}{dt}(VR^2) \tag{5}$$

$$\frac{dz}{dt} = w \tag{6}$$

$$\frac{dx}{dt} = v_x \tag{7}$$

Time, t, along the centerline plume path has been used as the independent variable (*i.e.*, ds = Vdt = increment of distance along the plume path). The meanings of other variables are given in the Nomenclature (*see* page 104). The first five equations describe conservation of mass, moisture, energy, vertical momentum and horizontal momentum, the latter ignoring solid-body drag effects. Equations (6) and (7) are needed to relate plume rise, z, and downwind distance, x, to the variable t. In Equation (1), v_e is an entrainment velocity which is determined by the dominant mechanism of mixing between plume and environment. In the initial phase of plume rise, $v_e = \alpha|w|$. It will be supposed that α is a constant, although there is strong evidence that this supposition is invalid close to the efflux point (*see*, for example, Hoult and Weil[13]).

Equations (1–7) do not form a closed set and the conditions $q_p = q_{sp}$ if $\sigma > 0$ can be used to close them. A more consistent approach is to incorporate supersaturation as an additional variable, but Wigley[10] has shown that the above equilibrium approximation is an excellent one for cooling tower plumes.

b. *The droplet growth (or decay) equation*

For droplet growth (decay) by condensation (evaporation) droplet radius, r, is related to supersaturation, E, by

$$\frac{dr}{dt} = \frac{\rho_e K^*}{4\pi r} (1 + .22\, a\sqrt{Re}) \left(E + \frac{K_2}{r^3} \right) \qquad (8)$$

where $$Re = \frac{2\rho_e\, r\, V_{dr}}{n_e}$$

is the droplet Reynolds number. V_{dr} is the magnitude of the droplet velocity relative to the environment, ρ_e is environment density and n_e (a function of temperature) is the environment viscosity. Note that the environment may be plume or atmosphere depending on the droplet location. The constant K* depends on temperature,[10] the K_2 term accounts for solute effects, and curvature effects have been neglected. The constant K_2 is of the order 5n where n is the number of g-moles of solute in the droplet. For seawater $K_2/r^3 \backsim 0.01$ so that, for droplets with concentrations of the order of seawater or higher the solute term in Equation (8) is significant. The factor $1 + .22\, a\sqrt{Re}$ in Equation (8) is a ventilation factor due to Kinzer and Gunn (*see*, for example, Byers[14]). The term a is an empirically determined function of droplet Reynolds number: $Re \leqslant 1$, $a = 0$; $Re \geqslant 100$, $a = 1$; and for $1 < Re < 100$,

$$a = 2(Re - 1) \quad : \quad Re \leqslant 2.1$$

$$a = 1 + .71523(2 - \log(Re)) \quad : \quad Re > 2.1$$

Equation (8) is a standard cloud physics result which may be derived by reference to works such as those by Byers[14] and Fletcher.[15] Droplet growth will be discussed in detail in a later section. Coalescence effects will also be considered in this later section.

c. *The trajectory equation for a droplet.*

The equation of motion for a droplet can be written in the form

$$m \frac{dV_{-d}}{dt} \approx mg + 6\pi n_e r\, V_{-dr} \frac{C_D Re}{24} \qquad (9)$$

where V_{-d} is the droplet velocity, m is the droplet mass and C_D is the droplet drag coefficient. C_D can be predicted very accurately for droplets up to 1000μm (0.1 cm) radius using

$$C_D = \frac{24}{Re} (1 + 0.144\, Re^{0.684}) \qquad (10)$$

Droplet radius and mass are, in general, constantly changing quantities so Equation (9) is coupled to the previous equations.

Droplet trajectories may be classified into five groups depending primarily on their endpoints: *i.e.*, the point where their mean vertical velocity reaches (effectively) zero. This may occur either by collision with the ground or by evaporating to a size so small that they no longer follow a ballistic trajectory, but are dispersed by atmospheric turbulence. The five classes are shown in Figure 8.1.

For "large" times, the vertical component of the droplet relative velocity tends to the terminal velocity, and the horizontal component tends to zero. The terminal velocity is, of course, dependent on droplet size and is therefore a constantly changing quantity.

Equation (9) can be solved analytically if a few approximations are made. By so doing it can be shown that the relaxation time for the approach of the relative-velocity vertical component to the terminal velocity is approximately

$$\tau_Z = \frac{V_T}{2g}$$

where V_T is the terminal velocity. For droplets in the drift droplet size range this relaxation time is very small (for example, for droplets of 300μm radius, τ_Z is about 0.13 seconds: for smaller droplets τ_Z is even smaller).

This is an important result since it means that by the time drift droplets have risen to the tower top they will be in dynamic equilibrium with their environment and will therefore be moving at terminal velocity relative to that environment. This gives us a simple way to determine droplet initial conditions at tower top. More importantly, it means that the only drift droplets which can escape from the tower are those whose terminal velocity is less than the tower-gas exit velocity.

d. *Droplet diffusion*

One available particulate diffusion model is the "tilted plume model". In particulate deposition problems, particles can be considered in three separate groups. Very small particles (radius $\leqslant$ 10μm, terminal velocity $\lesssim$ 1 cm/sec) are dispersed solely by the turbulence of the environment and their deposition rates from a point source can be calculated using simple Gaussian diffusion theory. Large particles (radius $\geqslant$ 100μm, terminal velocity $\gtrsim$ 100 cm/sec) are virtually unaffected by environmental turbulence in the atmosphere and their deposition rates can be calculated using the droplet trajectory equations alone. Between these limits both diffusion and trajectory effects are important and one can calculate deposition rates using a modified Gaussian diffusion model, the tilted plume model. This model assumes that droplets are dispersed by atmospheric turbulence according to Gaussian diffusion, but about the droplet trajectory for a single droplet. The tilted plume model includes the small particle group as a special case. The delimiting sizes quoted above are those given by Slade.[16] The limits are, in fact, not well defined and will vary with the level of atmospheric turbulence (or, more correctly, with the level of turbulence of the droplet environment, which will be the cooling tower vapor plume for a considerable part of the droplet's life). In the present model, we have used 50μm as the cut off radius between diffused and nondiffused droplets.

To apply the tilted plume model to the dispersion of the small-droplet plume the droplet trajectory equation is needed to predict the trajectories of test droplets of different (and variable) sizes. These predictions rest heavily on the droplet growth equation, which determines how the size of a droplet varies, and on the plume equations, which determine the environment velocity field for droplets within the vapor plume. Because of the importance of the environment velocity field on droplet trajectory, we have used a numerical solution of the plume equations. By using such a numerical solution we can incorporate into our model the following improvements over other drift deposition models:

1. The effect of latent heat release on plume rise is automatically included, and
2. Plume vertical velocity, relative humidity and length are modeled realistically.

For a point source of strength Q (mass units per unit time) the concentration at a point (X, Y, Z) in the absence of any boundaries is given by

$$C = \frac{Q}{2\pi U \, \sigma_Y \, \sigma_Z} \exp\left\{-\frac{1}{2}\left(\frac{Y^2}{\sigma_Y{}^2} + \frac{Z^2}{\sigma_Z{}^2}\right)\right\} \tag{11}$$

where U is the windspeed and σ_Y and σ_Z are dispersion coefficients in the Y and Z directions which depend on X and on atmospheric stability. This result applies strictly only to dispersion about a horizontal path, but it is frequently applied to tilted trajectories to yield quite reasonable results. It should also be noted that C is a time-averaged concentration (over a time interval of 10–60 minutes) rather than an instantaneous value.

The total mass flux past some point X is

$$Q = U \int_{-\infty}^{\infty} \int_{-\infty}^{\infty} C \, dZ \, dY \tag{12}$$

If the plume centerline at point X is at a height H above ground level then the mass flux above ground level is

$$U \int_{-\infty}^{\infty} \int_{-H}^{\infty} C \, dZ \, dY \tag{13}$$

The difference between (12) and (13) must be the amount of material deposited between the point source and X. The rate of deposition at a point (X, Y) at ground level must therefore be

$$DR = \frac{\delta}{\delta X}\left\{U \int_{H}^{\infty} C \, dZ\right\} \tag{14}$$

in mass units per unit area per unit time. Note that C involves σ_Y and σ_Z which are functions of X and, for a tilted plume, H is also a function of X.

Using (11), (14) can be evaluated to yield

$$DR = \frac{Q}{\sqrt{2}\,\pi\,\sigma_Y}\left[e^{-\xi^2}\left\{\xi\,\frac{\delta ln\sigma_Z}{\delta X} - \frac{1}{\sqrt{2}\,\sigma_Z}\,\frac{\delta H}{\delta X}\right\} + \frac{\sqrt{\pi}}{2}(1 - \operatorname{erf}\xi)\,\frac{\delta ln\sigma_Y}{\delta X}\left\{\frac{Y^2}{\sigma_Y{}^2} - 1\right\}\right]\exp\left(-\frac{1}{2}\,\frac{Y^2}{\sigma_Y{}^2}\right) \tag{15}$$

Here, erf is the error function and ξ is a dimensionless variable which occurs commonly in pollutant dispersion work, *i.e.*,

$$\xi = \frac{H}{\sqrt{2}\sigma_Z}$$

Equation (15) differs from other formulas in the literature, but reduces to a more familiar form in certain limiting cases. For large ξ (corresponding roughly to $X < 20$ km) the $ln\sigma_Y$ derivative term is negligible. Deposition rate is therefore controlled by the $ln\sigma_Z$ and H derivative terms. For small H, and provided that V_T is not too small, the $\frac{\delta H}{\delta X}$ term dominates and (15) reduces to

$$DR = \frac{Q\,V_T}{2\pi\,\sigma_Y\,\sigma_Z}\,e^{-\xi^2}\,\exp\left(-\frac{1}{2}\,\frac{Y^2}{\sigma_Y{}^2}\right) = V_T C_o \tag{16}$$

which is the usual deposition rate formula for particulate clouds (*see*, for example, Slade,[16] p. 203): C_o is the ground level concentration.

Equation (16) is not applicable in the present case because we are particularly interested in deposition rates for droplets which evaporate completely at some appreciable distance above ground (*i.e.*, H is not small, but V_T is).

For constant H and $\xi \gtrsim 1$ Equation (15) reduces to

$$DR \simeq HUC_o\,\frac{\delta ln\sigma_Z}{\delta X}$$

Since $\frac{\delta ln\sigma_Z}{\delta X} \simeq \frac{b}{X}$ (where b varies between about 0.2 and 0.8 depending on stability and on X) this becomes

$$DR \simeq \frac{HUC_o}{2X}$$

A very similar result has been derived by Csanady,[17] using a development which incorporates the effect of an absorbing boundary on the Gaussian profile. Future work should be based on Csanady's approach.

A standard result of diffusion theory is that the peak value of C_o occurs at about $\xi = 1$ and has the value

$$C_{om} = \frac{Q\sigma_Z}{2\ e\pi UH^2\ \sigma_Y}$$

Hence the peak deposition rate (for time-scale of order 10–60 minutes) is

$$DR_m = \frac{.0293\ Q\sigma_Z}{XH\sigma_Y} \tag{17}$$

occurring at $\xi \simeq 1$ (a more complete calculation shows that DR_m occurs at $\xi = 1.02$ to 1.66 depending on stability). Equation (17) provides a useful check on numerical calculations of deposition rate. Equation (17) does not work, of course, when $H = 0$ (the condition $\xi > 1$ fails in this case), but it does show that, for droplets that do not reach ground level, only those which evaporate close to ground level (*i.e.*, small H) are liable to contribute significantly to deposition rate. Computed results verified this important conclusion.

e. *Deposition rates for large droplets*

Large droplets ($\gtrsim 50\mu m$ radius) are relatively unaffected by atmospheric turbulent dispersion. These droplets are the ones which contribute the most to deposition rates mainly because they reach ground level much closer to the towers than do smaller droplets. The main mechanisms leading to a spread of large drift droplets of any given radius at ground level are:

1. The fact that these droplets are initially spread across the tower-top cross section,
2. Wind direction fluctuations associated with time-scales greater than of order 60 minutes (effectively eliminated by sector-averaging),
3. Spatial velocity variations across the plume cross section, and, possibly,

4. Turbulence within the plume not associated with atmospheric turbulence.

For the whole spectrum of droplet sizes, however, the main dispersing "mechanism" is the difference in trajectories between droplets of differing sizes. Incorporation of other factors does not seem feasible, or useful, at present.

To determine sector-average deposition rates for large droplets in the initial radius range r_1 to r_2 ($r_1 < r_2$) we first computed the distance at which test droplets of these radii hit the ground. If these distances are x_1 and x_2 from the tower then the mean deposition rate for the region within a sector of angle θ between x_1 and x_2 is

$$DR = \frac{2\,Q_{12}}{\theta(x_1^2 - x_2^2)} \tag{18}$$

where Q_{12} is the mass flux for droplets in the given range. By taking this value as representative of the distance ½ ($x_1 + x_2$) from the tower, and by using fairly small droplet radius ranges we can obtain a smooth deposition-rate-versus-distance-from-tower profile. Although this method does not account for downwind-distance spread of droplets due to mechanisms 1 and 3 above, the effect of these mechanisms tends to cancel out over much of the deposition zone because droplets in a given size range, which are spread away from the source are balanced by those in the next highest size range, which are spread towards the source.

DROPLET GROWTH

In cooling tower plumes, droplets can grow by two mechanisms—condensation and coalescence. Droplet size can also decrease by two mechanisms, evaporation and breakup. In any general model of the behavior of droplets in cooling tower plumes, all these mechanisms should be considered. It is sensible first to examine droplet growth analytically (and approximately) to determine the absolute and relative importance of these mechanisms. Droplet breakup can be ignored since this is only likely to occur in a very turbulent environment and for droplets somewhat larger than 1 mm in diameter. Droplet breakup may also be caused by shattering due to the expansion forces, which operate when a droplet freezes, but this is a relatively rare occurrence and may be neglected. This leaves the mechanisms of coalescence and condensation/evaporation to be considered.

Coalescence

It will be assumed that it is the larger droplet that grows by collecting a smaller droplet. Whenever droplets move with different velocities, they can collide and coalesce. However, not all droplets on collision trajectories will actually collide: smaller droplets will tend to be deflected around the larger droplet, in some cases enough to avoid collision. The collection efficiency, C_E, is defined as the fraction of droplets of radius r_1 whose centers lie within the volume swept out by a droplet of radius r_2 ($> r_1$), which are ultimately collected by it. C_E depends on r_1 and r_2 in a complex and imprecisely known manner.[15] Droplets with radii below about 18μm have $C_E \simeq 0$. The work of Wigley[10] indicates that nearly all cooling tower plume droplets other than drift droplets fall into this size range. Furthermore, the density of drift droplets is so low that we can ignore drift-drift collisions. This leaves only drift-nondrift collisions to consider.

To determine how rapidly a given droplet will grow by coalescence, consider a droplet of radius r falling with terminal velocity u through a uniform cloud (*i.e.*, condensed plume) of droplets of radii r_1 with terminal velocity u_1. If σ is the liquid water content of the plume (g/g) and ρ_p is the plume density, then, following the method by Fletcher[15] (p. 161, Equation 7.1), we have

$$\frac{d}{dt}\left(\frac{4}{3}\pi r^3 \rho_w\right) = \pi r^2 (u - u_1)\sigma\rho_P C_E$$

then

$$\frac{dr}{dt} = \frac{(u - u_1)\sigma C_E}{4}\,\frac{\rho_P}{\rho_W} \tag{19}$$

where ρ_W is the density of the collecting drop (*i.e.*, $\simeq$ 1gm/cc).

In applying Equation (19) we note first that the results of Wigley[10] indicate that nondrift droplets have mean radii up to 10μm. This means that for most droplets in a moderate length cooling tower plume the collection efficiency is < 0.5. For some droplets in such plumes, or for most droplets in short visible plumes it will be zero. Wigley's results also show that for short plumes $\sigma_{max} \backsim 0.3$ gms/kg and for longer plumes $\sigma_{max} \backsim 1.0$ g/kg. Using these values Equation (19) gives

$$\left(\frac{dr}{dt}\right)_{max} \simeq 0.45\mu m/sec$$

for a 50μm droplet and

$$\left(\frac{dr}{dt}\right)_{max} \simeq 0.59\mu m/sec$$

for a 300μm droplet.

The conditions leading to such extreme values (*i.e.*, high σ, and suitably large nondrift droplets to yield high values of C_E) will persist only for a fraction of the time that a drift droplet remains in the plume. If they persisted for 20 seconds, then a 50μm radius droplet could grow to 50.9μm and a 300μm radius droplet to 311.8μm by coalescence alone. Such conditions will persist for considerably less than 20 seconds in general. Allowing for the much longer time period during which less severe conditions prevail we can assign an upper limit to drift droplet radius increase by coalescence alone of 10%. In the light of the uncertainties in our knowledge of initial drift droplet size distributions we can, therefore, safely neglect the effects of coalescence on droplet growth within the condensed plume. For a more extensive discussion of coalescence effects see Foster.[20]

Condensation/Evaporation

The growth of nondrift droplets in cooling tower plumes has been discussed by Wigley[10] using well-established results from cloud physics. The general equation for growth of a droplet of radius r [Equation (8)] has been given in an earlier section.

If we let A be the ventilation factor, and ignore the solute term, then Equation (8) becomes

$$\frac{dr}{dt} \approx \frac{A\rho_e K^* E}{4\pi r} \qquad (8a)$$

Typical values for A are around 1.6 for 50μm radius droplets and 3 for 300μm radius droplets. Using the values of K* given by Wigley[10] and an estimate of an *extreme* value of E for a typical cooling tower from the same paper gives, after integration,

$$r^2 \approx r_o^2 + 6At \qquad (20)$$

where r_o is droplet radius at time t = 0, and r and r_o are in microns. This shows that it would take approximately 30 seconds for a 50μm radius droplet (A $\simeq$ 1.6) and approximately 500 seconds for a 300μm radius droplet (A $\simeq$ 3) to increase in radius by 5% by condensation alone. Since the extreme conditions on which these times are based only persist within the plume for very much shorter times, the growth of drift droplets within the plume by condensation can safely be neglected. Similarly, we also neglect the condensation-growth of droplets inside the tower between the mist eliminators and tower-top, since supersaturation here will most likely be very much less than the extreme of .02.

For drift droplets with very high salt concentrations the term involving K_2 in Equation (8) cannot be neglected. Such droplets can grow even when E = 0. However, the neglect of condensation-growth within the tower and the visible plume is also justified in these circumstances in view of the other uncertainties in the model.

Following the same arguments as above we can justify the neglect of evaporation of drift droplets within the undersaturated portion of the visible plume. (In any case neglect of evaporation is a conservative approximation.) However, we cannot neglect evaporation of droplets in the free atmosphere after they fall out of the plume, nor in the invisible section of the plume itself. For these cases droplet size decay is also controlled by Equation (8a), although E is negative. Using typical values for ρ_e and K* leads to

$$r^2 \approx r_o^2 - 3A(100\text{-RH})t \tag{21}$$

where RH is the environment relative humidity. From Equation (21) we can estimate the time it takes for a drift droplet to evaporate to the turbulence dispersion size (∼ 10μm radius). The results are shown in Table 8.1 with the times in seconds.

Table 8.1. Times (seconds) for Droplets to Evaporate to Turbulent Dispersion Size.

RH \ r_o	50%	70%	90%
50μm	10	15	45
300μm	285	475	1420

The variation in fall times is very large. Because of the difference in terminal velocities for droplets of 50μm and 300μm radius the variation in fall distances is even larger. Large droplets will fall to ground level without significant evaporation under almost all conditions, whereas small droplets will evaporate completely after only short fall distances under almost all conditions. For any given relative humidity there is a sharp cut off droplet size above which all droplets hit the ground, and below which all droplets evaporate far above ground level. This result is illustrated in Figure 8.2, (valid at 25°C), which has been constructed using an initial height of 300 m.

A similar conclusion was reached by Laskowski[18] after extensive numerical simulation of drift droplet dynamics.

DOWNWASH AND MULTIPLE SOURCE EFFECTS

Downwash

When the wind speed, U, is sufficiently large the bluff-body wake of the cooling tower may trap the plume and cause it to "downwash" into the wake. According to work by Overcamp and Hoult,[19] downwash is one of the most important factors in determining whether or not drift droplets rain out close to the cooling tower. Therefore, it is important to model downwash as realistically as possible. We have done this by incorporating an empirical relationship for the vertical component of the wake velocity field into the plume equations. In other words, we embed the plume in a simulated wake-region velocity field and modify the plume equations accordingly.

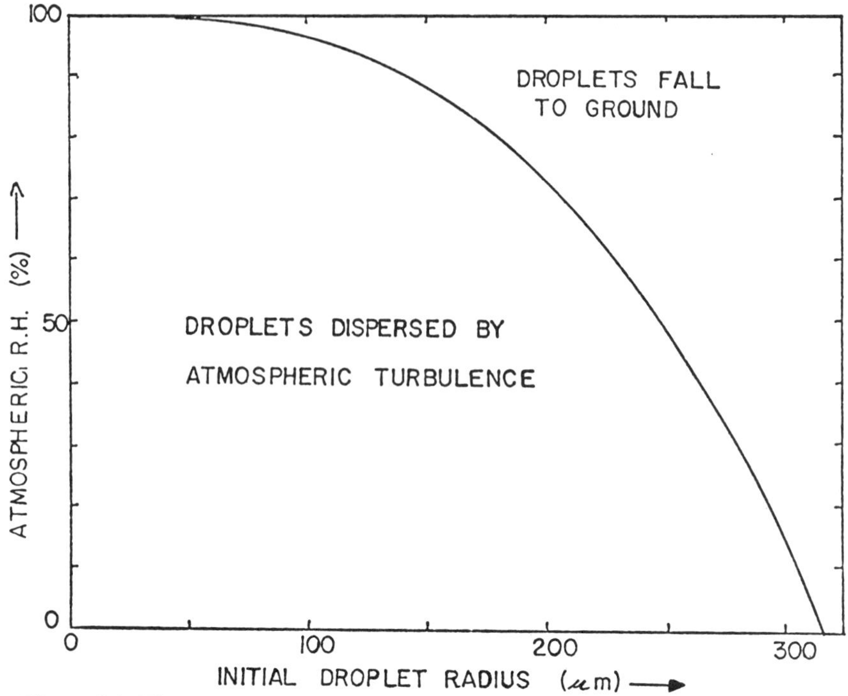

Figure 8.2. The effect of environment relative humidity and initial droplet size on the behavior of drift droplets released from a height of 300 m. All droplets to the left of the curve evaporate before reaching ground level.

Downwash has two main effects on plume trajectory. Close to the tower, plume rise is substantially reduced and may even be negative. A few hundred meters downwind, where the strength of the wake is reduced, the plume's buoyancy takes over as the dominant plume-rise mechanism and the plume "recovers" to a more normal, although delayed, plume rise. The overall effect has been described by Slawson *et al.*[11] in terms of the two-thirds law:

$$\text{no downwash: } z = A\, x^{2/3}$$

$$\text{with downwash: } z = A_1\, x^{2/3}$$

where $A_1 < A$. The reduction in A accounts for the observed reduction in overall plume rise due to downwash. Our numerical simulations of downwash produce essentially the same results as the empirically modified two-thirds law of Slawson *et al.*

The necessity for this type of modelling of downwash effects is twofold. First, there is a marked reduction in plume vertical velocity in the downwash region (plume vertical velocity may even be negative for large enough wind speed to exit velocity ratios) and this allows large drift droplets to fall out of a downwashed plume much sooner than they would otherwise do so. Second, for small drift droplets which remain within the plume far downwind and are diffused to ground level by atmospheric turbulence, it is essential to be able to predict plume rise reasonably accurately and to include in such a prediction the effect of downwash in reducing plume rise.

Multiple Source Effects

When plumes from separate sources merge, their combined flux of buoyancy exceeds that of the individual plumes and so the potential rise of the combined plume is greater than for either individual plume. However, if the plumes merge a long way downwind from their sources the enhanced rise due to merging may be quite small.

The model assumes that once two plumes come into contact merging is quite rapid. If the ambient wind is normal to the line joining the tower centers then the plumes will first touch when their radii equal one-half of the tower centers' separation (D/2). If the ambient wind is parallel to this line (the other extreme case), then the radius of the upwind tower's plume at the merge point will be somewhat greater than D/2 while the radius of the downwind tower's plume will be correspondingly less than D/2. Thus, in both cases, the combined flux of buoyancy for the merged plume should be well approximated by twice the flux of buoyancy for an individual plume when its radius is D/2. Although this is not a true representation of the complicated dynamics of plume merging it is a much better approximation than assuming a double source to be replaced by a single source with double the initial flux of buoyancy. With the present model we can account for the effect of premerging dilution on plume variables such as vertical velocity.

Simulation of realistic examples shows that premerging dilution has a significant effect on the overall plume rise. In stable atmospheric conditions final rise is, as one would expect, somewhere between the single tower plume rise and the rise expected if merging were assumed to occur at the source point. The precise quantitative effect of merging on plume rise depends on initial plume conditions, atmospheric conditions, and on the separation of the cooling towers.

APPLICATION OF MODEL TO TVA'S CANDIDATE SITE OF PHIPPS BEND

The cooling tower drift model consists of the numerical solution of Equations (1) to (9), with the tilted-plume Gaussian dispersion equation added to the system of equations. The equations are solved using a standard Runge-Kutta method.

Multiple source effects and the influence of latent heat on plume rise are both included in the plume model. Downwash is also incorporated by embedding the plume in an appropriate wake velocity field. Although the model predicts condensed plume lengths automatically these results are not given in this application. For drift droplets which evaporate to sizes with negligible terminal velocity the model assumes that dispersion is by atmospheric turbulence and calculates deposition rate using the modified tilted (Gaussian) plume model. For larger droplets which impact with the ground before reaching this size, deposition rate is calculated using droplet trajectories, the initial droplet-size spectrum and geometrical considerations.

Input data needed to run the program are:

1. Atmospheric temperature, relative humidity and velocity profiles and tower-top pressure. The program automatically calculates lapse rates and stability classes. Wind speeds below tower top are estimated using a log-linear velocity profile.
2. Tower design data: tower exit radius, height, efflux velocity temperature and site layout.
3. Initial drift droplet size spectrum and the total drift mass flux.
4. Probabilities associated with the various meteorological groupings characterized by items in 1 above.
5. Topographical data.

Some of the pertinent tower design information for the model is listed in Table 8.2.

Table 8.2.

	Tower Type	
Item	*Mechanical Draft*	*Natural Draft*
Towers/unit	2	1
No. of cells/tower	16	—
Tower length	600 ft	—
Tower width	40 ft	—
Tower height	70 ft	600 ft
Tower exit diameter	30 ft (each cell)	353 ft
Heat load/unit	7.9985×10^9 Btu/hr	7.9985×10^9 Btu/hr
Annual average exit velocity	32 ft/sec	9.7 ft/sec
Annual average drift loss/unit	68 gpm	48 gpm
Cycles of concentration	2.0	2.0
Exit temperature (January)	104° F	98° F

The drift mass flux used was a normalized value of 5 kg/hr/unit for natural draft towers and 25 kg/hr/unit for mechanical draft towers. The initial drift droplet size spectra used were based on the data given in Table 8.3.

Table 8.3. Drift Droplet Size Spectra.

Size range (μm)	10–60	60–120	120–180	180–225	225–325	325–580	>580
Mass fraction	.51	.07	.03	.04	.11	.20[a] .24[o]	.04[a] NIL[o]

[a]For mechanical draft towers; [o]for natural draft towers.

Figures 8.3 and 8.4 show the resulting monthly mean deposition rate isopleths for both natural draft and mechanical draft cooling system alternatives. The most striking differences in the deposition

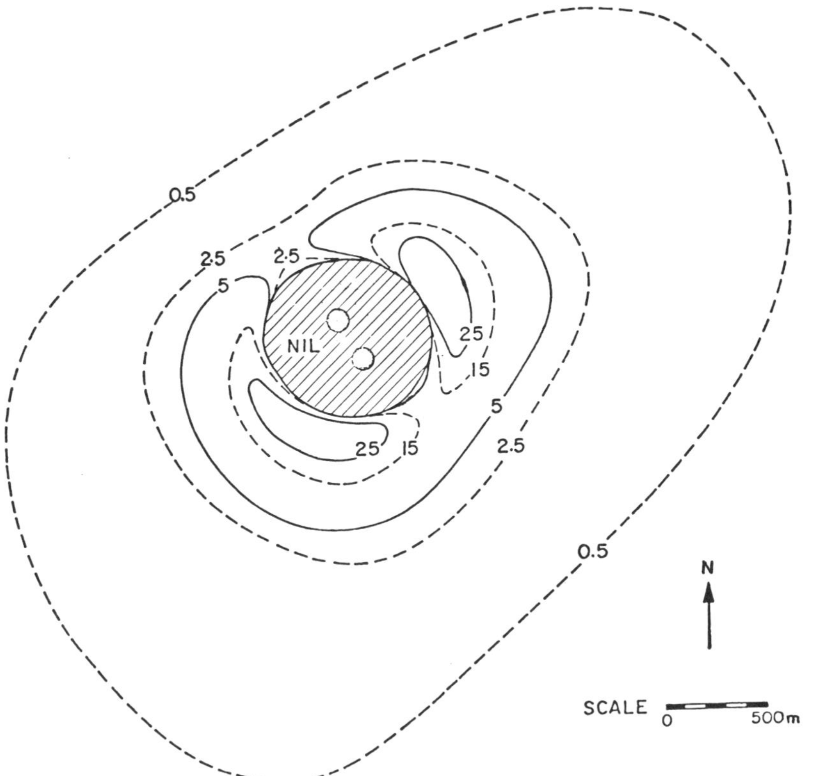

Figure 8.3. Phipps Bend—July. Natural draft towers. Deposition rate in mg/m²/day.

rate isopleths of Figures 8.3 and 8.4 are, first, the difference in magnitude, and, second, the difference in deposition pattern. The nil zone around the natural draft towers is due to the greater height of release along with the smaller drop sizes escaping from the tower compared to the corresponding mechanical draft tower parameters. On the natural draft tower the larger droplets have a terminal velocity greater than the average exit velocity and are assumed not to escape out of the tower. This results in a reduced mass flux and significantly altered droplet size spectrum in the natural draft case. The region of significant deposition extends further from the mechanical draft tower than from the natural draft towers. This is due entirely to the larger mass flux from the mechanical draft towers. In the case of the mechanical draft towers, the 840 mg/m^2/day deposition rate zone occurring immediately adjacent to the towers is a result of averaging a more detailed prediction throughout the immediate vicinity of the towers. This was done because of the very complex wind field generated by the wind, tower and plume interaction.

The far field deposition rates from the mechanical draft towers are approximately five times that for the natural draft towers. This can be compared with the initial mass flux ratio between mechanical and natural systems: *i.e.*, 25 kg/hr–5 kg/hr. In the far field, the ef-

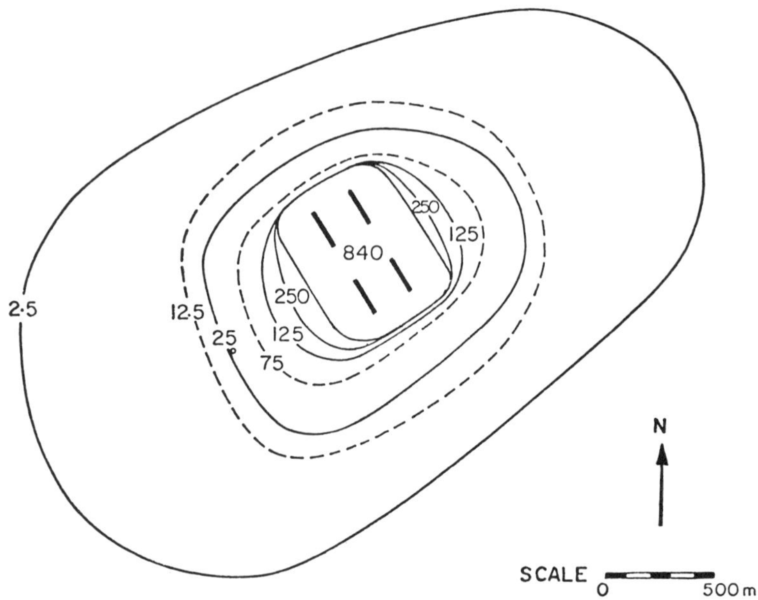

Figure 8.4. Phipps Bend—July. Mechanical draft towers. Deposition rate in mg/m^2/day.

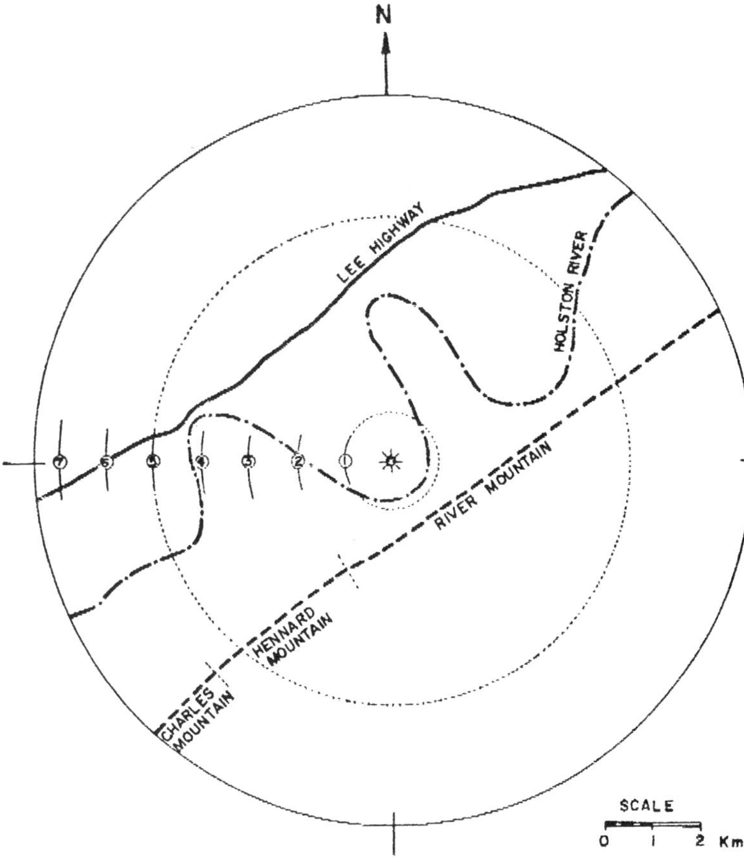

Figure 8.5. Phipps Bend site.

fect of the initial droplet size distribution becomes insignificant: only the small droplets contribute to the deposition.

The deposition rate pattern for both tower systems is elongated in the ENE-WSW direction due to topographically controlled winds prevailing at this particular site.

The influence of terrain irregularities on deposition rate is minor. The only noticeable change caused by terrain is for the region to the southeast of the Phipps Bend site (Figure 8.5) where the NE/SE trending ridge of River Mountain rises to some 500 ft above plant grade. The effect of this topographic irregularity is to increase the (already small) deposition rate on the northwest flank of River Mountain by approximately 25% over the rates shown in Figures 8.3 and 8.4. This increase tends to displace the deposition rate isopleths a few hundred meters further out from the source. On the lee or southeast flank of River Mountain slightly lower deposition rates could be expected. However, in both places the change in deposition rate is insignificant when compared with uncertainties in the model and in environmental conditions.

Conclusions

Some of the general conclusions obtained from the application of the model are listed below. These apply to all drift studies and are in accord with the conclusions of other physically realistic models.[18,20,21]

1. Drift deposition rates due to droplets which evaporate to atmospheric-turbulence-dispersion size are negligible compared with deposition rates produced by droplets which impact the ground prior to reaching this size.
2. For these important large drift droplets, atmospheric stability and temperature have only minor effects on deposition rate, especially when calculating monthly mean rates.
3. Because of (2) the main atmospheric parameters affecting drift deposition rate are wind speed and relative humidity.

APPENDIX A: WASHOUT

Drift material may be removed from the cooling tower plume by rainfall and deposited on the ground. The physical mechanism operating is coalescence of raindrops and drift droplets. The process is known as "washout." If a washout coefficient Λ is defined to be the area fraction swept out by raindrops per unit time (Λ is therefore a function of rainfall rate) then the washout rate (mass per unit area per unit time) is

$$WR = \Lambda \int_{h_1}^{h_2} C \, dz \qquad (A1)$$

where h_2 and h_1 are the upper and lower plume boundaries and C is the concentration of drift material (dissolved in drift droplets) in the plume (*see*, for example, Slade,[16] p. 218).

The usual approach for washout determination is to use a Gaussian plume model to find C and then integrate Equation (A1). For simplicity and to obtain an order of magnitude estimate, the simple top-hat distribution associated with the entrainment theory of plume behavior will be used. If Q (mass units per unit time) is the source drift-mass flux, then for a single tower, the mean drift-mass concentration in the plume at distance x downwind is approximately

$$\overline{C} = Q \frac{R_o^2}{R^2} \frac{1}{\pi R^2 U} \tag{A2}$$

where R is the plume radius (initially, R_o) and U the wind speed. The mean plume depth is $\pi R/2$ so that the mean washout rate across the plume is

$$WR = \frac{\Lambda Q R_o^2}{2UR^3} \tag{A3}$$

The sector-mean washout rate at distance x for a 45° sector is therefore

$$\overline{WR} = \frac{pQR_o^2 \Lambda}{2UR^2(R_o + x \tan \pi/8)} \tag{A4}$$

where p is the rainfall probability for the sector. The ratio R/R_o can be estimated as a function of x using results from elementary plume-rise theory:

$$R = R_o + \alpha z$$

$$z \approx 2.3\, l^{1/3}\, x^{2/3}$$

where $\alpha(\approx 0.3)$ is the entrainment parameter and l is the buoyancy length. Equation (A4) becomes

$$\overline{WR} \approx \frac{pQ\Lambda}{2U(R_o + x \tan \pi/8)} \left(1 + \frac{0.7}{R_o} l^{1/3} x^{2/3}\right)^{-2} \tag{A5}$$

When x = 0, $\overline{WR}$ has its maximum value, specifically

$$\overline{WR_m} = \frac{pQ\Lambda}{2UR_o} \tag{A6}$$

For large x, $\overline{WR}$ varies as $x^{-7/3}$:

$$\overline{WR} \approx \frac{pQ\Lambda R_o^2 U}{0.4\, F_o^{2/3}} x^{-7/3} \tag{A7}$$

In deriving Equation (A7) we have used the fact that $l = F_o/U^3$ where F_o (m^4/sec^3) is the initial flux of buoyancy.

Equations (A5), (A6) and (A7) do not account for the downwind depletion of washoutable material and should be multiplied by a depletion term such as $\exp(-\Lambda x/U)$ to allow for this. In addition, the washout region must be displaced downwind from the source to allow for the rainfall trajectory from plume level to ground level. Thus, for example, the maximum washout rate region will occur from tens to hundreds of meters downwind, the precise distance depending on wind speed and tower height.

To evaluate washout for the Phipps Bend site we need an estimate of the washout coefficient, the buoyancy flux (which determines l), the sector-mean wind speed associated with rainfall and the rainfall probabilities for each wind direction sector.

For moderate sized droplets (radius $\geqslant 50\mu m$), Λ is independent of droplet size and is approximately a linear function of rainfall rate. A reasonable estimate of monthly mean washout rate can therefore be found by using a value of Λ (*i.e.*, $\bar{\Lambda}$) corresponding to the *mean* rainfall rate (*i.e.*, approximately monthly rainfall divided by number of hours of rain).

The results of applying the necessary tower design, meteorological and rainfall data pertinent to the Phipps Bend site to Equation (A7) gives a sector mean washout rate of approximately 0.4 mg/m²/day for both natural and mechanical draft towers at a distance of 200 m from the source. The maximum predicted (sector mean) washout rates are 12 and 0.8 mg/m²/day for mechanical and natural draft towers respectively. The large difference in the predicted values of maximum washout rate is due mainly to the greater mass concentration of drift from a mechanical draft tower. The washout rates predicted above are both negligible when compared with the predicted drift deposition rates in the absence of washout.

ACKNOWLEDGMENTS

This work was supported by the Tennessee Valley Authority.

NOMENCLATURE

C_p = specific heat capacity at constant pressure

g = acceleration due to gravity

G = atmospheric moisture gradient $= \dfrac{\delta q_a}{\delta z}$

L = latent heat

l = buoyancy length scale $= \dfrac{g w_o R_o^2 \Delta T_o^*}{U^3 T^*_{ao}}$

N^2 = Vaisala frequency squared $\simeq \dfrac{g}{T_a} \dfrac{\delta\theta_a}{\delta z}$

q = specific humidity (g/g)

R = plume radius

s = distance along plume centerline trajectory

t = time

T = temperature (°K)

U = wind speed

V = centerline plume speed
v_e = entrainment velocity
v_x = horizontal component of V
w = vertical component of V
WR = washout rate
Λ = washout coefficient
x = horizontal plume coordinate measured from tower top
z = vertical plume coordinate measured from tower top
α = entrainment parameter
σ = liquid water content (g/g)
θ = potential temperature
r = droplet radius
ρ = density
n = viscosity
E = supersaturation
Re = droplet Reynolds Number
V_{-dr} = droplet velocity relative to environment
V_{-d} = droplet velocity
m = droplet mass
C_D = droplet drag coefficient
C = concentration
Q = source strength
DR = deposition rate (mass/unit area/unit time)
σ_y, σ_z = Gaussian dispersion coefficients
C_E = coalescence collection efficiency

NOTE. Subscripts a,e,o,p and s are used for atmospheric, environment, initial plume, and saturated values of variables (v_e is an exception). Superscript* is used to indicate virtual temperature. Prefix Δ is used to indicate the excess of a plume variable over its atmospheric value. For example, q_{spo} is the initial (*i.e.*, $t = x = z = s = 0$) value of the plume's saturated specific humidity, and $\Delta T_o^* = T_{po}^* - T_{ao}^*$ is the initial virtual temperature excess of plume over environment.

REFERENCES

1. Morton, B. R. "Buoyant Plumes in a Moist Atmosphere," *J. Fluid. Mech.* **2**, 127-144 (1957).
2. Morton, B. R., G. I. Taylor and J. S. Turner. "Turbulent Gravitational Convection from Maintained and Instantaneous Sources," *Proc. Roy. Soc.* **A-23**, 1 (1956).
3. Csanady, G. T. "Bent-over Vapor Plumes," *J. Appl. Meteorol.* **10**, 36-42 (1971).
4. Wigley, T. M. L. and P. R. Slawson. "On the Condensation of Buoyant, Moist, Bent-over Plumes," *J. Appl. Meteorol.* **10**, 253-259 (1971).
5. Wigley, T. M. L. and P. R. Slawson. "A Comparison of Wet and Dry Bent-over Plumes," *J. Appl. Meteorol.* **11**, 335-340 (1972).
6. Hanna, S. R. "Rise and Condensation of Large Cooling Tower Plumes," *J. Appl. Meteorol.* **11**, 193-199 (1972).

7. Slawson, P. R. "Buoyant Moist Bent-over Vapor Plumes at a Pilot Stack Facility," *C.E.P. Tech. Manual, A.I.C.E.* 87-90 (1972).
8. Richards, J. M. "The Stability of Wet and Dry Bent-over Plumes," *J. Appl. Meteorol.* **12**, 133-139 (1973).
9. Weil, J. C. "The Rise of Moist, Buoyant Plumes," *J. Appl. Meteorol.* **13**, 435-443 (1974).
10. Wigley, T. M. L. "Condensation in Jets, Industrial Plumes and Cooling Tower Plumes, *J. Appl. Meteorol.* **14**, 78-86 (1975).
11. Slawson, P. R., J. H. Coleman and J. W. Frey. "Some Observations on Cooling Tower Plume Behavior at the Paradise Steam Plant," *Proc. U.S.A.E.C. Cooling Tower Symposium* (1974).
12. Wigley, T. M. L. and P. R. Slawson. "The Effect of Atmospheric Conditions on the Length of Visible Cooling Tower Plumes," *Atmos. Environ.* **9**, 437-445 (1975).
13. Hoult, D. P. and J. C. Weil. "Turbulent Plume in a Laminar Cross Flow," *Atmos. Environ.* **6**, 513-531 (1972).
14. Byers, H. R. *Elements of Cloud Physics* (Chicago, Illinois: University of Chicago Press, 1965).
15. Fletcher, N. H. *The Physics of Rainclouds* (New York: Cambridge University Press, 1962).
16. Slade, D. H. (ed.) "Meteorology and Atomic Energy," U.S. Atomic Energy Commission, Division of Technical Information (1968).
17. Csanady, G. T. *Turbulent Diffusion in the Environment* (Boston: Massachusetts: D. Reidel Publishing Co., Inc., 1973).
18. Laskowski, S. M. "A Mathematical Transport Model for Salt Distribution from a Salt-water Natural-Draft Cooling Tower, Part II," A.M.S. Symposium on Atmospheric Diffusion and Air Pollution, Santa Barbara Preprints, (September, 1974) pp. 412-419.
19. Overcamp, T. J. and D. P. Hoult. "Precipitation in the Wake of Cooling Towers," *Atmos. Environ.* 5, 571-765 (1971).
20. Foster, P. M. "Droplet Growth Inside and Outside Cooling Tower—II," *Atmos. Environ.* **8**, 393-402 (1974).
21. Hanna, S. R. "Fog and Drift Deposition from Evaporative Cooling Towers," *Nucl. Safety,* **15**, 190-196 (1974).

9

Relating Emissions to Air Quality

Steven R. Hanna
Air Resources, Atmospheric Turbulence and Diffusion Laboratory,
National Oceanic and Atmospheric Administration,
Oak Ridge, Tennessee

INTRODUCTION

When modeling diffusion on regional scales (length scales greater than about 10 km), it is necessary to account for topography and changing meteorological conditions along the trajectory of the pollutant. Many sources and receptors are usually included in the model. Currently, a regional diffusion model is being developed which will be solved numerically on a large digital computer. This model will first be used to estimate diffusion in the Tennessee Valley. However, at the same time that the model is being developed, it is necessary to study existing emissions and air quality data. Hopefully, certain patterns between emissions and air quality will be revealed.

In this chapter, data from the 95 counties in Tennessee are analyzed. These data were collected by the state in 1970 to comply with federal regulations and develop the Air Pollution Control Implementation Plan (1972).[1] Average annual emissions of suspended particles, SO_2, CO, NO_x, and hydrocarbons are published.[2] In the four major metropolitan areas (Memphis-Shelby Co., Nashville-Davidson Co., Chattanooga-Hamilton Co., and Knoxville-Knox Co.), more detailed emissions and air quality were obtained. Air quality data for suspended particles and SO_2 were collected in about one-third of the counties. In almost all cases, air monitoring instruments were intentionally located near the pollution sources. Consequently, there are very few stations which represent rural areas. In 1970, only a handful of CO, NO_x, hydrocarbon, and ozone data were collected. More data are currently being collected but are not in a form suitable for analysis. The 95 counties are listed in Table 9.1, along with their areas and populations, and suspended particle emissions.

A simple urban dispersion model[3,4] states that concentration, $\chi(\mu g/m^3)$, is directly proportional to area emissions, $Q(g/m^2sec)$, and inversely proportional to wind speed, U(m/s):

$$\chi = C \frac{Q}{U} \tag{1}$$

where C is a dimensionless parameter, roughly equal to the region length divided by the depth of the pollutant cloud. This formula is intended for use in urban areas, but can be applied to regional data to determine rough relationships between emissions and air quality. The Tennessee data are used to work backwards in Equation (1) and estimate the parameter C. The correlation between published emissions, Q, and concentrations, χ, is also studied.

Suspended Particles Data Analysis

Suspended particles are studied first because there are more data available for this pollutant. Furthermore, chemical transformations are generally unimportant since most suspended particles are inert. Dry deposition and precipitation washout are the main mechanisms for removal of suspended particles from the atmosphere. The emissions data for suspended particles are listed in Table 9.1. So that the data can be compared on a common basis, emissions per unit area are given. Emissions per capita are calculated and listed in the table to determine whether it is possible to estimate suspended particles emissions as a function of population. The emissions rate for each county is plotted as a function of population in Figure 9.1. There is a great deal of scatter among the data. For counties with populations less than 10,000, the relation is accurate within about a factor of two.

$$\text{Emissions} = 8 \times 10^{-4} \text{ g/sec} \times \text{population}$$

But for counties with populations in the middle of the range no such relationship is valid, since emissions are dependent on the type of industry in the county rather than on the population. For example, among counties with populations of about 30,000, emissions range over three orders of magnitude. In the four metropolitan areas, however, the relation

$$\text{Emission} = 3 \times 10^{-2} \text{ g/sec} \times \text{population}$$

is accurate within about a factor of three. Apparently the degree of industrialization in large Tennessee cities in fairly uniform.

Table 9.1. 1970 Suspended Particles Emission for the Counties of Tennessee.

County	*Area (km^2)*	*Population*	*Suspended Particle Emissions (g/s)*	*Emissions per capita (mg/ person)*	*Emission per unit area (g/s km^2)*
Anderson	858	60,300	160	2.65	.186
Blount	1472	63,744	99.0	1.56	.0672
Bradley	856	50,686	16.3	0.317	.019
Campbell	1154	26,045	35.9	1.38	.031
Carter	891	43,259	354	8.18	.397
Claiborne	1137	19,420	11.1	5.73	.00976
Cocke	1085	25,283	22.8	0.893	.0210
Grainger	722	13,948	7.34	0.518	.0102
Greene	1570	47,630	22.0	0.461	.0140
Hamblen	397	38,696	380	9.82	.957
Hancock	589	6,719	3.80	0.576	.00645
Hawkins	1229	33,757	2466	72.6	2.00
Jefferson	702	24,940	13.9	0.547	.0198
Johnson	750	11,569	6.74	0.576	.00899
Loudon	606	24,266	16.4	0.691	.0271
McMinn	1106	35,462	53.0	1.50	.0479
Meigs	488	5,219	6.91	1.32	.0142
Monroe	1690	23,475	51.6	2.19	.0305
Polk	1110	11,669	34.4	2.94	.0310
Phea	798	17,202	11.0	0.63	.0138
Roane	897	38,881	581	14.9	.648
Sevier	1530	28,241	11.5	0.403	.00752
Sullivan	1060	127,329	1634	12.8	1.54
Unicoi	473	15,254	7.20	0.461	.0152
Union	543	9,072	4.38	0.490	.00807
Washington	827	73,924	53.8	0.720	.0651
Bledsoe	1033	7,643	5.64	0.749	.00546
Coffee	1110	32,572	61.8	1.90	.0557
Cumberland	1738	20,733	21.7	1.04	.0125
Fentress	1275	12,593	32.9	2.62	.0258
Franklin	1416	27,289	218	8.01	.154
Grundy	917	10,631	6.62	0.634	.0072
Marion	1295	20,577	873	42.3	.674
Morgan	1380	13,619	17.2	1.27	.0125
Overton	1128	14,866	13.9	0.922	.0123
Pickett	404	3,774	2.45	0.662	.00606
Putram	1037	35,487	18.3	0.518	.0176
Scott	1392	14,762	14.4	0.979	.0103
Sequatchie	699	6,331	7.11	1.12	.0102
Van Buren	651	3,758	7.03	1.87	.0108
Warren	1124	26,972	26.4	0.979	.0235
White	978	16,355	124	7.57	.127
Bedford	1233	25,039	7.2	0.288	.00589
Cannon	695	8,467	22.5	2.65	.0324
Cheatham	781	13,199	6.85	0.518	.00878
Clay	596	6,624	12.7	1.93	.0213
DeKalb	711	11,151	62.0	5.56	.0872

Table 9.1 (Continued)

County	*Area (km^2)*	*Population*	*Suspended Particle Emissions (g/s)*	*Emissions per capita (mg/ person)*	*Emission per unit area (g/s km^2)*
Dickson	1241	21,977	24.6	1.12	.0198
Giles	1585	22,138	21.2	0.950	.0134
Hickman	1562	12,096	4.09	0.346	.00262
Houston	514	5,853	3.28	0.547	.00638
Humphreys	1358	13,560	3661	270.	2.70
Jackson	817	8,141	3.83	0.461	.00469
Lawrence	1623	29,097	5.67	0.196	.00349
Lewis	730	6,761	2.59	0.374	.00355
Lincoln	1485	24,318	12.9	0.518	.00869
Macon	778	12,315	37.1	3.02	.0477
Marshall	965	17,319	28.1	1.61	.0291
Maury	1582	44,028	1318	29.7	.833
Montgomery	1381	62,721	17.8	0.288	.0129
Moore	317	3,568	2.48	0.691	.0078
Perry	1051	5,238	4.58	0.864	.00436
Robertson	1219	29,102	155	5.33	.127
Rutherford	1566	59,428	26.3	0.432	.0168
Smith	817	12,509	48.4	3.89	.0592
Stewart	1204	7,319	7.63	1.04	.00634
Sumner	1367	56,284	248	4.41	.181
Trousdale	292	5,155	3.11	0.605	.0107
Wayne	1892	12,365	5.56	0.461	.00293
Williamson	1520	34,423	9.59	0.276	.00631
Wilson	1452	36,999	30.1	0.806	.0207
Benton	1002	12,126	27.5	2.28	.0274
Carroll	1526	25,741	9.87	0.374	.00647
Chester	730	9,927	4.26	0.432	.00583
Crockett	689	14,402	8.84	0.605	.0128
Decatur	863	9,457	3.66	0.374	.0042
Dyer	1355	30,427	12.5	0.403	.00923
Fayette	1802	22,692	10.3	0.461	.00572
Gibson	1554	47,871	57.9	1.21	.0373
Hardeman	1680	22,435	13.3	0.605	.00792
Hardin	1502	18,212	45.4	2.51	.0302
Haywood	1330	19,596	8.96	0.461	.00674
Henderson	1319	17,360	7.49	0.432	.00568
Henry	1452	23,749	231	9.73	.159
Lake	428	8,091	5.52	0.691	.0129
Lauderdale	1220	20,271	8.35	0.403	.00684
Madison	1435	65,774	36.9	0.547	.0257
McNairy	1457	18,369	7.92	0.432	.00544
Obion	1424	29,936	10.3	0.346	.00723
Tipton	1175	28,001	15.6	0.547	.0133
Weakley	1474	28,827	40.3	1.41	.0273
Shelby	1930	722,111	1460	0.0576	.756
Davidson	1300	447,877	754	0.0490	.580
Hamilton	1408	254,236	478	0.0547	.339
Knox	1300	276,293	1385	0.150	1.07

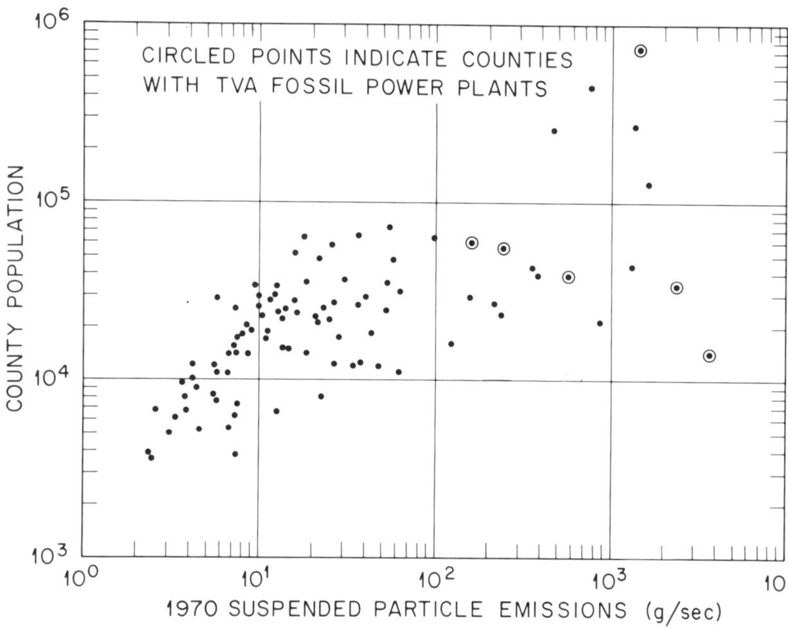

Figure 9.1. County emission rates (g/s) of suspended particles as a function of population (1970 data).

The map of Tennessee in Figure 9.2 contains the county by county emissions and air quality data for suspended particles. It is seen that the Tennessee Valley region in the eastern part of the state has the highest emissions per unit area. Average emissions over the state are .17 g/km^2 sec. Observed annual average suspended particle concentrations for some counties are also given on the figure. In the counties where there was more than one monitoring station, the number given is the average over all the stations. It is interesting that while the county emissions vary over three orders of magnitude, the observed concentrations vary only by a factor of three, from 50 $\mu g/m^3$ to 154 $\mu g/m^3$. The small variation in concentrations is partly due to the continual presence of a "background" amount of suspended particles, whose origin is natural and man-made sources upwind of the region. The annual average background of suspended particles in Roane County was found by Hanna[5] to equal about 40 $\mu g/m^3$, which agrees reasonably well with the minimum observed in this study. Two stations in suburban Davidson County registered annual average suspended particle concentration of about 40 $\mu g/m^3$.

Emissions and air quality are not well correlated in Figure 9.3, which suggests that the air monitoring stations were not situated so that their measurements would represent the regional average. In-

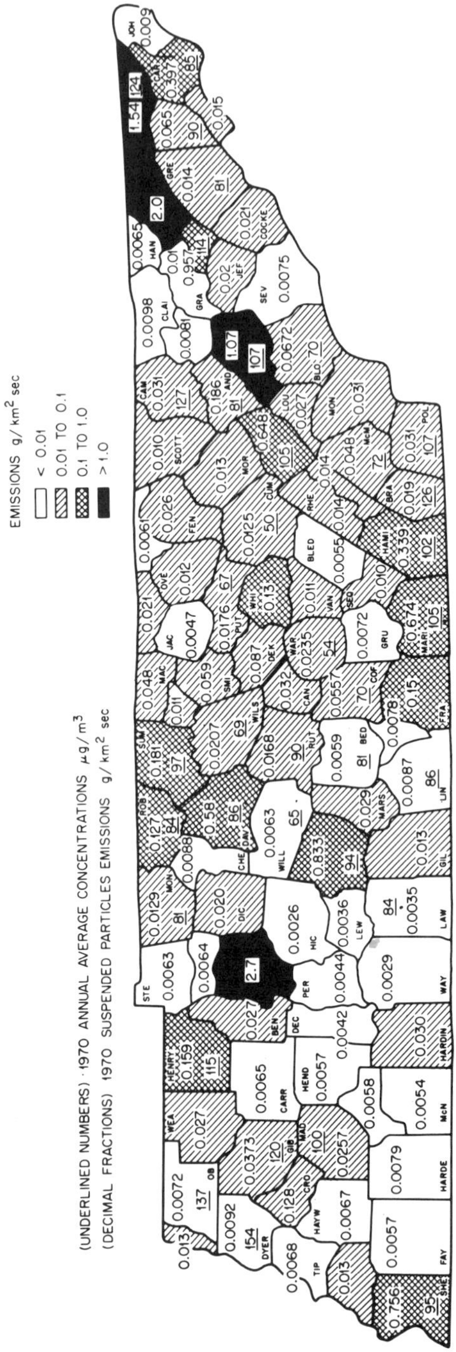

Figure 9.2. County emission rates (g/km²s) and air quality data ($\mu g/m^3$) for suspended particles (1970 data).

stead, most of the monitoring stations were intentionally located close to large pollution sources. Consequently the three highest concentrations were recorded in counties with very low total emissions. To record a truly representative regional average using a single monitor, it is necessary to locate the monitor far enough from large point sources that they do not strongly influence the measurement.

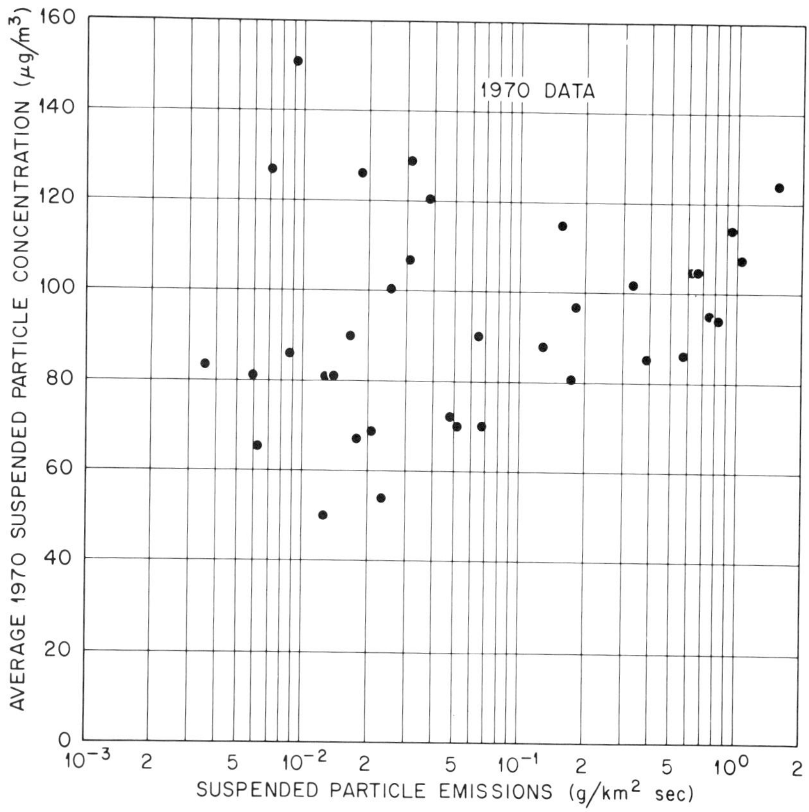

Figure 9.3. County emission rates (g/km^2s) as a function of air quality ($\mu g/m^3$) for suspended particles (1970 data).

If it is assumed that the background concentration of suspended particles is 40 $\mu g/m^3$, then the average contribution to the suspended particle concentration due to sources in the region is about $50\mu g/m^3$. Thus the dimensionless parameter C in Equation (1) can be calculated to equal

$$C = \frac{\chi U}{Q} = \frac{5 \times 10^{-5}\ g/m^3 \times 3\ m/s}{1.7 \times 10^{-7}\ g/m^2 s} = 880$$

In urban areas, Gifford and Hanna[3] find that C equals about 220. If we interpret C as the ratio of the region length to the depth of the pollution cloud, and assume that the depth of the pollution cloud is about 1 km, then the region length is about 880 km. This is roughly the length of Tennessee along the dominant wind direction.

Carbon Monoxide Data Analysis

Carbon monoxide (CO), which is the most inert of the three substances studied in this report, is emitted mainly from cars and trucks. Tennessee emissions of CO during 1970 averaged 63 g/km^2 sec and were more evenly distributed than emissions of suspended particles. County CO emissions plotted in Figure 9.4 vary by a factor of only about 20 and are better correlated with population than suspended particle emissions. Again, the Tennessee Valley region has significantly higher emission rates than the rest of the state. The only two available 1970 measurements of CO concentration are in Memphis (1.6 ppm) and Nashville (1.5 ppm). Since these air quality data are taken near highways in large cities, they are not representative of the region. Using Equation (1) and assuming that C equals 880, it is possible to predict the expected annual average contribution of Tennessee to the total concentration of CO:

$$\chi = 880 \frac{Q}{U} = 880 \frac{.62 \times 10^{-6}\ g/m^2 s}{3\ m/s} = 182 \frac{\mu g}{m^3}$$

or .15 ppm

The background concentration of CO is approximately equal to this value.[2] Thus the total regional CO concentration is expected to be about .3 ppm.

Sulfur Dioxide Data Analysis

Chemical transformations to sulfates cause sulfur dioxide (SO_2) to be a nonconservative pollutant. Its average annual emission rate over Tennessee is .37 g/km^2 sec. About two-thirds of the SO_2 emissions in Tennessee is from the TVA power plants, whose tall stacks and buoyant plume rise cause the pollution to be discharged at an effective height of from 200–500 m above the surface. These power plants were designed so that local, ground level air quality criteria for SO_2 are not exceeded. However, the great amounts of SO_2 emitted by these plants contribute to the gross flux of SO_2 over the state.

County emissions and air quality observations of SO_2 are listed in Figure 9.5. The heaviest SO_2 emissions are from the Tennessee Valley. The highest (by far) annual SO_2 concentration is 240 $\mu g/m^3$ in Copperhill in Polk County. This monitor is located too close to a large smelter to be of use in estimating a regional average. With the exception of this figure, the average SO_2 concentration in Tennessee is

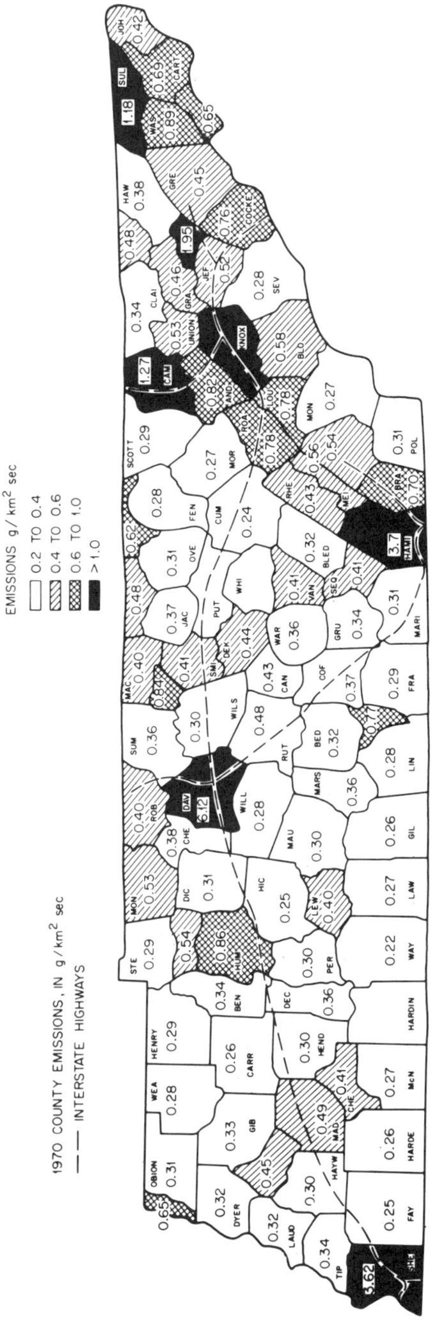

Figure 9.4. County emission rates (g/km^{2}s) of carbon monoxide (1970 data).

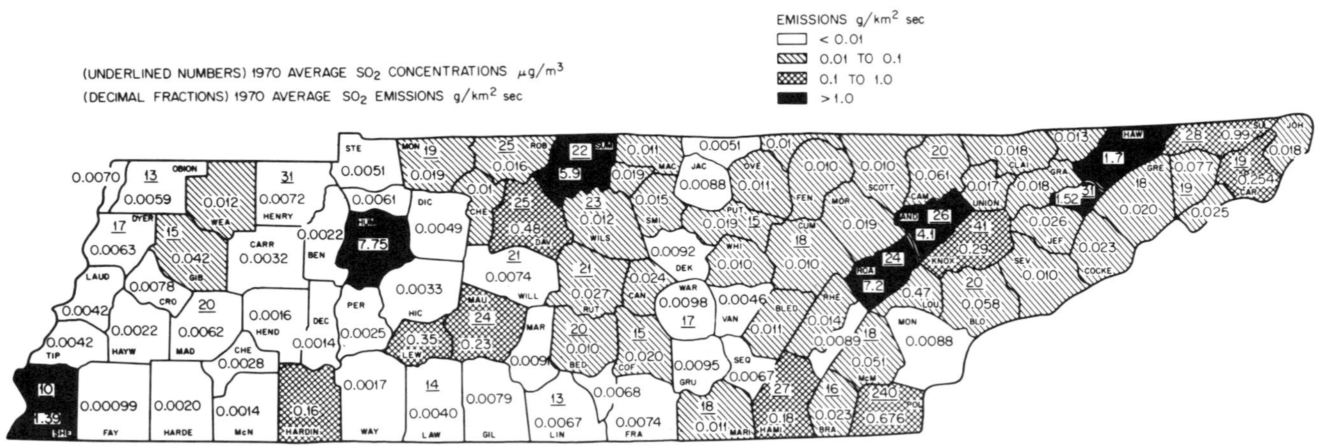

Figure 9.5. County emission rates (g/km^{2}s) and air quality data (μg/m^3) for SO_2 (1970 data).

21 $\mu g/m^3$. As is true for suspended particles and CO, the air quality monitors were usually placed in towns or close to large sources, so that the resulting observations are not representative of regions. Consequently there is poor correlation between observed concentrations and emissions. Instrumental errors may also contribute to this lack of correlation. For example, it is difficult to believe that the lowest SO_2 concentration (10 $\mu g/m^3$) would occur in Shelby County, in which Memphis is located.

From these data, the parameter C is calculated to have the value

$$C = \frac{\chi U}{Q} = \frac{21 \times 10^{-6}\ g/m^3 \times 3\ m/s}{.37 \times 10^{-6}\ g/m^2 s} = 170$$

This number is about a factor of five less than the value for C calculated for suspended particles. Gifford and Hanna[3] found the same difference using pollution data from many urban areas, and attributed part of the difference to the fact that SO_2 is generally emitted much higher in the atmosphere than suspended particles. Consequently ground level concentrations of SO_2 are relatively low. Chemical transformations further remove SO_2 relative to suspended particles.

Removal by dry deposition and precipitation washout is also important. Hanna[5] developed the following rough approximation for accounting for deposition and washout:

$$\chi = C \frac{Q}{U} \Big/ \left(1 + C \frac{V_d}{U} + \frac{\Lambda \Delta x}{U}\right), \tag{2}$$

where V_d(m/s) is the dry deposition speed, $\Lambda(s^{-1})$ is the washout coefficient, and Δx is the region length. Letting C, Δx, and V_d equal 880, 800 km, and .001 m/s, respectively, we find that dry deposition can decrease the flux of pollutant over Tennessee by about 25%. During rainy periods, it can be assumed that Λ equals 10^{-4} s^{-1}, and consequently about 95% of the flux pollutant over Tennessee is carried to the surface by rain.

Further Comments

While the emissions data for NO_x and hydrocarbons are fairly complete, there are very few air quality measurements of these substances. A revised emissions inventory is currently being compiled, and there are now more extensive and more accurate air monitoring instruments in operation. However, it is clear from this study that the air monitoring instruments do not represent regional air quality, defined as the average ground level pollutant concentration over a county. Instead, the monitoring instruments were usually placed in the largest town in the county, close to major pollution sources. A few instruments should be located in rural or suburban areas, or at least sufficiently far from major sources that they do not dominate the reading at that station.

The annual average concentrations of SO_2 and suspended particles, reported on Figures 9.2 and 9.5, do not vary much across the state. Thus, from Equation (1), the wind speed governs the general level of air pollution in a region on a day-to-day basis. The main use of a regional diffusion model would be to account for perturbations on this background due to local sources.

ACKNOWLEDGMENTS

The assistance of D. Neeley and B. Barnard of the Tennessee Division of Air Pollution Control is greatly appreciated. This research was performed under an agreement between the National Oceanic and Atmospheric Administration and the U.S. Atomic Energy Commission.

REFERENCES

1. Air Pollution Control Implementation Plan, Tennessee Department of Public Health, Div. of Air Poll. Cont., Cordell Hull Bldg., Nashville, Tennessee (1972).
2. Air Quality Criteria for Carbon Monoxide, NAPCA Publication No. AP-62 (Washington, D.C.: U.S. Govt. Printing Office, 1970).
3. Gifford, F. A., Jr. and S. R. Hanna. "Modeling Urban Air Pollution," *Atmos. Environ.* 7, 131-136 (1973).
4. Hanna, S. R. "A Simple Model of Calculating Dispersion from Urban Area Sources," *J. Air Poll. Cont. Assoc.* 21, 774-777 (1972).
5. Hanna, S. R. "Application of a Simple Dispersion Model to a Rural Industrial Region," Presented at the Third Annual Ind. Air Poll. Cont. Conf., Knoxville, Tennessee (March 29-30, 1973).

Section III
Monitoring Power Plant Emissions

10

Design of Air Monitoring Surveys Near Large Power Plants

Terry L. Miller
Enviro-Measure, Inc., Knoxville, Tennessee

Kenneth E. Noll
Environmental Engineering Group, Department of Civil Engineering, The University of Tennessee, Knoxville

In recent years, ambient air monitoring has become an important part of many environmental impact statements to satisfy the requirements of the National Environmental Policy Act.[1] Air monitoring systems are an essential element of supplementary control systems for large stationary sources. This has served to make air quality surveys more complex, requiring comprehensive planning to assure that prescribed objectives can be attained in the shortest possible time and at the least cost. Adequate planning and good survey design are also required because the actual monitoring is an expensive and time-consuming undertaking requiring skilled personnel and sophisticated analytical equipment.

This chapter will first provide a discussion of the role of air monitoring in environmental quality assessment and the planning and management process. A detailed methodology for the selection of air monitoring sites near point sources is then provided. A well-designed air quality survey will address the following items in the order listed:[2,3]

1. Set the objectives of the air monitoring investigation,
2. Determine the physical parameters to be measured,
3. Set the network specifications including the location of air monitoring stations; the duration of the study and sampling schedules; and the air sampling method to be used,

4. Set the specifications for the individual stations in the network including the equipment needed to conduct the study, the method and frequency for equipment calibration, and the data recording methods, and
5. Determine the type of data analysis to be performed and method of data reporting.

It should be recognized that these steps are interdependent and that a properly designed survey would consider each step in terms of the effect on the other parts of the survey design.

THE ROLE OF MONITORING AND MODELING IN AIR QUALITY ASSESSMENT

To determine the environmental impact of a new or existing project, provide routine source surveillance or operate an intermittent air pollution control program, it is necessary to estimate air pollution concentrations either by monitoring or modeling or by a combination of these methods. The interrelationship between monitoring and modeling, which can lead to an optimum design for an air quality survey, is illustrated schematically in Figure 10.1.

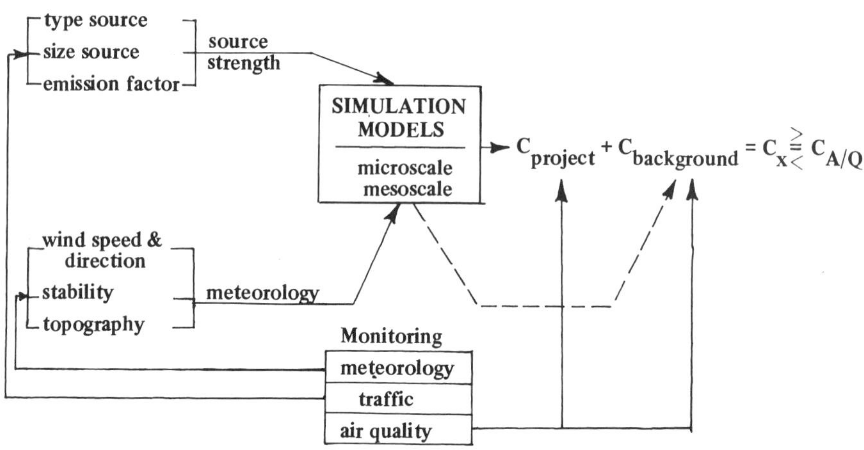

Figure 10.1. Interaction of modeling and monitoring to assess air quality impact.

As shown in the figure, the purpose of mathematical models is to quantitatively combine the effects of source strength and meteorology to describe the resulting ambient air pollution concentration. Source strength is affected by a number of variables including the size of the source, variable emission rates, and the efficiency of air pollution control equipment employed. Meteorology is affected by wind speed and direction, atmospheric stability, inversion height, and terrain features. Useful mathematical models must be able to account for all these parameters.

Air pollution models vary in complexity from simple microscale dispersion models to sophisticated mesoscale, multi-source models describing transport, dispersion, and photochemical reactions of pollutants.

Microscale models are used to estimate the ambient pollution levels near a single source or project, $C_{project}$. *Mesoscale models* are used to estimate the area wide impact of a proposed source or project, or the background concentrations, $C_{background}$, due to other sources.

Ambient air pollution concentrations occurring downwind of a source consist of two components: pollution contributed directly by the source and background. In most analyses, these components should be determined separately. The total air quailty impact, represented in Figure 10.1 by the concentration C_x, is equal to the sum of the background plus the concentration contributed by the project under study. Whenever other major sources are nearby, their contribution of pollution must also be added to the project contribution. The objective of many air quality surveys is to determine the value of C_x as accurately as possible and to compare that concentration to the National Ambient Air Quality Standards (NAAQS).[4]

The role of monitoring is to *measure* ambient pollution concentrations, meteorological parameters, and/or source strength parameters. Air monitoring at carefully selected sites provides a direct measurement of background concentrations. Measurements of meteorological or source strength parameters can be used to verify model input data or validate air pollution dispersion models. Validating microscale models requires source strength, meteorology, and both microscale and mesoscale (background) air pollution concentration measurements. Validating mesoscale models requires that source strength, meteorology, and air pollution measurements all be representative of area wide conditions.

Setting Objectives and the Initial Air Monitoring Survey Planning Process

Figure 10.2 provides an overview of the important decision points required in the initial phases of planning a survey. In general, the broad objectives of any air quality monitoring study are to determine the extent of the existing air pollution problem, and to validate any mathematical models or assumptions used to estimate the future impact of a new source or change in emissions from an existing source on air quality. More specific objectives may include such things as: (a) checking for compliance with ambient air quality standards at critical locations, (b) determining when and where the worst case background pollutant concentrations presently occur within the impact area, (c) validating or calibrating a mesoscale model to allow an accurate prediction of future background concentrations, and (d) validating or calibrating a microscale model to allow an accurate estimate of the air quality impact of a proposed facility or change in emissions from an existing facility. Consideration should also be given to the potential exposure of people to air pollutants. If sensitive receptors such as children, the elderly, or the sick, are to be exposed to additional pollutant concentrations after the completion of a proposed project, it might be important to document the present levels of exposure or sensitive populations so that the future levels of exposure can be more accurately estimated.

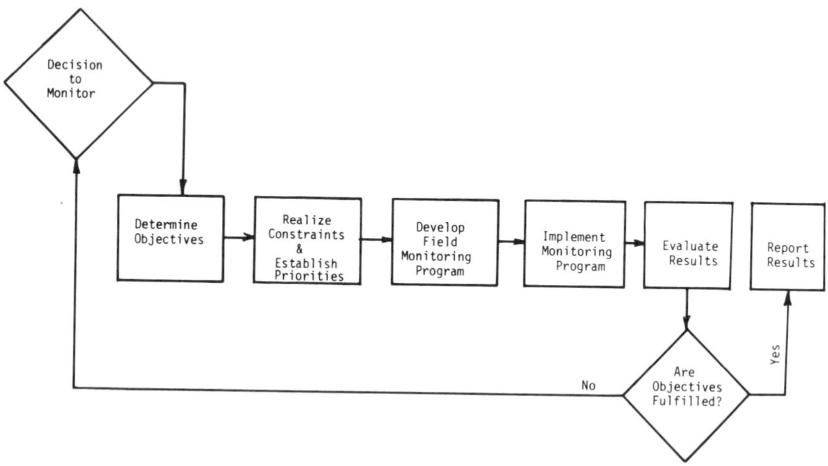

Figure 10.2. The planning and management process for an air quality survey.

Once the objectives of the air quality study are set, it is possible to develop and implement a field monitoring program which will fulfill the objectives. Working within the constraints of money, manpower, and time, priorities can be set in terms of the objectives which can realistically be met within the cost constraints imposed upon the air quality study. If available resources are severely limited, an air quality monitoring study might not be advisable at all, because a poorly planned, poorly financed, and hurriedly conducted air sampling survey will not provide data which can be easily evaluated in terms of the survey objectives.

AIR MONITORING NETWORK SPECIFICATIONS

The total of all air monitoring stations, meteorological stations, calibration equipment, and data acquisition equipment required to meet the total objective of an air quality survey represents the air monitoring network.

To understand the interrelationships between all of the component parts of the network and allow decisions to be made about the number and type of each piece of equipment and the interdependence of the equipment in meeting the survey objectives, a set of specifications for the network must be developed early in the planning process.

Air monitoring network specifications should include the number of sites to be monitored, air pollution and meteorological measurements required at each site, the duration of the study, and manpower requirements. Network specifications should be determined in light of known limitations of physical, engineering, economic, and human factors as well as limitations due to completion deadlines or time delays in equipment procurement. Time, manpower, and budget constraints may significantly affect the feasibility of utilizing specific techniques or equipment to conduct the study.

Considering all the stated sampling requirements and available resources, the types of air quality and meteorological monitoring equipment to be employed can be identified including the total number of samplers and analyzers for each type of measurement to be performed. To accomplish this, available sampling methods should be evaluated. For example, the number of continuous analyzers versus semi-automatic samplers versus manual sampling methods should be investigated. Obviously, the needed mobility of air sampling equipment should be considered and such factors as the high capital costs of continuous analyzers versus the high operating costs of semi-automatic or manual methods should be optimized for cost effectiveness where possible.

The calibration system to be used to ensure accurate data from all the analyzers should be considered next. Each station can be equipped with calibration hardware, or calibration procedures can be utilized where either portable instruments are calibrated at a central laboratory, or permanent instruments are calibrated with a portable calibration device. The choice of needed calibration equipment will depend on the number of stations, distance separating the stations, and the frequency at which calibrations are performed. A typical integrated calibration system might consist of sophisticated dynamic calibration apparatus located at a central laboratory for performing periodic primary calibrations, with several portable calibrators or span gas bottles used for frequent on-site single-point calibration checks.

Once the air monitoring and calibration specifications have been fixed, the requirements of the data acquisition system can be evaluated. The data rates and quantities generated by the monitoring network should be determined as a whole to calculate the data handling capacity needed for the overall sampling network. Data recording equipment, which is ideal for a single monitoring station, might be too expensive for use at numerous monitoring stations. Frequently it is less expensive to replace a variety of special devices with a central general-purpose control element, even though the speed and flexibility requirements of the system do not require it. Data telemetry and central data processing systems might represent the most efficient approach for monitoring networks with many remote sampling stations employing continuous air pollution analyzers. Conversely, for single station applications, strip chart recorders with manual data reduction may represent the most cost effective data collection method.

The network design will provide the number of air monitoring stations to be employed and set the functional sampling objectives of each station. At this point, air monitoring station specifications can be developed for each monitoring site in the network. Station specifications include the setting of sampling objectives for each station and the selecting of compatible hardware components for the station resulting in an integrated air monitoring sampling station design.

There are nine discrete component elements of air monitoring station design: the sampling probe and tubing, sample preconditioning, sample collectors, continuous air pollution analyzers, calibration equipment, flow control and flow meters, air movers, data acquisition equipment, and equipment shelters (Figure 10.3).

The sequence of selection of station equipment should be arranged so that primary components are chosen before supportive or supplementary hardware. For example, equipment shelters should be designed after the total amount of equipment to be housed in the shelter has been determined; thus, the shelter will be properly sized, with adequate electrical wiring, plumbing, instrument mounting, and storage space as required by hardware components.

SYSTEMATIC METHOD FOR SELECTING AIR MONITORING SITES NEAR POINT SOURCES

One of the most critical elements of air monitoring survey design is the location of sampling sites. Criteria for site selection has not been well defined in the literature, consisting mostly of subjective general recommendations. It is the objective of this section to present methods of applying available mathematical dispersion models to assist in site selection decisions.

Choosing the "right" location for the "right" monitoring objective requires insight into the nature of air pollution emission, transport, and dispersion from different types of sources. Because many variables must be considered, site selection is one of the most complex and critical elements in the design of an air quality monitoring survey. If the wrong sites are chosen or if a critical site is missed, no amount of accurate data will allow the objective of the study to be fully realized. Variables to be considered regarding the transport and dispersion of pollutants include relative locations of pollution sources (inside and outside the study area), sensitive receptors, the effects of topography, source size and configuration, and meteorol-

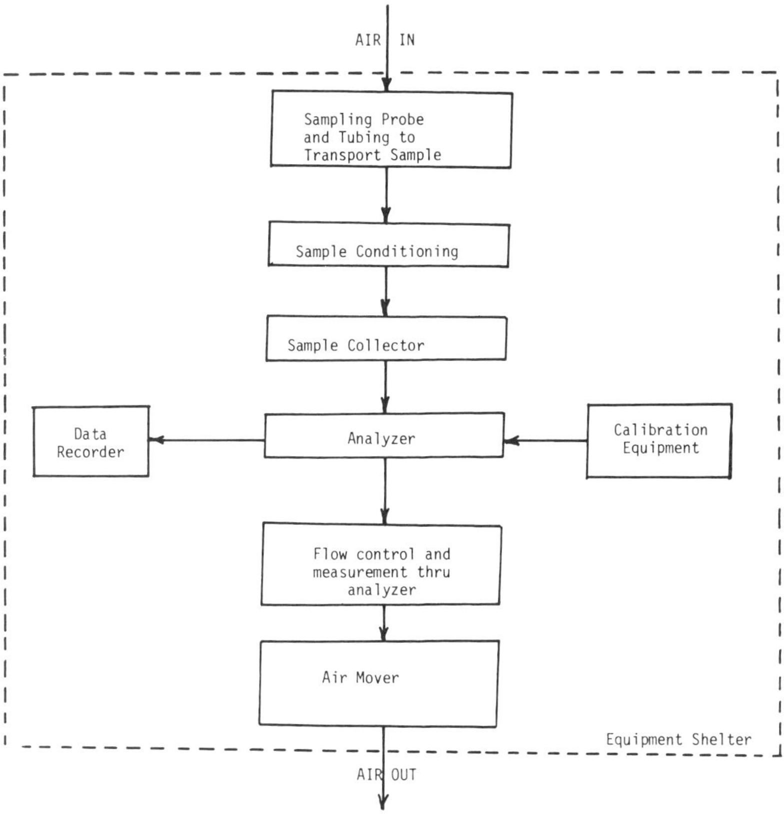

Figure 10.3. Hardware components of an air quality monitoring station.

ogy. Each site chosen should fulfill specific objectives, such as determining area wide "background" concentrations, measuring the ambient concentrations within the microscale influence of a nearby source, and determining the mean, minimum, and maximum pollution concentrations near sensitive receptors.

Before air monitoring sites can be effectively selected, an understanding of the spatial distribution of pollution concentration is important. It is useful to define three separate air pollution regimes—

the microscale regime, the mesoscale regime, and the macroscale regime as follows:

Microscale: The microscale air pollution regime represents a relatively small air mass exhibiting large variations in ground level air pollution concentrations. This phenomenon usually occurs very near sources of air pollution when the rate of increasing atmospheric dispersion with downwind distance is very great.

Mesoscale: The mesoscale regime represents a "community size" air mass exhibiting fairly homogeneous ground level air pollution concentrations such as the ambient concentrations which result within urban areas, from the emission of relatively small quantities of air pollutants from a large number of ground level sources (*i.e.*, automobiles, residential and commercial space heating furnaces, and even numerous small industrial sources). These "local background" concentrations can vary considerably at different locations within an urban area.

Macroscale: A macroscale regime is represented by "regional background" air pollution concentrations, which can be fairly homogeneous over linear distances from *tens to hundreds of kilometers.* Large variations in pollutant concentration indicate the presence of mesoscale and microscale air pollution regimes which are superimposed upon the regional background air pollution concentration regime.

Pollution levels in the proximity of major sources consist of three component parts—the microscale concentration (contributed directly by the nearby source), the mesoscale concentration (the local background concentration due to area wide sources of pollutant emissions), and the macroscale concentration (the regional background concentration due to distant pollution sources). Frequently, the macroscale pollution concentration is so low that this term can be ignored, leaving only two components to any ground level observation—the local background air pollution concentration plus the microscale concentration due to a nearby source. Air monitoring sites located outside the microscale regime measure only mesoscale and macroscale concentrations (background). Sampling sites located within the microscale regime measure the concentration due to the combined microscale, mesoscale, and macroscale regimes.

Selecting Monitoring Sites Near Point Sources

Whenever air quality near a specific air pollution source is to be monitored, mathematical dispersion models should be employed to assist in determining optimum air sampling sites. Numerous point source models are available,[5,6] which can be used to calculate where the maximum ground level concentrations are expected to occur. Models can also be used to determine ground level concentration profiles at various distances from the source to determine the extent of the impact area.

The following sections describe a general methodology for selecting air monitoring sites near point sources. The basic approach is to determine where the maximum concentrations are expected to occur using point source atmospheric diffusion models, then locate monitoring sites in these areas. Measured concentrations can then be compared to NAAQS.

The methodology also includes (a) determining the size of a representative sampling area which can be covered by a single station (coverage ratio), (b) determining the minimum number of sites needed to insure a desired number of observations of C_{max} (a statistical approach), and (c) selecting the optimum number and location stations required to meet monitoring objectives with a high probability of success.

Ground level pollution concentrations near point sources are highly dependent on meteorological conditions which affect plume rise and the rates of transport and diffusion of pollutants. Before selecting air monitoring sites the meteorological conditions for which monitoring data is desired must be specifically delineated. Locations where the maximum concentrations occur under these conditions can then be determined using diffusion models. The diffusion of air pollutants from point sources is complex due to the number of variables interacting to affect nearby ground-level concentrations (*i.e.*, wind speed, wind direction, atmospheric stability, inversion heights, plus emission rates, plume rise, and topographic effects). Therefore, the meteorological conditions that produce maximum pollution levels are not always easy to determine. However, an organized step-by-step approach can be used to determine the locations where maximum levels occur and the likelihood that an air monitoring station located at these points would in fact measure the maximum values.

Using Model Results to Delineate Potential Zones for Monitoring

One method of selecting potential zones for monitoring is to present the results of diffusion calculations as shown in Figure 10.4. This shows the distances downwind from the source where maximum concentrations occur under different meteorological conditions. A station located in Zone (1) would be needed to monitor during low wind speed and unstable conditions; Zone (2) is for monitoring during both trapping and inversion breakup conditions; Zone (3) could monitor during conning at critical wind speeds and mean wind speeds under neutral conditions; Zone (4) could monitor both mean wind speed-neutral conditions and low wind speed-stable conditions; in some cases sites in Zone (5) may be desirable to document that ambient levels are less outside the impact area. The maximum concentration can be recorded in each box as well as the probable frequency of occurrence of various wind speeds, stabilities, mixing heights, and wind directions.[4] This data can then be used to establish priorities in choosing monitoring sites. Monitoring locations where maximum concentrations occur frequently should probably receive a higher priority than locations where C_{max} occurs infrequently.

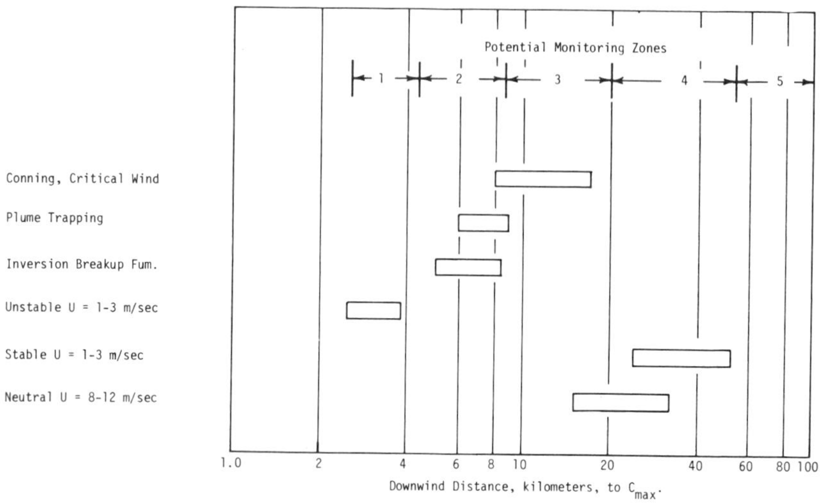

Figure 10.4. Identifying potential zones where C_{max} may occur.

Calculating C_{max} and X_{max}

Two calculation steps are required to determine C_{max} and its location downwind, X_{max}. The first step is to calculate the plume rise. This can be done using the nomograph presented in Chapter 5, where the plume rise under either neutral or stable conditions can be estimated for various wind speeds. Plume rise, Δh, plus stack height, h_s, equals the effective stack height, h_e, used in diffusion calculations. Most authors agree that estimating plume rise under unstable conditions is questionable due to lack of data and uncertainties in defining air motions. For this reason, a conservative solution is recommended; that is, to underestimate plume rise for neutral conditions and wind speeds = 3 m/sec, which is TVA's reasonable minimum $\bar{u}$.

The second step is to determine the distance downwind where the maximum concentrations occur under each meteorological condition of interest. This can be determined for the coning plume, trapping plume, and inversion breakup fumigation by using other nomograms in Chapter 5.

Diffusion calculations do not predict accurately the precise location where maximum concentrations occur. Therefore, a more realistic solution to the models may be to calculate where maximum concentrations occur for a narrow range of conditions rather than a precise point. For example, rather than determine where the maximum concentration occurs for neutral stability at the mean wind speed, $U = 5.7$ m/sec, it would be more realistic to determine X_{max} for wind speeds from say 4–7 m/sec. The frequency of occurrence of wind speeds between 4–7 m/sec could be determined from historical meteorological summaries such as STAR.[7]

Determining the Optimum Number and Location for Air Monitoring Stations

The number of air monitoring stations required to determine the total impact of a source can be quite large. Four or five zones might require data in a single direction downwind. If this number is multiplied by the number of directions from which the wind can blow, then a very large number of monitoring sites seems necessary. In practice, such a large number of sites is usually out of the question so that the survey objectives must be compromised by monitoring in only a few selected downwind directions. Which downwind directions should be monitored can be determined from an examination of historical wind direction/frequency data, and a consideration for the potential receptors in each direction.

Even after a downwind direction has been selected, all of the maximum ambient air pollution concentrations which occur at different locations for different meteorological conditions can not be measured because of the large number of sampling locations which would be required. However, if air monitoring stations are selected to measure a range of C_{max} values (tolerance range) then the number of stations required is reduced. Thus, monitoring locations can be chosen for the purpose of measuring ambient levels within some tolerance range (*i.e.*, within 10% of C_{max} or within 20% of C_{max}, etc.). The acceptable tolerance range will have an effect on the number and location of stations for air monitoring; the greater the tolerance, the fewer stations will be required. An acceptable tolerance range should be based on an analysis of the maximum expected levels compared to air quality standards. If maximum levels are expected to be near standards then the tolerance should be small. Conversely, low concentrations need not be resolved as accurately as high concentrations, so that a greater tolerance range may be acceptable.

Diffusion models can be used to determine the size, shape, and extent of the representative area about a sampling station which corresponds to a given tolerance range. Knowing the size and shape of the representative area allows guidelines for minimum spacing between stations.

From curves of normalized concentration, C U/Q, versus distance downwind, x,[6] (Figure 10.5) it is apparent that the point where the maximum ground level concentration occurs is also where the rate of change in the concentration with respect to downwind distance is at a minimum. This means that concentrations are fairly uniform where C_{max} occurs. This figure can be used to determine the distance toward and away from the source within which the concentrations deviate from C_{max} only by some minimum amount (such as 10%). Figure 10.6 shows the distance from the source, X_{10}, within which concentrations are within 10% of C_{max} when the peak occurs at X for different stability classes.

Figure 10.6 also provides an upper limit to the number of downwind monitoring sites needed. If sites are to be located to provide complete coverage, the maximum number of downwind stations required can be determined graphically as shown in Figure 10.7. The figure illustrates the same boundaries as in Figure 10.6, and each "step" represents the distance which can be covered by a single monitoring site. For example, if all C_{max} values are expected to occur within 1–10 km, then, from Figure 10.7, the maximum number of downwind stations needed in one direction is 4–5. Note that under unstable atmospheric conditions the boundaries in Figure 10.6 are narrower such that more "steps" are required to cover the same distance shown in Figure 10.7.

In addition to the downwind extent of the "representative area," it is also important to know the crosswind extent. The distance, Y_{10}, within which concentrations are within 10% of C_{max} can be calculated from the equation

$$(1\text{–}0.10) = \exp - \frac{1}{2}\left(\frac{Y^2_{10}}{\sigma_y^{\,2}}\right) \tag{1}$$

which is derived from the general Gaussian plume equation.[9] This reduces to

$$Y_{10} = 0.92\ \sigma_y(x) \tag{2}$$

Values of Y_{10} for various distances downwind, and stability classes are presented in Figure 10.8 (σ_y's are Pasquill-Gifford values[7]).

Values of X_{10} and Y_{10} determine the size and shape of the "representative area" under C_{max} conditions. Values of X_{10} tend to exceed Y_{10} by an order of magnitude for most stabilities. This results in an impact area which is elliptical in shape, (an ellipse is used here to approximate what is actually a "teardrop" shape). An example of

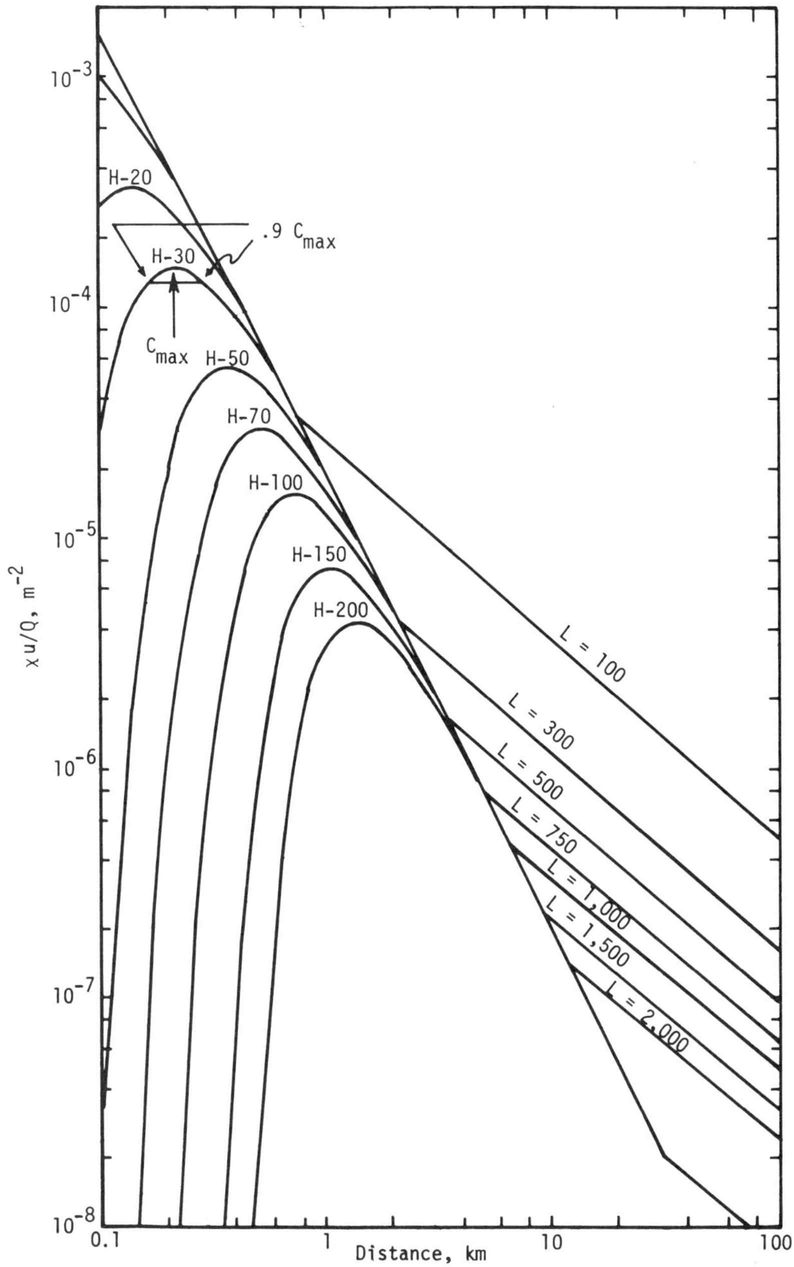

Figure 10.5. $\chi u/Q$ with distance for various heights of emission H and limits to vertical dispersion (L), B stability.[10]

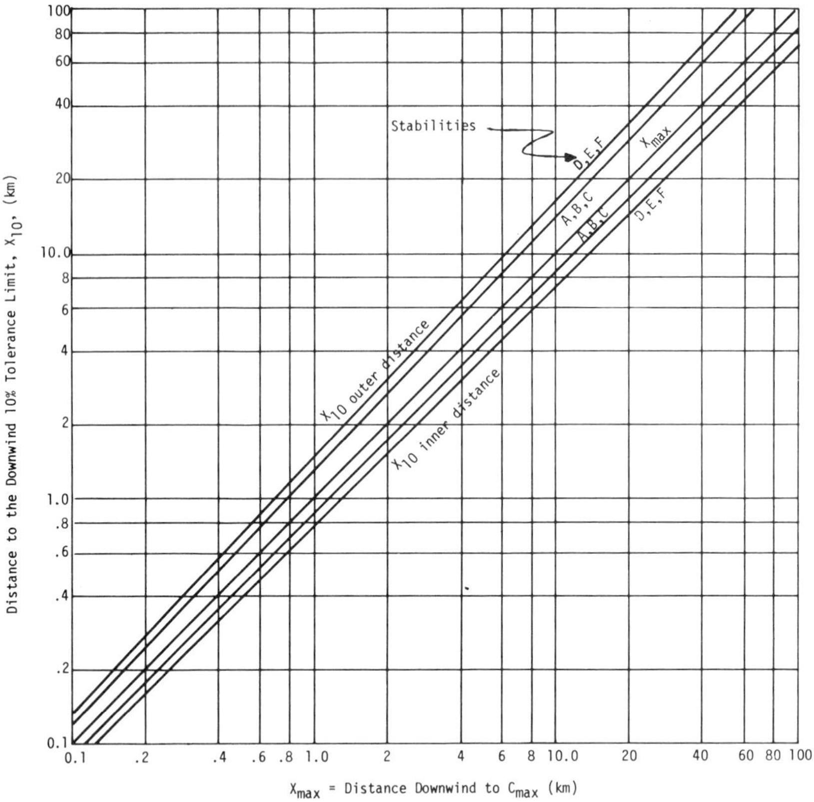

Figure 10.6. Graphical method for determining X_{10} versus distance downwind to X_{max}. (X_{10} = distance from the source to 0.9 C_{max} = $|X - X_{max}|$).

an impact area downwind of a source is illustrated in Figure 10.9.

The number and location of monitoring sites can now be determined from (a) the size of the "representative area" under C_{max} conditions, (b) the size of the potential zone requiring monitoring (from Figure 10.4), and (c) the frequency of occurrence of meteorological conditions causing C_{max} values in each zone. The first step is to determine the ratio of the representative area of a monitoring station located in the center of a potential zone to the area encompassed by the zone (the zone is "donut" shaped if wind direction is not accounted for). If X_{10} is less than the radial dimension of the zone, then the proportion of the zone covered by a single monitoring station can be calculated from

$$CR_S = \frac{\left(\frac{X_{10} + Y_{10}}{4}\right)^2 \pi}{(X_{Z_0} - X_{Z_i})^2 \pi} \qquad (3)$$

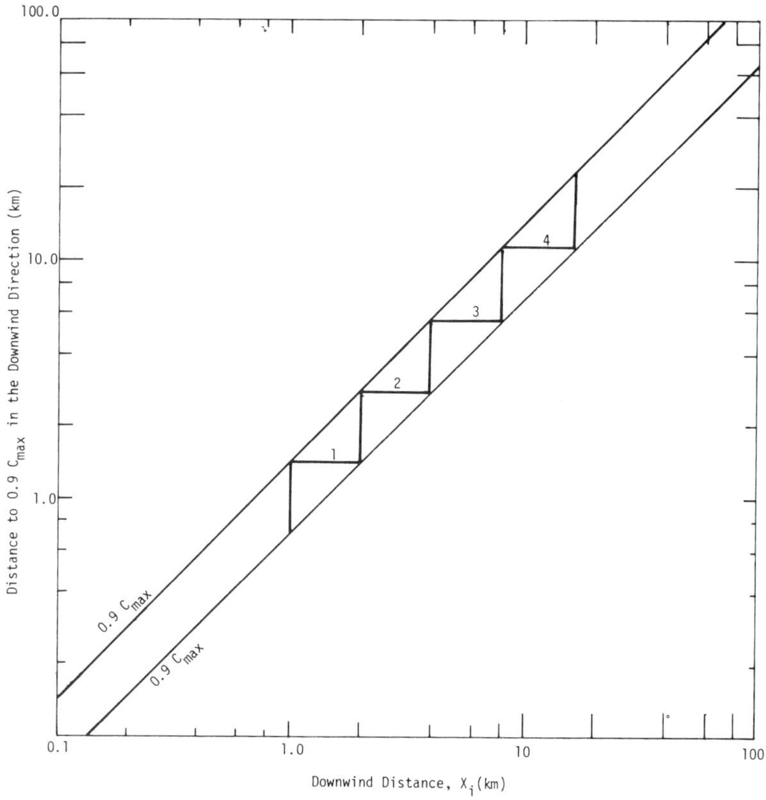

Figure 10.7. Illustrating the number of sites needed to maximize coverage between 1.0–10 km for D stability.

where CR_S = the coverage ratio for one station (without range for wind direction)
X_{Z_i} = inside radius of circular zone, and
X_{Z_0} = outside radius of circular zone.

To calculate coverage ratio for a single wind direction, multiply the CR_S from Equation (3) by the number of wind sectors the "donut" shaped zone is divided into. Figure 10.9 depicts a representative area and a potential zone, and shows the coverage ratio calculation. Coverage ratios can be calculated for a single station, CR_S, or for all stations combined. The sum of all station coverage ratios equals the total coverage ratio, CR_T, for the network.

To calculate the probability that one station will measure the maximum concentration n times when the meteorological conditions causing C_{max} occur N times per year requires use of the binomial distribution function.[8]

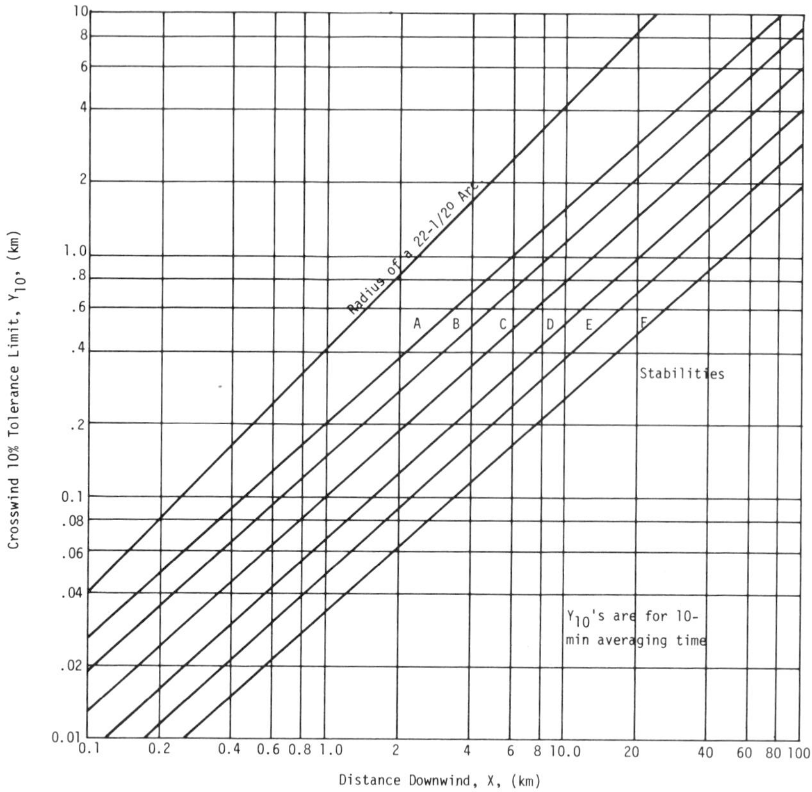

Figure 10.8. Y_{10} versus distance downwind for six stabilities.

$$P = \frac{N!}{n!(N-n)!} CR^n (1-CR)^{N-n} \tag{4}$$

For large values of N, the probability, P, or coverage ratio, CR, is best determined by using the normal approximation to the binomial distribution and solving for the statistic, Z. The probability associated with various Z values can be determined from standard statistical tables. Z is calculated using the equation

$$Z = \frac{n-NP}{(NP(1-P))^{1/2}} \tag{5}$$

One solution of Equation (5) is given in Figure 10.10. This figure can be used to determine the total coverage ratio, CR_T, required to allow n successful measurements of C_{max} with 99% confidence given the frequency of occurrence, N, each year of the meteorological event producing C_{max}.

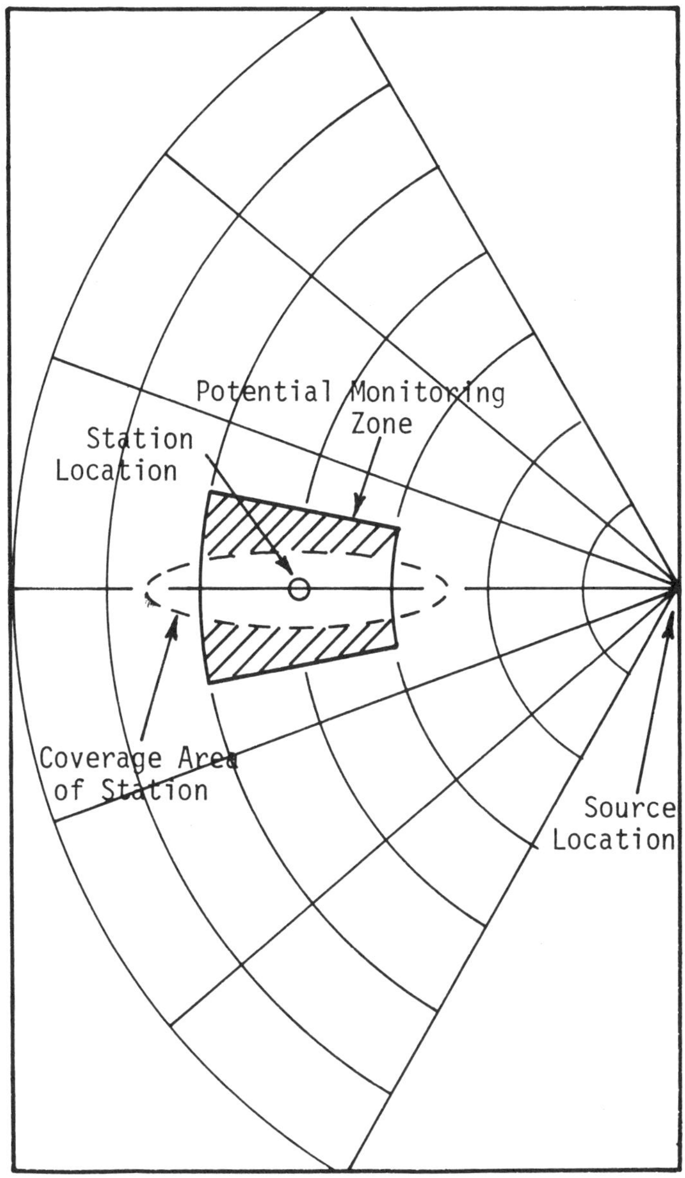

Figure 10.9. Illustration of the coverage area of a single station downwind of a point source. (Shaded area represents the area within the potential monitoring zone not covered by the station. The ratio of the coverage area to the total area of the potential zone equals the coverage ratio.)

After solving the total coverage ratio required, the number of stations needed in the crosswind direction can be calculated.

$$N_S = \frac{CR_T}{CR_S} \tag{6}$$

where N_S = number of stations needed,
CR_S = coverage ratio of each station, and
CR_T = total coverage ratio required to achieve desired results.

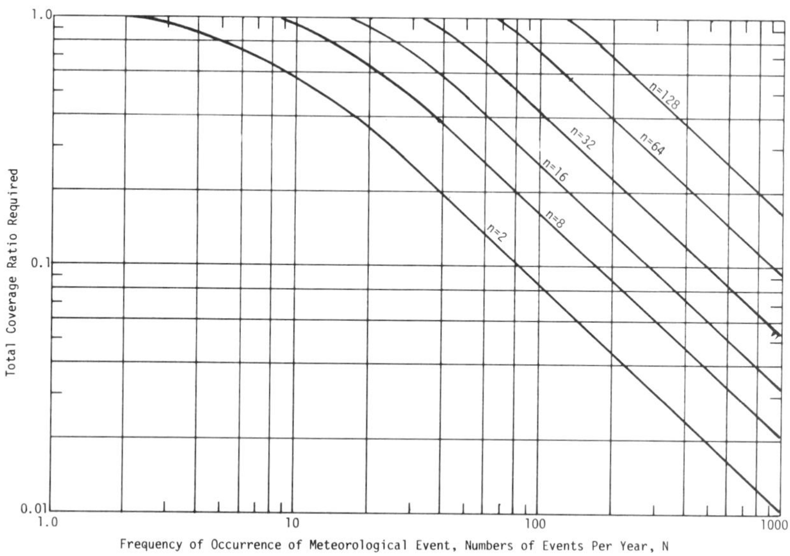

Figure 10.10. Coverage ratio required to observe C_{max}, n times, with 99% confidence, versus the occurrences, N, of meteorological phenomena causing C_{max}.

Simplified Method for Station Location

The previous analysis indicates that the area where maximum ground-level concentrations occur is narrow in the Y-direction and long in the X-direction. For this reason, the likelihood that a monitoring station will observe maximum levels is more dependent on its circumferential location about the source than its radial distance from the source. For this reason, simplified estimates of CR for $X_{10} > (X_{Z_0} - X_{Z_i})$ are sufficient in most cases. CR is then determined by dividing $Y_{10}(x)$ by either the circumference of a circle (for all wind directions) or a single arc about the point source (for one wind direction) having a radius equal to the downwind distance, x.

Using this simplified method, Figure 10.11 gives coverage ratios for one station located with a single 22½° wind sector plotted against downwind distance, x, for various stabilities. Using this figure, coverage ratios for single stations can be determined assuming they are properly located in the downwind direction, x.

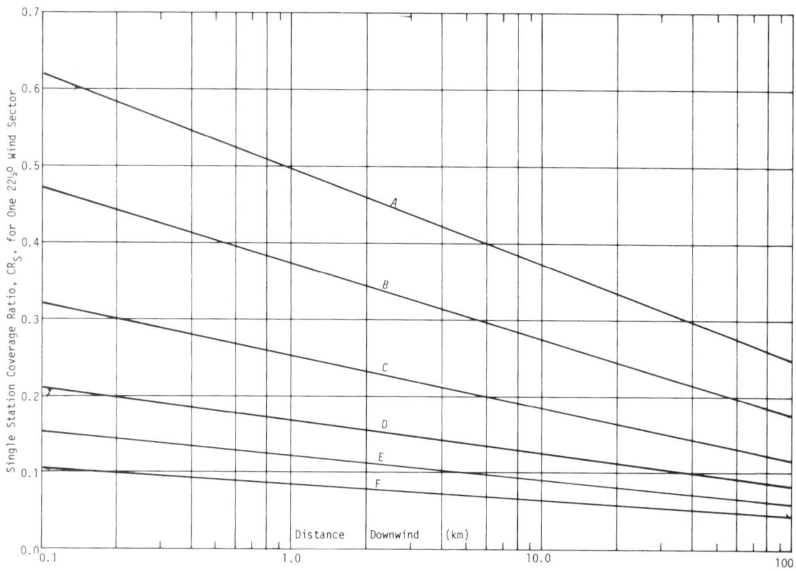

Figure 10.11. Single station coverage ratios, CR_s, versus distance downwind for six stability classes. (CR_s values are for 10-min averaging times.)

The Effect of Averaging Time on Coverage Ratio

Coverage ratios are highly dependent on the extent of the representative sampling area in the Y-direction, which is in turn dependent on the averaging time of the sample. Several authors (*i.e.*, Hino,[9] Turner, *et al.*)[6] have pointed out the relationship between σ_y and sample averaging time, usually expressed as the exponential relationship

$$\sigma_y \propto (t)^{-.05} \tag{7}$$

where t = averaging time in minutes. The physical cause of increasing σ_y with averaging time is due to the effects of meandering wind direction.

Because CR $\propto Y_{10} = .92\ \sigma_y(x)$, the coverage ratio varies with averaging time. The values of Y_{10} given in Figure 10.8 are for 10-minute average concentrations. Correction factors, CF, which can be used to increase Y_{10} and subsequently CR_S to account for averaging times different from 10 minutes, can be calculated from the equation

$$CF = \frac{(10)^{-0.5}}{(t)^{-0.5}} \; .316 \; t^{0.5} \quad (8)$$

Typical values are:

Averaging Time	*CF*
1-hr	2.4
3-hr	4.2
10-hr	7.7

Can One Station Observe C_{max} Twice

The number of stations required to observe C_{max} at least twice can be determined using Figure 10.10; however, what is the likelihood that C_{max} will be observed twice by the same station (appropriately applicable to checking compliance with NAAQS)? This probability is solely dependent on the coverage ratio of each station and not affected by the number of stations used. Figure 10.12 can be used to determine the coverage ratio required by one station to ensure that C_{max} is observed at least twice, given the number of times the meteorological event causing C_{max} occurs, N, and the desired confidence level. The results can be used to indicate if monitoring some meteorological phenomena should be attempted at all. For example, if the coverage ratio of a monitoring site is known equal to (CR = .2) and the expected number of occurrences of meteorology causing C_{max}, N = 10, then the probability that C_{max} will be measured at least twice is only 50%. Hence, there is only a 50:50 chance that any monitoring station will observe two occurrences of C_{max} even though it occurs 10 times. In such a case, it is probably inadvisable to attempt to monitor C_{max} under this condition at all. In light of the expense of air monitoring, sites should not be chosen to monitor C_{max} when the probability of success is low. Figure 10.12 shows how many times a meteorological event should occur to be 90% certain that C_{max} will be observed twice at one site for a range of values of CR_S. For example, for CR_S = .2 the meteorological event should occur at least 20 times each year to be 90% certain; for CR_S = 0.1 the same event must occur 43 times per year.

SUMMARY

Techniques for evaluating the number and location of air monitoring sites needed to monitor values of C_{max} have been presented. The various techniques can be integrated into a systematic approach to site selection. The steps are outlined below.

1. Evaluate historical meteorological data to determine prevailing meteorology and the frequency of occurrence of adverse condition for air pollution dispersion. Choose ranges of meteorological

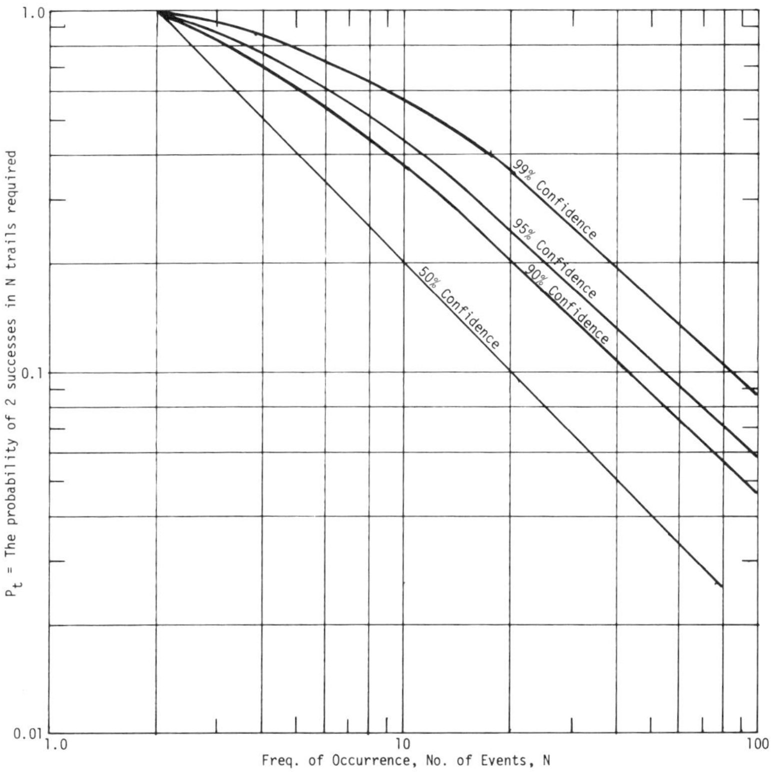

Figure 10.12. The probability required to successfully monitoring C_{max} twice if the appropriate meteorological event occurs N times (for various confidence levels).

conditions under which air monitoring may be desirable and directions from source where conditions occur most frequently.

2. Using point source models (computer models, nomographs, or general solutions presented herein) determine C_{max} and X_{max} under each range of meteorological conditions. From X_{max} identify potential zones where air monitoring stations may be needed.
3. Locate potential sites in the downwind directions where C_{max} values occur most frequently. Note that different C_{max} values occur most frequently. Note that different C_{max} values occur under different types of meteorology all of which may not occur most frequently in the same direction. Hence, this step does not simply mean to locate sites downwind of the prevailing wind direction.
4. Using Figure 10.6 determine the proper spacing and num-

ber of sites needed in the X direction to provide adequate coverage of each monitoring zone.

5. Determine the single station coverage ratio, CR_S, for the distances downwind and stability classes chosen in steps 1–3.
6. Calculate the total coverage ratio, CR_T, required by the network for each meteorological condition of interest and desired number of observations, n. Divide CR_T by CR_S to determine number of stations needed in the crosswind direction.
7. Finally calculate the probability of successful observations of C_{max} for each station, and if this probability is very low then some stations *may* be eliminated.

It is important to remember that the figures given are based on a tolerance range for C_{max} of 10%. If a different tolerance range is desired, then new curves should be prepared from the basic equations presented in the text. The steps in the site selection process remain the same.

The site selection procedure presented herein is designed for locating sites having the highest probability of successfully measuring maximum ground-level pollutant concentrations occurring near point sources. In practice, other objectives may be equally important, such as monitoring pollutant levels near sensitive receptors. Therefore, the sites identified using the procedures presented here should also be evaluated to see if certain site locations can be altered in order that multiple objectives might be served by the same site or if additional sites are needed.

REFERENCES

1. National Environmental Policy Act, U.S.C., Public Law No. 91-190 (1970).
2. Noll, Kenneth E. and Terry L. Miller. *Design of Highway Air Monitoring Surveys — Vol. I*, (Federal Highway Administration, 1975).
3. Miller, Terry L. and Kenneth E. Noll. *Highway Air Quality Monitoring Manual*, (Federal Highway Administration, 1975).
4. "National Primary and Secondary Air Quality Standards," *Federal Register*, **36**, No. 84 (April 30, 1971).
5. Slade, David H., (Ed.). *Meteorology and Atomic Energy — 1968*, (Oak Ridge, Tennessee: U.S. Atomic Energy Commission, Office of Information Services, 1968).
6. Turner, D. Bruce. *Workbook of Atmospheric Dispersion Estimates*, U.S. EPA Publication No. AP-26 (1970).
7. STAR Program: Seasonal and Annual Wind Distribution by Pasquill Stability Class (6), National Climatic Center, Federal Bldg., Asheville, North Carolina.
8. SAS (Statistical Analysis System) designed by Anthony Barr and James Goodnight, Dept. of Statistics, North Carolina State University, User Manual Published by Student Supply Stores, NCSU, Raleigh, NC, August 1972.
9. Hino, M. *Atmos. Environ.* 2, 149 (1968).

11

Ambient Monitoring at Tennessee Valley Authority Power Plants

Auburn E. Owen, Jr. and **Roy May**
Tennessee Valley Authority, Muscle Shoals, Alabama

INTRODUCTION

TVA has maintained surveillance over air quality since its first large coal-fired generating plant began operating in the early 1950s.[1] This surveillance program has grown with the TVA system, which now includes 12 coal-fired plants in Tennessee, northern Alabama, and southwestern Kentucky. With its plants widely dispersed over a large geographic area, TVA has had an opportunity to monitor a wide range of meteorological and topographical conditions. The diversity of plant characteristics (such as total capacity, number and size of units, emission rates, number and heights of stacks) has added to the scope of TVA's monitoring experience.

Air quality monitoring documents ambient air quality to assure compliance with applicable state and federal air quality standards, to enable selection and operation of adequate measures for controlling air quality, to provide data to support applied research programs, and to supply information that may contribute to air quality criteria for refining and improving standards and regulations. The air quality surveillance program consists primarily of four interrelated activities for monitoring meteorological conditions, source emissions, ambient air quality levels, and biological effects.

METEOROLOGICAL MONITORING

Meteorological measurements have been made for more than two decades, generally in two phases—before and after operations began. At least one year before plant startup, data are obtained to define the local meteorological conditions that will affect the transport and dispersion of emissions and to provide a basis for selecting sites for permanent air quality monitors. After plant startup, permanent meteorological facilities provide data that may be used to investigate the actual dispersion of stack effluent and to determine the effect of the plant on ambient air quality.[2,3]

Installation configurations, monitoring instruments, and data recording and display systems have become increasingly complex as a

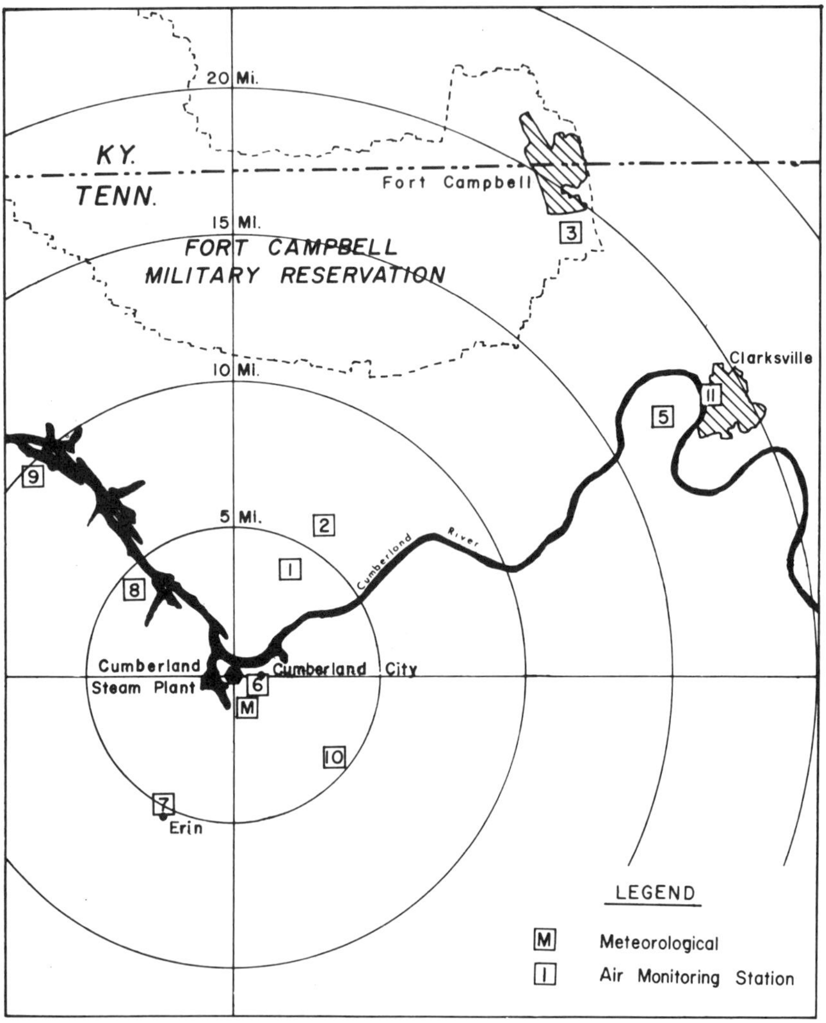

Figure 11.1. Air monitoring network—Cumberland Steam Plant.

result of technological advancements in equipment design and expanded requirements for environmental monitoring. At many of the early power plants, instruments were mounted on plant structures or on relatively low towers (50–60 m tall) and included only analog chart records of dry-bulb temperature and wind speed and direction. Today, facilities at the larger power plants include high-speed digital computer recording and processing systems and instruments at multiple levels on towers 100–120 m tall for measuring wind speed and direction, dry- and wet-bulb temperatures, solar radiation, total radiation, rainfall, and atmospheric pressure (Table 11.1).

SOURCE EMISSIONS MONITORING

Emissions from the power stations are determined mainly by routine analyses of the fuels that are consumed. For each unit, daily samples of coal are combined into weekly composite samples, and a proximate analysis is performed. The resulting information, along with information about plant operating conditions and about the efficiency of the control devices, can be used to determine emission rates of sulfur dioxide and particulates, the emissions of greatest interest from coal-fired power plants. More recently, monitors are being installed for direct measurement of sulfur dioxide emissions and smoke opacity within the stack.

Stack gases are also occasionally sampled to determine emissions of other materials such as nitrogen oxides, mercury and other trace elements, and fluorides.

AMBIENT MONITORING

TVA maintains stations that monitor ambient air quality in the general area of each power plant. At some plants, ambient air quality monitoring was recently expanded to include nitrogen oxides and oxidants; at others, monitoring has been extended to include emissions from local sources other than the TVA plant. Ambient concentrations of sulfur dioxide and particulates, the emissions of principal interest, are monitored most extensively. There are 195 stations for sampling gases or particulates (Table 11.2). As with meteorological monitoring, intensive air monitoring is begun at a plant site about one year before plant startup. The data obtained are used to determine baseline levels of ambient air quality in the area. Figure 11.1 shows such a monitoring network for the recently completed Cumberland power plant. After operation begins, the same large number of monitoring stations is operated for several years until the range of effects of the plant operations on ambient air quality is defined. Afterwards, the number of stations and the amount of monitoring may be reduced. If plant operations change significantly, if capacity

Table 11.1. Meteorological Monitoring Facilities. TVA Coal-fired Power Plants, January 1, 1975.

Plant Area	*Meteorological Stations*	*Wind System*	*Temperature Difference System*	*Hygro-thermograph*	*Rain Gage*	*Dew Point*	*Solar Radiation*	*Total Radiation*	*Atmospheric Pressure*
Allen	1	1	3	—	—	1	—	—	—
Bull Run	1	1	3	—	1	1	1	1	1
Colbert	—	—	—	—	—	—	—	—	—
Cumberland	1	1[a]	3	—	1	1	1	1	1
Gallatin	1	1	3	—	—	1	—	—	—
John Sevier	1	1	—	—	—	—	—	—	—
Johnsonville	1	1	2	1	—	—	—	—	—
Kingston	1	2[a]	3	—	—	—	—	—	—
Paradise	1	2[a]	3	—	1	1	1	1	1
Shawnee	1	1	2	1	—	—	—	—	—
Watts Bar	1	2[a]	4	—	1	1	1	1	1
Widows Creek	2	2[a]	3	2	—	—	—	—	—
Total	12	15	29	4	4	6	4	4	4

[a]Includes Bivane.

Table 11.2. Air Quality Monitoring Facilities. TVA Coal-fired Power Plants, January 1, 1975.

Plant Area	*Air Quality Stations*	*Sulfur Dioxide*	*Nitrogen Dioxide*	*Ozone*	*Particulate*		*Soiling Index*	*Lead Peroxide (Sulfation)*	*Fluoride*	*Corrosion Effects*	*Special Precipitation Sensor*
					Suspended	*Settled*					
Allen	4	3	1	—	3	3	1	—	—	—	—
Bull Run	41	4	—	—	3	5	—	36[a]	—	—	—
Colbert	3	4	—	—	3	3	1	—	—	1	—
Cumberland	72	12	1	1	5	10	1	62	—	—	4
Gallatin	6	4	—	—	4	5	—	—	—	—	—
John Sevier	3	3	—	—	2	3	—	—	—	—	—
Johnsonville	12	4	—	—	4	4	—	5	6	—	—
Kingston	9	10	—	1	3	3	1	—	—	1	—
Paradise	10	11	1	—	5	7	1	—	—	—	—
Shawnee	6	6	—	1	4	4	—	—	2	—	—
Watts Bar	3	3	—	—	1	2	—	—	—	—	—
Widows Creek	25	10	—	—	5	6	1	17	—	7	—
Total	194	74	3	3	42	55	6	120	8	9	4

[a]Some stations sited in adjacent Kingston power plant area.

increases, or if unusual conditions occur, intensive monitoring is resumed.

Although sulfur dioxide is now being measured by both conductometric and coulometric monitors, the conductometric monitors are being progressively replaced by instruments based on coulometry. Sulfation levels are determined with lead peroxide candles or Hueystyle platelets. Suspended particulates are measured with high-volume samples; settled particulates are measured with a dust deposit gage approved by the American Society for Testing and Materials.

Sulfur Dioxide Monitoring

Ambient sulfur dioxide is monitored by the Philips PW 9700 SO_2 monitor. This monitoring method is based on a coulometric principle, whereby measurement of SO_2 is accomplished by titration with bromine and a feedback amplifying system that controls the current for regeneration of the bromine. Hourly average SO_2 data are collected at each monitoring station by analog strip chart recorders and simultaneously relayed to a computer-controlled paper punch machine and teletype printer at the central environmental data station. SO_2 data, along with meteorological parameters, are displayed at the environmental data station for use in plant operational control and are ultimately converted to computer-generated summary printouts for other uses. All ambient SO_2 is logged on an hourly average basis, the 24-hour day beginning with midnight to correspond with daily summaries of plant emissions and meteorological parameters.

Suspended Particulate Monitoring

Suspended particulates are collected by the standard, EPA-approved monitoring method.[4] Standard high-volume samplers collect air-aspirated particulates on 8 by 10-inch fiberglass filters that are weighed before use. The samplers operate for 24-hour periods from midnight to midnight on each sixth day. Filter samples are equilibrated for 24 hours in the laboratory below 50% relative humidity and then weighed to determine weight of the particulates by comparison with the weight measured before the filters were used. The weight of the particulates is divided by the volume of air sampled and expressed in $\mu g/m^3$. Filter samples are stored indefinitely for analyses that may be requested.

Settled Particulate Monitoring

Settled particulate samples are collected and analyzed by the method approved by the American Society for Testing and Materials.[5] The sampling gage is a polyethylene container 7¾ inches in diameter and 10 inches high mounted on a steel pole about 8 feet above ground. Collecting gages are placed in the field at the beginning of each calendar month and remain there for 30 days (plus or minus three days). The collected sample is analyzed for the content of water-insoluble and water-soluble particulates and for ash residues. Results are expressed as tons per square mile of total particulates (soluble plus insoluble) and of filter ash residues. When requested, the approximate fly ash content of the settled particulate samples is determined from the insoluble portion of the filter ash residue, by the same method used for the suspended particulate samples. Filter ash residues are stored for two years.

Particulate Analyses

TVA monitors concentrations of airborne particulates primarily to assess the contribution of fly ash emissions from a power plant to the total levels of particulates in the air around the plants. If a sample exceeds state or federal standards, or if complaints are received regarding excessive particulates, that sample can be analyzed for fly ash content to assess the contribution of the power plant to the total amount of particulates in the area. All samples for suspended particulates that exceed the EPA primary standard of 260 $\mu g/m^3$ are analyzed microscopically for fly ash content; those that exceed 150 $\mu g/m^3$ from some power plants located near urban centers are also analyzed. A 2-inch-wide strip of the fiberglass filter sample is weighed and ashed in a low-temperature vacuum oven to destroy organic matter and weighed again. The percentage of noncombustible material in the particulate sample is calculated from the weight lost by the ashed strip and the fraction of the area of the total filter. A portion of the ash is dispersed in a drop of mounting oil on a microscope slide. A cover slide is placed over the oil droplet, and the particles are allowed to settle on the slide. All particles within a counting graticule field (Whipple micrometer disc) are classified as fly ash (spherical) or not fly ash (nonspherical) at 400X. Particles greater than 0.5 μm in diameter can be classified at this magnification. Three or more fields are counted until 300–500 particles have been classified. The ratio of fly ash particles to the total number of particles counted approximates the fly ash content of the ash residue.[6] The fly ash content for the total particulate samples is then calculated from the percentage of ash residue and the percentage of fly ash in the residue.

The ash residues of the high-volume filter samples and settled particulate gage samples that exceed standards give some indication of whether the high results could be due to power plant emissions or to organic substances. Samples high in fly ash content usually have a low weight loss when they are ashed. However, the fly ash content should be determined, to assess the composition of the noncombustible material.

The ratio of the number of fly ash particles to the total number of particles gives only an approximation of the percentage by weight of fly ash. The average particle size and densities for fly ash particles and for particles that are not fly ash are assumed to be similar. Bloss[6] identified the principal components of a portion of the ash residue that are not fly ash and found that densities of these substances ranged from 1.6–2.6 g/cm^3. Densities of fly ash particles range from 2.0–2.9 g/cm^3.

The diameter of fly ash particles and of other suspended particulates can vary widely. However, most suspended particulates are less than 80 μm in diameter, and the fly ash collected by high-volume filters is less than 40 μm in diameter. Therefore, the percentage of fly ash can more accurately be determined by classifying the particle sizes of the components and calculating the fly ash content from the mean weighted diameters of the fly ash and non-fly ash particles.[7] The particle numerical ratio method tends to give 10–25% higher results that the particle size method but it requires less time. Although the ratio method is less accurate, it gives a good indication of the contribution of the fly ash from the power plant to the total suspended particulates.

Biological Surveillance

Biological surveillance is maintained around all fossil fuel power plants to observe acute effects, if any, on terrestrial vegetation. Surveillance before operations begin includes land use studies to determine the nature and condition of crops and forest species in proposed plant area. Forest surveillance plots are established, along with experimental gardens, to determine typical growth characteristics of local plant species and to document significant ecological factors. After plant operations begin, routine biological surveys are made at least once each growing season. Season field surveys are made, along with special botanical studies in growth plots, to determine effects on local plant species. Results of these field investigations are correlated with ambient air quality data. Biologists plan to increase the limited studies they have been conducting of potential chronic effects on vegetation around TVA power plants.

DATA HANDLING

During the early phases of the monitoring program, data from the meteorological and continuous recording ambient sulfur dioxide monitors were recorded on analog charts and abstracted by hand. As the program became larger and uses for the data became more complex, data from the meteorological facilities at plants were placed on paper punch tape and processed by the TVA Computing Center for subsequent summary analyses. At the plants such as Cumberland Steam Plant, SO_2 data from the monitoring stations were converted to transmit SO_2 data by radio telemetry. These data are now entered with the meteorological data on paper punch tape for later computer analysis and summary. With the implementation of TVA's new sulfur dioxide emission limitation (SDEL) program, monitors for all aerometric data from all 12 steam plants are being converted to telemetry and automatic data logging systems. At a number of steam plants, the SO_2 and meteorological monitors will operate on a real-time basis to provide the continuous monitoring of ambient SO_2 levels needed to assure compliance with state standards. Data are summarized and submitted monthly and quarterly to the state air quality control agencies.

REFERENCES

1. Gartrell, F. E. "Monitoring of SO_2 in the Vicinity of Coal-fired Power Plants — TVA Experience," *Symposium on the Air Pollution Problem, Proc. of the Amer. Power Conf.* **27**, 117-123 (1965).
2. Tennessee Valley Authority and Public Health Service. "Full-scale Study of Dispersion of Stack Gases, Summary Report," Chattanooga, Tennessee (August 1963) and Parts I-IV, Chattanooga, Tennessee (June 1965).
3. Montgomery, T. L., J. M. Leavitt, T. L. Crawford and F. E. Gartrell. "Controlling Ambient SO_2," *J. Metals* **25**, No. 6 (1973).
4. "Appendix B — Reference Method for Determination of Suspended Particulates in the Atmosphere," *Federal Register* **36**, No. 84 (April 30, 1971).
5. "Recommended Standard Method for Continuing Dustfall Surveys," APM-1, Revision 1, *J. Air Poll. Cont. Assoc.* **16**, 372 (1966).
6. Bloss, F. Donald. "Microscopic Study cf Contents of Fly Ash Gages Near Kingston, TVA" (July 12, 1956).
7. McCrone, Walter C. and J. G. Delly. *The Particle Atlas Edition Two*, Vol. I (Ann Arbor, Michigan: Ann Arbor Science Publishers, Inc., 1973).

12

Airborne and Mobile Ground Level Measurements of Pollutants in the Sphere of Influence of a Coal-Fired Electric Generating Station

N. Thomas Stephens
Department of Civil Engineering, Virginia Polytechnic Institute and State University, Blacksburg

Alvin N. Bird, Jr. and **Joseph D. McCain**
Southern Research Institute, Birmingham, Alabama

INTRODUCTION

An accurate assessment of the influence of atmospheric emissions from a particular source on the existing background levels of atmospheric pollutants is a necessary input for both technical and administrative decision-making processes. The results of such an assessment could influence such decisions as plant siting, facilities expansion, process design, equipment modifications, and fuel substitutions. In any event, data concerning both actual and potential contributions of emissions to ambient levels are desirable.

The objectives of the study described in this chapter were to design and evaluate a rapid technique for mapping the background levels and source contributions of sulfur dioxide and particulate matter within the sphere of influence of a coal-fired electric generating station, and to provide data for evaluation of predicted concentration values based on diffusion and dispersion models. While numerous relationships have been derived to predict stack emission dispersion, the utility of these equations is limited due to the inability to accurately describe the complex interaction of factors such as topography, atmospheric stability, emission rate, stack height, wind speed, and rate of pollutant decay for a particular situation.

The predominant factors which are responsible for the variability in ground level concentrations also present the major difficulty in a sampling program. On a long-term average basis, wind velocity and direction are predominantly within some definite range, and one may set up fixed location samplers to obtain these average concentrations. On the short-term basis required for rapid evaluation purposes, however, conditions are not sufficiently constant to assure adequate coverage by a few fixed sampling stations.

To minimize the limitations inherent in the fixed location sampling method, instrumentation mounted in a light aircraft and similar equipment mounted in ground vehicles were used in this study to provide simultaneous plume measurements in the air and on the ground. These measurements were then used to study transport and diffusion in the plume and to contribute data for evaluation of the influence of the power plant on the surrounding area.[1]

STUDY SITE AND EQUIPMENT

A pulverized-coal, coal-fired generating station of 1500 Mw capacity located in the Southeast was selected for the study. Airborne combustion effluents from the facility were discharged from three stacks ranging in elevation from 250–500 feet above ground level. During the week of the tests, the plant load was between 800 and 960 Mw. The plant is located in relatively flat terrain, and is isolated from major population and industrial centers which might contribute significant quantities of the atmospheric constituents of interest in the study.

For the ground level measurements, two panel truck-type vehicles were equipped with continuous recording sulfur dioxide analyzers, portable generators, and radio gear. In addition, one of the vehicles contained an experimental model of a condensation nuclei counter designed by Southern Research Institute. The sulfur dioxide analyzers, which were equipped with strip chart recorders, were of the electroconductivity type described by Hocheiser.[2]

For the airborne measurements, a Cessna 172 aircraft, capable of operating at slow speeds and low altitudes, was equipped with analytical instrumentation for sulfur dioxide and particulate matter in addition to specialized temperature, altitude, and turbulence recording equipment. Airborne sulfur dioxide concentrations were also measured with an electroconductivity analyzer. This device was similar to the ground level instruments, but with a response time on the order of 2 seconds as compared with approximately 15 seconds for the ground instruments. The instrument is small and easily adapted to limited space and weight situations. The SO_2 analyzers were calibrated with commercially available permeation tubes and the output of the permeation tubes was checked by chemical analysis.[3,4]

The difference in time constants of the two analyzers had very little influence on the interpretation of results of the test because of the difference in speed between the aircraft and the ground sampling vehicles. That is, the ground vehicle moved through the plume more slowly than the aircraft so that the slower response of the Davis analyzer still provided about the same ratio of response time to sampling time as the airborne unit. In addition to SO^2 concentration data, other parameters that were measured and recorded included temperature, altitude, and particulate matter concentration in the size range from 0.5–10 μ in diameter. The particle counter was a Bausch & Lomb Model 40-1. The aircraft instrumentation package is described more completely by Stephens and McCaldin.[5]

Wind speed and direction at 350 feet above ground level and the vertical temperature gradient to this elevation were obtained from the plant meteorological station.

SAMPLING PROCEDURE

Based on the prevailing meteorological conditions, the sampling teams were dispatched as early in the morning as practical to locate the plume and set up a simultaneous sampling pattern at a convenient location. Radio contact was maintained between all stations to coordinate the sampling efforts.

Ground level sampling consisted of measuring background concentrations outside the plume, and then making a series of crosswind traverses through the plume at the time the aircraft was executing crosswind traverses overhead at the same location. The aircraft traverses were made at different altitudes to measure variations in SO_2 with height. On each test one traverse was made near tree top height for comparison with ground level measurements. As soon as data had been collected at one location, the ground vehicle was relocated further downwind for the next series of sampling traverses.

In addition to the crosswind sampling pattern, the aircraft was flown downwind along the longitudinal axis of the plume. Wind velocity at plume altitude was determined by timed upwind and downwind flights over a fixed distance at constant air speed. The vertical temperature lapse rate and apparent depth of the mixing layer were measured and recorded periodically.

The aircraft made from two to nine passes through the plume at various altitudes from 100 feet above ground level to a few thousand feet above ground level. In most cases an attempt was made to obtain a sufficient number of passes through the plume within a short enough time frame to be able to define the plume characteristics with some confidence. The plume cross sections were obtained at distances ranging from about 1.5-22 miles downwind of the plant.

The choices of downwind distances for the simultaneous air-ground sampling were dictated by the availability of reasonably straight roads running at approximately right angles to the wind direction at the time of each set of sampling runs. As a result of changes in the wind direction, sampling took place throughout an arc of about 180° during the five-day sampling period.

BACKGROUND CONCENTRATIONS OF SULFUR DIOXIDE

Background SO_2 concentrations at ground level in the vicinity of the plant were obtained during four of the five days on which testing took place. The SO_2 analyzers have automatic hourly calibration cycles or check cycles which were used to obtain background measurements. During part of this cycle, a filter removes the SO_2 from the sample air. Comparison of the responses obtained from the background (outside the limits of the plume) with the response obtained from SO_2 filter in place provided the desired data on the background concentrations.

Background concentrations ranged from 0.032–0.092 ppm during four days of testing with an average value of 0.048 ppm and a median value of 0.042 ppm over distances from 2.5–20 miles from the plant. Since the high readings were all experienced during the first day of sampling within 5 miles of the plant, it is possible that these values were influenced by plant emissions and do not represent true background measurements.

PARTICLE CONCENTRATIONS IN THE PLUME

Unfortunately, few usable data were obtained on particle concentration in the plume. The ground-based condensation nuclei counter was an experimental device that had not been calibrated at the time of these tests; hence only relative concentration values were obtained with this device. However, these measurements did show that particles in the condensation nuclei size range, about 0.01–0.1 μm, were still present in the plume at distances up to 20 miles downwind from the source. When the airborne particle counter was used, the concentrations of particles in the 0.5–1 μ size range often exceeded the capacity of the particle counter (one million particles per cubic foot). The particle counter was not equipped with a dilution device so the high concentrations that were encountered in the plume "saturated" the counter when it was set to detect all particles larger than 0.5 μ in diameter. An attempt was made to obtain some indication of the concentration of particles in the plume by setting the particle size threshold at 1 μ. The concentration of particles greater than 1 μ was much lower than the concentration between 0.5 and 1 μ but it was still possible to detect the plume with the particle counter. Some representative particle counter data have been included in Table 12.1.

SULFUR DIOXIDE CONCENTRATIONS IN THE PLUME

Sulfur dioxide data from the aircraft and ground level measurements were used to plot concentration contour diagrams. This particular series of contour diagrams shows the effect of dilution and growth of the plume as the distance from the plant increases from 5 miles (Figure 12.1) to 20 miles (Figure 12.3). Of particular interest is the plume configuration under the prevailing meteorological conditions.

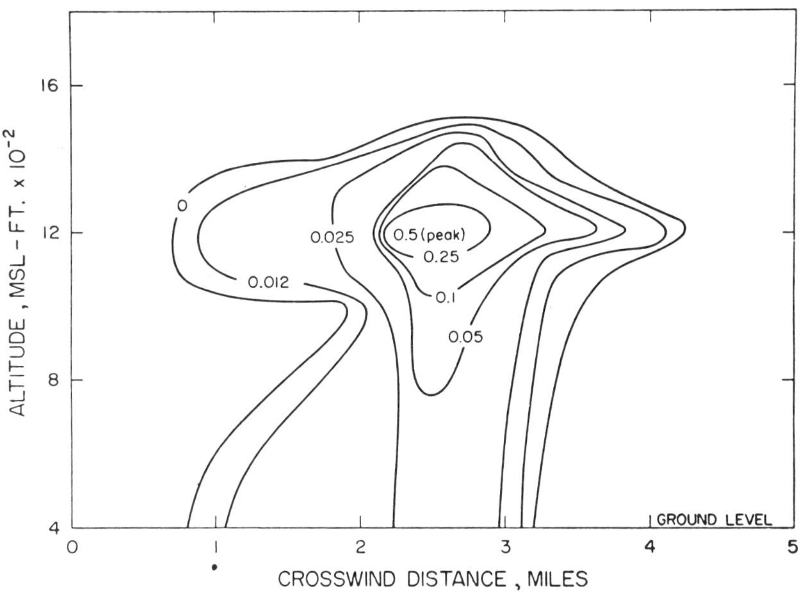

Figure 12.1. Sulfur dioxide profile 5 miles downwind.

The temperature lapse rate from 900–1300 feet MSL (mean sea level) was isothermal and stable conditions prevailed, and at an altitude of 1600 feet there was a noticeable decrease in air roughness which was attributed to reaching the maximum mixing height for that particular time period.

Thus the plume profile illustrated in Figure 12.1 shows the influence of this stable layer existing in the isothermal region, and the effect of a greater degree of turbulence, particularly from ground level (400 feet MSL) to the base of the isothermal layer at 900 feet. During

Table 12.1. Representative Field Test Data. Airborne and Ground Level Sampling.

Time: 0850 — Wind Speed at Stack Height: 15 mph

Location: 5.0 miles downwind — Wind Speed at Plume Height: 16 mph

Ground Level Data			*Aircraft Data*				
SO_2 Peak (ppm)	*SO_2 Avg. (ppm)*	*Plume Width (miles)*	*SO_2 Peak (ppm)*	*SO_2 Avg. (ppm)*	*Particulate Peak 1 μ (counts/m³)*	*Altitude (ft MSL)*	*Lapse Rate (°C/100 m)*
0.200	0.106	1.5	0.162	0.088	50,000	600	—1.1
0.222	0.100	1.6	0.236	0.106	80,000	850	
0.439	0.159	1.4	0.248	0.141	60,000	1,400	
			0.385	0.157	90,000	1,900	
			0.423	0.170	70,000	1,700	
			0.102	0.054	40,000	1,000	

*Note: All concentration values reported excluding background. Particle background count 125,000 counts/ft^3. SO_2 background approximately 0.05 ppm.

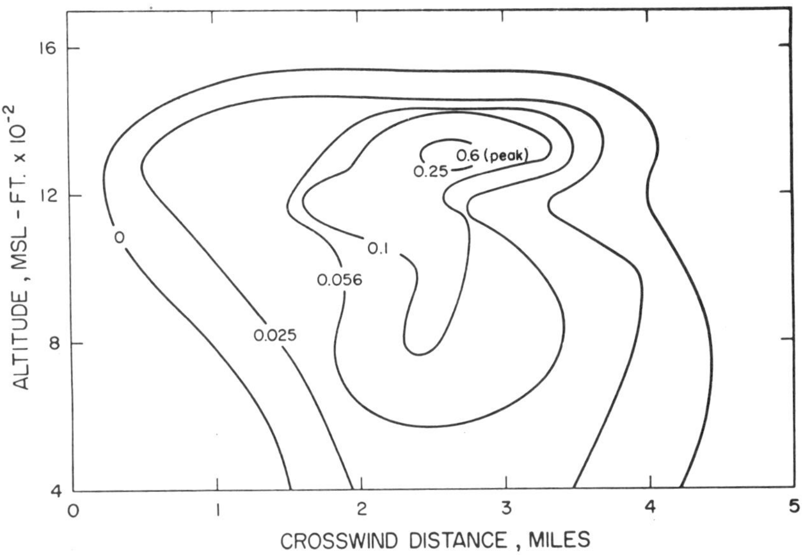

Figure 12.2. Sulfur dioxide profile 8 miles downwind.

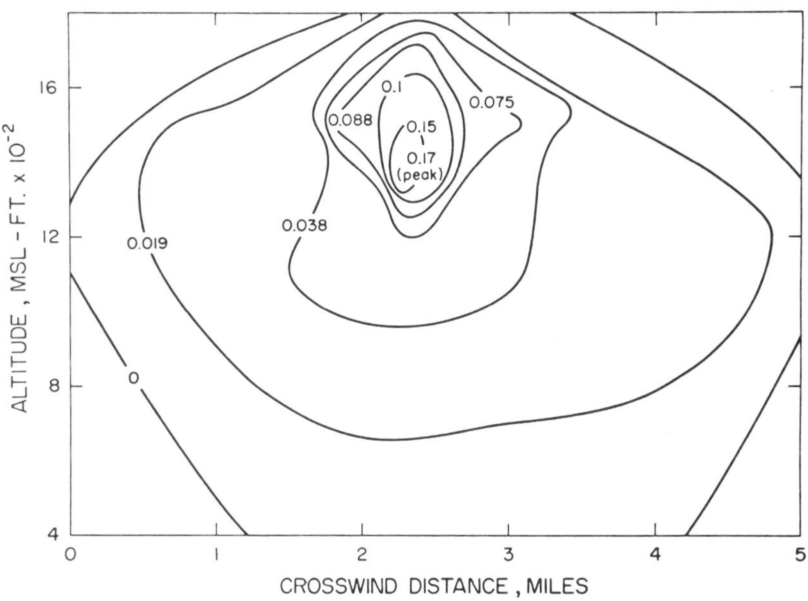

Figure 12.3. Sulfur dioxide profile 20 miles downwind.

this test period, the wind speed at plume altitude was about 20 mph, and at stack height it ranged from 14–17 mph.

Plume spread in the vertical and horizontal dimensions as a function of downwind distance were illustrated in Figures 12.1, 12.2, and 12.3. The data used for these figures were obtained from traverses made at approximately 5, 8 and 20 miles downwind from the plant.

These data may be compared to data obtained the previous day at the same location under differing meteorological conditions. As noted in Figure 12.4, the plume cross section is greater in the vertical dimension than under more stable conditions seen in Figure 12.1. Measurable concentrations are noted from ground level to over 3000 feet, and the top of the mixing layer was judged to be approximately 3200 feet. For this set of traverses, the existing lapse rate from 600–3000 feet was near adiabatic, with a wind speed at plume elevation of 7 mph. Wind speed measured at stack height also averaged about 7 mph during this period of time.

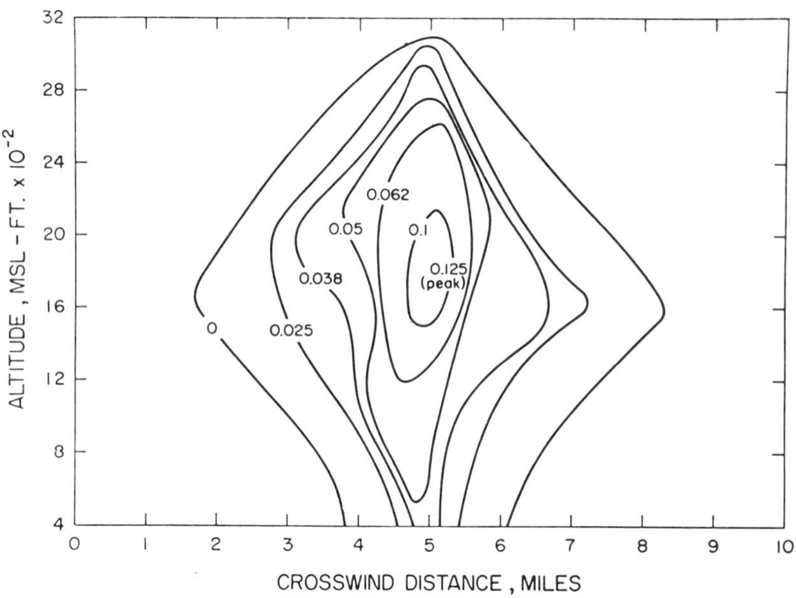

Figure 12.4 Sulfur dioxide profile 5 miles downwind.

Some appreciation for the degree of variability of atmospheric stability may be gained by looking at the results shown in Figure 12.5. A profile was obtained two days later at the same 5-mile downwind location as the one illustrated in Figure 12.4. The temperature lapse rate from 600–2000 feet was also nearly adiabatic; however, from 2000 feet to 3600 feet conditions were isothermal. There was evidently a much greater degree of turbulence at lower altitudes on this date as illustrated by the broken nature of the plume. The major difference in conditions recorded in the two days data was a wind speed that was two times higher on the day that a greater degree of turbulence was present.

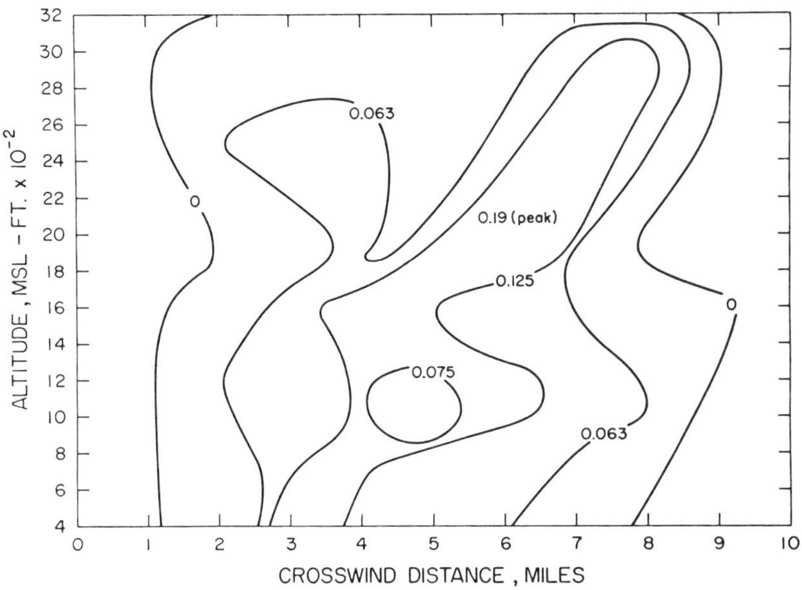

Figure 12.5. Sulfur dioxide profile 5 miles downwind.

During stable atmospheric conditions, one would anticipate the least amount of Plume spread, particularly if the wind direction is fairly constant. The reduced plume spread and reduced dilution in comparison with a more unstable atmospheric regime will result in higher values of pollutant concentration in the plume. This restricted plume spread in the horizontal dimension is illustrated in Figure 12.6.

The SO_2 concentration data given in Figures 12.1 through 12.3 were used to construct the contour diagram of a horizontal cross section through the plume at about 1400 feet. The zero concentration line indicates the points at which the SO_2 concentration in the plume was approximately equal to the background concentration.

Representative data for these plots was given in Table 12.1.

COMPARISON OF AIRCRAFT AND GROUND LEVEL DATA

Data from airborne and ground based SO_2 analyzers were used to compare concentrations measured during low altitude traverses of the plume at the same location and time as mobile ground level data

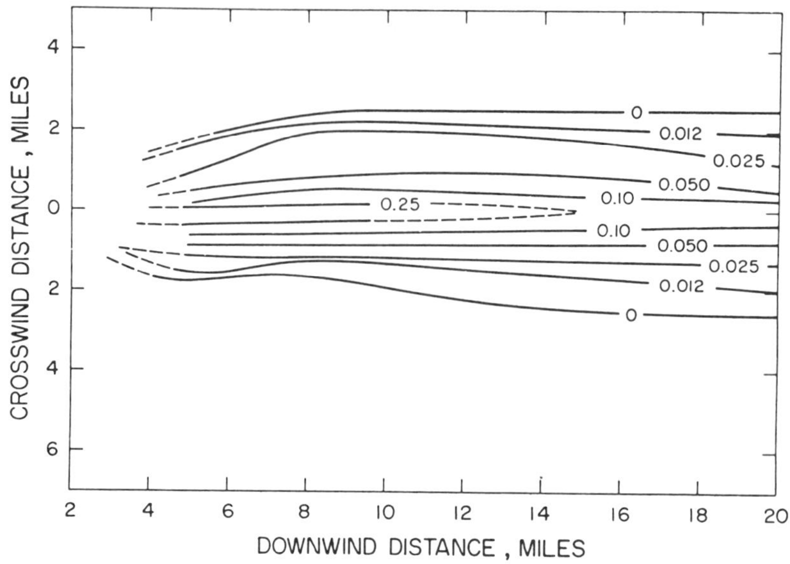

Figure 12.6. Sulfur dioxide isopleths at plume altitude of 1400 ft MSL.

were obtained. The results are illustrated in Figure 12.7. From these data it would appear that in cases where suitable flying conditions prevail, ground level measurements would not be required to establish ground level concentrations.

UTILIZATION OF THE DATA

One key factor to remember in reviewing these data is that the concentrations measured are "instantaneous" rather than "long-term" averages. When one deals with a stationary fixed location monitor, the data recorded are averaged over some relatively long time period, usually 24 hours. Since the atmospheric motions are characterized by temporal and spatial variations, the concentration recorded by continuous or time-averaged monitors becomes subject to the same type of fluctuations. It may well be, for example, that high concentrations were experienced when the wind carried the plume over the station, but that the frequency of occurrence of this situation was low, so the average concentration for that time period would reflect very low values. The converse could also be true.

Compared to data obtained in the "average" sense, the data recorded in this study represent much higher values since every effort was made to measure plume concentrations. The usefulness of these data is more in the realm of assessment of maximum expected short-term values rather than annual mean value situations.

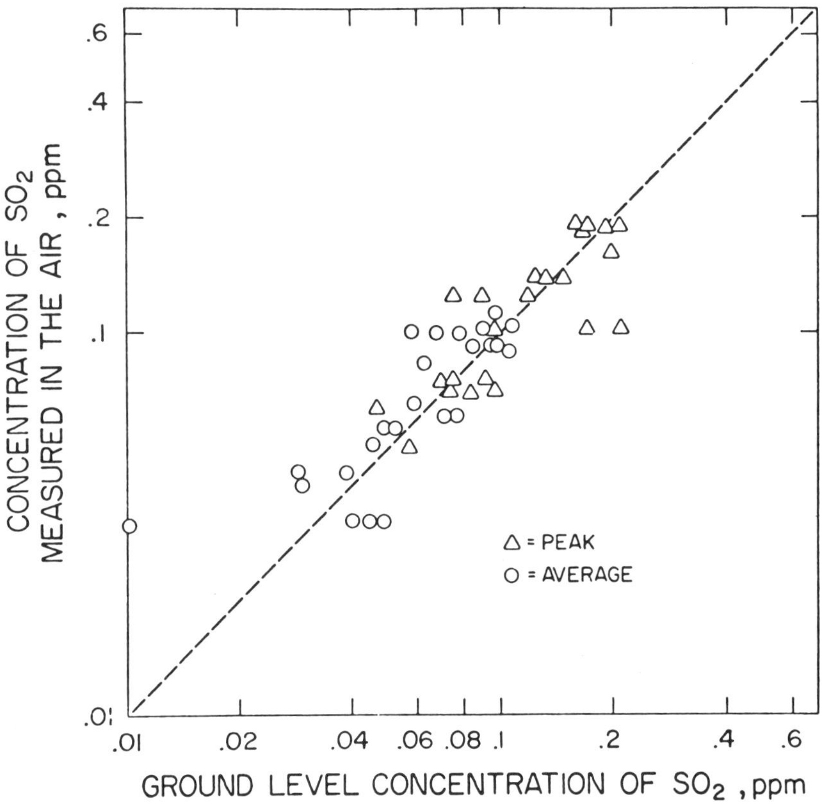

Figure 12.7. Comparison of low-altitude aircraft and ground-level measurements.

When one is concerned with acute exposure or maximum concentration values expected, predictive models based on long-term statistical trends are less useful; hence, it is desirable to have short-term averaged information.

LOCATION OF MAXIMUM CONCENTRATIONS

One such comparison of short and long-term data is illustrated in Figure 12.8, where maximum instantaneous and crosswind average concentrations as a function of downwind distance are compared to a predictive model.[6] The model is based on half-hour averages, and is shown for two wind speeds that were representative of the speeds encountered during the week-long sampling program. One of the main points to be noted is the downwind location of the maximum expected concentration. Each experimental point on the plot represents the maximum value observed for the downwind distance during the week of sampling. Background concentrations of SO_2 (approximately 0.042 ppm) are not included in the values.

EXPERIMENTAL DISPERSION COEFFICIENTS

The field measurements may also be used for developing plume dispersion coefficients which are specific for a particular topographic and stack discharge situation. Since the plume width and depth to background levels were measured, the horizontal and vertical dispersion coefficient can be determined based on these measurements.

A common definition of plume boundary is that point at which the concentration drops to a value of one-tenth of the maximum centerline concentration. Using this convention, the plume boundary will be a distance from the centerline of 2.15 times the standard deviation of plume spread, σ (dispersion coefficient) if we assume a Gaussian

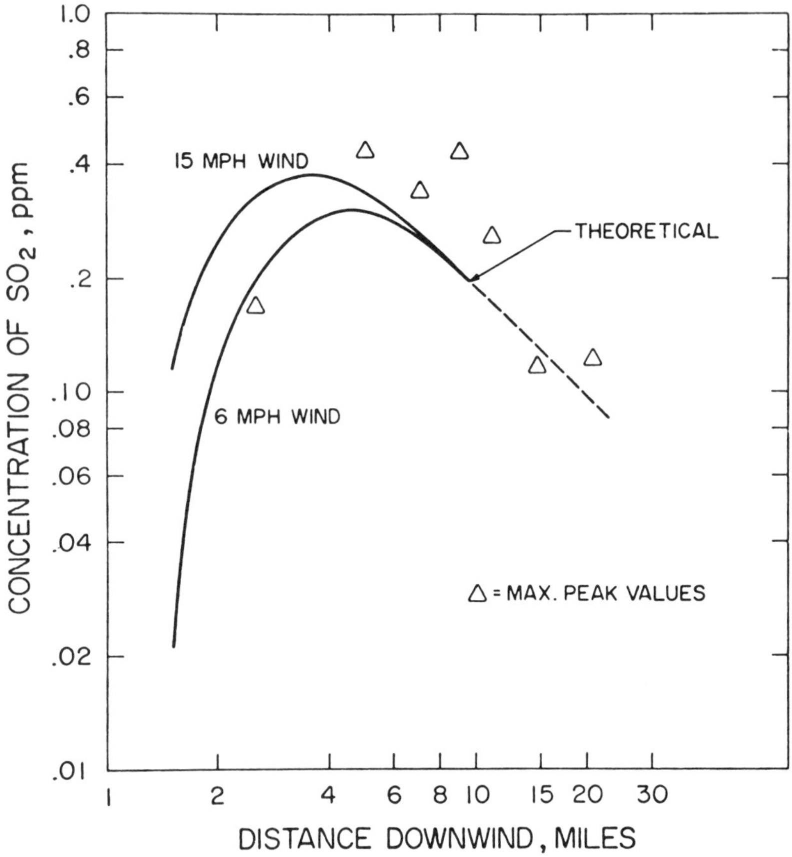

Figure 12.8. Predicted and measured.

distribution. With this assumption, values for σ_y and σ_z, respectively, the horizontal and vertical dispersion coefficients for the sampling period illustrated in Figures 12.1 through 12.3 have been calculated. Again one must remember that these are "instantaneous" values of plume spread. These experimentally determined coefficients are compared with values representative of a 10-minute sampling time in Figures 12.9 and 12.10. With these dispersion coefficients, one may predict concentration values for comparable averaging intervals by methods described by Turner.[7] It should be noted in Figure 12.10 that the vertical parameter is lower than would be expected from the horizontal parameter because of the restricted movement in the inversion layer.

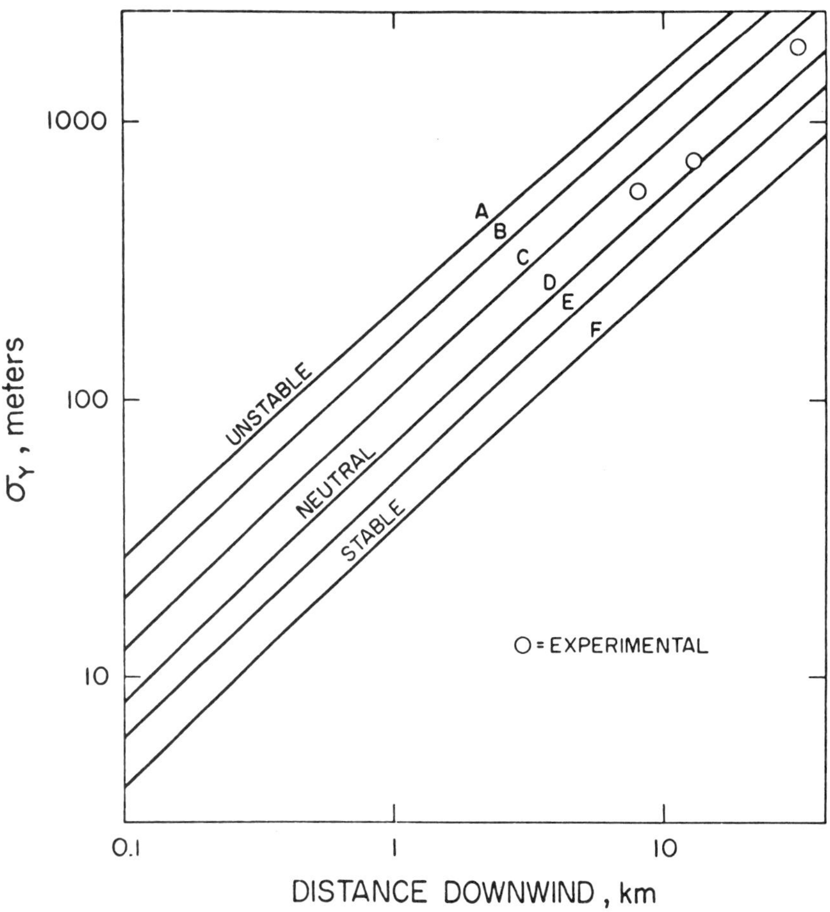

Figure 12.9. Horizontal dispersion coefficient as a function of downwind distance.[4]

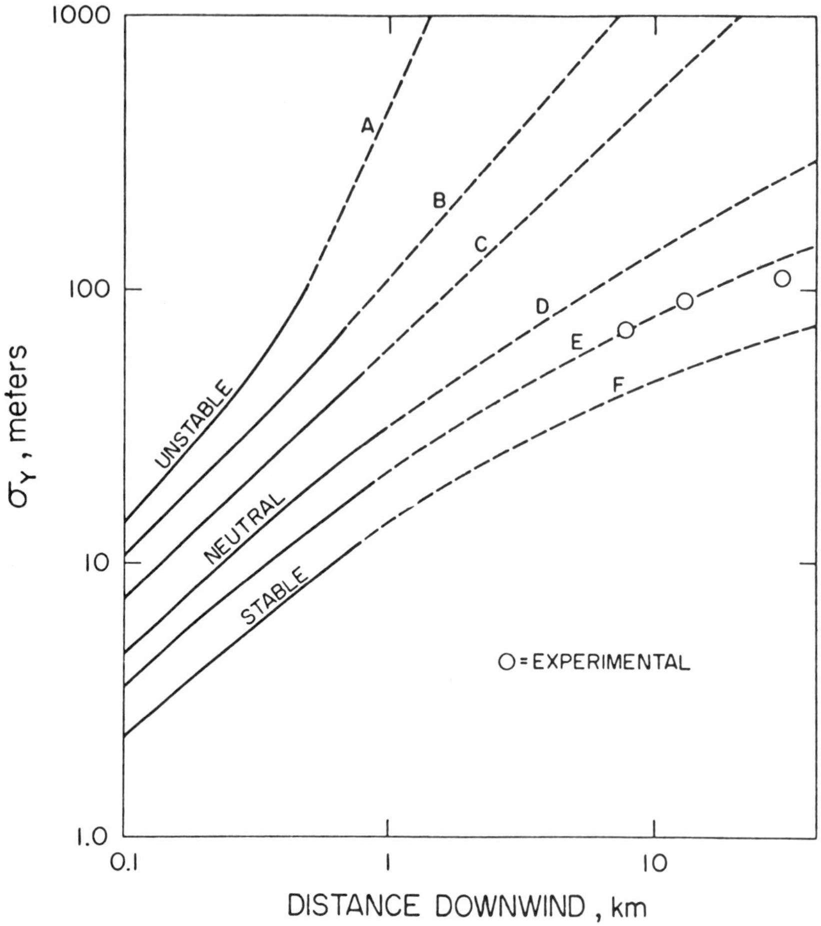

Figure 12.10. Vertical dispersion coefficient as a function of downwind distance.[4]

SUMMARY

The results of these tests demonstrated the usefulness and limitations of both airborne and ground sampling techniques. The aircraft affords rapid access to areas where ground vehicles cannot be used and provides certain data for use in validating plume dispersion models that are not available from the ground. However, bad weather can severely curtail a sampling program based exclusively on aircraft sampling.

The relatively short period of time that the aircraft spends in the plume during any one crosswind pass makes it necessary to use instrumentation with a fast response. The SO_2 analyzer used in the aircraft had a noise level that was considerably higher than that of the ground-based instruments. The respective figures are about 0.3 mv for the airborne instruments and 0.02 mv for the ground level instruments. This results in a noise level equivalent to about 0.03 of SO_2 for the aircraft data and about 0.001 ppm for the ground level data. This higher noise level may be inherent in obtaining the necessary rapid response, but it increases the difficulty in interpreting the aircraft data. For example, it was difficult to locate the plume edges accurately from the aircraft data when the measured concentration of SO_2 was low.

Several further problems were encountered in interpreting the aircraft data. After a pass through the plume, the output of the analyzer slowly returns to background level. This effect made it difficult to determine both the location of the plume edge and the correct background concentration level. A second problem resulted from an apparent shift in the instrument response to background concentrations with changes in altitude. This effect might be due to real changes in SO_2 background, a change in the background concentration of CO_2, or sensitivity of the instrument to changes in pressure or temperature.

Some of the limitations of the ground vehicle sampler are apparent; however, it was possible to locate the plume, measure its concentration and dimensions, and track its motion. For some tests, it would seem that one or more ground vehicles could provide useful data without aid of the aircraft. Better ground level meteorological equipment might be required for this work.

ACKNOWLEDGMENTS

The authors would like to thank Southern Services, Inc. of Birmingham, Alabama for support of this project as a part of their continuing environmental evaluation and management program.

The airborne measurements were provided under contract by Dr. R. S. Sholtes, who at the time of these tests was employed by the University of Florida.

REFERENCES

1. Bird, Alvin N., Jr. and Joseph D. McCain. "Airborne and Ground-Level Measurements of SO_2 and Particulate Emission from a Coal-Fired Electric Generating Station," Final Report to Southern Services, Inc., Birmingham, Alabama (March, 1971).
2. Hochheiser, Seymour. "Methods of Measuring and Monitoring Atmospheric Sulfur Dioxide," PHS Publ. No. 999-AP-6 (1964).
3. Analytical Instrument Development, Inc., 250 So. Franklin Street, West Chester, Pennsylvania.

4. Barium perchlorate titration of bubbler sample.
5. Stephens, N. Thomas and Roy O. McCaldin. "Attenuation of Power Station Plumes as Determined by Instrumented Aircraft," *Environ. Sci. Technol.* 5, No. 7 (July 1971).
6. McNair, Larry. Southern Services, Inc., Birmingham, Alabama, private communication.
7. Turner, D. Bruce. "Workbook of Atmospheric Dispersion Estimates," PHS Publ. No. 999-AP-26 (1969).

13

A Field Evaluation of Ambient Sulfur Dioxide Monitors

R. W. Garber, J. W. Hamby, G. E. Titcomb and **T. L. Montgomery**
Tennessee Valley Authority, Muscle Shoals, Alabama

INTRODUCTION

The Clean Air Act Amendments of 1970 and resultant state regulations have caused increased interest in ambient air monitoring and monitoring instruments. Instrument manufacturers have responded rapidly to this interest and provided a multitude of instruments to meet many air monitoring needs. The EPA[1] has published specifications for ambient air monitoring instruments to assure uniformity of different groups that conduct atmospheric monitoring. Although these specifications indicate certain minimum requirements and suggest certain methodologies, EPA has not formally approved any specific instrument.
the many available instruments to use.

TVA has monitored the ambient air around its coal-fired power plants for many years,[2-6] and recently designed new monitoring networks. The primary requirements of any monitor to be used in this new monitoring program system are given in Table 13.1. These specifications are very similar to those of the EPA.[1] Table 13.2 shows the criteria, in order of importance, that were used in this instrument evaluation. To determine which of the available ambient sulfur dioxide monitors to use in its improved ambient air monitoring program, TVA field tested some selected instruments.

DESIGN OF THE TEST

Instrumentation

The instruments evaluated are listed in Table 13.3. All instruments except those that operate on the principle of conductivity utilize detection methods approved by the Environmental Protection Agency

Table 13.1. Requirements for Ambient Sulfur Dioxide Monitors.

Parameter	*Requirement*
Ranges	0–1 ppm, 0–10 ppm
Minimum detectable sensitivity (0–1 ppm range)	0.01 ppm
Rise time, 90%	5 minutes
Fall time, 90%	5 minutes
Calibration drift	± 1% per day
Zero drift	± 1% per day
Precision	± 2% per day
Unattended operation period	7 days
Noise	± 0.5% of full scale
Interference equivalent[a]	0.01 ppm
Operating temperature range	60–80° F
Linearity	2%

[a]1 ppm of interfering agent has equivalent reading to 0.01 ppm sulfur dioxide.

Table 13.2. Evaluation Criteria Ambient Sulfur Dioxide Monitor Evaluation.

1. Accuracy of measurement as compared to reference method.
2. Compliance to the Ambient Air Monitoring Requirements as specified by the Environmental Protection Agency according to the *Federal Register*, **36**, No. 158, 15486 (August 14, 1971), and any subsequent EPA requirements.
3. Serviceability and maintenance (the instrument must have a maintenance-free operation period of at least 168 hours of continuous service, equivalent to 7 days, 24 hours/day), excluding time required for zero and span calibration.
4. Ease of operation and adaptability to conventional dynamic calibration methods.
5. Ease of use of instrument output with automatic data processing systems.
6. Availability of parts and manufacturers servicing representative.

for measuring sulfur dioxide. Because a large amount of data was previously collected with instruments that operate on the principle of conductivity, two conductivity type monitors were included in the test for comparison. Two Gelman Model 24008 sequential samplers were used to take one-hour samples for analysis by the Pararosaniline method as reference. The strip chart recorders used to record the output of the monitors are listed in Table 13.3.

One requirement peculiar to flame photometric sulfur analyzers such as the Bendix and Meloy Total Sulfur Analyzers is the need of gaseous hydrogen for the flame. To eliminate the troublesome and possibly hazardous task of hauling and changing large cylinders of compressed hydrogen gas, electrolytic hydrogen generating systems were used to supply the fuel for these instruments. Table 13.3 also indicates the type of hydrogen generator used with each instrument. Note that the inclusion of the hydrogen generator to the flame photometric analyzers greatly increased the complexity and the required maintenance of these systems. The hydrogen generators were recommended and proved by the manufacturer of each monitor.

Table 13.3. Sulfur Dioxide Monitors Evaluated.

Manufacturer	*Model*	*SO_2 Detection Principle*	*Recorder*	*Peripheral Equipment*
Bendix Corporation	8300	Flame photometry	Esterline-Angus Speed-Servo II	Trienco 911PHG Hydrogen Generator
Calibrated Instruments, Inc.	Wostoff V-35	Conductivity	Wostoff No. 660	
Leed & Northrup Co.	Autometer 64251-AL SO_2	Conductivity	Leed & Northrup Speedomax G Model S 60000	
Meloy Laboratories, Inc.	SA185	Flame photometry	Esterline-Angus Speed-Servo III	Toledyne Isotopes HG-500 Hydrogen Generator
Philips Electronics, Inc.	PW9700	Coulometry	Esterline-Angus Speed-Servo II	
Technicon Instruments Corp.	Air Monitor IV	Colorimetry	Technicon 69A-TC-T51-1PH760	

Test Site

The instruments were evaluated under field conditions at an ambient air monitoring site 1.3 miles south-southeast of TVA's Kingston Steam Plant near Kingston, Tennessee.

A special instrument shelter was built to provide space to position the six monitoring instruments and peripheral equipment for easy maintenance and repair. The shelter was provided with 110-V 60-hz power and with heating and air conditioning to maintain the temperature in the shelter within the comfort zone (65-80° F). The floor plan of the monitoring shelter is shown in Figure 13.1.

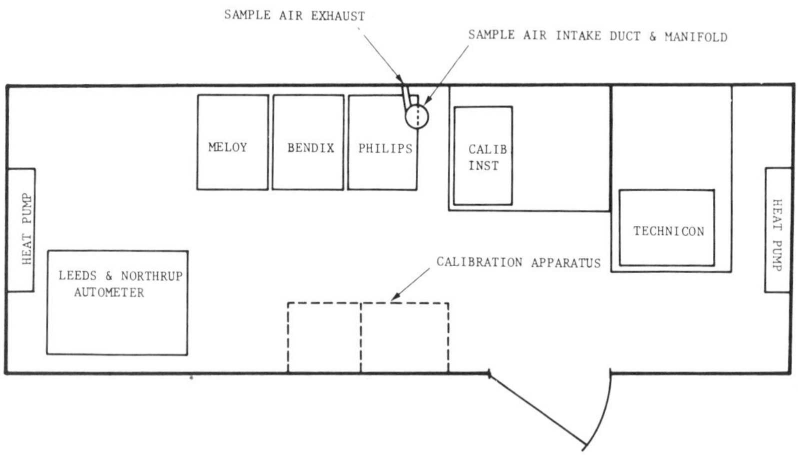

Figure 13.1. Air monitor evaluation shelter—instrument layout.

Experimental Design

To ensure that the instruments were tested properly they were installed and operated as prescribed by the manufacturers. All instruments were operated on the same concentrations range (*i.e.*, 0-1 ppm SO_2 in air).

Ambient air samples were taken above the roof of the monitoring shelter by means of the sampling duct shown in Figure 13.2. A blower at the base of the sampling duct provided a flow rate of about 400 l/min. The air monitors sampled the air flowing through the large duct by means of individual teflon sampling lines connected to the manifold on the duct. The sequential samplers were also connected into the same air sampling system with teflon lines.

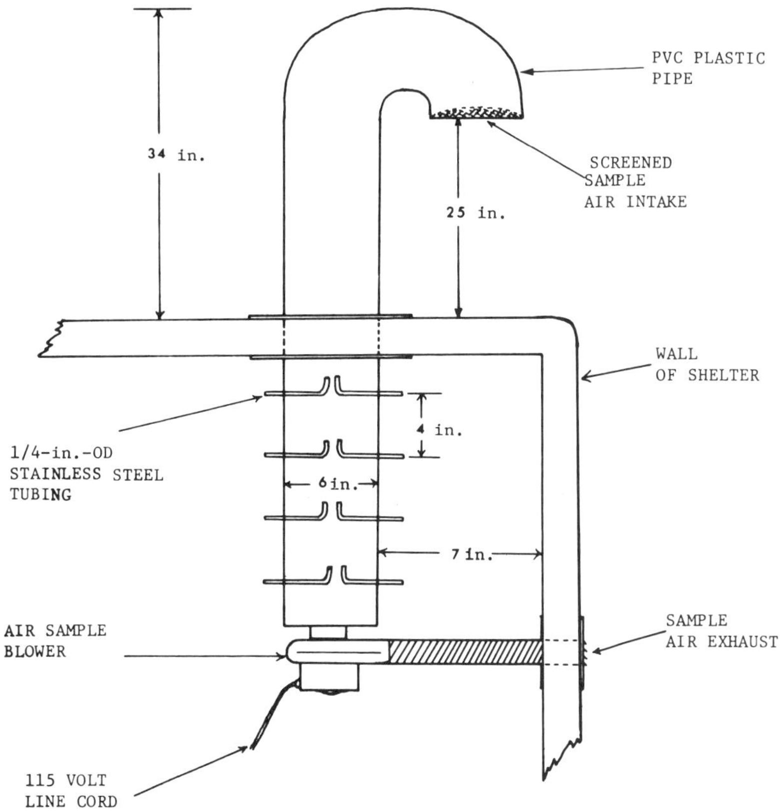

Figure 13.2. Air sampling duct and manifold.

Instrument Calibration and Zero Check

Before atmospheric sampling was started, all instruments were dynamically calibrated. Analytical Instrument Development, Inc. and Metronics Associates, Inc. sulfur dioxide permeation tubes were used as a source of known concentration of sulfur dioxide for calibrating the monitors. The calibrations apparatus (Figure 13.3) operates at ambient temperature but, because of its insulation and large mass of water, small fluctuations in ambient temperature have no effect on the permeation tubes during the 60 minutes required for calibration. The rate at which sulfur dioxide was emitted from the permeation tube was determined from emission rate versus temperature data. Such data were provided by the manufacturers and confirmed gravimetrically. The concentration of sulfur dioxide was calculated from the known emission rate and the rate at which

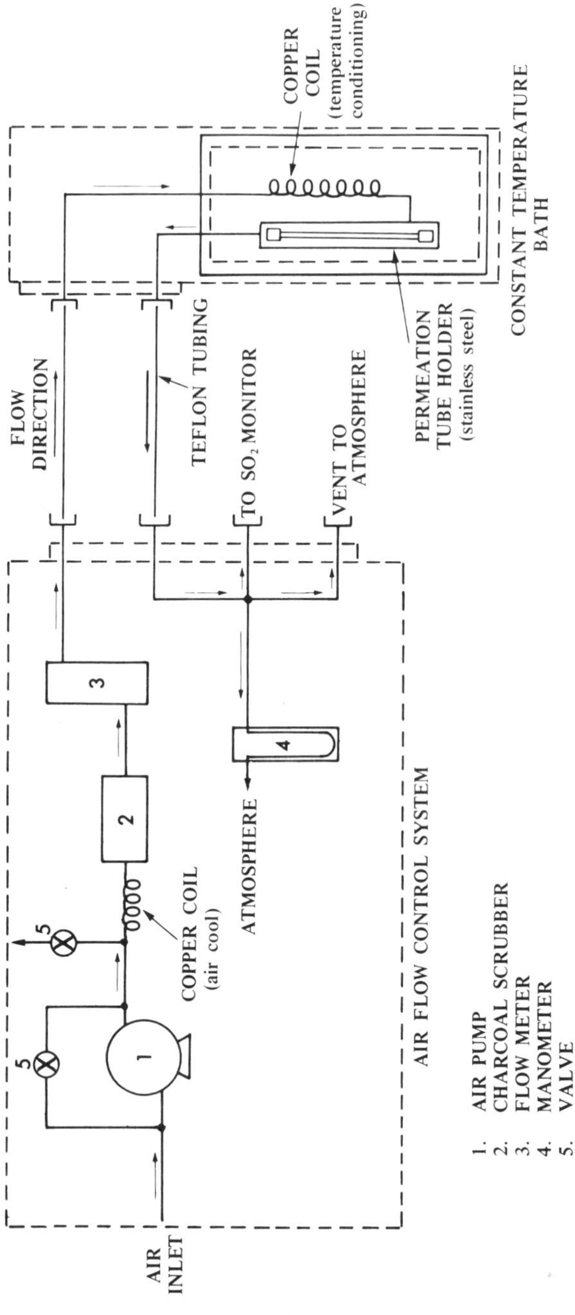

Figure 13.3. Sulfur dioxide monitor calibration apparatus.

charcoal-filtered air flows over the permeation tubes. Charcoal-filtered air was used as a dynamic check of the zero indication on each monitor. The recorders of all instruments were zeroed at + 5% of full scale so that zero fluctuations in both positive and negative directions could be observed. Because only one calibration apparatus was available and each calibration required more than one hour, only a single point calbiration and a zero check were made for each monitor.

Test Phases

The instruments were evaluated in three phases over a period of about five months. This report is principally concerned with the first two phases of the test, which lasted about four months.

Phase I—The first phase of the evaluation (30 days) was intended to determine the operational stability of the monitors, and the minimum maintenance schedule that would allow reliable measurements. During this period no maintenance or adjustments were made except those necessary to keep the instruments operational. The instruments were checked at least twice each week and their operational status was determined. No adjustments were made unless the instrument zero indication was in error by more than ± 20% as determined by dynamic tests. Air flow was corrected as required.

Hourly duplicate bubbler samples were taken from midnight Sunday to noon Friday only by means of the Gelman sequential samplers. Duplicate analyses were performed on each sample by the standard Pararosaniline analysis method. The Pararosaniline method was calibrated by use of a sodium metabisulfite calibration solution.

Phase II—The second phase of the instrument evaluation, which lasted 86 days, was designed to test the overall performance characteristics of the instruments under typical field conditions of use and maintenance. All instruments were operated continuously with normal, routine maintenance and were checked for proper operation twice weekly.

The sequential samplers were operated on demand from midnight Sunday until midnight Friday. The Calibrated Instruments monitor was used to start the sequential samplers whenever the monitor registered 0.04 ppm or higher. So that data from the sequential samplers would be one-hour average values, they were set to operate at one-hour intervals after being triggered for as long as the trigger value continued to be registered. As during Phase I, both sequential samplers operated simultaneously, and duplicate Pararosaniline analyses were made on each sample. All other instruments were operated continuously seven days a week.

Phase III—The third phase of the instrument test, which lasted 35 days, was designed to determine the effect of instrument sensitivity upon performance and maintenance and to gain more experience with these instruments particularly with diagnosing and repairing malfunctions. The instruments were operated continuously, and the range was changed from 0–1 ppm to 0–5 ppm. The sequential samplers were not used during this final phase.

RESULTS AND DISCUSSION

Calibration and Zero Drift

The calibration drift for an ambient air monitor should be less than that which would produce a standard deviation of 10% of the mean concentration.[7] The mean one-hour average concentration of ambient sulfur dioxide measured at the test site during the entire test was 0.025 ppm. Thus, a drift of less than 1% per week of the full-scale calibration will not adversely affect the precision and accuracy of the measurements.

To check the instrument zero and calibration drift, the instruments were calibrated and checked for zero indication twice a week. The monitors were calibrated dynamically by means of the sulfur dioxide permeation tube method. The calibration data appear in Table 13.4. Instrument zero was dynamically checked with activated charcoal-filtered air (zero air). The results of the zero check are shown in Table 13.5.

Table 13.5 shows the stability of the zero indication for each instrument. The zero indication for each instrument was read directly from the strip chart recorder. In order that both positive and negative zero fluctuation could be observed the instruments were set to read + 5 chart divisions of 5% of full scale with zero air. The greatest zero fluctuation was observed with the Calibrated Instruments monitor and the smallest zero fluctuation was observed with the Meloy monitor.

The results of the calibration experiments were subjected to further study by linear correlation analysis and direct comparison against input concentrations (Table 13.6). Linear correlation analysis indicated in most cases the expected one-to-one relationships between the applied concentration and the measured value is valid to a very high degree of correlation: *e.g.*, typical slope 0.97 with a correlation coefficient of 0.99. A direct comparison by taking the ratio of the measured concentration by the applied concentrations shows the expected one-to-one relationship, *e.g.*, typical mean ratio of 1.00 with a standard deviation from the mean of 0.03. In only one case is lack of linear correlation indicated, and this is attributed to the very short concentration range of available calibration measurements. It should be noted that the Bendix monitor was in operation for only 7 of the possible 27 attempted calibration experiments. The analysis of the

Table 13.4. Instrument Calibration Data. Ambient SO_2 Monitor Evaluation. Concentration of Sulfur Dioxide (ppm)

Calibration Test No.	*Calibration Range*	*Conc. Applied*	*Values Measured by Instruments*[a]						*Mean Reading*	*Difference (Mean Reading and Applied Conc.)*	*Standard Deviation*
			Bendix	*Calibrated Instruments*	*Leeds & Northrup*	*Meloy*	*Philips*	*Technicon*			
1	0.40-0.54	0.40		0.49	0.44	0.37	0.40	0.42	0.42	0.02	0.05
2	,,	0.42		0.43	0.43	0.39	0.42	0.43	0.42	0.00	0.02
3	,,	0.42		0.44		0.43	0.43	0.44	0.43	0.01	0.01
4	,,	0.43		0.46	0.44	0.40	0.35		0.41	0.02	0.05
5	,,	0.44		0.48	0.43		0.46		0.46	0.01	0.03
6	,,	0.46		0.47	0.39		0.45		0.44	0.03	0.04
7	,,	0.51		0.54	0.53	0.47	0.52		0.52	0.00	0.03
8	0.55-0.69	0.55		0.55		0.53	0.50		0.53	0.02	0.03
9	,,	0.55		0.55	0.24		0.45		0.50	0.14	0.07
10	,,	0.55		0.58	0.55		0.57		0.57	0.02	0.02
11	,,	0.60		0.64	0.56		0.60	0.63	0.61	0.01	0.04
12	,,	0.63		0.64	0.60		0.63		0.62	0.00	0.02
13	,,	0.64		0.66	0.60		0.64	0.68	0.65	0.00	0.03
14	,,	0.66		0.32	0.70		0.66	0.69	0.68	0.03	0.02
15	,,	0.67		0.67	0.65	0.68	0.68	0.66	0.67	0.00	0.01
16	,,	0.67	0.62	0.66	0.60		0.65	0.66	0.64	0.03	0.03
17	,,	0.69	0.54		0.64		0.67	0.72	0.64	0.04	0.08
18	0.70-0.85	0.70		0.71		0.70	0.70	0.63	0.69	0.01	0.04
19	,,	0.70		0.67			0.68	0.71	0.69	0.01	0.02
20	,,	0.72		0.16		0.70	0.74	0.67	0.70	0.02	0.04
21	,,	0.72	0.73		0.71		0.73		0.72	0.00	0.01
22	,,	0.73	0.64	0.73			0.73	0.76	0.72	0.01	0.05
23	,,	0.73	0.64	0.74	0.73		0.74	0.80	0.73	0.02	0.06
24	,,	0.73	0.69	0.72	0.72		0.75	0.73	0.72	0.01	0.02
25	,,	0.74	0.69	0.74	0.68		0.73	0.74	0.72	0.02	0.03
26	,,	0.75		0.76	0.73		0.76	0.77	0.76	0.00	0.02
27	,,	0.84		0.83	0.79		0.82	0.80	0.81	0.02	0.02

[a]Exceptional values greater than 3 σ from mean reading are excluded from calculations.

Table 13.5. Instrument Zero Data. Ambient SO_2 Monitor Evaluation.

	Readings on Zero Gas (ppm)[a]					
Calibration Test No.	*Bendix Sulfur Analyzer*	*Calibrated Instruments SO_2 Analyzer*	*Leeds & Northrup Autometer*	*Meloy Sulfur Analyzer*	*Philips SO_2 Analyzer*	*Technicon Air Monitor IV*
1		0.06	0.03	0.05	0.05	0.03
2		0.05	0.03	0.05	0.05	0.05
3		0.05		0.05	0.07	0.02
4		0.05	0.03	0.05	0.04	
5		0.02	0.02		0.05	
6		0.06			0.06	
7		0.05	0.05	0.05	0.05	
8		0.05		0.05	0.05	
9		0.05	0.02	0.05	0.05	
10		0.06	0.01		0.04	
11		0.01	0.01		0.06	0.01
12		0.07	0.01		0.06	
13		0.03	0.01		0.06	0.04
14		0.02	0.02		0.03	0.04
15		0.06	0.02	0.04	0.06	0.00
16	0.06	0.05	0.03		0.04	0.02
17	0.07		0.04		0.05	0.04
18		0.05		0.05	0.06	0.05
19			0.05		0.06	0.03
20		0.10		0.05	0.06	0.03
21	0.05		0.05		0.05	
22	0.07	0.04			0.05	0.04
23	0.06	0.06	0.04		0.04	0.03
24	0.07	0.06	0.04		0.05	0.05
25	0.06	0.05	0.04		0.05	0.05
26		0.06	0.04		0.05	0.05
27		0.03	0.01		0.05	0.02
Maximum	0.07	0.10	0.05	0.05	0.07	0.05
Minimum	0.05	0.01	0.01	0.04	0.03	0.00
Mean	0.06	0.05	0.03	0.05	0.05	0.03

[a]100 units equal to full scale on recorder — all recorders set on 0.05 unit for zero gas initially.

Table 13.6. Analysis of Calibration Data.[a]

	Linear Correlation		*Comparison-Measured Value to Applied Concentration*	
Instrument	*Slope Dimensionless*	*Correlation Coefficient Dimensionless*	*Mean Ratio Dimensionless*	*Standard Deviation Dimensionless*
Bendix	0.92	0.47[b]	0.93	0.05
Calibrated Instruments	1.02	0.99	1.02	0.03
Leeds & Northrup	0.96	0.97	0.98	0.06
Meloy	0.97	0.99	0.96	0.04
Philips	1.00	0.99	1.00	0.03
Technicon	1.01	0.96	1.01	0.05

[a]Exceptional values greater than 3σ from mean ratio are excluded from the calculations.
[b]Data available only over very limited range of concentrations.

calibration data indicates that most of the instruments tested performed fairly well with respect to calibration stability.

The calibration data scatter and zero fluctuation observed with some of the monitors is believed to be due to impurities in the reagent and/or improperly mixed reagent. These instruments are extremely sensitive to some of the impurities commonly present in distilled water used for the reagents as well as slight inaccuracies in the formulation of the reagent mixture. Therefore, extreme care must be used in preparation and handling of these reagents if data scatter and zero fluctuation is to be avoided.

Ambient Sulfur Dioxide Data

The frequency of occurrence of measured concentrations is shown in Table 13.7. Table 13.8 is a similar frequency table for the sequential sampler measurements.

Table 13.7. Frequency of Occurrence of Sulfur Dioxide Concentrations. Ambient SO_2 Monitor Evaluation.

Range One-hour Conc. (ppm)	*Number of Occurrences*					
	Bendix	*Calibrated Instruments*	*Leeds & Northrup*	*Meloy*	*Philips*	*Technicon*
0.00	603	1781	2074	1707	2094	1720
0.01–0.05	139	489	396	233	468	348
0.06–0.10	51	90	133	74	139	98
0.11–0.15	18	51	39	27	42	38
0.16–0.20	16	32	48	27	39	39
0.21–0.25	10	19	19	8	21	14
0.26–0.30	14	20	20	15	23	16
0.31–0.35	8	10	12	7	11	7
0.36–0.40	9	7	10	9	19	11
0.41–0.45	4	11	10	2	2	7
0.46–0.50	5	7	9	8	6	8
0.51–0.55	1	2	5	4	0	5
0.56–0.60	1	4	6	2	4	3
0.61–0.65	1	3	1	1	2	2
0.66–0.70	2	3	5	5	1	5
0.71–0.75	3	0	1	0	0	1
0.76–0.80	0	2	0	2	1	0
0.81–0.85	0	0	0	1	0	1
0.86–0.90	2	1	1	0	0	0
0.91–0.95	0	0	0	0	0	0
0.96–1.00	0	0	0	0	0	0
Lost Record[a]	2089	444	186	844	104	653
Total	2976	2976	2976	2976	2976	2976

[a]Lost Record—Number of hours SO_2 not measured due to instrument malfunction, instrument service, calibration or zero check, etc.

Table 13.8. Frequency of Occurrence of SO_2 Concentrations. Gelman Sequential Sampler Data. Ambient SO_2 Monitor Evaluation.

Range One-hour Conc. (ppm)	*Number of Occurrences*		
	Sequential Samplers		
	B	*A*	*A B Average*
0.00	254	311	246
0.01–0.05	328	263	251
0.06–0.10	45	42	39
0.11–0.15	23	21	20
0.16–0.20	13	13	12
0.21–0.25	10	11	8
0.26–0.30	9	8	8
0.31–0.35	5	9	4
0.36–0.40	6	8	5
0.41–0.45	5	3	2
0.46–0.50	5	8	5
0.51–0.55	4	5	4
0.56–0.60	2	4	2
0.61–0.65	1		
0.66–0.70	3	3	1
0.71–0.75	4	1	
0.76–0.80			
0.81–0.85	3	1	1
0.86–0.90		2	
0.91–0.95	1	1	1
0.96–1.00			
>1.00		5	
Lost Record[a]	106	98	

[a]Lost Record—All A and B values are averages of two measuremenets per sample. If values differed by a factor of two or more of each other, no A and/or B value was calculated yielding lost record. If A and B differed by a factor of two or more for a given sampling period a lost record resulted.

The distribution of the sulfur dioxide concentrations recorded by the continuous instruments and the sequential samplers cannot be compared quantitatively because of the differing operating schedules of the two systems. The continuous monitoring instruments were operated as continuously as was possible. The sequential samplers were operated continuously from midnight Sunday to noon Friday during the first 30 days of the test, then as demanded from midnight Sunday to midnight Friday during the second phase of the test.

Trends in the two sets of data were compared. Data from all continuous monitors were similar: about 75% of the values are zero, and about 18% of the values fall between 0 and 0.05 ppm. The highest one-hour average concentration recorded was about 0.9 ppm as measured by the Bendix, Calibrated Instruments, and Leeds and Northrup instruments, 0.85 ppm as measured by the Philips instrument. The instruments performed similarly except for duration of operability.

Except for the zero distribution, data from the sequential samplers follow a trend similar to those from the continuous instruments, although the data from the sequential samplers appear to be slightly skewed toward the higher values. According to recent work by Blacker *et al.*,[8] this may be an effect of the sodium metabisulfite solution calibration method used for Pararosaniline analysis.

TVA defines *lost record* (LR) as time when SO_2 is not measured due to such things as instrument malfunction, calibration, and zero check in reported data. The total lost record data of the continuous instruments is shown in Table 13.7. After adjustment to account for the time for each calibration and zero check, the minimum number of hours that each instrument was inoperable is shown in Figure 13.4. Examination of the causes of lost record is very informative and is possibly indicative of problems that may be expected.

The Bendix instrument was inoperable when it was received because of severe instability; the Trienco hydrogen generator supplied by Bendix was also inoperable because of a fouled palladium hydrogen purification filter. The Bendix instrument was returned to the factory for repair. The factory representative visited and repaired the Trienco hydrogen generator. The Bendix system started operation after 76 days of lost record. Soon after startup, the hydro-

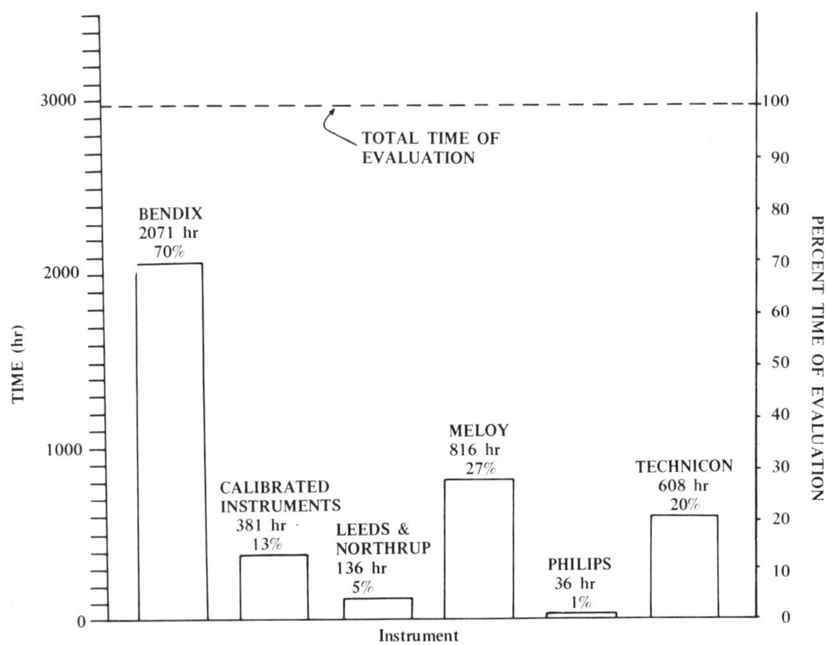

Figure 13.4. Lost record due to instrument malfunction.

gen purification system of the Trienco hydrogen generator fouled again and caused a hydrogen overpressure cutoff. The system went back into service after another six days of lost record.

The Calibrated Instruments sulfur dioxide analyzer lost about 400 hours of record due to problems of reagent fouling, a pump breakdown, and other minor malfunctions.

Although the Leeds and Northrup Autometer used was nearly 20 years old, all of its lost record time was accounted for by changing the time reagent and other minor maintenance. These instruments have undergone multiple modifications by TVA over the nearly 20-year service time which have improved their reliability.

Part of the lost record time for the Meloy instrument was caused by a leak in a valve in the air line. The instrument was exchanged for another of the same model by the manufacturer. The new instrument developed some instability that is believed to be attributable to hydrogen pressure changes caused by the hydrogen generator. The Teledyne Isotopes hydrogen generator later became inoperable because of icing of the oxygen vent. The oxygen backpressure forced the potassium hydroxide electrolysis reagent through the hydrogen line into the Meloy instrument, where it severely damaged the sulfur analyzer. The lost record amounted to about 700 hours.

A break in the vacuum seal and a rupture of the vacuum-pump diaphragm accounted for most of the lost record for the Philips analyzer. Normal maintenance accounted for the remainder.

Most of the 400 hours of lost record for the Technicon analyzer was caused by difficulties encountered during startup. A broken proportioning pump accounted for another approximately 170 hours of lost record. Normal maintenance and minor problems accounted for the remainder.

All of the instruments evaluated were quite easily calibrated with conventional dynamic calibration methods. Therefore, no distinction could be made between the monitors evaluated on the basis of items 4, 5, and 6 in the evaluation criteria, Table 13.2.

CONCLUSIONS

The main purpose of this study was to evaluate several ambient sulfur dioxide monitors to determine which would prove most accurate and reliable in unattended field operation. In terms of accuracy, the instruments were comparable, and no clear separation was possible on this basis. The lost record associated with instrument malfunctions shows some measure of the reliability of each complete system.

Another reason for such an instrument test was to discover problems associated with particular instruments in field use and to design methods of coping with these problems. The regular, twice weekly checkups of the monitors revealed many such problems and thus provided valuable experience for foreseeing and eliminating many instrument breakdowns.

This instrument evaluation also provided comparative data for the newer agent-specific pollutant monitors and the older conductivity analyzers. These data will prove useful in correlating measurements made with the newer instruments with those of the past. More comparative data will have to be collected at various monitoring sites to ensure the validity of the correlation and to determine any regional dependence for such a correlation.

This evaluation was made under typical field conditions. The mean one-hour average sulfur dioxide concentration measured during the test was 0.025 ppm. The highest one-hour average sulfur dioxide concentration measured by a continuous monitor was 0.90 ppm. During more than 75% of the test period, no detectable sulfur dioxide was present in the atmosphere at the test site.

The following conclusions are drawn from the results of this evaluation:

1. The Leeds and Northrup (conductometric) and the Philips (coulometric) analyzers had less than 5% lost record and required the least maintenance and operated most reliably in the test.

2. The Meloy flame photometric analyzer had relatively lower measurements than the other instruments and during its operation it had the most stable zero.

3. The greatest zero fluctuation was observed in the Calibrated Instruments SO_2 analyzer. As discussed earlier this is probably caused by impure or improperly mixed reagent.

4. The measurements of all of the instruments were very similar and comparable except for slightly lower measurements obtained from the Meloy analyzer.

REFERENCES

1. *Air Quality Surveillance* **40** C.F.R., Part 51.17 (1973).
2. Montgomery, T. L., *et al.* "A Simplified Technique Used to Evaluate Atmospheric Dispersion of Emissions from Large Power Plants," *J. Air Poll. Cont. Assoc.* **23**, 388-394 1973).
3. Tennessee Valley Authority. "Full-Scale Study of Dispersion of Stack Gases—A Summary Report," Chattanooga, Tennessee (1964).
4. Tennessee Valley Authority. "Full-Scale Study of Plume Rise at Large Electric Generating Stations," Muscle Shoals, Alabama (1968).
5. Carpenter, S. B., *et al.* "Principal Plume Dispersion Models: TVA Power Plants," *J. Air Poll. Cont. Assoc.* **21**, 491-495 (1971).

6. Huang, Chao-Ming, *et al.* "Atmospheric Interactions in Coal-Fired Power Plant Plumes," Paper No. 72-70, 65th Annual Meeting of the Air Pollution Control Association, Miami Beach, Florida (1972).
7. Stevens, Robert K., *et al.* "Field Evaluation of Sulfur Dioxide Monitoring Instruments," U.S. Environmental Protection Agency, Raleigh, North Carolina (1972).
8. Blacker, J. H., *et al.* "Note on an Evaluation of the Reference Method for Determination of Sulfur Dioxide in the Atmosphere (Pararosaniline Method), **23**, 525-527 (1973).

14

Calibration Procedure for Continuous Air Monitoring Instruments

Kenneth E. Noll and **John Walling**
Environmental Engineering Group, Department of Civil Engineering,
The University of Tennessee, Knoxville

INTRODUCTION

The monitoring of atmospheric levels of air pollutants is an integral part of the air pollution control program for any air pollution control agency and for many industries. This is because air monitoring is usually the most direct method of determining the magnitude of an air pollution problem. The monitoring data can be used to (a) compare with air quality standards, (b) determine the extent of control activities required, and (c) evaluate the effectiveness of the control effort.

A key part of an air monitoring program is instrument calibration. Calibration determines the relationship between the instrument measurement and the true value of a specific pollutant. Present-day air monitoring instrumentation is complex and subject to electronic drift, as well as time variation in such things as sensor and reagent response. Therefore, not only is an initial calibration of new instruments required, but periodic recalibration is necessary to assure the accuracy and validity of the monitoring data obtained over a period of time.

The purpose of this paper is to explain basic air monitoring instrumentation calibration procedures for continuous gas analyzers. This includes the identification of static and dynamic calibration of the instrument, the use of a dilution panel, the use of the reference method, and the generation of known gas concentrations. A specific example is provided to demonstrate the typical procedures and calculations required to calibrate an air monitoring instrument.

GENERAL CALIBRATION PROCEDURE

The basic sequence for calibration of continuous air monitoring instruments is as follows:

1. Precalibration instrument check; replacement of old reagent, calibration of the rotameters, and performance of a static electronic calibration.
2. Preparation of a standard absorbing reagent and working standard solution.
3. Preparation of a static calibration curve of absorbance versus the concentration of the specific pollutant of interest in micrograms or microliters.
4. Generation of a known concentration of gas for delivery to dilution panel from either a permeation tube, ozone generator, or certified bottle gas.
5. Provision for a steady gas flow to the instrument and the reference colorimetric method (dynamic calibration).
6. Preparation of a dynamic calibration curve and adjustment of the instrument for a direct ppm response.

The general procedure for the calibration of air monitoring instruments provides both a static and dynamic instrument response evaluation. The static electronic calibration, listed under the precalibration check, is used to check the electronic response of the instrument and is usually accomplished by providing a known signal or using a material having the same effect on the sensor as the pollutant of interest. The dynamic calibration uses a known gas concentration to establish the instrument response and provides information the response of all components of the instrument. Dynamic calibration allows the instrument to be adjusted to provide any relationship between the observed and the measured concentrations desired (usually a direct readout in ppm).

DYNAMIC CALIBRATION PROCEDURE

Figure 14.1 shows a flow diagram for the dynamic calibration procedure. Dynamic calibration is performed with an artificially generated pollutant concentration. A major component of this dynamic calibration of an instrument is the simultaneous measurement of the gas stream supplied to the instrument and a reference method. The part of the dynamic calibration procedure that allows this simultaneous measurement is the dilution panel.

Dilution Panel

Special dilution panels are usually used to provide accurate concentrations of gas pollutants as artificial atmosphere for calibration. The dilution panel allows a concentrated stream of a specific pollu-

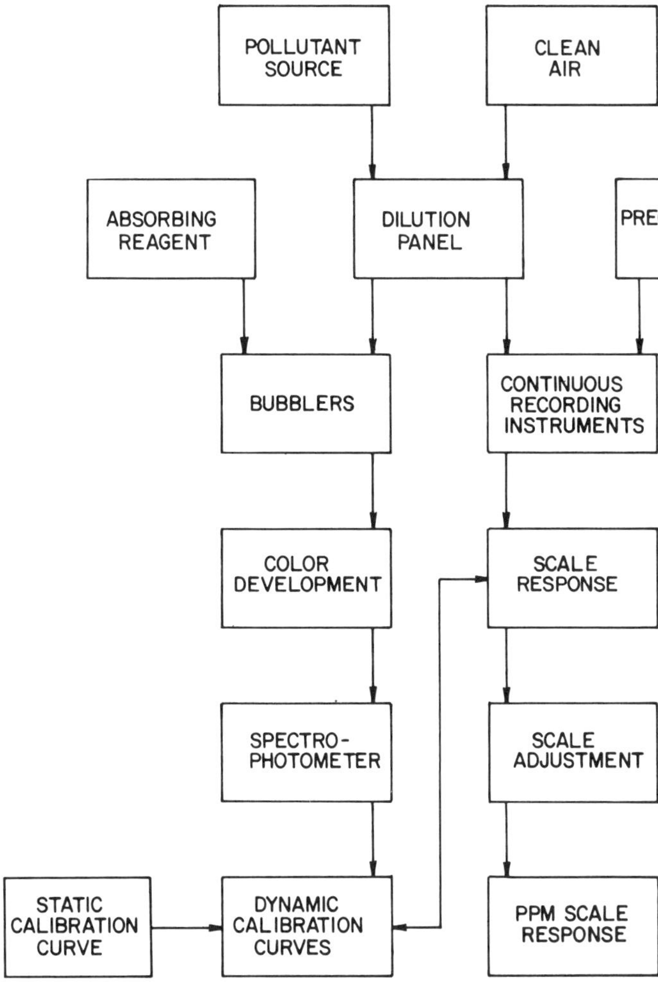

Figure 14.1. Dynamic instrument calibration.

tant gas to be diluted with clean air in varying amounts. Figure 14.2 shows a schematic diagram of a typical dilution panel. There are two inlet connections, one for a concentrated source of a desired pollutant gas and the other for diluent air. The diluent air is cleaned by columns containing activated carbon and drierite. Each air flow is accurately measured by rotometers with micrometer adjustments before being combined in a mixing chamber. The diluted mixture then passes simultaneously to the reference method and to the analyzer, the equivalent method. A wide range of concentrations of the calibration gas can be produced by varying the flow ratios into the mixing chambers. What is equally important is that a uniform concentration can be produced if the flow rates are kept constant. A micrometer adjustment is also used for the rotometer measuring the flow to the reference method.

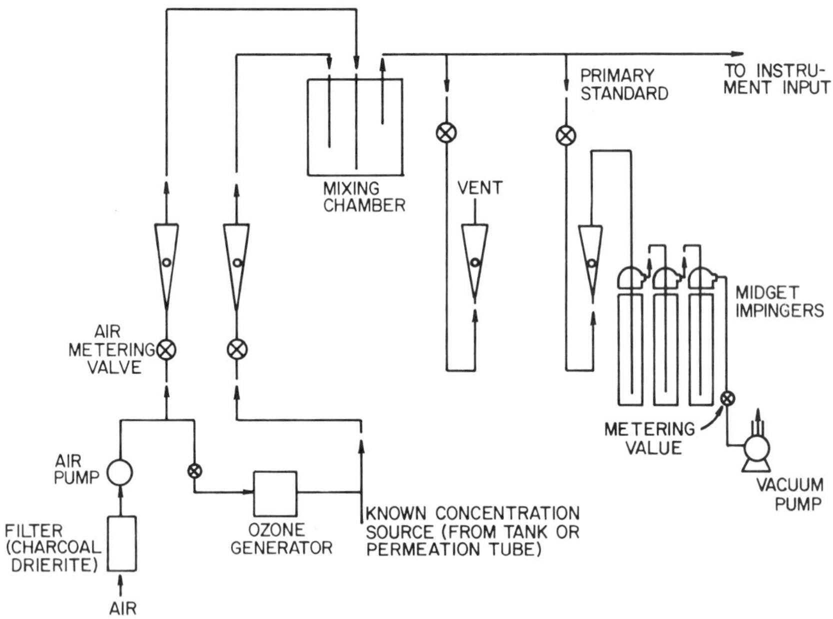

Figure 14.2. Dilution panel.

With the dilution panel it is possible to vary the sample concentration, the sampling time, or the flow rates to the instrument and the reference method.

Reference Method

The reference method can be a specific type of instrument or a wet chemical colorimetric method. For many of the major gaseous air pollutants (NO, NO_2, O_3, SO_2), it is the wet chemical colorimetric method.[1] In the colorimetric method, the gas stream is bubbled through an absorbing reagent. The color developed in the reagent follows Beer's law, and a spectrophotometer is used to measure the degree of absorption due to the color development. Since the color development in the bubbler is a function of the amount of specific pollutant, one must find a way to relate the two. This is accomplished by the production of a static calibration curve.

Figure 14.3 shows a diagram of the static calibration procedure. A working standard solution, an absorbing reagent, and sometimes an indicator are combined to produce the static curve (Figure 14.4). A working standard solution is a substance with a precisely defined or calculated response when added to the absorbing reagent. The nature of the working standard solutions used in dynamic calibration, as well as the absorbing reagent for NO, NO_2, O_3, and SO_2, is shown in Table 14.1.[2] Since the volume of the working standard solution can be used to calculate the concentration of the specific pollutant (in μg or μl), the static calibration curve is obtained by varying the amounts of the working standard solution added to the absorbing reagent; and determining the spectrophotometer response. In Figure 14.4 the absorbing reagent was sodium tetrachloromercurate which was used in the pararosaniline method for SO_2 determination. Sodium metasulfite in water is the standard sulfite solution. When standardized to 0.0123 normality using 0.01 normal iodine solution, then 1 ml is equivalent to 150 μl of SO_2 at STP.[3] To construct the curve shown in Figure 14.4 (left scale), varying amounts of the standard solution were added to the absorbing reagent, and the absorbance was measured. This curve provides a relationship between SO_2 collected in the absorbing reagent and absorbance units.

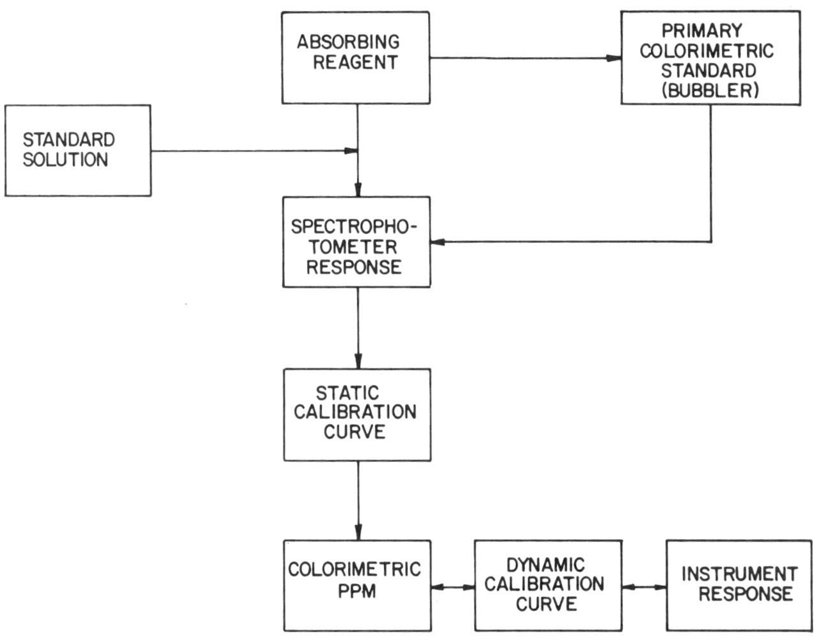

Figure 14.3. Colorimetric primary calibration standard.

Table 14.1. Calibration Standards for Analyzer and Reference Methods.

	Analyzer Standards			*Reference Method*		
Pollutant	*Calibrating Solution (Static)*	*Primary Standard*	*Calibrating Gas (Dynamic)*	*Detection Principle*	*Working Standard Solution*	*Absorbing Reagent*
NO	$NaNO_2$	$NaNO_2$	NO	Colorimetric (azo-dye)	$NaNO_2$	NaOH
NO_2	$NaNO_2$	$NaNO_2$	NO_2	Colorimetric (azo-dye)	$NaNO_2$	NaOH
O_3,O_x	$I_2(I_3^-)$	$KH(IO_3)_2$	O_3	Colorimetric (KI)	I_2	KH_2PO_4 KI
SO_2	H_2SO_4	$KHC_8H_4O_4$	SO_2	Colorimetric (pararosaniline)	$Na_2S_2O_5$ or Na_2SO_3	Potassium tetrachloro- mercurate (TCM)
	Na_2SO_3 $Na_2S_2O_5$	$KH(IO_3)_2$ or KIO_3	SO_2	Colorimetric (pararosaniline)	$Na_2S_2O_5$ or Na_2SO_3	

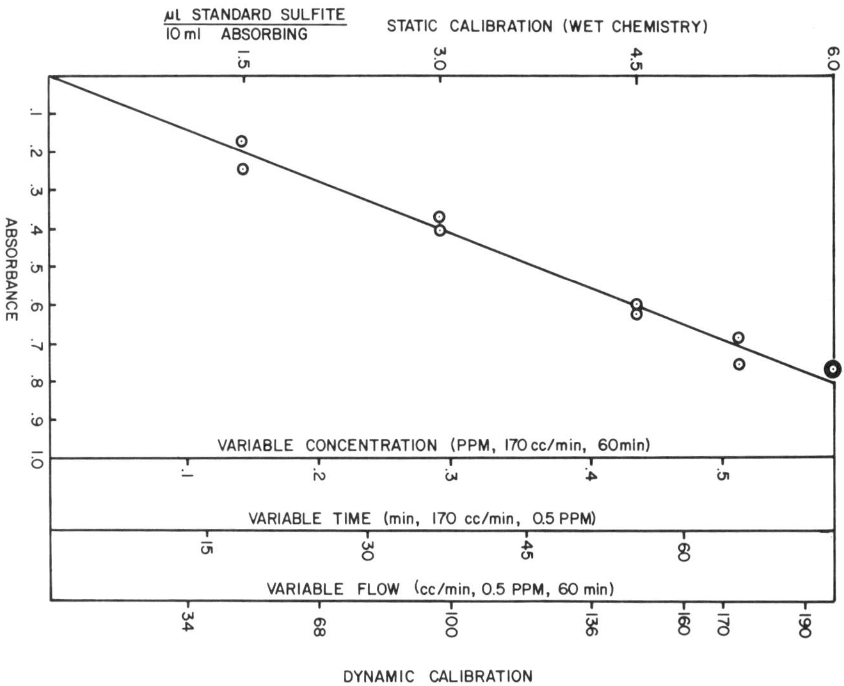

Figure 14.4. Sulfur dioxide calibration curve.

Dynamic Calibration Curve

To obtain the concentration of a gas bubbled through the absorbing reagent, a dynamic calibration curve must be produced as shown in Figure 14.4 (right scale). The absorbance of the scrubber reagent can be converted to a concentration of pollutant in the gas stream introduced to the bubbler by use of the static calibration curve. The calculation of the actual concentration in ppm is a function of the concentration input from the dilution panel, the flow rate, and the sampling time. A dynamic calibration curve can be constructed by varying any of these three variables as shown on the right scale in Figure 14.4. For example, a dynamic calibration curve can be obtained by maintaining a constant flow rate and sampling time, and varying only the input concentration. The equivalent concentration due to the color development in the bubbler is determined by measuring the absorbance of the reagent and using the static calibration curve, the flow rate, and the sampling time.

As a number of different gas mixtures are allowed to flow through the dilution panel the response of the continuous monitoring instrument is compared to the reference method concentration as demonstrated in Figure 14.5. An important feature of the instrument response curve is that it has a linear relationship to the reference method. The scale response of field monitoring instruments may be adjusted to provide the readout of concentration directly in ppm. Depending on the type of instrument, the instrument parameters which may be adjusted to change the scale response of the instrument included air flow, liquid flow, electric gain, and sensor response. Figure 14.5 shows how the equivalent method's concentration is affected by an adjustment. When the instrument is adjusted to give a direct readout in ppm that agrees with the reference method for different concentrations, the instrument is calibrated.

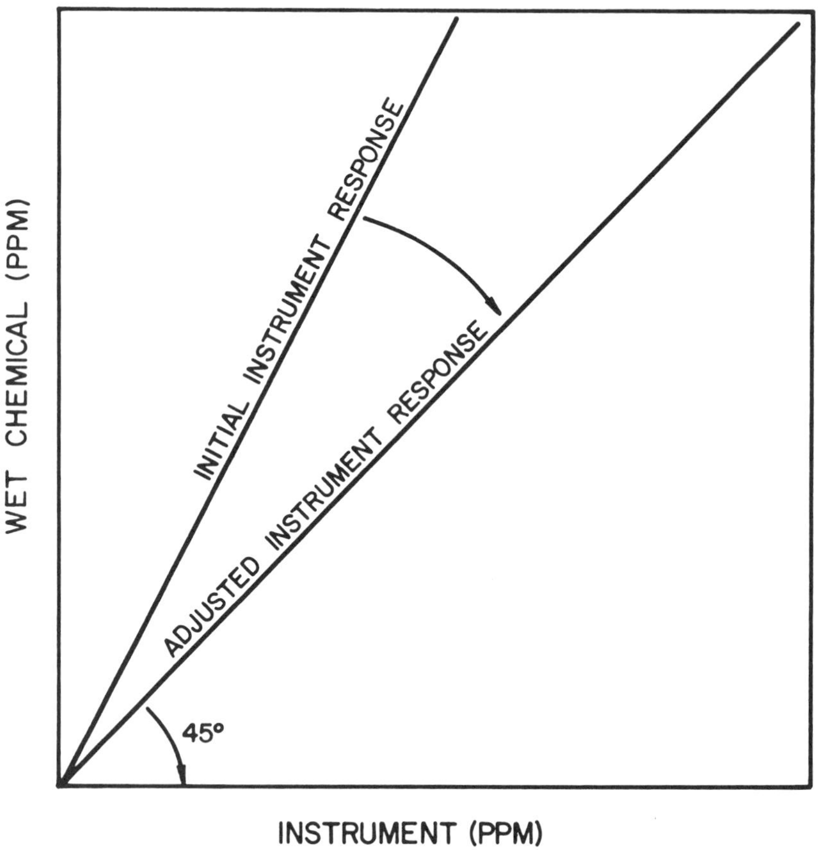

Figure 14.5. Adjustment of continuous instruments to allow a direct readout (ppm).

Sources of Known Gas Concentrations

The source for the specific pollutant gases used to calibrate an air monitoring instrument can be a permeation tube, a certified gas bottle, or a special gas generator as in the case of ozone.

Permeation Tubes

A permeation tube consists of a liquefied material in a sealed, permeable Teflon tube. The permeation of the liquefied material through the tube walls can be controlled and determined accurately by collecting weight data losses over a period of time. Temperature control is critical, as slight changes ($\pm 0.2°C$) can cause measurable changes in the gas permeation rate.[4]

Permeation tubes are presently available for NO_2, SO_2, and hydrocarbons. A special dilution apparatus (Figure 14.6) is usually required to provide satisfactory temperature control. The main feature of such a dilution instrument is the constant temperature bath for the permeation tube. Clean air is obtained by filtration through charcoal and drierite. The clean air is then split with a small amount flowing over the permeation tube as a carrier gas and the remainder directed to a dilution tube. Dilution air and permeation air are controlled by adjustments of rotometers which control the air stream flow rate. The permeation tube dilution apparatus is portable and is well suited to field application.

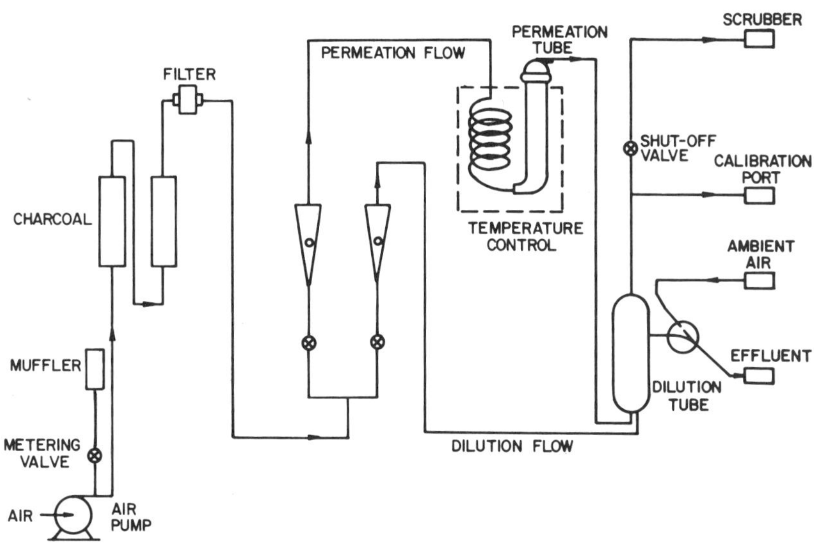

Figure 14.6. Permeation calibration instrument.

Certified Gas Bottles

Mixtures of stable gases can be prepared to exact concentrations in pressurized cylinders and can be used as known gas concentrations. Carbon monoxide and hydrocarbon gases are stable for several months; SO_2, NO_2, NO are less stable and require a reference wet chemical check when used to calibrate air monitoring instruments. The dilution panel is usually used for this purpose.

Ozone Generator

Ozone is usually generated by the use of an ultraviolet light source just prior to use. The concentration of ozone is varied by controlling the lamp current and thus the intensity of the light source or by controlling the flow rate of dilution air over the lamp.

The ozone generator, if well-calibrated, may be used in the field to supply ozone directly to instruments with good accuracy. However, the ozone generator may also be attached to a dilution panel where a referee concentration may be monitored simultaneously with the instrument being calibrated.

EXAMPLE CALCULATION

Table 14.2 illustrates three methods of obtaining the concentration of SO_2 for a typical set of calibration data. The three methods are: (a) the calculation of the concentration based on the permeation rate, (b) the measurement of the concentration by the reference pararosaniline method, and (c) the measurement of the concentration by the continuous instrument. The calculated permeation value was 0.336 ppm, while the reference method and the instrument both showed 0.332 ppm.

A permeation apparatus (Figure 14.6) with an SO_2 permeation tube was used to supply the pollutant gas. The following equation provides the effluent gas concentrations:

$$C = \frac{R\ (22.4)\ Ta}{Mw\ (273)\ (F_1 + F_2)}$$

C = output concentration in ppm
R = gravimetric permeation rate of the permeation tube in ng/min
Ta = temperature of sample in °K
F_1 = dilution flow rate over permeation tube in cc/min
F_2 = dilution flow to the dilution tube cc/min

Table 14.2 shows the calculated concentration for one set of flow conditions. The permeation rate was determined by successive weighings of the permeation tube to be 1423.69 ng/min.

The flow from the permeation apparatus was input to the dilution panel so that the 0.336 ppm concentration could be simultaneously

Table 14.2.

A. Permeation Apparatus

Dilution — 1600 cc/min. — F_2

Permeation — 46 cc/min. — F_1

$$C = \frac{(R)\ (22.4)\ (Ta)}{Mw\ (273)\ (F_1 + F_2)} = 0.336 \text{ ppm}$$

R = Geometric Mean Permeation Tube Weighings — 1423.699 ng/min.

Ta = Temperature of Sample — 303 °K

Mw = Molecular Weight of SO_2 — 64 g/g mole

B. Continuous Instrument

Range — (0–.5) ppm

Zero — 5.5%

Final — 72%

Actual Vacuum — 170 cc/min

Instrument Response — .3325 ppm

C. Wet Chemistry

	Absorbance	*Bubbler*
	0.445	1st
	0.015	2nd
	0.00	3rd
Total	0.46	

From Static Calibration Curve (left scale of Figure 14.4), .46 Absorbance is equal to 3.39 μl. Time of the Run — 60 min. Volume of flow — 170 cc/min (60 min) = 10.2 l. ppm = 3.39 μl/10.2 l = .33235 ppm. Since the dynamic calibration curve's scale was calculated previously, one can read the .33235 ppm from the right scale Figure 14.4 directly.

measured by the reference method and the continuous instrument. Since the permeation apparatus gave the concentration range desired, further dilution by the dilution panel was not needed.

In this case the continuous instrument to be calibrated operated on the coulometric principle. The continuous instrument adjustment for direct readout in ppm could only be made by an air flow adjustment. Therefore, after a constant concentration was obtained from the permeation apparatus, the flow of the continuous instrument was adjusted so that the readout corresponded to the calculated permeation concentration (Part B of Table 14.2). If a number of different concentrations are established from the permeation apparatus, a plot similar to Figure 14.5 of the permeation concentration versus the instrument's response can be obtained. For this example an adjustment of the instrument's air flow rate to 170 cc/min was needed to give a ppm response corresponding to the calculated permeation concentration.

At this point in time and with the instrument still operating, a reference sample was taken through three bubblers for a reference method check. Table 14.2 shows the measured value of absorbance

(.46) for a flow rate of 170 cc/min and a 60-minute sample period. Since a static calibration curve (left scale of Figure 14.4) and a dynamic calibration curve (right scale of Figure 14.4) were previously established, one can read the actual concentration of 0.332 ppm from the dynamic calibration curve by knowing the total absorbance of (.46) for the bubblers.

Since the calculated permeation concentration, the reference method concentration, and the analyzer response all agree, the instrument is calibrated and ready to be used in the field for monitoring SO_2 levels.

SUMMARY

Periodic calibration of air monitoring instruments is required to provide meaningful information on the true level of air pollution. The techniques for calibration have been well defined and involve the simultaneous measurement of an arificially polluted atmosphere by the instrument to be calibrated and by a standard reference method.

Present calibration techniques involve the use of permeation tubes, certified gas bottles and ozone generators to produce artificial polutants of known concentration. A dilution panel is required to (a) allow simultaneous sampling, (b) verify permeation rates, (c) check concentrations in certified bottles, and (d) verify ozone generation rates.

Instruments are usually adjusted so that the scale response represents the true concentration in ppm. This is accomplished by setting of the air and/or liquid flow rate or adjustment of the electronic response of the instrument to be calibrated so that the reference method and instrument record the same concentration.

REFERENCES

1. Environmental Protection Agency. *Federal Register* **36**, 8186–18210 (1971).
2. Tokiwa, Y. and E. R. DeVera. "Outline for Workshop on Static and Dynamic Calibration of Continuous Analyzers," Proc. 13th Methods Air Poll. Ind. Hyg. Studies, California Department of Public Health (June 1972).
3. *Methods of Air and Analysis,* Intersociety Committee, Publication (Washington, D.C.: American Public Health Association, 1972).
4. *Field Operations Guide for Automatic Air Monitoring Equipment*, National Technical Information Service, APTD - 0736, (Washington, D.C.: U.S. Dept. of Commerce, 1971).

15

Sampling and Analysis of Trace Metals in the Atmosphere

Gary D. Rawlings and **Hal B. H. Cooper, Jr.**
Texas A&M University, Department of Civil Engineering,
Environmental Engineering Division, College Station

INTRODUCTION

Air pollution control authorities have been monitoring gaseous and particulate concentrations in urban environments for several years to assess their potential impacts on ambient air quality and possible adverse effects upon humans, plants, animals, and materials. It has been recognized that the physical size of the particles in the atmosphere was very important, since the degree of penetration into the lower respiratory system increases as the particle size decreases. It has only been recently that the chemical composition of particles and the presence of trace hazardous materials have been given serious consideration in air pollution research.

Determination of the respective concentrations of specific trace metals in the atmposhere has only recently become of interest. The presence of elements such as mercury, arsenic, selenium, cadmium, chromium, vanadium, beryllium, lead and copper in urban atmospheres may be significant because of their potential for reaching the lower portions of human respiratory systems, for acting as potential catalytic agents for atmospheric photochemical reactions, and transmission through with subsequent contamination of the food chain following removal from rainfall. It is particularly important to identify the specific chemical and physical forms in which the specific trace metals may be present in the atmosphere in terms of their potential effects.

Most of the trace metals in urban atmospheres are probably present as particulate matter, but mercury can also exist in either elemental or organic vapor forms in addition to the inorganic particulate form. The same may also be true for other metals such as arsenic and cadmium, though to a lesser extent than for mercury, where the potential for substantial errors can exist when attempting to determine trace metal concentrations in the atmosphere by suspended particulate measurements alone. In addition, the trace metals are normally present in extremely low concentrations in the atmosphere, which requires sophisticated sample collection and analytical techniques such as emission spectroscopy, x-ray fluorescence, neutron activation, atomic absorption spectrophotometry, and spark source mass spectrometry.

SAMPLE COLLECTION

It is necessary to use considerable care in the selection of a sample collection system suitable for the metal or metals of interest. Most of the metals can be collected by means of conventional particulate removal methods but special techniques may be required for mercury sample collection.[1] The major concern with collection of metal-containing particulate matter is particle size, where particles of less than 2 μ in diameter are difficult to collect but tend to have preferentially large trace metal contents. Lee[2] reports that lead and vanadium are the predominant metals associated with particles between 0.4 and 0.6 μ in diameter in urban atmposheres. Major collection techniques for particulate trace metals include inertial impaction, liquid impingement, filtration, and electrostatic precipitation, while metallic screen amalgamation, solid absorption, and liquid absorption can be used for metals in the vapor form.

Inertial Impaction

Inertial impaction is used for particle collection because particles moving in or on an air stream tend to follow their original direction of motion when the streamlines of air flow are deflected by a target. The result is that particles colliding with and being collected on an impaction plate. The major types of inertial collection systems include single-stage impactors, cascade single-stage impactors, multistage or cascade impactors, dry impingers, and centrifugal collectors. Single-stage impactors and centrifugal collection devices are not normally effective to collect particles for trace metal analysis because of their low collection efficiency for particles less than about 5 μ in diameter. In addition, single-stage impactors usually require a sticky coating on the impactor plate to aid in collecting the particles, which can result in an interference with analytical techniques.

Cascade impactors are multiple-stage devices, which are frequently used for collection of particulate samples for subsequent trace metal analyses, and have two major advantages. First, the particles are collected in a dry state so that fewer sample interferences from the collection system exist for the subsequent chemical analysis. Second, the particles can be separated by specific size ranges from 0.1–20 μ microns where each stage can be individually analyzed to determine the specific metal concentrations as functions of particle size. A schematic diagram of a cascade impactor is presented in Figure 15.1.

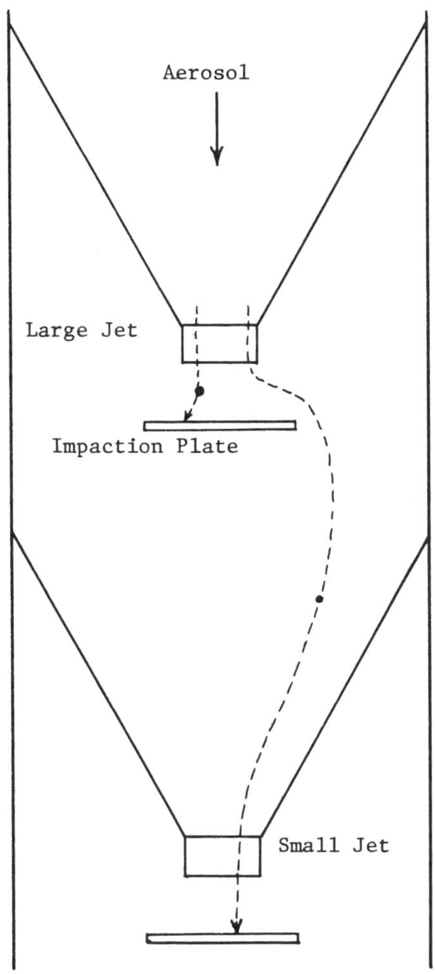

Figure 15.1. Diagram of a typical cascade impactor.

Liquid Impingement

Liquid impingement can be used for collection of particulate matter where subsequent wet chemical analyses can then be made for specific trace metal constituents. The air is drawn through a conventional Greenburg—Smith or midget impinger containing a known volume of liquid into which particles larger than approximately 1 μ in diameter can be collected.[3] The technique is suitable for determination of overall mass metal concentrations without respect to particle size or chemical form except in terms of soluble and insoluble materials in water. The technique may also cause subsequent problems with sample deterioration or alteration following collection, or potential interferences with analytical techniques.

Filtration Methods

Filtration provides a very useful means for collection of particulate materials for subsequent analysis for specific trace metal constituents because of its simplicity, low cost, and ease of sample storage. The storage of filter pads from suspended particulate measurements by air pollution control agencies will provide a readily available source of historical information regarding ambient air level trends of specific trace metals in the future. The particle collection efficiency of filtration is dependent upon the gas flow rate, the filter surface area, the filter material employed and its porosity, the particle size distribution of the air being sampled, the sampling time employed, and the mass of particulate material previously collected.[4]

The basic types of filter materials include cellulose fiber, glass fiber, membrane, nucleopore, polyester fibers, and mixed fiber filters. Cellulose and glass fiber filters are useful for trace metal analysis of aerosols because of their durability and low trace metal background content. Metal-free filters are especially important when nuclear analytical techniques are used, to reduce the background radiation level. The respective levels of elemental impurities for three different types of filter media used in ambient air sampling for trace metal constituents is presented in Table 15.1.[6] A summary of the penetration characteristics of submicron particles through several different types of filter media is presented in Table 15.2.[6] The basic filtration system employed in ambient air sampling is illustrated in Figure 15.2.

Membrane-type filter media are best suited for particle size analysis studies because their porosities can be carefully controlled and usually ranges from 0.01–10 μ. The major disadvantages of membrane filters are that deposited particles tend to slough off when more than one layer of particles is collected and they are not as durable as other media. However, the filter can be readily dissolved in re-

Table 15.1. Background Impurity Levels in Filtration Materials used in Ambient Air Sampling for Trace Metal Constituents.[6]

	Elemental Impurity Level — μg/cm²		
Element	*Glass Fiber*	*Organic Membrane*	*Silver Membrane*
Arsenic	0.08	—	—
Beryllium	0.04	0.0003	0.2
Bismuth	—	<0.001	—
Cadmium	—	0.005	—
Cobalt	—	0.00002	—
Chromium	0.08	0.002	0.06
Copper	0.02	0.006	0.02
Iron	4	0.03	0.3
Manganese	0.4	0.01	0.03
Molybdenum	—	0.0001	—
Nickel	<0.08	0.001	0.1
Lead	0.8	0.008	0.2
Antimony	0.03	0.001	—
Silicon	7000	0.1	13
Tin	0.05	0.001	—
Titanium	0.8	2	0.2
Vanadium	0.03	0.0001	—
Zinc	160	0.002	0.01

agent grade acetone or alcohol to separate the particles from the filter media.

Electrostatic Precipitation

Electrostatic precipitation can be used for collecting trace metal-containing particulate matter from ambient air samples. The technique has the advantages of being able to collect particles smaller than possible by filtration or impaction under certain conditions, and that the particles can be collected free from a collecting medium.[7] Electrostatic precipitation differs from other particle collection mechanisms in that electrical forces act to separate the particles from the gas stream instead of inertial, impaction, centrifugal or diffusion forces. The result is that less lower gas pressure drop is passing through the collector. Particles can be deposited and collected on microscopic glass slides, filter papers, thin metal sheets, or other thin, flat surfaces. The technique is not suitable for collecting all types of particulate matter, but is probably suitable for most metal-containing inorganic particles. (A typical electrostatic precipitation sampler is shown in Figure 15.3.)

Vapor Collection

Certain metals such as mercury can exist in forms other than the particulate which requires the use of other sampling techniques, of

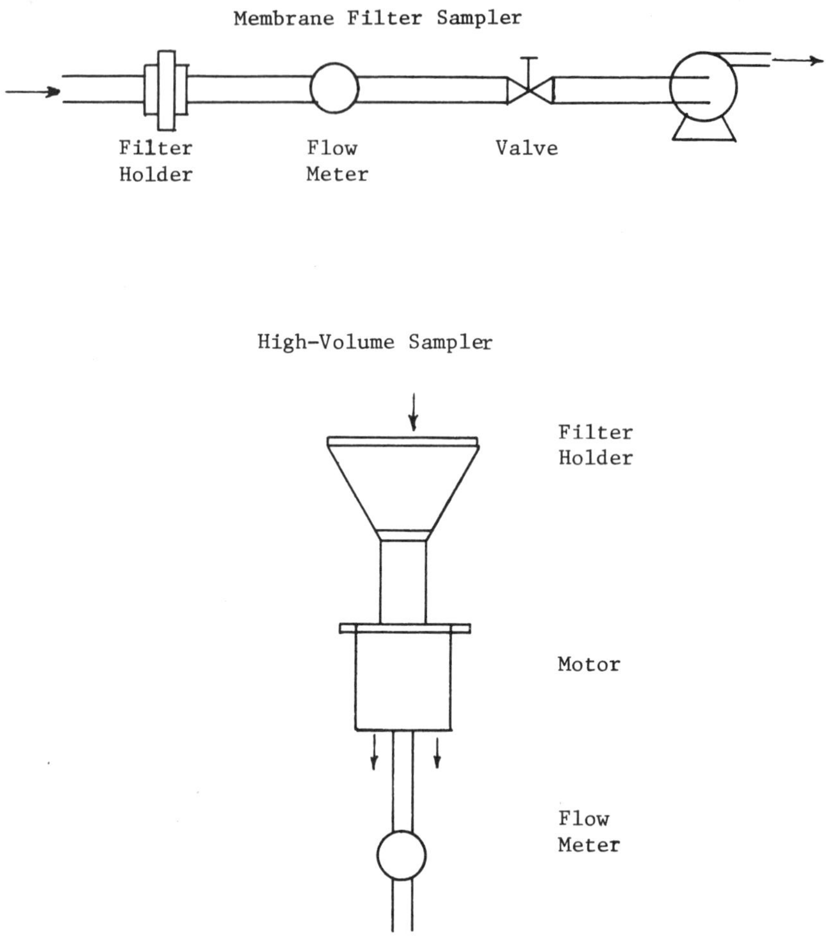

Figure 15.2. Filtration sample collection system.

which metallic grid amalgamation and liquid absorption are the most common. Foote[8] describes the use of gold screens for collecting elemental mercury vapor from ambient air by formation of a mercury–gold amalgom. The gold screen is then heated to drive off the mercury collected, which is then analyzed by means of atomic absorption spectrophotometry. The collection and analytical system is illustrated in Figure 15.4. Total mercury is analyzed by passage of all materials through a pyrolysis furnace to convert inorganic and organic mercury to the elemental form, followed by collection on a gold screen, followed with analysis by atomic absorption spectrometry.

Table 15.2. Penetration of Submicron Particles Through Commonly Used Filter Media in Percent by Weight.[6]

		Filter Media				
		Whatman 41	*MSA 1106B*	*Delbag Microsorban*	*Millipore AA*	*Gelman AM-4*
Flow Rate	*Particle Size*	*Cellulose Fiber*	*Glass Fiber*	*Polystyrene Fiber*	*Membrane Filter*	
FPM						
5	0.025	16.0	1.0	1.8	<0.1	<0.1
	0.30	89.0	0.001	—	0.002	—
50	0.025	18.0	<0.1	2.8	<0.1	0.3
	0.30	67.0	0.005	—	0.02	—
100	0.025	25.0	<0.1	2.8	<0.1	0.1
	0.30	9.0	0.005	—	—	—

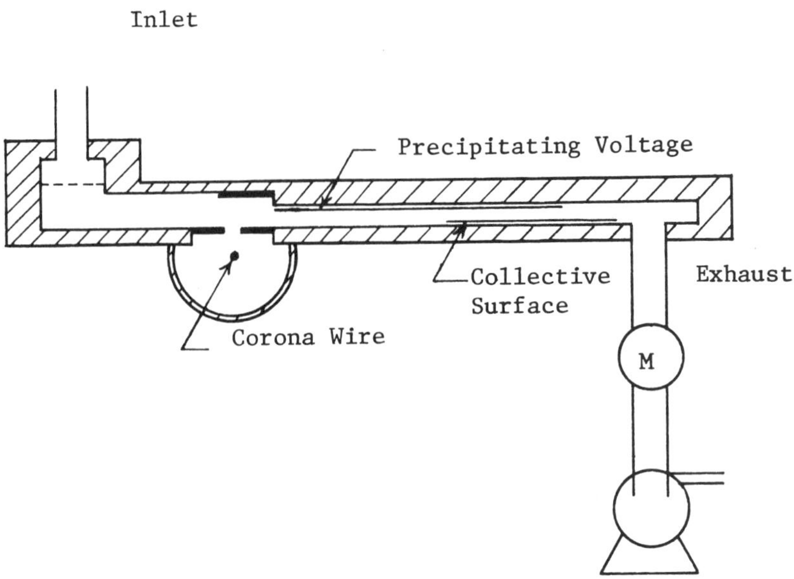

Figure 15.3. Electrostatic precipitator aerosol sampler.[7]

Rawlings and Cooper[9] report on the development of a technique for collection of organic mercury compounds in urban atmospheres by means of liquid absorption. The organic mercury compounds are collected by drawing air samples through two straight tip midget impingers in series containing diethyl ether at a rate of approximately 500 ml/min. The impingers are contained in a bath of dry ice to prevent evaporation of the ether during the sampling period. Diethyl ether samples containing the dissolved organic mercury compounds are then injected into a gas chromatograph where the individual components are separated by means of a spiral column. The individual mercury compounds are then converted to elemental mercury by means of pyrolysis at elevated temperatures in the presence of a copper oxide catalyst, and the mercury detected by means of atomic absorption spectrophotometry. The system is illustrated in Figures 15.5 and 15.6.

Other Methods

Collection systems have also been developed to measure trace metal constituent levels in rainfall samples and soil samples. Rainfall samples have been collected and analyzed for specific trace constituents,[10,11] and studies have also been made of the effectiveness of rainfall for scavenging particles from the atmposhere.[12] Soil samples

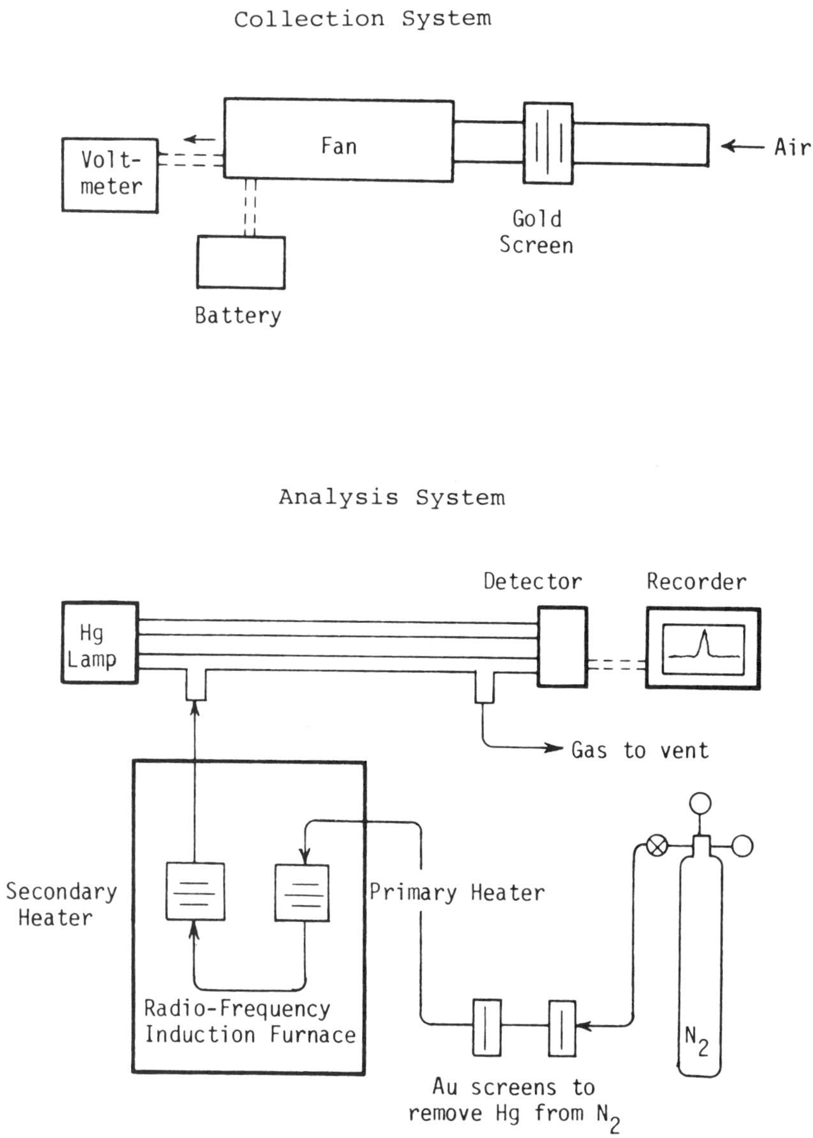

Figure 15.4. Sampling and analysis for elemental mercury.

can also be used to provide an indication of atmospheric fallout or rainout of particulate materials containing trace metal constituents near specific sources.[13]

ANALYTICAL TECHNIQUES

Trace element analysis is a separate area of analytical chemistry with problems peculiar to those of very low concentrations. The concentration ranges of most trace metals present in urban atmos-

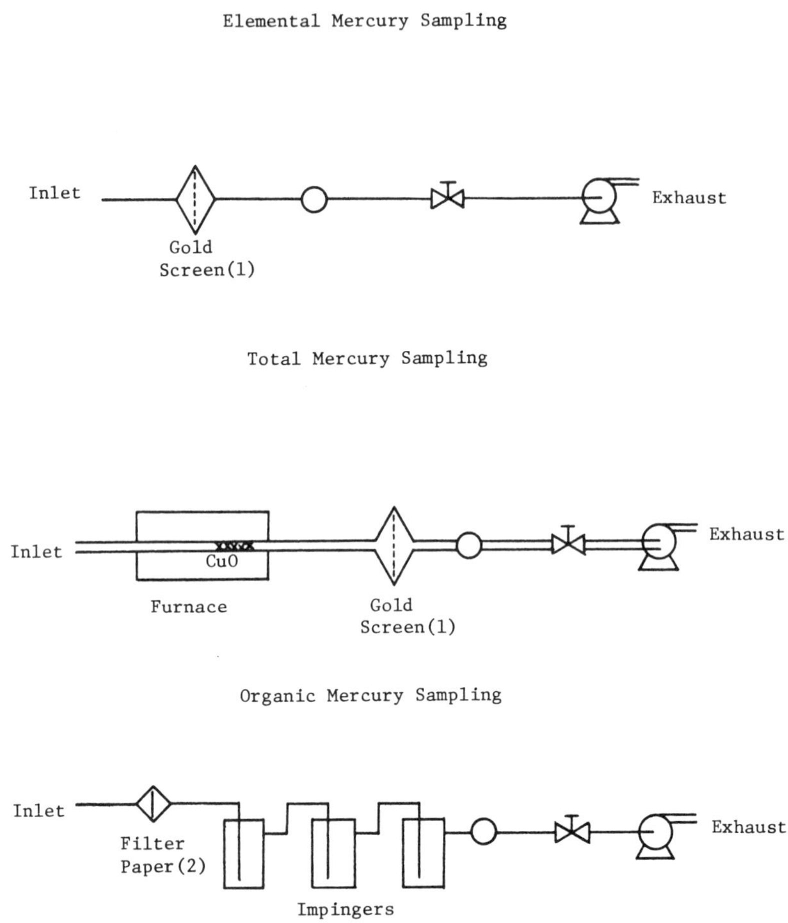

(1) Gold screen analyzed as shown in Figure 4.

(2) Filter paper analyzed by Hatch and Ott procedure (17).

Figure 15.5. Mercury sampling systems.

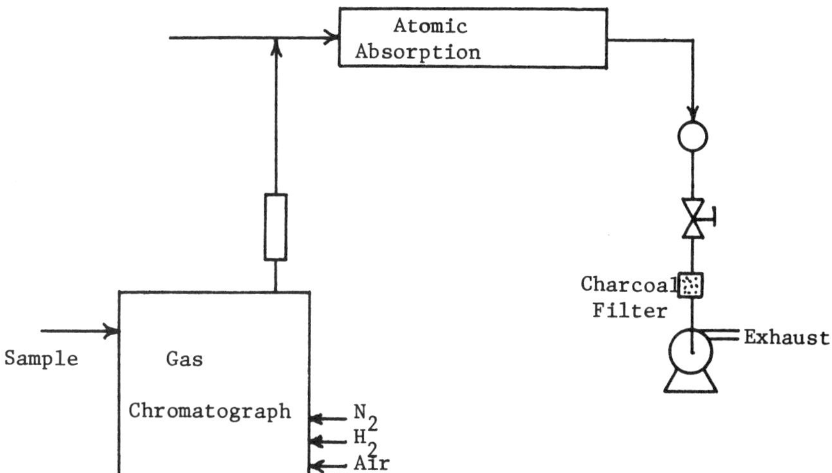

Figure 15.6. Organic mercury analysis system.

pheres are normally in the range of one to 1000 ng/m^3 of air.[14] The most commonly employed methods for measuring trace metal concentrations of samples collected from ambient air include atomic absorption spectrophotometry, neutron activitation analysis, and x-ray fluorescence, as listed in Table 15.3.

Table 15.3. Analytical Techniques for Trace Element Analyses in Ambient Air Samples.[15]

Analytical Principle	*Specific Method*
Optical absorption	Flame atomic absorption spectrophotometry
	Flameless atomic absorption spectrophotometry
Optical emission	Flame emission spectrography
	X-Ray fluorescence
	Charged particle induced x-ray fluorescence
Electrical analyses	Polarography
	Voltommetry
	Anodic stripping
Nuclear measurements	Mass spectrometry
	Neutron activation analyses

Atomic Absorption Spectrophotometry

Conventional flame atomic absorption has been used to measure trace metals in aqueous samples for many years. In recent years, this technique has been used in the analyses of trace metals in atmospheric particulates. One reason for the widespread use of this analytical technique is because most analytical laboratories have atomic absorption systems. However, the method by itself suffers from several disadvantages, such as large consumption rate of sample solutions, insensitivity to extremely low concentrations, and potential chemical interferences from other materials present.

Sample preparation is necessary to obtain satisfactory results for atmospheric trace metal analyses. Since most trace metal samples are collected on filter paper, one procedure of sample preparation normally involves removal of organic matters from the particulate sample and conversion of the metals to soluble salts by ashing the samples with subsequent dissolution into nitric acid. High-temperature ashing of filter samples results in the loss of the following metals by volatilization including mercury, lead, zinc, cadmium, tin, selenium, arsenic, antimony, and thallium. Hwang[5] has found that low temperature ashing can significantly increase analytical sensitivities and reduce the amount of metals lost. Ashing is not required when membrane filters are used because the filter material can be dissolved in reagent grade acetone or alcohol.

On of the more sensitive atomic absorption techniques is the semiflame solid phase sampling boat technique.[16] This method employs placing the sample in a tantalum boat and electrothermal heating of the sample in an enclosed chamber to produce an atomic vapor sample for subsequent introduction to the detector system. The method increases the analytical sensitivity of the conventional atomic absorption spectrophotometric system by 10–50 times.

Flameless atomic absorption techniques have received increasing attention in the environmental field because of their extremely low sensitivity and fewer interferences. Hatch and Ott[17] increased the sensitivity of mercury detection by three orders of magnitude using a flameless atomic absorption technique. Other flameless techniques include the tantalum ribbon atomizer, heated graphite atomizer, and carbon rod atomizer. Advantages of this method are excellent sensitivity at the nanogram and picogram levels with very small sample volumes. The complexity of the system requires a highly trained technician for successful analyses. The relative sensitivities of the respective atomic absorption spectrophotometric analytical techniques for several different metals are presented in Table 15.4.

Nuclear Activation Analysis

Nuclear methods are particularly well suited for measuring trace metals in the atmosphere because they require no wet chemical or sample pretreatment which can contaminate the sample or result in the loss of material. The sensitivities obtainable vary depending on the metal, but range from 1–10 ng/m^3 of air to as low as 10,000 mole/m^3. Nuclear methods stand apart in their relative freedom from interferences, particularly because of the development of solid state detectors and larger multi-channel analyzers.

Table 15.4. Analytical Sensitivities of Atomic Absorption Spectrophotometry for Trace Metal Analyses in Ambient Air.[5,16]

Element	*Conventional Flame AAS (μg/ml)*	*Flameless AAS (ng)*	*Sample Boal Flamless AAS (ng)*
Aluminum	1.0	—	0.4
Chromium	0.07	0.04	0.006
Copper	0.03	0.02	0.003
Manganese	0.01	0.02	0.003
Molybdenum	0.3	—	3
Nickel	0.05	0.04	3
Lead	0.06	0.03	—
Vanadium	1	1	0 6
Zinc	0.01	0.004	—

The sensitivity of thermal neutron activation analyses depends on the reactor neutron flux, irradiation time, sample size, and counting time. Reactor fluxes range from 2 to $5x10^{12}$ neutrons/sec/cm^2. If a broad spectrum of trace metals (33 elements) is to be determined, then the irradiation and cooling periods for analyses of 33 separate elements is presented in Table 15.5.

Table 15.5. Irradiation, Cooling, and Counting Schedule for Multi-element Analyses by Neutron Activation.[20]

Irradiation Time	*Cooling Period*	*Counting Time*
5 min	3 min	400 sec
5 min	15 min	1000 sec
2–5 hr	20–30 hr	2000 sec
2–5 hr	20–30 days	4000 sec

The purpose of the first short irradiation period is to activate only the short lived radionuclides and results in little interference from the other larger metals. Often long irradiation periods, the intermediate and long cooling periods allow the short lived radionuclides to die away. If high volume glass fiber filter pads are used to collect the particles, the analytical sensitivity can be increased if trace metal-free paper is used. Cellulose-base filters generally have lower background concentrations of trace metals than do fiber glass filters.[17]

X-Ray Fluorescence

X-Ray fluorescence is gaining increasing popularity as an analytical system for atmospheric trace metals. The technique has a sensitivity equivalent to neutron activation analysis, but the long cooling periods are eliminated. X-Ray fluorescence has the added advantage of being capable of detecting both metallic and nonmetallic elements. X-Ray fluorescence and atomic absorption equipment are of equal sophistication and are of approximately equal initial and operating expense. The respective analytical sensitivities of neutron activation and x-ray fluorescence for determination of trace metals present in the particulate matter in ambient air is presented in Table 15.6.

Table 15.6. Analytical Sensitivities of Neutron Activation and X-Ray Fluorescence for Trace Metals in Urban Atmospheres in ng/m^3.

Element	*Thermal Neutron Activation*[19,20]	*X-Ray Fluorescence*[21]
Aluminum	8	200
Silver	1	45
Gold	0.1	14
Copper	5	12
Iron	20	31
Manganese	0.6	42
Nickel	20	42
Sulfur	5000	84
Vanadium	2	9.8
Zinc	1	9

The sensitivity of the x-ray fluorescence also depends on the length of the irradiation period, but it is generally shorter than for nuclear activation analysis. In addition the sensitivity will vary depending on the source of x-rays. For example, 45 ng/m^3 of silver can be detected using Fe-55 x-rays, but only 70 ng/m^3 can be detected using Gd-153 x-rays. A recent development is the use of secondary x-rays as an excitation source. The purpose of this is to obtain purer, monoenergetic x-rays in which to activate the sample. This improvement will further increase the sensitivity of the x-ray fluorescence technique.

CONCLUSIONS

A number of methods exist for collection and analysis of trace metal concentrations in urban atmospheres but there are significant problems with sufficient analytical sensitivities. Filtration is an excellent method for collecting large quantities of particles, but are not effective for particle size analyses. Electrostatic precipitation is the most efficient means of collecting submicron particles, but is limited in the quantity which can be collected. Cascade impactors are about the best collection system where particle size analyses in terms of trace metal content are involved. However, the particles must be dry or the collector will clog and there is a greater chance of losing some of the sample during weighing and transfer operations. Liquid absorption and metallic amalgomation may be suitable for collection of metals which may be present in the vaporous instead of the particulate form. The relative advantages and disadvantages of several methods for trace metal analyses are presented in Table 15.7.

Table 15.7. Summary of Advantages and Disadvantages of Trace Metal Analytical Systems for Ambient Air.

Technique	*Advantages*	*Disadvantages*
Flame atomic absorption spectrophotometry	Easy to use Relatively inexpensive Good for routine work	May not be sensitive enough Much sample pretreatment needed Requires large sample volume
Flameless atomic absorption spectrophotmetry	More sensitive than flame type Little sample needed	Highly trained technician needed More costly
Neutron activation analysis	Most sensitive for metals No sample pretreatment needed Little sample reuired Good for routine work	Most expensive, however equipment can be rented Highly trained technician needed Dry samples are better Radiation safety procedures needed Too long to analyze for some metals
X-Ray fluorescence	Very sensitive Detects other elements than metals with good sensitivity Not too expensive Good for routine work Rapid multi-element analysis Sample media flexible	Requires some radiation safety Highly trained technician needed

REFERENCES

1. Cooper, H. B. H., R. S. Foote, and G. D. Rawlings. "Measurement of Mercury Vapor Concentrations in Urban Atmospheres," Presented at the 1973 Annual Meeting of the Instrument Society of America, Environmental Sensing and Applications Session, Houston, Texas (October 15, 1973).
2. Lee, R. E., S. S. Goranson, R. E. Enrione, and G. B. Morgan. "National Air Surveillance Cascade Impactor II., Size Distribution Measurements of Trace Metal Components," *Environ. Sci. Technol.* 6, No. 12, 1025-1030 (1972).
3. *Air Sampling Instruments for Evaluation of Atmospheric Contaminants,* 4th ed., American Conference of Governmental Industrial Hygienists, Cincinnati, Ohio (1972).
4. Cooper, H. B. H. and A. T. Rossano. *Source Testing for Air Pollution Control,* (New York: Environmental Science Services Corp., 1971).

5. Hwang, J. Y. "Trace Metals in Atmospheric Particulates and Atomic Absorption Spectroscopy," *Anal. Chem.* **44**, No. 14, 20A-27A (December 1972).
6. Posner, S. "Air Sampling Filter Retention Studies using Solid Particles," *Proc. Seventh Annual Atomic Energy Commission Air Cleaning Conf.*, Brookhaven National Laboratory, Washington, D.C. (October 1961).
7. Liu, B. Y. H., K. T. Whitby, and H. H. S. Yu. "Electrostatic Aerosol Sampler for Light and Electron Microscopy," *Rev. Scient. Instr.* **38**, No. 1, 100-102 (1967).
8. Foote, R. S. "Mercury Vapor Concentrations Inside Buildings," *Science* **177**, No. 4048, 513-514 (1972).
9. Rawlings, G. D. and H. B. H. Cooper. "Identification of the Major Chemical Forms of Mercury in Urban Atmospheres," Special Report for the Texas Engineering Experiment Station, Texas A&M University, College Station, Texas (January 24, 1974).
10. Drozdova, V. M. and E. P. Makhonko. "Content of Trace Elements in Precipitation," *J. Geophys. Res.* 75, No. 18, 3610-3612 (1970).
11. Lazrus, A. L., E. Lorange, and J. P. Lodge. "Lead and Other Metal Ions in United States Precipitation," *Environ. Sci. Technol.* 4, No. 1, 55-58 (1970).
12. Peirson, D. H., P. A. Cawse, L. Salmon, and R. S. Cambray. "Trace Elements in the Atmospheric Enivornment," *Nature*, **241**, No. 4, 252-256 (1973).
13. Lagerwerff, J. V. and A. W. Specht. "Contamination of Roadside Soil and Vegetation with Cadmium, Nickel, Lead and Zinc," *Environ. Sci. Technol.* **4**, No. 7, 583-586 (1970).
14. Lee, R. E. and D. J. Von Lehmden. "Trace Metal Pollution in the Environment," *J. Air Poll. Cont. Assoc.* **23**, No 10, 853-857 (1973).
15. Filby, R. H. "Problems in the Analysis of Trace Metals," In *Heavy Metals in the Environment*, Seminar Conducted at Oregon State University, Water Resources Research Institute, Corvallis, Oregon (January, 1973).
16. Donega, H. M. and T. E. Burgess. "Atomic Absorption Analysis by Flameless Atomization in a Controlled Atmosphere," *Anal. Chem.* **42**, No. 13, 1521-1524 (1970).
17. Hatch, W. R. and W. L. OH. "Determination of Sub-Microgram Quantities of Mercury by Atomic Absorption Spectrophotometry," *Anal. Chem.* **40**, No. 14, 2085-2087 (1968).
18. Brar, S. S., D. M. Nelson, J. R. Kline, P. F. Gustafson, E. L. Konabrocki, C. E. Moore and D. M. Hattori. "Instrumental Analysis for Trace Elements Present in Chicago Area Surface Air," *J. Geophys. Res.* 75, No. 15, 2939-2945 (1970).
19. Dams, R., J. A. Robbins, K. A. Rahn, and J. W. Winchester. "Nondestructive Neutron Activation Analysis of Air Pollution Particulates," *Anal. Chem.* **42**, No. 8, 861-866 (1970).
20. Kuykendall, W. E. "Neutron Activation Analysis of Air Pollution Particulates," Preliminary Report, Activation Analysis Research Laboratory, Texas A&M University, College Station, Texas (1972).
21. Payne, J. S. "X-Ray Fluorescence Analysis of Air Pollution Samples," Special Report, Texas State Department of Health, Air Pollution Control Services, Austin, Texas (March 7, 1972).

16

Principles and Applications of Cooling Tower Drift

Carl B. Wilber, Thomas B. Carlson and **Frederick M. Shofner**
Environmental Systems Corporation, 1212 Pierce Parkway,
Knoxville, Tennessee

INTRODUCTION

Increased projected usage for evaporative or wet cycle cooling systems necessitates the evaluation of their environmental effects. Inherent to the operation of evaporative cooling towers is the entrainment of droplets, known as drift or carryover, into their effluent air. These droplets contain the same dissolved minerals as the circulating cooling water often at similar concentrations. With the advent of large salt water cooling towers, increased attention has been focused on the possible effects this brackish water drift will have on the environment surrounding the towers. To aid in evaluating these effects, it is prudent to characterize not only the source, but also areas downwind of the source in both pre- and postoperational studies.

This chapter will treat the basic parameters measured, their interpretation, and the equipment used for their acquisition. A case study of a large hyperbolic natural draft brackish water cooling tower is examined with regard to specific instrumentation and data acquisition techniques employed.

DEFINITIONS AND TERMINOLOGY

Discussion of Appropriate Terminology

The essential source characterization data are droplet density distribution, mineral composition for each droplet size range, and absolute droplet velocity at each measurement point. Accurate measurement of this parameter together with a measurement of the droplet velocity allows calculation of the total drift fraction emitted. The droplet density distribution is measured by two complementary techniques: the PILLS System and the Sensitive Paper (SP) device. This fundamental parameter is defined by

$$\frac{\Delta N_{ij}}{\Delta d \bullet V_s} = \frac{\text{number of drift particles of mean diameter } d_i}{\text{unit size range } (\Delta d_i, \mu m) \text{ x unit volume } (m^3)} \tag{1}$$

where i indicates the particle size range (d_i, $d_i + \Delta d_i$) and j is a specific measurement point.

Directly following from the droplet density distribution is the mass density distribution defined as:

$$\frac{\Delta \chi_{ij}}{\Delta d} = \frac{\pi}{6} \bar{d}_i^3 \, s \frac{\Delta N_{ij}}{\Delta d \bullet V_s} \quad [g/\mu m \bullet m^3] \tag{2}$$

where s is the specific mass of the drift particles and $\bar{d}_i$ is the mean diameter of the droplets in the range Δd_i.

A summation of the applicable size ranges yields the total mass density χ_j where

$$\chi_j = \Sigma_i \, \Delta\chi_{ij} = \Sigma_i \frac{\Delta \chi_{ij}}{\Delta d} \bullet \Delta d_i$$

$$= \frac{\pi}{6} \, s \, \Sigma_i \, d_i^3 \frac{\Delta N_{ij}}{\Delta d \bullet V_s} \bullet \Delta d_i \quad [g/m^3] \tag{3}$$

With the knowledge of the absolute droplet velocity, $\bar{u}_{ij}$, (which is simply the vector difference between the air velocity at the jth measurement point and the settling velocity of the particle), the mass flux distribution is derived as

$$\frac{\Delta^2 D_{ij}}{\Delta d \bullet \Delta A} = \frac{\Delta \chi_{ij}}{\Delta d} \bullet u_{ij} \quad [g/\mu m \bullet m^2 \bullet sec] \tag{4}$$

It is important to note that the absolute droplet speed u_{ij} is the component normal to the efflux plane over which measurements are made. This velocity can be measured at the tower throat for a natural draft tower. For an induced tower, measurements are conveniently made just above the fan stack.

With this convention, the drift mass flux is then

$$\frac{\Delta D_j}{\Delta A} = \sum_i \frac{\Delta^2 D_{ij}}{\Delta d \cdot \Delta A} \cdot \Delta d_i$$

$$= \frac{\pi}{6} s \sum_i \bar{d}_i^3 \left(\frac{\Delta N_{ij}}{\Delta d \cdot V_s} \right) \Delta d_j \; [g/m^2 \cdot sec] \quad (5)$$

Thus, from Equation (5) the *total* mass emission is simply

$$D = \sum_j \frac{AD_j}{\Delta A}$$

$$= \sum_j A_j \left[\frac{\pi}{6} s \sum_i \bar{d}_i^3 u_{ij} \left(\frac{\Delta N_{ij}}{\Delta d \cdot V_s} \right) \Delta d_i \right] \quad [g/sec] \quad (6)$$

It has been the custom to express drift emission rate D as a percentage Δ of circulated water rate R (in g/sec)

$$\Delta(\%) = \frac{D}{R} \cdot 100 \quad (7)$$

This suffices for a general picture of drift emission, although it is of interest to define a specific drift fraction,[1] δ, which is sometimes employed for measurements made in the vicinity of the drift eliminators. For the jth location, the specific drift fraction is given by

$$\delta_j = \frac{\Delta D_j}{\Delta A} \frac{\Delta R_j}{\Delta A} \text{ x } 100\% \quad (8)$$

which is simply the mass flux near the drift eliminators divided by the water loading per unit area of drift eliminators.

It is important to note that the equations presented here evolve either directly or indirectly from the droplet density distribution. Both the PILLS and SP techniques measure this parameter directly and conveniently provide a valuable cross check on measurement accuracies within their overlapping ranges of response.

A third device, the isokinetic sampling (IK) tube complements the PILLS and SP-acquired data by yielding the mineral mass flux. Evolving from this measurement is the apparent drift water mineral concentration (χ_{kj}) at the jth location within the plane of measurement. In the present art, it is assumed[1] that the drift droplet concentration is nearly equal to the basin water concentration C_k. Thus, the apparent drift water mass concentration can be expressed approximately as

$$\chi_{kj} = (M_{kj} - M_{Bk}) \; / \; C_k \, V_s \; [g/m^3] \quad (9)$$

where M_k is the mass of chemical element k in μg stripped from the IK tube, M_{Bk} is the average background value of the tube for the element k (also in μg), and V_s is the volume of air sampled (in m^3). Note that C_k defined previously is given in ppm = $\mu g/ml$.

The total drift emission across the plane of measurement is found using the following equation:

$$D = \sum_j u_j \chi_j A_j \quad [g/sec] \tag{10}$$

It is emphasized that this is only an approximation due to the assumptions that the droplet velocity is equal to the air velocity and that the chemical concentration of all droplets sampled is the same as that of the basin water.

DRIFT MEASUREMENT INSTRUMENTATION

Source Characterization

The PILLS System

The PILLS System is composed of a solid state GaAs laser source and solid state signal conditioning electronics and employs a light scattering method of droplet sizing. One current instrument incorporates three scattering volumes ranging in size from approximately 0.03 cm^3 to 2.5 cm^3. This type of arrangement provides both redundancy (and therefore accuracy of interpretation) of data as well as suppression of background fog levels.

The operational characteristics of the PILLS System are illustrated in Figure 16.1. A single drift droplet present in one of the scattering volumes at the instant of irradiation from the laser scatters light into one of the detectors and produces a voltage signal. This voltage signal, Δv, is related to the diameter of the drift droplet in the scattering volume through the relationship

$$\Delta v = K d_i^2 \tag{11}$$

where Δv is the average voltage response to a particle of diameter d_i and K is an experimentally determined calibration constant. By measuring the distribution of voltage pulses and producing a voltage height histogram one arrives at the droplet size distribution through Equation (11).

The measured voltage signal, which is directly related to the diameter of the measured drift droplet, is stored by incrementing the appropriate channel of an on-line minicomputer or pulse height analyzer. The voltage distribution obtained from the sequential measurement of an ensemble of drift droplets is obtained from the number of counts accumulated in each channel of the pulse height analyzer. Combining this measured voltage distribution with the known response function of the instrument gives the drift droplet density distribution, p(d).

The Sensitive Paper Device

A chemically Sensitive Paper technique has been refined and calibrated to aid in the determination of the drift droplet density distri-

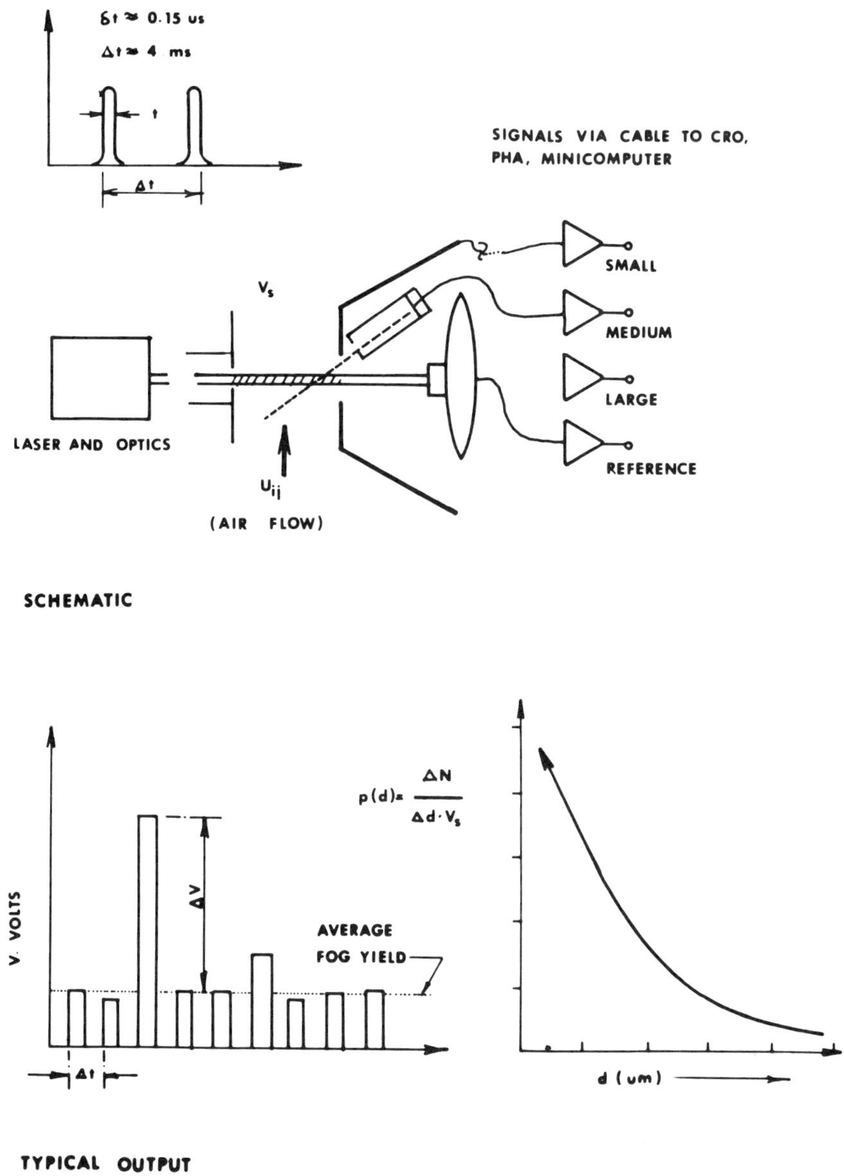

Figure 16.1. PILLS II-A operational characteristics PILLS IIA concept.

bution. When a droplet of water impinges on the sensitive paper, a stain, the size of which is a function of both the droplet diameter and its impingement velocity, is produced.

The SP device is calibrated by means of a vibrating orifice monodisperse water droplet generator that emits uniform water droplets of known diameter and simulates the dynamics of particle impingement. For large droplets, inertial forces tend to cause the stains to be larger for higher impingement velocities. Also, impingement angles other than normal can lead to elongation of the stains. For smaller particles, aerodynamic drag forces can cause the droplets to be accelerated transversely to the impingement direction or parallel to the paper surface. Small droplets, which do impinge on the paper, produce very uniform, circular stains indicating that inertial forces play a less significant role than surface tension forces.

Two sensitive paper geometries have been employed as collection elements for drift particles: a 47 mm disk and a 3 mm x 47 mm strip. We have conducted experiments and have established the collection efficiencies of the disk and strips based on impaction theory. It has been determined that the effective cut-off diameter for the disk occurs at about 16 μm and at about 5 μm for the strip.

Figure 16.2 shows a version of the Sensitive Paper device. Sensitive papers are mounted in electrically heated heads which open under centrifugal acceleration upon attainment of a peripheral velocity of 10 m/sec. The duration for which the head is opened and the papers exposed is controlled by electronic circuitry. Exposure time should

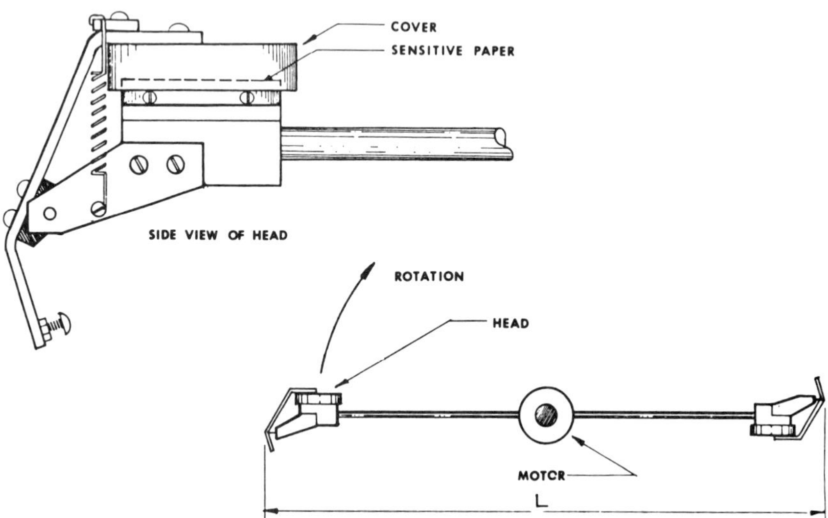

Figure 16.2. Sensitive paper device.

be adjusted such that we acquire a statistically significant sample while at the same time taking precautions not to overexpose the paper.

The Isokinetic Sampling (IK) System.

The heated glass bead isokinetic sampling tubes, shown in Figure 16.3, accumulate the mineral deposits from the drift droplets: the liquid water evaporates upon heating by the internal surfaces of the Pyrex tubing. A resistance wire heater is wrapped around the 25-mm diameter tube to elevate the temperature of the glass surface inside and to heat the air entering the tube such that heat is convected to its downstream portions.

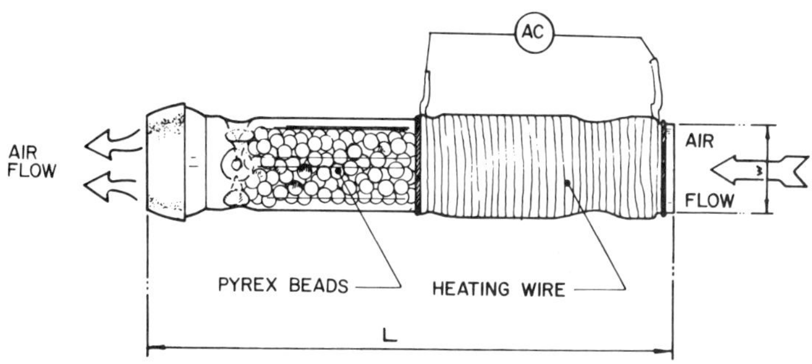

Figure 16.3. Isokinetic Sampler.

The air volume sampled, V_s, is measured with a standard positive displacement gas meter. An orifice manometer calibrated with the gas meter or commercial ball flow meters measure instantaneous volumetric flow rate. Pressure drops are sufficiently low and isothermal conditions are sufficiently approximated to permit direct use of the indicated sampled volume.

The Pyrex glass construction has proven reliable for field usage and is used due to lower cost of fabrication and, more important, lower background mineral contributions for a wider variety of minerals than other materials. These tubes have been successfully used for collection of calcium, magnesium, sodium, and chromium.

Ambient and Downwind Monitoring

Introduction

Before discussing ambient monitoring as related to pre- and post-operational cooling tower studies, it is again important to identify some concepts and terminology felt pertinent to the discussion of cooling tower impact. One approach often employed in the assessment of salt or brackish water cooling towers is based on the following rationale: if it can be shown that the levels of salt drift contributed by the tower are negligible compared to those levels which are already typical of the ambient environment, then the tower can be considered environmentally acceptable.[2]

In most ambient monitoring programs, two basic parameters are measured in an attempt to characterize atmospheric mineral loadings in pre- and postoperational cooling tower impact studies. These are near ground air concentration χ_a and deposition flux d_a. It is felt that χ_a is the more representative and basic parameter with regard to collection of minerals by vegetation and structures, since the amount of mass collected and retained depends strongly on the physical qualities of the drift as well as the characteristics of the surface on which they impinge.[3]

Collection of mass by simple deposition techniques (*e.g.*, pans, buckets) behaves as $D = d_a \, AT\tau$ where A is the vertically projected area of the collection surface, T is the exposure time and τ is the collection efficiency. Deposition values are sometimes inferred from more accurate airborne concentration measurements via a mean particle settling speed, V_d. Ambient measurements often reveal the value of V_d (which is the computed ratio of d_a to χ_a) to well exceed the settling velocity of particles under simple gravitational settling which are typical of the test environment. This is in part explained by the role of nongravitational mechanisms such as washout, surface impaction, electrostatic attraction and chemical interaction which are responsible for the reduction of χ_a.

Ambient Monitoring with the Airborne Particle Sampler (APS)

To meet the need for accurate airborne concentration data, the Airborne Particle Sampler (APS) was developed (Figures 16.4 and 16.5). APS employs the principle of inertial impaction[4,5] and was designed to collect airborne particlates independent of wind speed, wind direction, or relative humidities, typical of coastal regions.

Five distinct features make the APS a unique and field reliable device:

1. The device operates on one or two standard automotive batteries which facilitates placement of the system in any remote area selected for a sampling site. (It can also operate on AC where conveniently available although experience in several different multiple site programs has shown this to be rarely the case.)
2. Woven polyester meshes are used for collection elements due to their ease of handling, low chemical backgrounds and proven precision.
3. A 0.25-m diameter fan is responsible for moving air along the axis of rotation of the mesh arms so that the mesh sampling volumes do not overlap, a situation that might accur in calm wind conditions.

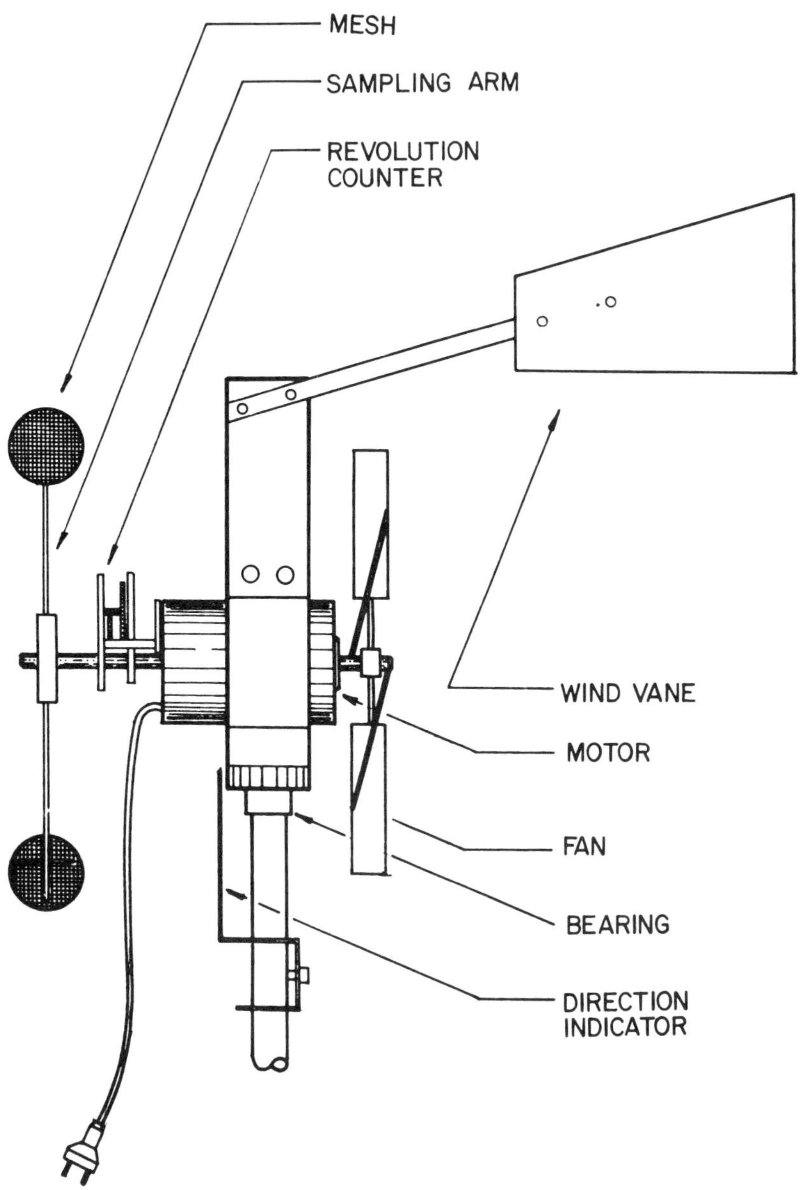
MESH
SAMPLING ARM
REVOLUTION
COUNTER
WIND VANE
MOTOR
FAN
BEARING
DIRECTION
INDICATOR

Figure 16.4.

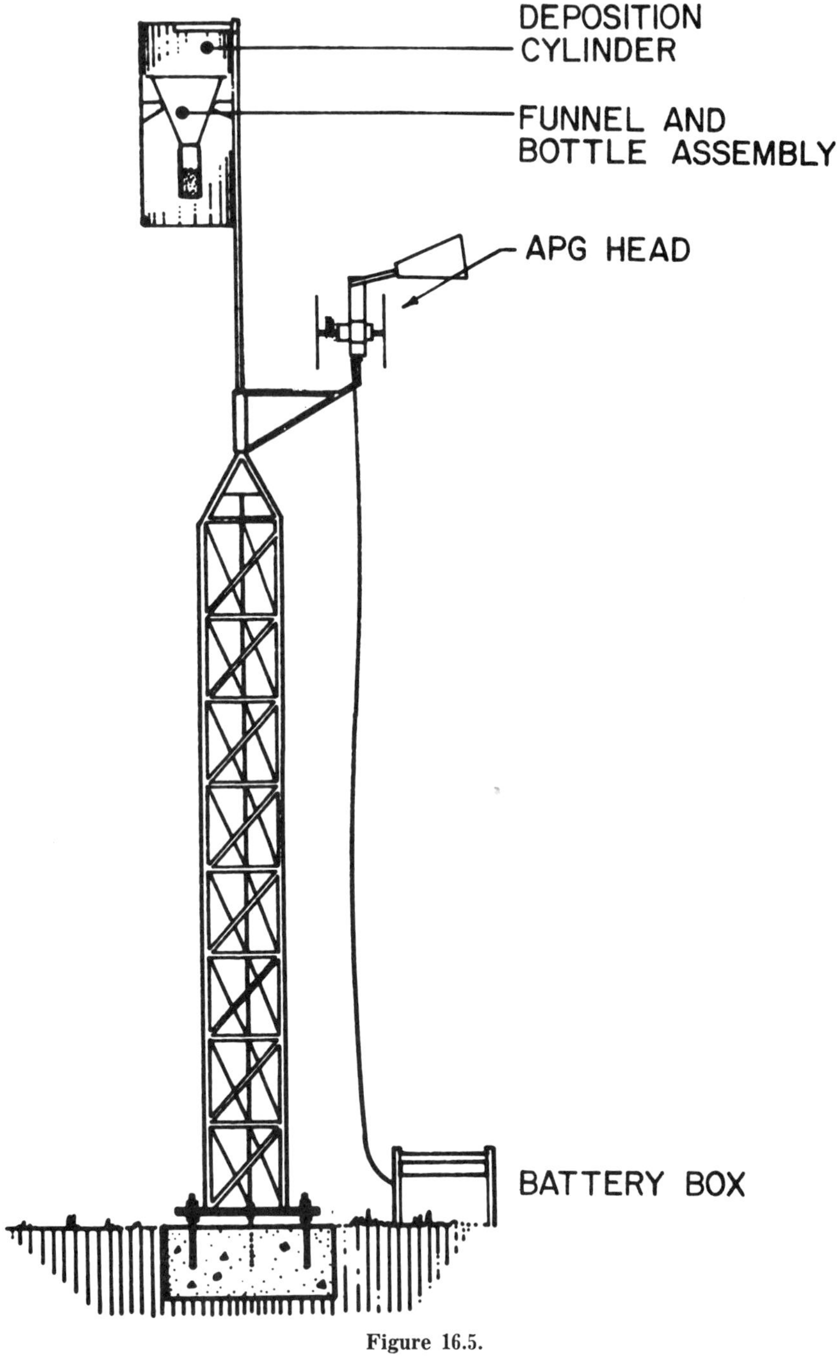
DEPOSITION CYLINDER
FUNNEL AND BOTTLE ASSEMBLY
APG HEAD
BATTERY BOX

Figure 16.5.

4. A wind vane is used to keep the bearing-mounted APS head in a direction such that the fan induced air and ambient wind vector are collinear.
5. Finally, a rotation counter permits accurate measurement of the total number of revolutions made during a testing period, thereby ensuring accuracy of the sampling volume of air.

The APS has been compared with other instruments and techniques (*e.g.*, single strand techniques, a high volume sampler, a cascade impactor, and an isokinetic sampler). Detailed results of some of these tests along with precision tests run between two simultaneously operated APS stations have been published.[6,7] In summary, the results of these comparative tests reveal the APS to be a viable collection device for pre- and postoperational cooling tower impact studies. Nearly three years of experience with the APS in (preoperational) ambient sea salt measurements have shown the combined characteristics of high sampling rate (≈ 100 m^3/hr) and low chemical background levels (*e.g.*, ∽1 μgNa) on the collection elements to be a very suitable and unique combination for this application.

EXEMPLARY COOLING TOWER MONITORING PROGRAM

The Chalk Point Cooling Tower Project[8] (CPCTP) is a research and monitoring endeavor to fully characterize the effluent plume of a large natural draft cooling tower. This continuous monitoring effort will provide data allowing the environmental impact of hyperbolic cooling towers to be determined. These data will provide information both with regard to phenomena such as fogging, precipitation enhancement, and downwind icing, and also with regard to possible effects on vegetation by the deposition of salts due to the brackish or saline water drift.

Power Plant and Cooling Tower Descriptions

This program will be carried out on a large natural draft cooling tower located in Chalk Point, Maryland, which represents a typical cooling tower installation. This tower is used to cool the circulating water of a 630 MW electric oil fired generator owned by the Potomac Electric Power Company. The cooling tower is 40 feet in height and 350 feet in base diameter. The throat diameter is 162 feet and is the location for most of the measurements, about 300 feet above the base. Design operation conditions for the Chalk Point tower are presented in Table 16.1.

Measurements and Test Plan

Measurements will be carried out within the tower shell to determine the drift characteristics and to relate these to tower operational parameters. As previously mentioned, the tower drift characteristics will be monitored with the PILLS, Sensitive Paper and Isokinetic Sampling devices. Simultaneous measurements will be carried out on the cooling tower water circuit and tower effluent parameters. That is, inlet and outlet circulating water flow rates, together with inlet and outlet water temperatures will be monitored along with tower effluent updraft velocity and temperature. These aforementioned parameters will be continuously recorded while the plant generated load and firing rate will be periodically acquired as needed. This comprehensive set of measurements will permit correlation of drift emissions with all pertinent operational parameters. Table 16.2 summarizes the measured parameters and the instruments with which they will be acquired.

Table 16.1. Tower Design Operating Conditions.

Circulating water flow rate	260,000 gpm
Warm water (inlet) temperature	120°F
Cooled water (outlet) temperature	90°F
Range (Inlet-Outlet)	30°F
Wet bulb temperature	78°F
Relative humidity	50%

The instrumentation package which includes the drift and tower effluent sensors will be suspended from two winch operated steel rigging cables enabling horizontal traverses along either of two diameters. Additionally, a walkway has been constructed below each set of pulleys for ready access to the lowered instrument package. This rigging concept permits a single operator to service the equipment as well as to traverse the chosen diameter.

Indicators of cable lengths, which will be calibrated when the tower is not in operation, will reveal the position of the instrument package without the necessity of making a visual inspection.

Initially, tests will be carried out to determine the positional dependence of the drift characteristics by making drift and related measurements along vertical and horizontal traverses. From these tests we will determine whether a symmetrical relationship exists between the two diameters and we will determine a representative altitude at which to operate. It is anticipated that initial results will show that measurements should be carried out at the throat of the tower.

After the initial complete characterization of the tower has been carried out, the testing schedule will be divided into intensive and continuous testing phases for a duration of at least one year. Intensive testing periods will consist of one week tests, one for each season of the year, and will utilize all the aforementioned drift and

Table 16.2. Parameters Monitored in the Chalk Point Program.

A. Drift	PILLS II-A ($50 < d < 1000$ μm)
	Sensitive Paper ($20 < d < 1000$ μm)
	Isokinetic Sampler (mass mineral concentration, χ)
B. Tower water circuit	Inlet, outlet, makeup water temperature (platinum RTD's)
	Circulating water rate, makeup water rate, blowdown rate
	Circulating water conductivity
C. Tower effluent other than drift	Updraft speed (Gill Propeller Anemometer)
	Wet bulb temperature (thermilinear element)
	Dry bulb temperature (thermilinear element)
D. Power plant parameters	Generation rate (MW)
	Firing rate

tower effluent measurement equipment. These tests will consist of a series of tower effluent measurements along a horizontal traverse at the altitude determined from the initial characterization phase.

In the intervening periods, a technician will conduct the continuous testing phase. These tests will be carried out at a representative altitude and will utilize all instrumentation with the exception of the IK and SP which are personnel intensive devices. The intensive tests will provide a complete data base for each of the four seasons while the continuous tests will provide data for virtually all tower operational conditions.

The Chalk Point Data Acquisition System

All continuously monitored sensors will be hardwired to the Chalk Point automated data acquisition system. Figure 16.6 shows the essential components. Analog data are transmitted via shielded cables directly to the buffering and signal conditioning electronics. The system 256 analog-digital (A/D) converter system translates these data lines from a parallel to digital serial format. Finally, an input-output (I/O) interface ensures compatibility between the A/D converter and the Texas Instruments (TI) Model 980 A minicomputer.

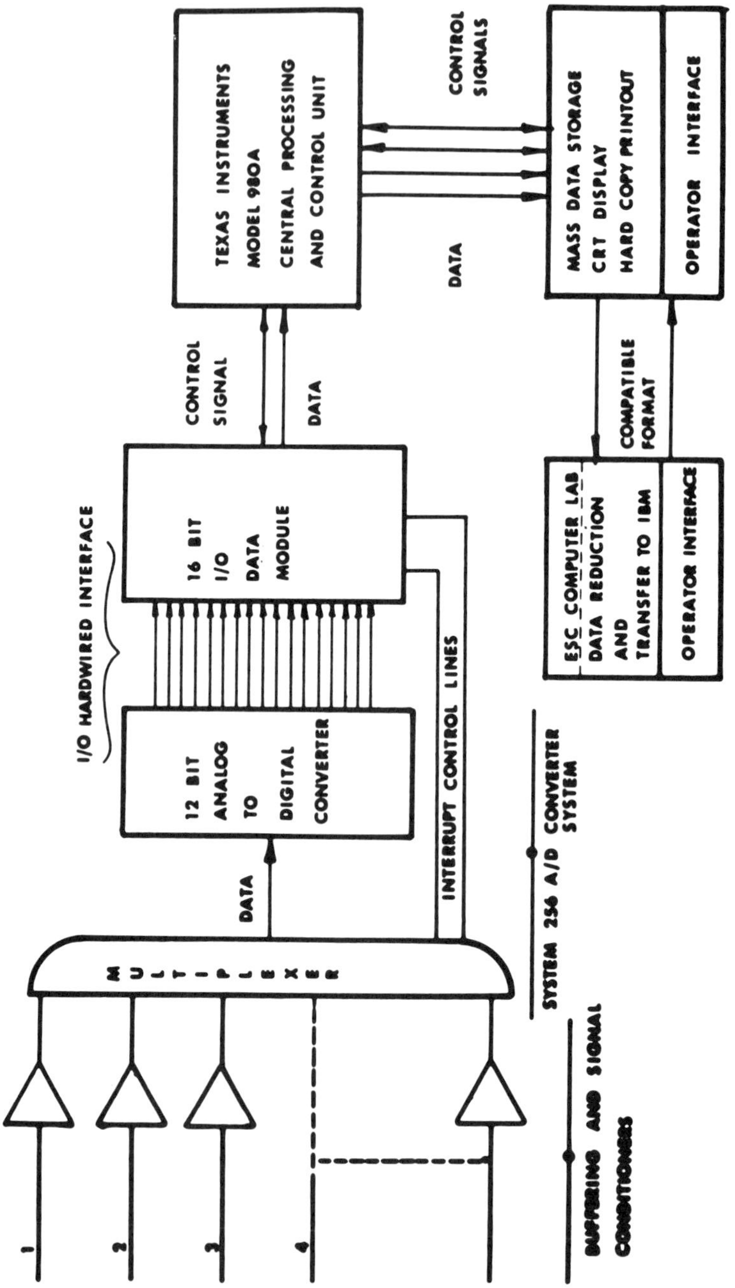

Figure 16.6. Chalk Point data acquisition system schematic.

The computer is programmed to provide preliminary data reduction via the software package developed specifically for this test program. The software controls the logging of the data from the various electronic sensor signals including the PILLS II-A System, tower effluent sensors and water circuit parameter sensors. Finally, the software controls the display of these data hourly, transfers the data to cassette magnetic tape and prints a hard copy of the PILLS voltage histogram upon operator command.

The software package was developed in "assembly language" and has the capability of sampling the various electronic sensors at different rates. For example, PILLS II-A is sampled at the rate of 300 Hz while the tower effluent and water circuit parameters are sampled once every 30 seconds since they vary slowly with time.

For each experimental run, the PILLS voltage histograms and computed 10-minute averages of the remaining continuous parameters are stored on magnetic cassette tape. An additional feature is that data from any sensor can be either displayed on an oscilloscope or can be printed on the teletype on call from the operator. The software and hardware comprising this system have been designed for ease of modification. For example, the system can interface up to 256 sensors and sample up to a rate of 50 kHz with minimal additional hardware cost and programming effort.

The Chalk Point data acquisition system provides automated data recording, preliminary data reduction, and formatting and storage of the raw data via the cassette magnetic tapes. ESC will further reduce these raw data at its home office computer facility and transfer the pertinent data to punched cards. Finally, these data will be made available to other program participants for use in their plume modeling efforts.

Ambient Salt Monitoring Program

The salt sampling program at Chalk Point, Maryland, is carried out at six stations located one, three, and six miles from the cooling towers at the power plant, in directions North, East, South and West. Each station location is indicated by the digit and the letter in the station name; thus station 1N is located one mile north of the cooling towers. The designations of the other five stations are 3N, 1E, 3E, 3S, and 6W. A map showing the locations of the power plant and the six salt sampling stations is presented in Figure 16.7.

Inspection of the ambient salt concentration levels measured thus far in the Chalk Point environs reveals an almost uniform level of airborne salts from natural sources independent of station location and prevailing meteorological conditions. Inland environments are expected to exhibit this uniform concentration of ambient salt concentrations. Measurements will continue for at least one year after

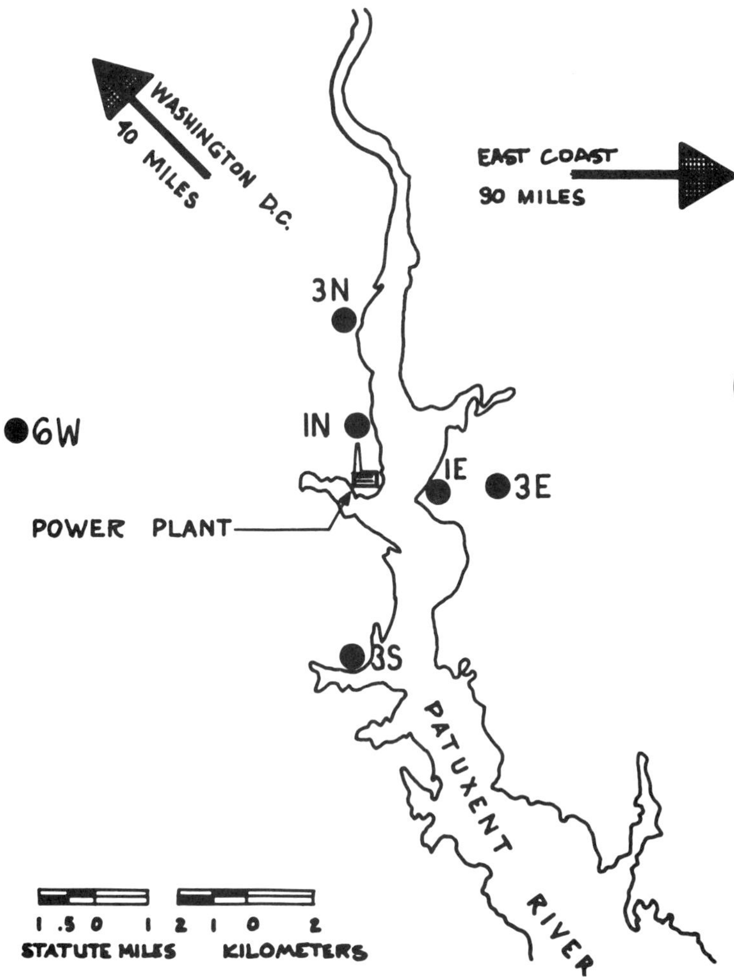

Figure 16.7. Location of APS Stations.

the plant is operational to determine possible salt concentrations contributed by the cooling tower.

SUMMARY AND CONCLUSIONS

1. Reliable drift measurement methods have been developed over the past four years. At the source, PILLS and SP techniques can be used with accurate results for determination of the droplet density distributions. The IK System can be reliably employed to

measure mineral mass flux attributed to drift effluent from a cooling tower.

2. For ambient monitoring as related to source contributions, the APS has been shown to be an accurate, field reliable device for characterization of ground level mineral concentrations both in pre- and postoperational cooling tower studies. The combination of low chemical background, easily handled collection media and its flexibility of placement (due to its battery operation) make the device suitable in these studies.
3. Drift measurement technology can be applied in a comprehensive and coordinated effort to assess the possible environmental effects of a salt or brackish water cooling tower. The Chalk Point Cooling Tower Project is an example of such an application. Here, tower effluent characteristics, tower water circuit and power plant parameters, as well as local meteorology are being monitored to assess the immediate local environmental effects of the natural draft tower of the Chalk Point site.

REFERENCES

1. Shofner, F. M., G. O. Schrecker, T. B. Carlson, and R. O. Webb. "Measurement and Interpretation of Drift Particle Characteristics," Presented at the "Cooling Tower Environment — 1974" ERDA Symposium Series, CONF-740302.
2. Wilber, K. R., F. M. Shofner, and M. J. Margetts. "Measurement and Characterization of Cooling Tower Drift," presented at the ASME Winter Annual Meeting, New York (November 17-22, 1974).
3. Wedding, J. B., R. W. Carlson, J. J. Stukel, and F. A. Bazzaz. "Aerosol Deposition on Plant Leaves," *Environ. Sci. Technol.* **9**, No. 2, 151-153 (1975).
4. Golovin, M. N. and A. A. Putnam. "Inertial Impaction on Single Elements," *Ind. Eng. Chem. Fund.* **1**, 264 (1962).
5. Langmuir, I. and K. Blodgett. "Mathematical Investigation of Water Droplet Trajectories," General Electric Research Laboratory Report, RL-225 (1945).
6. Schrecker, G. O., K. R. Wilber, and F. M. Schofner. "Prediction and Measurement of Airborne Particulate Concentrations from Cooling Device Sources and in the Ambient Atmosphere," Presented at the "Cooling Tower Environment—1974" ERDA Symposium Series, CONF-740302.
7. Wilber, K. R. "An Experimental Approach to the Evaluation of Mesh Collection Efficiencies Using the Mechanism of Inertial Impaction," Master's Thesis, University of Tennessee (June, 1974).
8. Carlson, T. B., J. Pell, and F. M. Shofner. "Case Study of the Environmental Effects of the Chalk Point Natural Draft Cooling Tower," Presented at the Cooling Tower Institute, Houston, Texas (February 11, 1975).

17

In Situ, Continuous Measurement of Particle Size Distribution and Mass Concentration Using Electrooptical Instrumentation

F. M. Shofner, G. Kreikebaum, H. W. Schmitt and **B. E. Barnhart**
Environmental Systems Corporation, 1212 Pierce Parkway,
Knoxville, Tennessee

INTRODUCTION

Accurate, reliable, convenient measurement of particles entrained in process streams constitutes a general and fundamentally important problem. Particle size distribution has often been ignored in the design and operation of large-scale processes, principally because of the general unavailability of suitable monitoring equipment. The unavailability of *in situ*, continuous monitoring equipment, in turn, has its cause in the difficulty of making the measurements in the hostile environments often associated with large process streams.

Four examples will help to illustrate the importance of particle size distribution monitoring. The first concerns design optimization of particulate control equipment (such as electrostatic precipitators or fabric filter "baghouses") servicing stationary sources (such as electric power generating plants). It is well known that the collection efficiency of all control equipment is sensitive to particle size; yet, most current specifications (including regulations) of control equipment performance are in terms of mass collection efficiency. Only recently has the performance in terms of fractional efficiency—collection efficiency as a function of particle size—begun to receive serious attention with regard to measurements on operating sources[1] and with regard to the design of new systems.[2] Performance versus control equipment size/cost considerations are highly sensitive to particle size distribution used for design.

It is clear how continuous monitoring can enhance understanding of the operation of control equipment and lead to better, more cost-effective design. But further, as in the second example, the operational parameters of the control equipment may be adjusted to yield optimum performance (*e.g.*, maximum mass collection efficiency, maximum equipment reliability, minimum energy consumption). Such adjustments could be made, automatically and continuously, on the basis of continuous particle monitors at the input and output of the control equipment.

Third, the performance of a process itself is, in some cases, dedetermined by the size distribution of the process stream particles. For example, the combustion efficiency of coal and fuel oil depends very sensitively upon the combustible particle size distribution. As in the first and second examples, it is desirable to continuously measure input particle size distribution for optimization of both design and operations.

As a fourth example, for advanced energy production techniques such as pressurized, fluidized bed, coal gasification, and magnetohydrodynamic processes, the presence of particles in the process stream can cause serious degradation of the energy conversion equipment, *i.e.*, to the turbines, heat exchangers, or MHD channels. The nature of such damage and the rate at which it occurs are functions of the particle size distribution, which should be monitored continuously. Even in conventional steam turbine energy conversion equipment, a similar problem exists with condensation droplets. Thus we see that *in situ*, continuous measurement of process particles is of fundamental importance for both design and operational optimization.

The advantages of *in situ* measurements are self-evident. In most cases, it is extremely difficult or impossible to withdraw a sample of the process stream particles without modifying the particle size distribution or the physical and chemical characteristics of the particles. McKenna and Weisberg report measurements on a spreader stoker boiler, in which they obtained a median diameter of roughly 8 μm in an *in situ* measurement as opposed to 40 μm from a "grab sample" followed by laboratory analysis.[3]

The advantages of continuous measurements are equally self-evident in terms of quantity of information available. An important consideration is whether such measurements cost or save money; in many cases, the latter is clearly demonstrated.

This chapter presents an overview of the PILLS family of electro-optic instrumentation, including a discussion of the basic physical parameters measured and interpretation of the data generated. A brief description of the PILLS IV and V instruments as applied to "fine particles" monitoring and data from the PILLS IV and V Systems is provided, and finally, some brief examples of the practical economic benefits of such data are given.

THE PILLS FAMILY

Principles of Operation

PILLS is an acronym derived from the basic principle of operation: *P*articulate *I*nstrumentation by *L*aser *L*ight *S*cattering. Referring to Figure 17.1, where the major elements of the system are identified, it is seen that, a beam of laser light is directed into the process medium containing the particles. The particles interact with the electromagnetic radiation and scatter light out of the beam in all directions; the intensity of the scattered light is directly related to the intensity of the original beam, the effective size of the particle(s), and the surface characteristics of the particle(s). For particles that are spherical, the scattering can be described by the well known but complex theory of Mie for arbitrary particle diameter d, wavelength of coherent radiation λ, and complex index of refraction $m = m_1 - jm_2$. A selected part of the scattered radiation is collected by an optical system and is focused upon a detector, which converts the photon flux into a flow of current. This electrical signal is then conditioned and presented to a conventional electronic readout device.

All of the present PILLS family of instruments measure single particles, except PILLS V, which responds to the scattered light from a large number of particles in its sampling volume. For a single par-

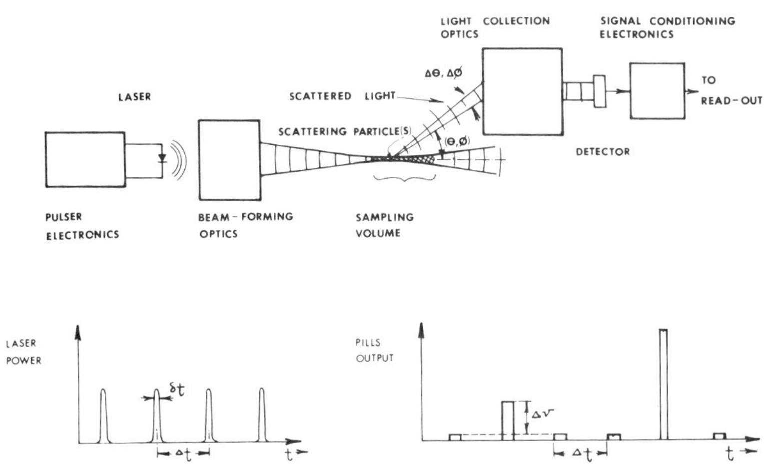

Figure 17.1. PILLS electrooptic block diagram.

ticle measurement, *e.g.*, with PILLS IV, the sampling volume must be such that the probability of occupancy by a particle in the size range of interest is small compared to unity. The sampling volume is periodically illuminated by bursts of laser power, as indicated in Figure 17.1. If there is a particle in the sampling volume at the instant of a laser pulse, the scattered light is collected, detected, and converted into an output signal, also shown in the figure. The voltage Δv (above a residual background voltage, caused either by stray light from the incident beam or by many smaller particles outside the size range of interest to the measurement) is related through calibration of the size of the particle. This voltage signal is stored as a count in the appropriate channel of a multi-channel analyzer or minicomputer. The recorded voltage distribution corresponds to the size distribution and concentration of the particles present and the particle size distribution and concentration may be obtained by applying instrument calibration data to the measured voltage distributions.

By proper electrooptical design, a wide range of particle sizes and concentrations may be measured by the light scattering technique. The PILLS technology is primarily based on the semiconductor, junction diode laser, which operates at $\lambda \approx 0.9\ \mu m$. This device yields intense bursts of optical power; it is extremely small in physical size, highly reliable, and among the most efficient of laser devices in conversion of electrical to optical power. The utilization of this device and other semiconductor electronics, including detectors, results in a compact and reliable electrooptical system suitable for measuring particle characteristics over a very wide range of sizes.

Given the selection of a semiconductor device operating at $\lambda = 0.9\ \mu m$ and semiconductor electronics, the balance of the design of PILLS instruments are concerned with the size of the sampling volume, its definition (whether by completely optical or by optical/mechanical means), and the angles and angular ranges for which scattered light is collected. In Figure 17.1, the polar angle θ is usually a few degrees for forward scatter designs and nearly 180° for backscatter designs. The azimuthal angular range $\Delta\phi$ is a few degrees for an off-axis design and essentially 360° for a so-called inline design.

For the PILLS V instrument, a mass concentration monitor, the sampling volume is chosen such that many particles are in it at all times. It follows that the pulses of radiation falling on this multiplicity of particles will produce an integrated scattering yield that is, for realistic particle concentrations, essentially the sum of individual scattering yields from each of the particles within the sampling volume. Accordingly, rather than a random distribution of voltage pulse heights, the PILLS V output is a constant voltage, as long as the total mass concentration within the sampling volume remains constant.

Representative Members of the PILLS Family

Table 17.1 enumerates the various members of the PILLS family, giving a brief, general description of each instrument type, along with typical system parameters, and exemplary applications.

Historically, the first major application of the PILLS technology was to the problem of measuring liquid aerosol droplets entrained in the air flows through evaporative cooling systems used for the rejection of waste heat from power plants. But one of the major unknowns surrounding cooling towers was the emission rate of "drift" droplets entrained in the air flows of these devices. These droplets, to first order, contain the same mineral concentration as the water being circulated and cooled by the tower. For coastal installations, this concentration is approximately twice the salt water concentration used for makeup. Thus, if appreciable quantities of these droplets are emitted, severe environmental effects could result. Data on the drift emission characteristics of these large cooling systems have been widely published.[4–7]

Figure 17.2 shows a PILLS II-A unit along with other instrumentation. The three sampling volumes for this instrument are located between the two large coaxial cylinders. This instrument will be used in the Chalk Point Cooling Tower Program where it will monitor drift emissions continuously for a year or more.[7] A similar instrument has generated data as part of the Turkey Point Program.[8]

Definition and Interpretation of Parameters Measured

The single particle instruments of the PILLS family basically produce a voltage pulse height histogram as recorded in the channels of a pulse height analyzer. As noted above, the number of counts in given voltage ranges are converted to the number of particles in given size ranges using the instrument calibration data determined by laboratory or field measurements with monodisperse particles. The transverse dimension d_t of the sampling volume and the gas velocity parallel to this dimension obey the inequality

$$u_{11}\,\Delta t > d_t \qquad (1)$$

No particle is counted twice and it follows that not only the particle size distribution is measured but the concentration of particles in a given size range as well. The total volume of process gas sampled is derived by recording the number of times the "PILL-BOX" sampling volume was sampled during a run.

Thus the PILLS single particle instruments measure the number

Table 17.1. The PILLS Family *In Situ*, Continuous, *P*articulate *I*nstrumentation by *L*aser *L*ight *S*cattering Systems, Environmental Systems Corporation.

		Typical System Parameters[a]			
Member	*General Description and Applications*	*Size Range* d'-d'' (μm)	*Sampling Volume* V (cm^3)	*Maximum Gas Temperature* T_{max} (°F)	*Maximum Pressure Differential*[d] Δp (in. H_2O)
I[b] II[b] II-A II-B	Forward scatter instruments for sizing liquid aerosols, especially droplets from cooling towers and spray nozzles. I, II, and II-A have 1, 2, and 3 independent sampling volumes. I and II off-axis. II-A combination of two off-axis and one in-line channels.	50–1000	2	130	30
III	Forward scatter, in-line. Measures atmospheric giant aerosols; fog; fine mists.	5–100	0.3	130	30
IV	Forward scatter; ratio of two in-line optical channels. Pollution control equipment performance evaluation; process monitoring.	0.2–3.0	2×10^{-7}	400	5
V	Mass concentration monitor for combustion, melting, grinding, etc., processes.	0.1–10 [c]	12	400	5

[a]The representative values given indicate range of applicability and actual design values for specific field instruments; these values should not be regarded as limits to the ranges of applicability.
[b]R&D instruments built to support ESC field measurement services.
[c]Range of usable response; this instrument yields the integrated response from many particles throughout its sampling volume.
[d]$\Delta p = p_{duct} - p_{atmos}$.

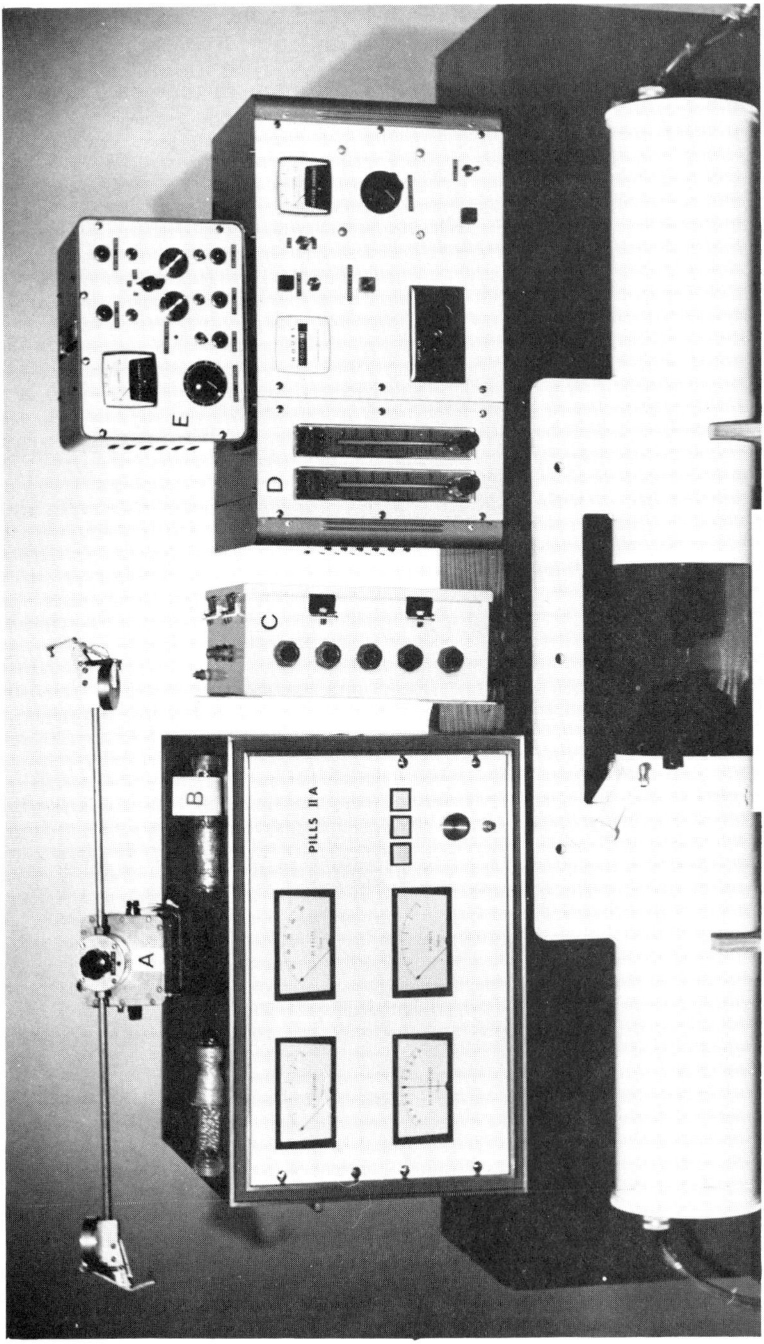

Figure 17.2. PILLS II-A system.

of particles in given size ranges, per unit volume of the process stream. Mathematically, we may write this as

$$\frac{\Delta C_i}{\Delta d_i} = \frac{\Delta N_i}{\Delta d_i V_s} = \frac{\text{number of particles of mean diameter } d_i}{\text{size range } (\Delta d_i, \mu m) \text{ x volume } (cm^3)}, \quad (2)$$

where the subscript i indicates the particle size range (d_i, $d_i + \Delta d_i$), ΔN_i is the number of particles in the size range Δd_i, V_s is the total volume sampled, and $\Delta C_i = \Delta N_i / V_s$ is the concentration of particles in the size range Δd_i.

The total mass concentration is derived from the particle density distribution according to

$$\chi = \frac{\Sigma}{i} \Delta\chi_i = \frac{\Sigma \Delta\chi_i}{i \ \Delta d} \cdot \Delta d_i$$

$$\sum_i \frac{\pi}{6} \bar{d}_i^3 \rho_i \frac{\Delta N_i}{\Delta d \cdot V_s} \cdot \Delta d_i \ [g/cm^3] \quad (3)$$

where ρ_i is the specific mass of the particles and d_i is the mean diameter of the particles in the range Δd_i.

The total mass emission crossing a surface A is given by

$$D = \sum_j \frac{\Delta D_j}{\Delta A} \cdot \Delta A_j = \sum_j \Delta A_j \left[\sum_i \frac{\pi}{6} \rho_i \ d_i^{-3} \ u_{ij} \left(\frac{\Delta N_{ij}}{d \cdot V_s}\right) \Delta d_i\right] [g/sec] \quad (4)$$

where ΔA_j represents appropriately chosen segments of the total surface area $A = \sum_j \Delta A_j$ across which the particles are transported, u_{ij} is the absolute i^{th} particle speed associated with the jth measurement station and is the velocity component normal to the efflux plane over which the measurements are made. Note that if it is assumed $u_{ij} = u_j$, the air speed normal to the measurement plane, then the total mass emission becomes

$$D = \sum_j u_j \ \chi_j \ A_j \qquad [g/sec] \quad (5)$$

This approximation is very good for fine particulates.

Whereas the single particle counters fundamentally yield the particle size distribution, the PILLS V instrument yields the mass concentration χ_j directly. Of course, it provides no information on particle size.

PILLS IV

Figure 17.3 is a block diagram of the PILLS IV System which is a forward scatter, ratio device. A single particle within the sampling volume at the instant of a laser pulse scatters light in all directions. Parts of the scattered light pattern are collected by an optical system and brought to a detector. Radiation from two optical channels, defined by central angles θ_1 and θ_2 are simultaneously detected, amplified, and conditioned and then input to a ratio circuit. The input from the θ_0 or main beam is used for synchronization purposes. The ratio of the two optical signals has several distinct advantages for particle sizing, especially for the highly important fine particulate size range from a few tenths to a few micrometers. These advantages are: (a) sizing independent of particle index of refraction; (b) sizing independent of instantaneous laser power; (c) sizing independent of position within the sampling volume; and (d) generation of a monotonic response in this size range. These advantages are all first order effects and have been investigated by Hodkinson,[9] Gravatt,[10] Carlson,[11] and many others. The electrooptic design permits complete optical definition of a sampling volume as small as $\approx 10^{-7}$ cm^3. Further, this system permits sizing particles as small as 0.2 μm.

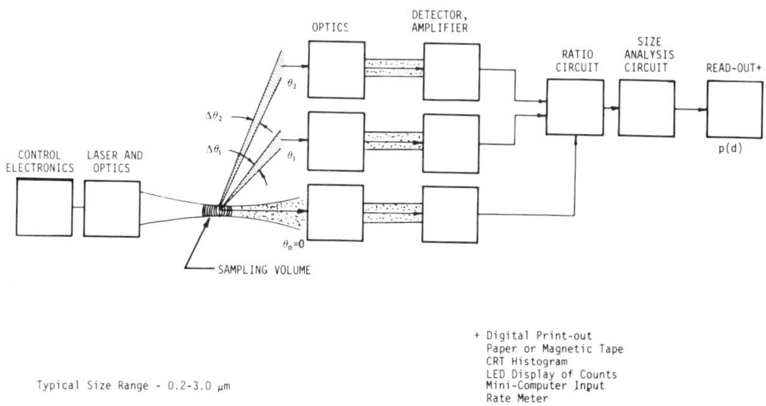

Figure 17.3. PILLS IV block diagram.

As seen in Table 17.1, the design size range for the PILLS IV instrument is 0.2–3.0 μm. In many applications, the bulk of mass emission is contained within this range. Also, for electrostatic precipitators, it appears that there is a minimum in the fractional efficiency in this size range. Thus the availability of an *in situ*, continuous monitor, which covers this important size range, will lead to better understanding and, following that, improved performance of the control equipment.

Figure 17.4 is a photograph of a PILLS IV unit which will be utilized by the Environmental Protection Agency to evaluate the fractional efficiencies of control equipment. The probe tube is 5 feet in length (from the center of the sampling volume to the support pack-

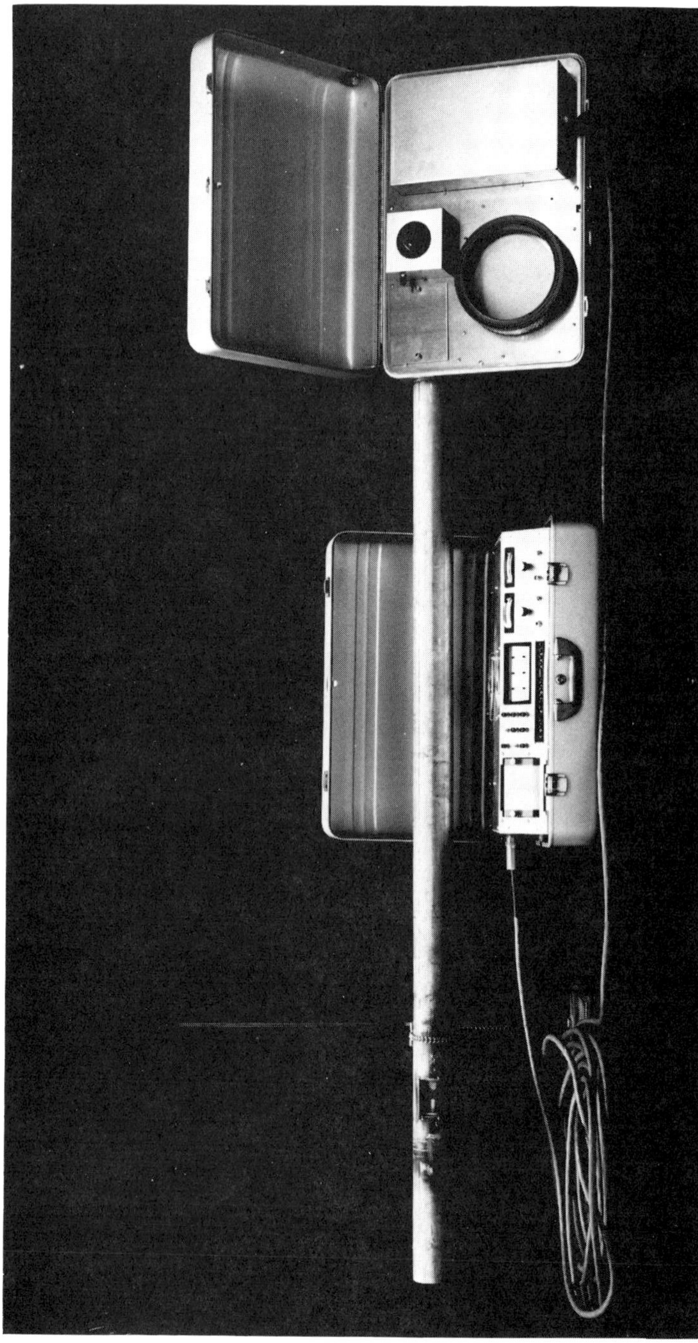

Figure 17.4. PILLS IV system.

age). The sampling volume is in the slotted section of the probe tube, midway between the conical snouts. The instrument system is completely self-contained, requiring only AC power at the site. The support package contains a filter/blower system for supplying purge/ cool air down the insulated probe tube to the sensor head.

An umbilical cable connects the sensor head/support package to the control/readout package, shown in the middle of the photograph. This units contains a pulse height analyzer with arbitrarily settable voltage limits, permitting 32-channel resolution of the voltage height histograms, to be recorded. These data are printed on a paper hard copy in the field. This permits a rapid evaluation of system performance and yields a written record of that performance. Additionally, a rate indicator circuit has been developed, which has provisions of meter and chart record readout.

The sampling volume location within the duct can be traversed to evaluate the homogeneity of the process stream or to evaluate "hot spots" found in the stream. For example, gas sneakage paths in precipitators can be quickly located and their severity evaluated. Faulty bags in a fabric filter may be identifiied similarly.

Figure 17.5 shows exemplary data obtained with a PILLS IV unit early in its development. These data are from the output of an electrostatic precipitator serving a coal-fired power plant.

PILLS V

The PILLS V System is a mass concentration monitor. It represents an extension of single particle sizing technology and provides a direct measure of the total mass emissions. More specifically, since for fine particulates $u_{ij} = u_j$, knowledge of the mass concentration and the air velocity in a process duct lead to the total mass emission rate. Of course, this information could be generated from particle size distribution data but the equipment is more complex and requires nontrivial computations for automatic readout.

Figure 17.6 is a photograph of a PILLS V instrument from an early production run. As with PILLS IV, the instrument contains a filter and blower system for supplying the purge/cool air. Thus only AC power is required. The filter is noted at the lower right-hand corner of the support box.

Exemplary data from the PILLS V System is shown in Figure 17.7. This is a chart record giving the mass concentration at the output of an electrostatic precipitator serving a coal-fired power plant. Note particularly the rapping puffs, indicating the temporal resolution of the instrument. In this particular experiment, significant variations in mass concentration were found as a function of position within the duct.

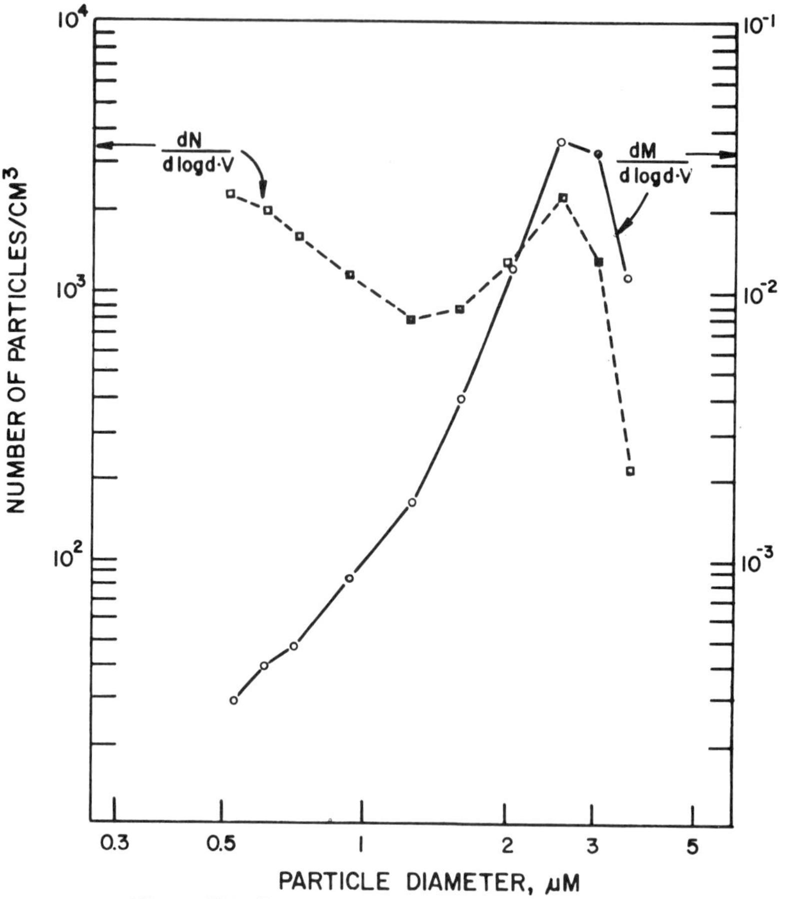

Figure 17.5. Exemplary particle size and mass distribution.

REFERENCES

1. McCain, J. D., T. M. Cushing, and W. B. Smith. "Measurement of the Fractional Efficiency of Pollution Control Devices," Presented at the 67th Air Pollution Control Association Meeting, Denver, Colorado (June, 1974).
2. Nichols, G. B. "Significance of Particle Size and Resistivity Testing for Precipitator Sizing," *Proc. of the 4th Annual Ind. Air Poll. Cont. Conf.*, Knoxville, Tennessee (March, 1974).
3. McKenna, J. D. and R. Weisberg. "A Pilot Scale Investigation of Fabric Filtration as Applied to Coal Fired Industrial Boilers," *Proc. of the 4th Annual Ind. Air Poll. Cont. Conf.*, Knoxville, Tennessee (March, 1974).
4. Shofner, F. M. "Development and Demonstration of Low-Level Drift Instrumentation," EPA Project No. 1610 GNK 10/71, Final Report to the Environmental Protection Agency (October, 1971).

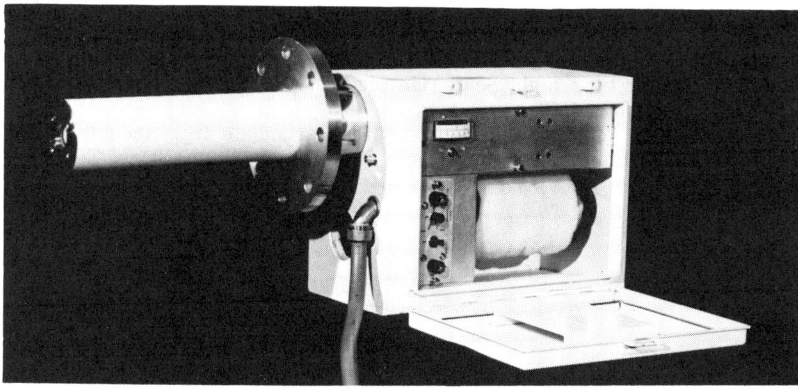

Figure 17.6. PILLS V system.

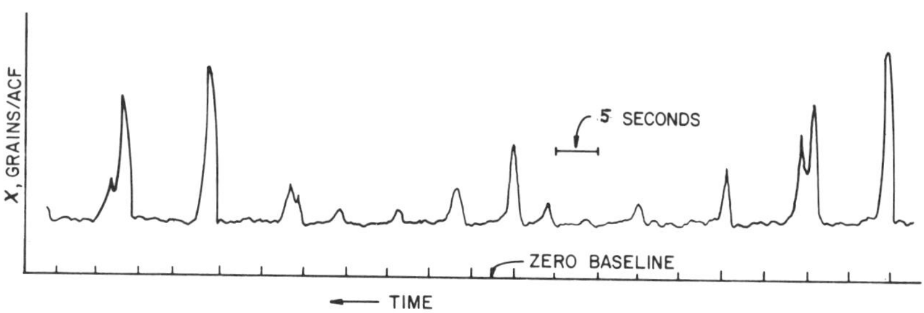

Figure 17.7. Exemplary PILLS V chart.

5. Shofner, F. M., G. O. Schrecker, T. B. Carlson, and R. O. Webb. "Measurement and Interpretation of Drift Particle Characteristics," Presented at the "Cooling Tower Environment — 1974" Symposium, University of Maryland, March 4-6, 1974 and published in the *Proc. of the Cooling Tower Environment Symp.*, University of Maryland (March 4-6, 1974).
6. Margetts, M. J. and F. M. Shofner. "Characterization of the Drift Emissions of A Natural Draft Cooling Tower and Examination of Sensitivity to Operational Parameter Variations," Presented at the Joint Power Generation Conference; ASME, IEEE, and ASCE, New Orleans, Louisiana (September 16-19, 1973).
7. Carlson, T. B., J. Pell, and F. M. Shofner. "Case Study of the Chalk Point Cooling Tower Project." Presented at the Cooling Tower Institute Annual Meeting, Houston, Texas (February 11, 1975).
8. Schrecker, G. O., R. O. Webb, D. O. Rutherford, and F. M. Shofner. "Drift Data Acquired on Mechanical Salt Water Cooling Devices," Draft final report to the Environmental Protection Agency, Contract No. 68-021365 (December, 1974).

9. Hodkinson, J. R. "Particle Sizing by Means of the Forward Scattering Lobe," *Appl. Optics* 5, No. 5 (May, 1966).
10. Gravatt, C. C., Jr. "Real Time Measurement of the Size Distribution of Particulate Matter by a Light Scattering Method," *J. Air Poll. Cont. Assoc.* **23**, No. 12 (December, 1973).
11. Carlson, T. B. "Design Considerations for Particulate Instrumentation Employing Laser Light Scattering," Ph.D. Dissertation, University of Tennessee (June, 1974).

18

Monitoring Industrial Sulfur Scrubbers by Flame Photometry

James D. Barden and **Daniel P. Lucero**
Meloy Laboratories, Inc., Springfield, Virginia

INTRODUCTION

One of the most important parameters in the utilization of equipment and apparatus to remove sulfur effluents from industrial waste streams is the efficiency of the process. A direct measurement of industrial scrubber efficiency is obtained by measurement and comparison of the sulfur concentration levels and mass flow rate of the scrubber input and output streams. The instrumentation commercially available to perform this task is widespread. For example, over 70 monitors utilizing 15 distinct principles of operation can be used. However, all possess operating characteristics that establish their performance limitations and define specific interface requirements with the sample stream. Many are limited by dynamic range of sample concentration response, interferent molecules in the sample, and/or sampling rate capacity. Others are confined in their response to a single sulfur species such as SO_2 and thus require special sample processing prior to analysis.

One of the more modern and direct techniques for the analysis of sulfur is by flame photometry or the flame photometric detector.[1,2] Over the past several years it has developed into a commonly used sulfur detection method.[3] The flame photometric detector (FPD) is responsive to all gas phase sulfur molecules in proportion to the number of sulfur atoms contained by the molecule.[4] For example, the FPD relative response to equal concentrations of SO_2, and CS_2 is 1 and 2 respectively. This feature permits the use of specific sulfur molecule scrubbers and sample stream diverting apparatus to obtain concentration level measurements of particular sulfur compounds in a sample.

Its operation is based on the chemiluminescence of an activated molecular sulfur species produced by a hydrogen hyperventilated diffusion flame.[5,6] As the activated sulfur species reverts to a lower energy state it emits a broad band of radiation with its maximum intensity at 3940Å. The FPD signal current output diminishes rapidly near 10 ppm as illustrated by Figure 18.1. The minimum detectable response is approximately 1 ppb. For most source and many process monitoring applications, the FPD upper concentration response limit is below the sample concentration. Therefore, the sample must be conditioned prior to analysis to ensure that sample concentrations are within the dynamic response range of the analyzer. It is equally important that sufficient air or oxygen enter the detector with the sample to maintain the flame. Both of these objectives are fulfilled by dynamically diluting the sample with clean air. Two basic dilution techniques are available to accomplish the sample conditioning: a) pneumatic dilution in which a sample stream is mixed with a clean air stream, and b) diffusion dilution where a small portion of the sample migrates through a membrane and is transported by a clean air stream to the analyzer. The methods of diffusion dilution are superior to pneumatic dilution in that a much greater degree of control over wide dilutions can be obtained. In addition, a degree of selective dilution of particular molecular species is obtained by the diffusion technique. It stems from the differences in the membrane material permeability coefficients for particular molecules. This feature is particularly important in anlyzing "S" in pure hydrocarbon streams.

SULFUR ANALYSIS SYSTEM

The sample dilution and analysis can be performed with the system depicted by Figure 18.2. It is composed of the FPD analyzer and a membrane diluter cartridge.

FPD Analyzer

An FPD analyzer is basically an FPD with appended supporting sample handling and signal conditioning subsystems.[7] Constant operating conditions are maintained at the FPD and photomultiplier tube to provide low-level noise and drift characteristics required for ppb sensitivity, and unattended long-term operation. The analyzer sample handling network is shown in Figure 18.3.

Hydrogen and sample air are continuously supplied to the analyzer at 140 and 200 ml/min respectively. The analyzer sample line is connected directly to the diluter cartridge. The pneumatic network

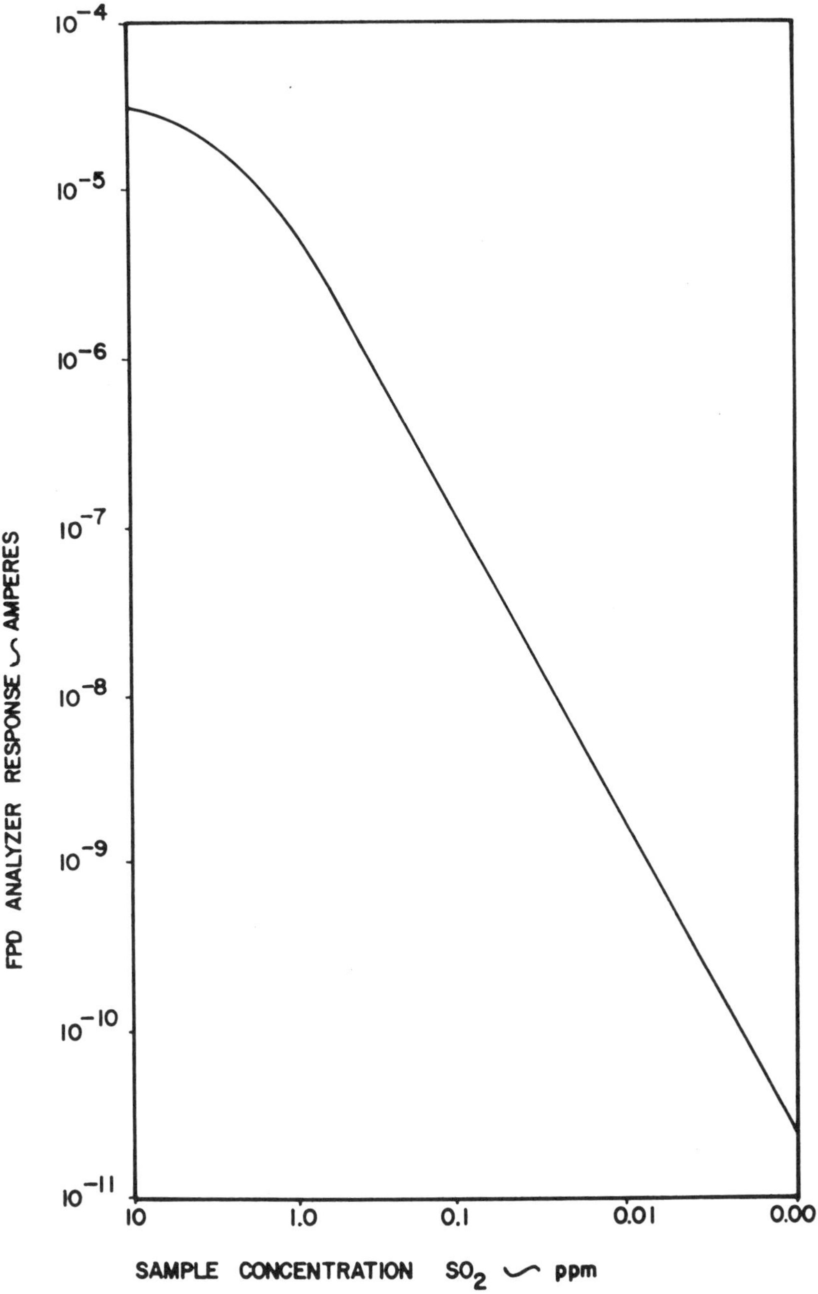

Figure 18.1. Flame photometric detector sulfur response.

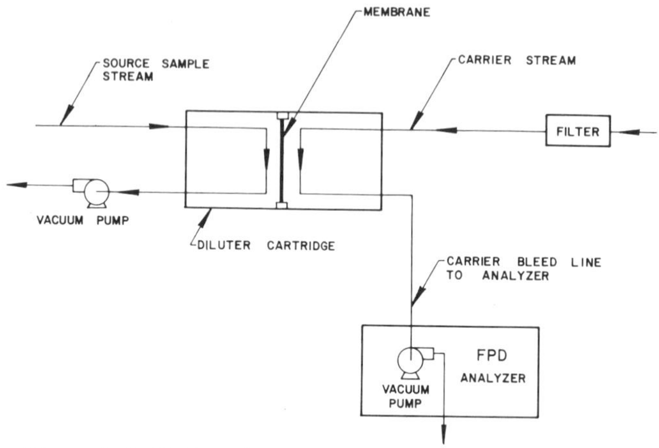

Figure 18.2. FPD analyzer and membrane diluter cartridge.

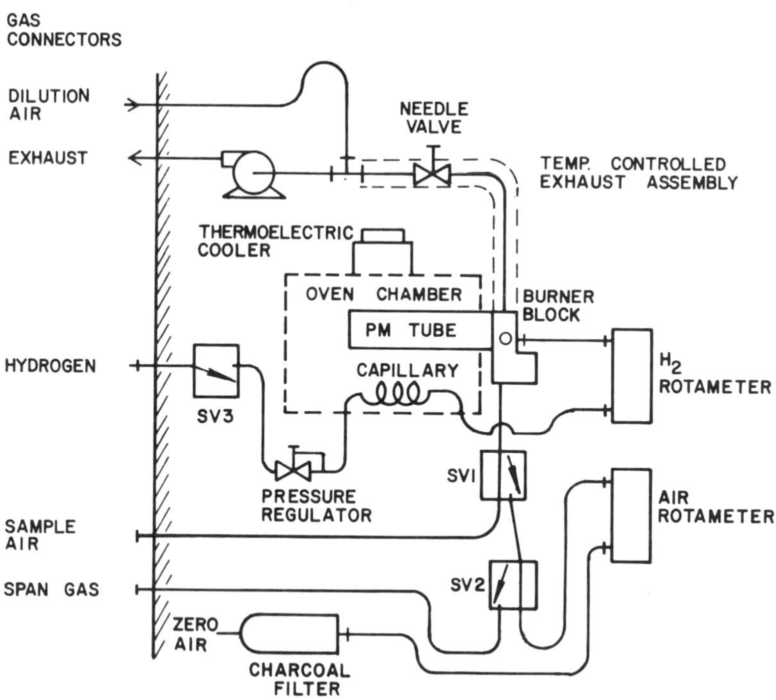

Figure 18.3. FPD analyzer sample handling network.

maintains precise control of the flow rate level of the gas streams leading to and away from the burner block. It is accomplished by applying direct controls to the hydrogen and exhaust gas streams and their attendant pneumatic elements. Since the exhaust gas flow rate is the sum of the hydrogen and sample gas flow rates, indirect control of sample flow rate is achieved by virtue of mass conservation. In addition, it is accomplished without introducing a single pneumatic element to the sample line and thus the chemical integrity of the sample stream is preserved for analysis.

Safety Features

The analyzer system also provides operational safety features. With an FPD a fire or explosion can possibly arise from two sources: a) the hydrogen gas cylinder supplying the FPD, and b) propagation of the flame from the FPD burner chamber to a surrounding explosive atmosphere. The FPD analyzer hydrogen line is equipped with a shutoff solenoid valve (SV3), as shown by Figure 18.3, which stops the hydrogen flow to the FPD in the event of a power failure or flame-out. If required, additional precautions can be introduced by utilizing an electrochemical hydrogen generator to supply the FPD burner.

The sample line carries air to the burner where it supports a hydrogen hyperventilated flame. Because of the lack of hydrogen in the burner tip, the flame cannot strike down the burner tip and out of the flame chamber. The flame cannot propagate through a leak since the flame chamber does not contain oxygen outside the boundary of the flame primary reaction zone. In the event of a leak, hydrogen will not flow out of the flame chamber; but the surrounding atmospheric gases will leak in since the flame chamber operates at a slight vacuum.

It is only during ignition of the FPD that a flame could propagate out of the flame chamber through the exhaust. The possibility arises in this condition because it is the only time that a combustible mixture of hydrogen and oxygen exists within the chamber. However, the flame cannot propagate out because the orifice diameter of the exhaust valve is much smaller than the hydrogen/oxygen flame-quenching diameter.[8] It acts as a flame arrester. During operation the exhaust gas is composed normally of 60, 160, and 80 ml of hydrogen, nitrogen, and water vapor respectively. Downstream of the exhaust valve it is mixed with 3500 ml of dilution air to yield a hydrogen concentration of 1.6% by volume. In the event the flame does not ignite, the exhaust gas ratio is 70 hydrogen and 100 air. After mixing with the exhaust dilution air the hydrogen concentration is reduced to 3.6%. In both cases the hydrogen concentration entering the vacuum pump and discharged from the analyzer exhaust is below the lower hydrogen explosion concentration limit (4.1%) in air. Thus, as a consequence of the characteristics of a diffusion flame, the vacuum operational mode of the FPD, and the design of the analyzer pneumatic network, propagation of the flame beyond the flame chamber boundary is remote.

Diluter Cartridge

A sample stream is drawn into and out of the diluter cartridge by a vacuum pump, as illustrated by Figure 18.2. It does not mix directly with the carrier gas stream entering the diluter and analyzer because they are separated by a membrane. However, a small portion of the sample stream diffuses through the membrane and into the carrier side of the cartridge. On the carrier side, the sample molecules permeating the membrane mix with the carrier stream and are transported to the analyzer.

The mass exchange across the membrane occurs by a diffusion process where the sample concentration or partial pressure difference across the membrane is the diffusion potential.[5] The SO_2 concentration level at the carrier exit is related to the SO_2 permeation flow rate through the membrane as follows: $[SO_2]_c = [Q_m/Q_c]\, 10^6$.

$[SO_2]_c$ = carrier stream SO_2 concentration level, ppm
Q_m = membrane SO_2 diffusion flowrate, ml/min
Q_c = carrier stream flowrate, ml/min

For a membrane whose lateral dimensions are much larger than its thickness, Q_m is related to the sample stream SO_2 concentration by the expression: $Q_m = \dfrac{60\, P_m A}{(\Delta X)} [P_{SO_2} S - P_{SO_2} C]$.

P_m = membrane material SO_2 permeability coefficient, cc/sec-cm^2 torr/cm
A = membrane cross-sectional diffusion area, cm^2
ΔX = membrane thickness, cm
$P_{SO_2} S$ = sample stream SO_2 partial pressure, torr
$P_{SO_2} C$ = carrier stream SO_2 partial pressure, torr

Under most operating conditions:

$$P_{SO_2} S >> P_{SO_2} C \qquad P_{SO_2} C \rightarrow 0$$

It is clear that this condition will prevail for the system of Figure 18.2 in considering that, at 100°C, a 5.6-cm diameter and 0.00254-cm thick TFE Teflon membrance will reduce a 1000 ppm SO_2 sample gas concentration to 0.56 ppm in a 200 ml/min carrier stream. The diluter is characterized by the dilution factor relating $[SO_2]_s$ and $[SO_2]_c$. The equation is dilution factor $(k_d) = \dfrac{(\Delta X)\, Q_c}{60 P_m A\, P_T\, S}$. Within practical limits, any parameter comprising k_d can be varied to obtain the dilution required.

Although it is not shown as expressed by k_d the membrane permeability coefficient varies logarithmically with temperature. Figure 18.4 shows the variation of k_d for a TFE Teflon membrane with temperature, area, and thickness.

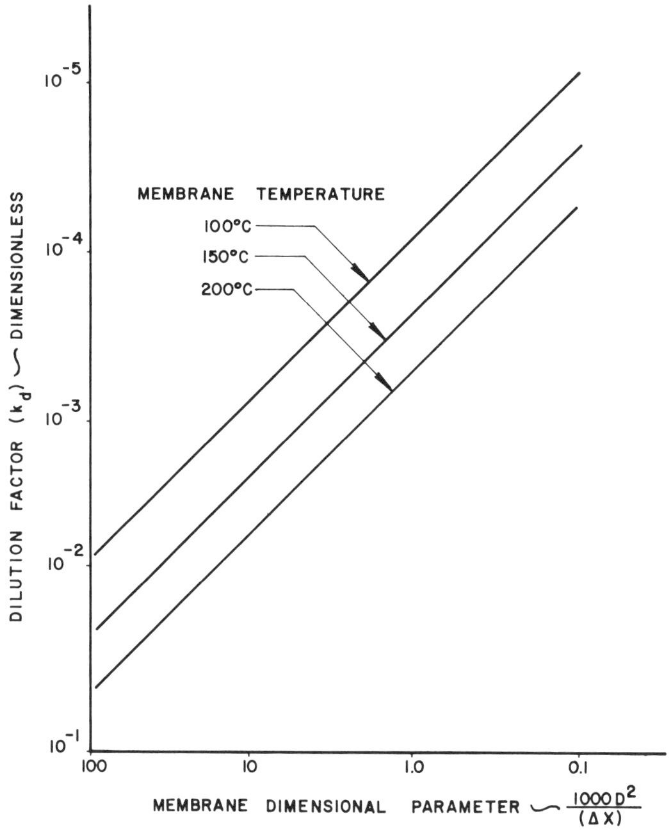

Figure 18.4. Diluter membrane dimensional parameter.

Figure 18.5 illustrates the essential elements and operation of the diluter cartridge. It functions as the membrane holder, heat exchanger and a temperature controller. The cartridge is divided into the sample and carrier sides. They are identical in all respects. Sample dilution occurs at the carrier gas membrane surface located in the center of the cartridge. Two identical open right circular cylinders with sealing flanges at each end are mated together to hold the membrane. Two additional cylinders closed at one end with a flange at the other end are inserted into the outer cylinder. Each inner cylinder is provided with gas inlet and outlet ports and a sealed electrical feed-through at the closed end. The feed-through supports a thermistor and a thermocouple junction suspended in the air stream near the membrane for temperature control and measurement. Heating elements are wrapped on the outer cylinder's external surface to provide power for the carrier air and sample streams. The internal surface of the outer cylinder is the heating surface or heat exchanger. As the gas streams enter at each end of the cartridge, they are distributed circumferentially and flow axially through the annular gap between the inner and outer cylinder towards the membrane. The heat exchanger ensures that the membrane is controlled at temperature levels from 80 to 210°C ± 0.1°C for carrier air and sample flow rates from 50–1500 ml/min at inlet gas temperatures from 0–200°C.

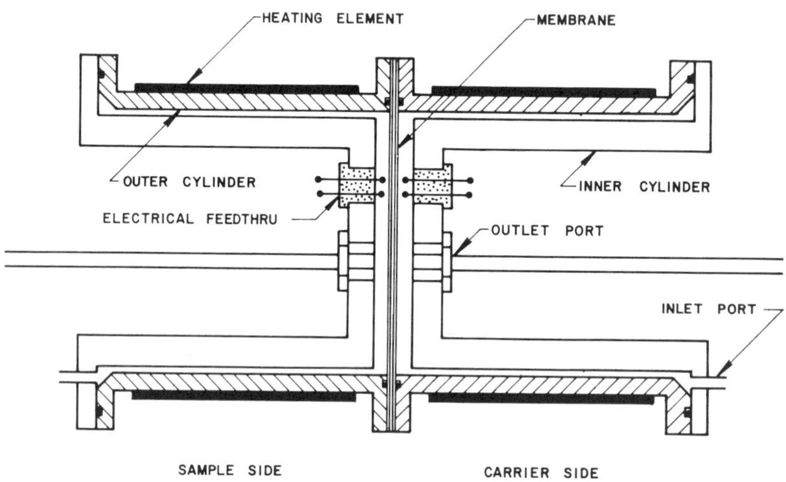

Figure 18.5. Diluter cartridge.

SYSTEM PERFORMANCE

The FPD sulfur and membrane diluter system performance was characterized experimentally.[9] Sulfur dioxide in air at concentration levels from 10–100,000 ppm and SO_2 in pure isobutane and propane at concentration levels 1–4 ppm were measured continuously with the experimental arrangements illustrated by Figures 18.6 and 18.7.

SO_2-in-Air Results

Figure 18.8 shows the relationship between the sample and carrier SO_2 concentration for diluter operation at 203°C, 140°C, and 106°C. The slope of the isotherms is equivalent to the k_d. Dilution factors of 101,000, 4975, and 1660 were attained at 106, 140, and 204°C respectively with 0.0065-cm thick Teflon membranes 7.6 cm and 1.3 cm in diameter. Long-term operations showed that the stability and accuracy of the analyzer/diluter system is equal to that of the analyzer itself. This indicates that the performance specifications of the

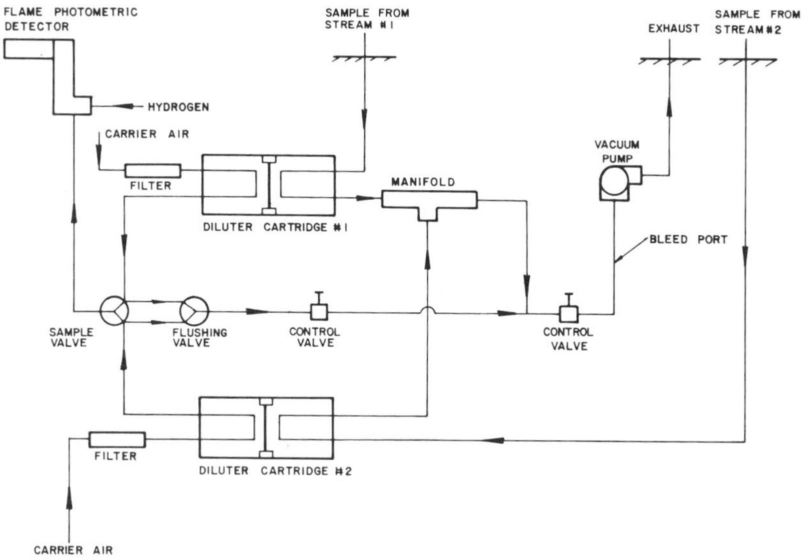

Figure 18.6. Flow schematic—citrate scrubber SO_2 analyzer system.

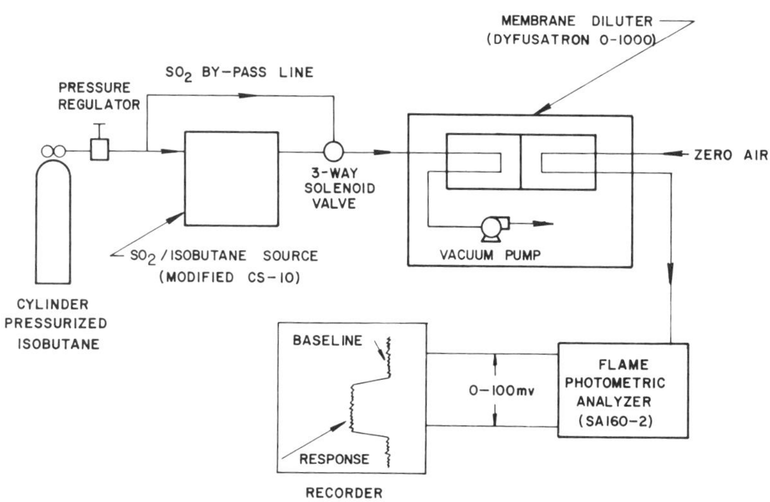

Figure 18.7. Sulfur in isobutane analysis system.

analysis system are established primarily by the analyzer. The diluter k_d established from the 101,000 ppm SO_2 source was wtihin +1% of the k_d established with the 4975- and 1660-ppm SO_2 sources.

SO_2-In-Isobutane and -Propane Results

The isobutane and propane gas samples generated by a permeation tube system ranged in SO_2 concentration from 1.3–3.5 ppm.

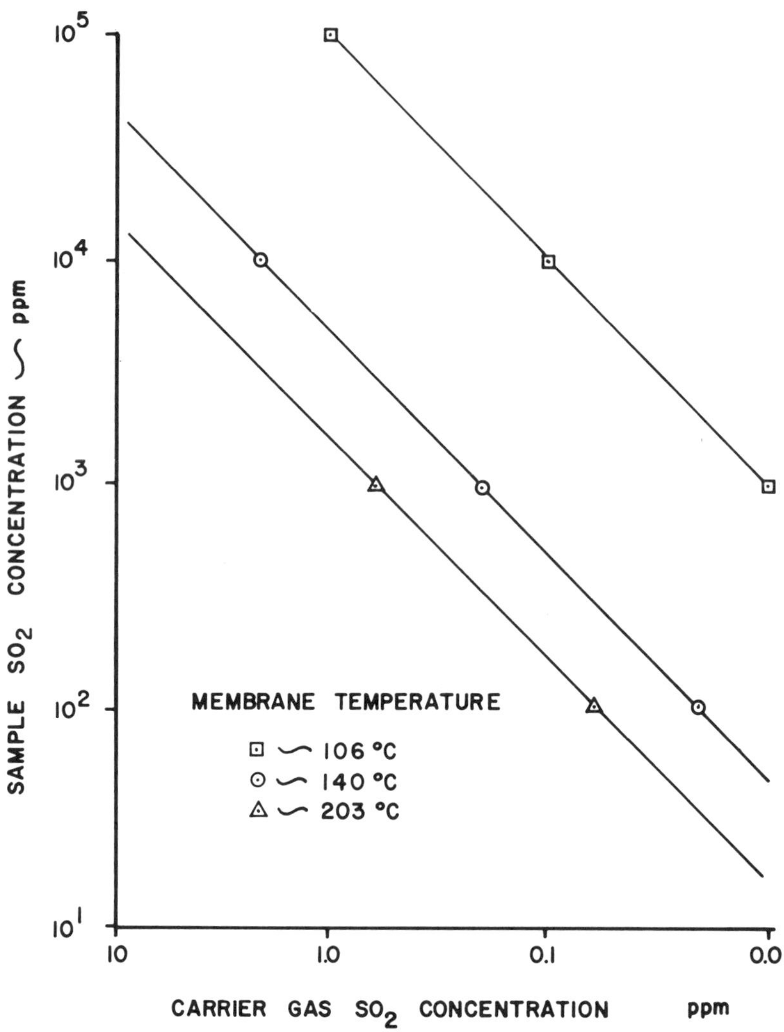

Figure 18.8. Diluter SO_2 sample/carrier relationship.

After dilution by the membrane, the corresponding SO_2 in air concentration range was 0.00137 to 0.00368 ppm.

Figure 18.9 shows the measured SO_2 concentration in these samples. A direct comparison of these data to the corresponding SO_2 in the hydrocarbon gas concentration determined gravimetrically is performed at identical isobutane flow rate conditions as presented by Figure 18.9. The isobutane and propane flow rate through the permeation tube device is the common parameter relating the gravimetric and FPD SO_2 concentration measurements. The maximum, minimum, and average discrepancy of the data are −8.38%, −0.38%, and −4.4%.

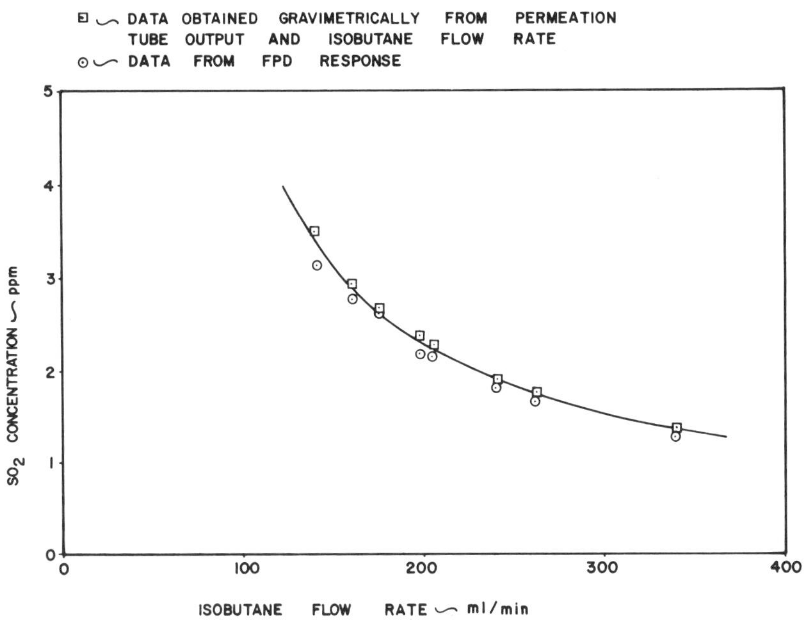

Figure 18.9. FPD analysis correlation.

Field Operation

An FPD sulfur analysis system is presently employed to monitor the performance of the U.S. Bureau of Mines citrate process to remove sulfur dioxide from smelter stack gas.[9] The citrate scrubber process is comprised of absorbing SO_2 in citrate solution, reacting the absorbed SO_2 with H_2S, filtering and melting the precipitated sulfur, and recycling the regenerated citrate solution to the SO_2 absorption step. Water produced by the sulfur precipitation reaction is evaporated by the hot gas in the absorption column. Preparation of H_2S from recycle sulfur by reacting sulfur vapor with natural gas and steam or aluminum catalyst was also examined. A pilot plant of the citrate type scrubber was constructed by the Bureau of Mines and is being operated jointly by the Bureau with Bunker Hill Company at the lead smelter in Kellogg, Idaho. The plant capacity is 1000 SCFM of 1.5% SO_2 with a sulfur yield of 1/3 ton/day. Figure 18.10 is a flow sheet of the plant citrate process.

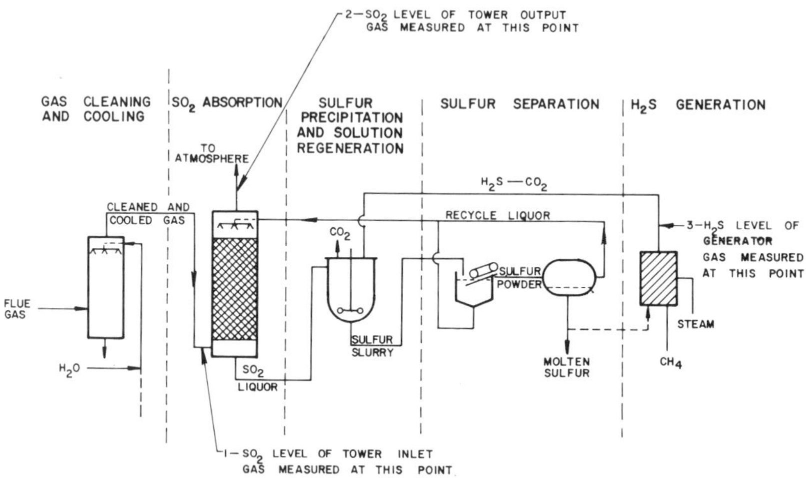

Figure 18.10. Generalized citrate process flow sheet.

Sulfur dioxide is absorbed from cooled and cleaned smelter gas in a 2.5-foot diameter by 30-foot-high packed SO_2 absorption tower using a 10 gpm countercurrent stream of citric acid-sodium citrate-sodium thiosulfate solution. The absorbed SO_2 reacts with H_2S to precipitate elemental sulfur. The citrate solution is regenerated for recycle while sulfur is removed from the slurry by oil flotation. Sulfur is melted in a heat exchanger, then separated as molten sulfur from the occluded citrate solution and oil in a settler at 35 psi and 135°C.

The SO_2 concentration level is measured at the gas inlet and outlet points of the absorption tower as shown by points 1 and 2 on Figure 18.10. In addition, a separate FPD sulfur analysis system continuously measures the H_2S concentration level of the H_2S generator as shown by point 1 in Figure 18.10.

Ambient air is utilized as the carrier gas. It is transported by two streams through sulfur gas scrubbers to the diluter cartridges. Each stream carries the source gas molecules which diffuse through the membrane and flows out of the diluter cartridges to the sampling valve. At this point, one of the carrier streams flows through the sampling valve to the FPD for analysis of sulfur. Simultaneously the other carrier stream is continuously flowing through the flush valve and source gas vacuum pump. A bleed valve at the flush valve is provided to adjust the carrier flush rate to its sampling flow rate.

The carrier streams corresponding to each sampling station are transported to a single FPD analyzer. It is accomplished by the sample switching valve, which can operate in a manual or automatic mode. The complete analysis cycle for both streams in the automatic mode is 20 minutes.

The diluter cartridge handling the input scrubber stream operates at one temperature and reduces the sample SO_2 concentration by a factor 10^5 or reduces 100,000 ppm range to 1 ppm. The diluter cartridge handling the output scrubber stream operates at two temperatures. At the high and low temperature levels, the sample SO_2 concentration is reduced by factors 1660 and 4975. The H_2S generator output level is measured continuously by a system similar to that depicted by Figure 18.2.

REFERENCES

1. Brody, S. S. and J. E. Chaney. U.S. Patent No. 3,489,498.
2. Brody, S. S. and J. E. Chaney. *J. Gas Chromatog.* **4**, 42 (1966).
3. Stevens, R. K. and A. E. O'Keeffe. *Anal. Chem.* **42** 143A (1970).
4. Stevens, R. K., A. E. O'Keefe and G. C. Ortman. *Environ. Sci. Technol.* **3**, 652 (1969).
5. Syty, A. and J. A. Deam. *Appl. Opt.* **7**, 1331 (1968).
6. Tewerson, A. and H. B. Palmer. *Proc. 13th Internat. Symp. On Combustion*, The Combustion Institute, Pittsburgh, Pennsylvania (1971), p. 99.
7. Lucero, D. P. and J. W. Paljug. "Monitoring Sulfur Compounds by Flame Photometry," Presented at the EPA/ASTM/NCAR Committee D-22 Meeting, Conference on Instrumental Monitoring for Ambient Air, University of Colorado, Boulder August 14-16, 1973. In *ASTM Standards Book* (Philadelphia, Pennsylvania: American Society of Testing Materials, 1973).
8. Spalding, D. B. *Proc. Royal Soc. (London)* **A 240**, 83 (1957).
9. McKinney, W. A., W. I. Bissen and J. B. Rosenbaum. "Design and Testing of a Pilot Plant for SO_2 Removal from Smelter Gas," Presented at the Annual AIME Meeting, Dallas, Texas, February 23-28 (1974).

Section IV
Intermittent Control Systems

19

Supplementary Control Systems for Attaining Ambient Air Quality Standards

George W. Reynolds
Woodward-Clyde Consultants, Scottsdale, Arizona

INTRODUCTION

This chapter discusses Supplementary Control Systems (SCS) as a means of complying with ambient air quality standards. Supplementary Control Systems have also been known as Intermittent Curtailment Systems (ICS), Closed Loop Systems, Variable Control Systems, and Sulfur Dioxide Emission Limitation Systems (SDEL).

The Environmental Protection Agency uses the term Supplementary Control System, or its abbreviation, SCS, as "an acceptable control measure only in cases where constant emission reduction measures are unavailable, and only until such measures become available."[1] However, official discussions of the SCS frequently refer to intermittent control, since this term most accurately characterizes the general nature of the system.

Supplementary Control Systems are systems by which the rate of pollutant emission is limited during periods when meteorological conditions are such that violations of short-term ambient air quality standards would occur without this limitation. In general, these intermittent emission limitations are accomplished by curtailing production and/or by shifting fuels.

ACCEPTABILITY OF SUPPLEMENTARY CONTROL SYSTEMS

A few general guidelines are apparent from EPA literature concerning the acceptability of a SCS system.

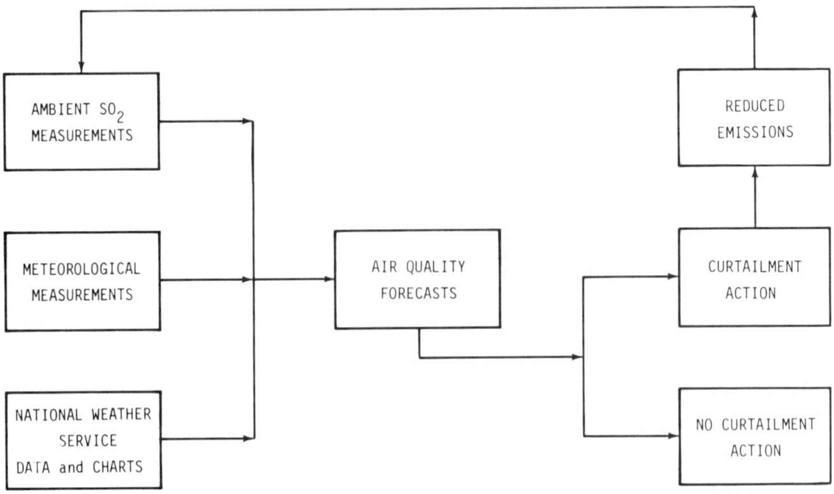

Figure 19.1. Information—action cycle.

1. SCS is not applicable for carbon monoxide, particulates, hydrocarbons, nitrogen oxides, or photochemical oxidants, because of the difficulties of source isolation.[2] Thus, of the six materials for which there are federal primary and secondary ambient standards, sulfur dioxide is the only contaminant for which emission control with an SCS, even as an interim measure, is presently recognized as reasonable by the EPA.
2. The source must be sufficiently isolated so "that the owner or operator is willing and able to accept full legal responsibility for maintaining the national ambient short-term standards throughout the area in which emissions from that source influence ambient air quality."[2] Thus, if there are other sources in the area, predicted ambient concentrations from the full combination of sources must be the basis for curtailment.
3. Use must be made of available continuous emission control technology. The reference here is primarily to the desulfurization of flue gas rather than the desulfurization of fuels, since the latter technologies are not expected to be adequate for widespread commercial use until the 1980s. When the supply becomes adequate, continuous use of low-sulfur coal could also be required before authorization of a Supplementary Control System is granted, even though "coal with low sulfur content reduces the efficiency of electrostatic precipitators."[1]
4. Timing is an important consideration. That is, "temporary use of intermittent controls in carefully selected circumstances will fa-

cilitate more rapid attainment of the current primary sulfur dioxide standards without the necessity for power plant shutdown, will allow the continued use of the nation's high-sulfur coal reserves while control technology, which will make it environmentally acceptable, is being installed, and will also allow time to increase the availability of low-sulfur fuels."[1]

5. A Supplementary Control System must exhibit a high degree of reliability. The key words, insofar as the reliability criterion is concerned, are forecasts and reaction. That is, the forecasts of ambient SO_2 concentration levels must be sufficient to avoid violations of short-term ambient standards.

If high concentrations occur when it is convenient to curtail, it may be reasonable to be very conservative in forecasting air quality and in the assignment of safety factors. For those cases in which curtailment is costly (in terms of safety, comfort, convenience, or dollars) less conservatism is acceptable, and more technical information, a larger staff, and a more sophisticated support facility are essential.

Reaction is an obvious requirement. To have an acceptable SCS, the operator must have a specific plan for actions in response to forecasts of ambient concentrations in excess of standards. This plan may vary with the magnitudes of excesses, the time of day, the season, or the phase of the operation. At any given time, the plan may call for partial curtailment, total curtailment, fuel switching, or some other corrective measure, but it must tell the proper authority precisely what to do, once this authority has received the forecast requiring reaction.

This leads to a vital statement. Obviously, violations can be avoided only with appropriate and timely reaction. Consequently, it is apparent that to have an officially acceptable system, it will be necessary for the company proposing an SCS to commit to a forecast-reaction program.

Thus, the decision to use a Supplementary Control System relies on the answers to several overlapping questions, not necessarily in the order given.

1. Has use been made of continuous control (hardware) technology?
2. Is further emission control needed? That is, will there be ambient violations without additional emission controls? The answer to this question may come from SO_2 monitoring data or from estimates of ambient concentrations based upon emission, meteorological, and topographic data.
3. Is the source sufficiently isolated so that the operator can take legal responsibility for ambient concentrations over an entire assigned area?

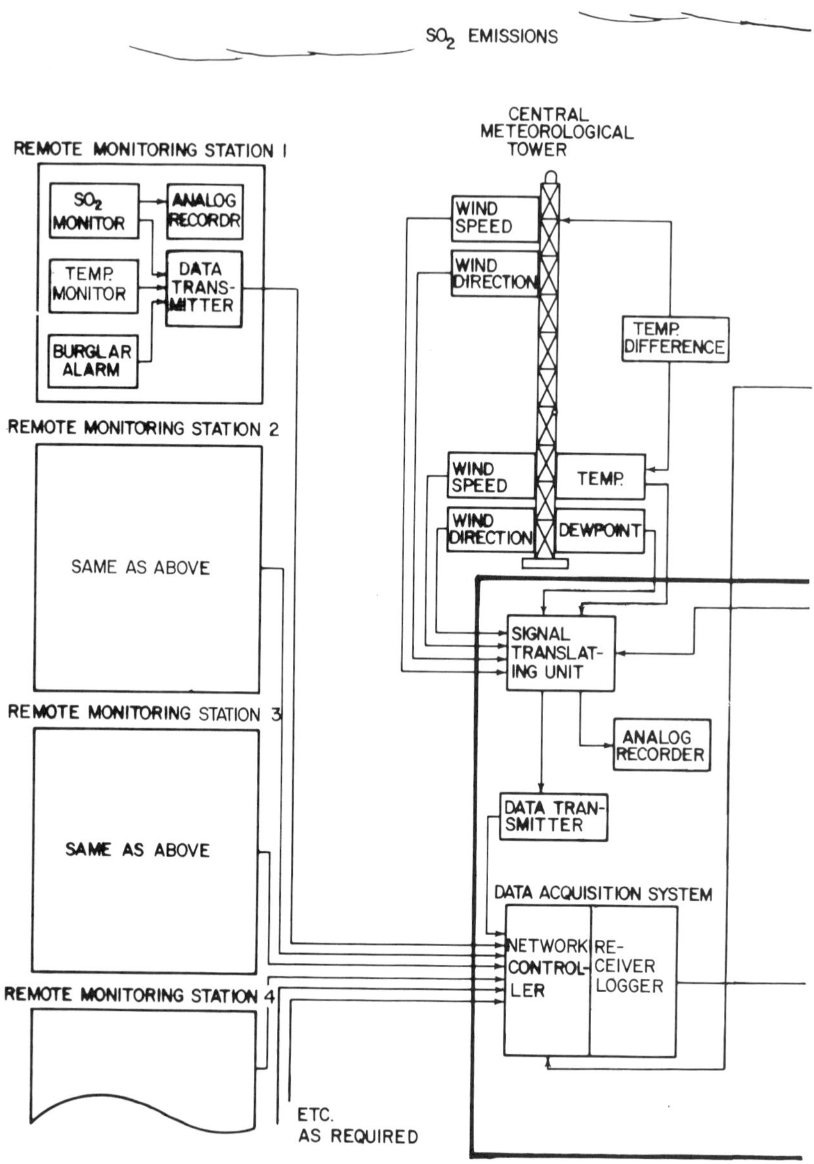

Figure 19.2. Components of a fairly sophisticated Supplementary Control System.

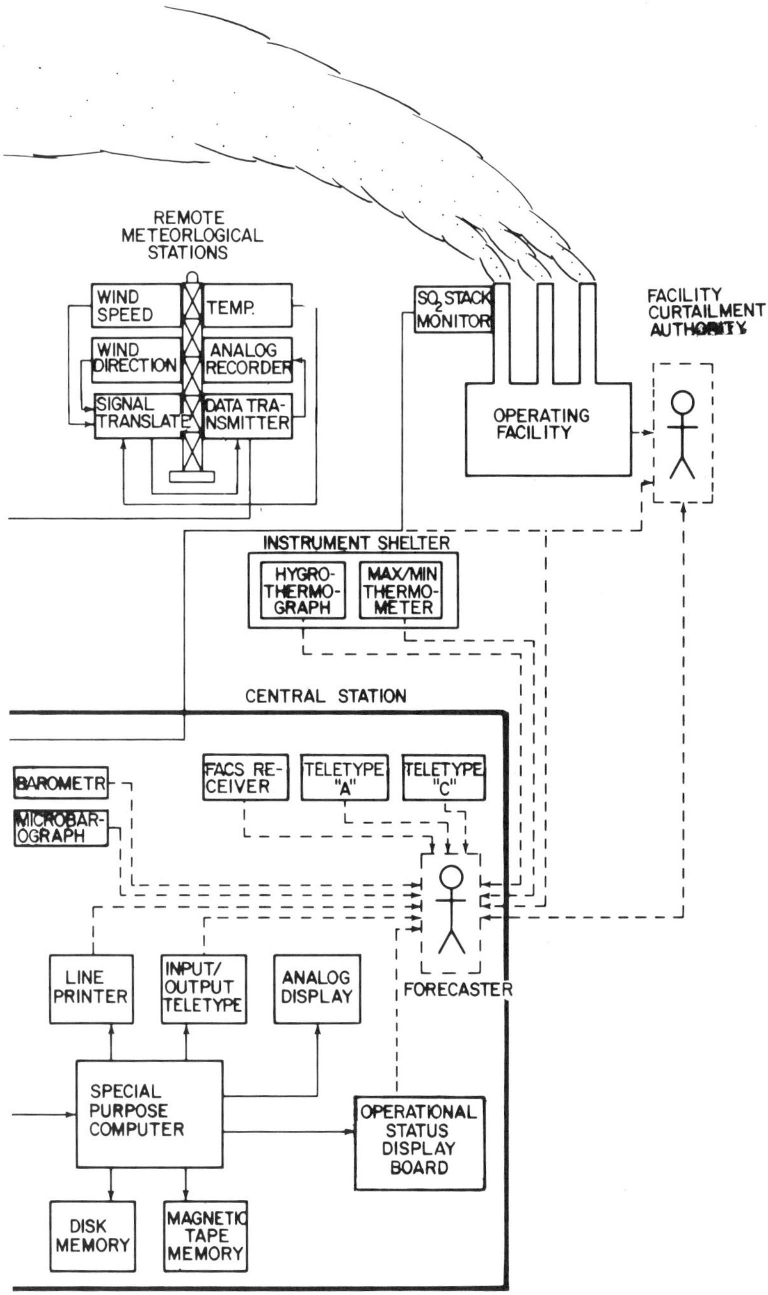
REMOTE METEORLOGICAL STATIONS
WIND SPEED
TEMP.
WIND DIRECTION
ANALOG RECORDER
SIGNAL TRANSLATE
DATA TRANSMITTER
SO_2 STACK MONITOR
FACILITY CURTAILMENT AUTHORITY
OPERATING FACILITY
INSTRUMENT SHELTER
HYGRO-THERMO-GRAPH
MAX/MIN THERMO-METER
CENTRAL STATION
BAROMETR
MICROBAR-OGRAPH
FACS RE-CEIVER
TELETYPE "A"
TELETYPE "C"
FORECASTER
LINE PRINTER
INPUT/OUTPUT TELETYPE
ANALOG DISPLAY
SPECIAL PURPOSE COMPUTER
OPERATIONAL STATUS DISPLAY BOARD
DISK MEMORY
MAGNETIC TAPE MEMORY

4. How often and by how much will curtailment be required? How much will these curtailments cost in terms of safety, convenience, special problems, and dollars? Both development and operational costs should be considered.

5. Can the source curtail by the amount required in response to forecasts of concentrations?

6. Can air quality be forecast with sufficient accuracy to permit reaction in time to avoid violations of ambient standards?

For some plants, some of these questions may have already been answered, and sometimes the answers are qualitatively evident to the professional who is experienced with air quality and SCS problems and procedures. For other locations, a substantial data collection and analysis program is required before a decision can be reached on whether to go SCS, and if so, on the type of system that is needed.

COMPONENTS OF AN SCS

The degree of sophistication and the components of Supplementary Control Systems will, by necessity, vary from System to System (Figure 19.2). Systems can include: a) SO_2 monitoring, b) meteorological monitoring from large towers, c) telemetering from monitoring sites to a central station, d) computer processing, manipulation, SO_2 concentration prediction, and system evaluation, requiring the development of specialized computer programs, e) on-site weather station, complete with facsimile and teletype facilities, and manned around the clock by highly experienced professional meteorologists, f) air quality forecast techniques specifically developed for the individual SCS and/or g) procedures for self-evaluation and updating the System.

For some operations, intricate systems have been required because curtailment is very costly, meteorological conditions change rapidly, violations can occur at any time of day, and topographical modifications of synoptic meteorological circumstances are significant. Local, on-site air quality forecasters are required for these Systems for accurate operational forecasts that will avoid requiring unnecessary curtailments.

On the other hand, under less complex circumstances it may be possible to eliminate one or more of these elements, such as the onsite specialized computer, and/or the meteorological tower. However, it is probable that a considerable study would be involved in determining, documenting and justifying the simplicity of this SCS requirement.

REFERENCES

1. *National Strategy for Control of Sulfur Oxides from Electric Power Plants* (Washington, D.C.: U.S. Environmental Protection Agency, 1974).
2. Quarles, John R., Jr. "Use of Supplementary Control Systems and Implementation of Secondary Standards," *Fed. Reg.* **38**, No. 178, 25697-25703 (1973).

20

Operational Supplementary Control for Industrial and Utility Applications

Vincent A. Rocco
Manager, Midwest Technical Center

Robert J. Merlino
Manager, Program Management Office
Environmental Research & Technology, Inc.
Concord, Massachusetts

INTRODUCTION

Much controversy exists concerning the role of Supplementary Control Systems in the complex task of balancing costs of control against adequacy of control of sulfur oxides. What constitutes adequate control is the basic subject of this controversy. At one extreme, adequate control requires zero anthropogenic sulfur molecules emitted into the atmosphere. At the other extreme, the status quo is considered adequate. An intermediate position is that adequacy is demonstrated when Primary and Secondary National Ambient Air Quality Standards (NAAQS) are achieved and maintained.

Some groups argue that we do not know all of the deleterious effects of airborne sulfur compounds and that sulfate aerosols may be more of a hazard to human health and welfare than sulfur dioxide. Therefore, the only prudent course is to reduce, as far as possible, the emissions of sulfur in any form into the atmosphere. Others argue that a) to spend billions of dollars for possible excessive sulfur oxide control on the basis of poorly documented evidence of the importance of other sulfur compounds is a luxury the country can ill afford now and is highly imprudent at any time; b) if economical control methods exist now through which present standards can be met, and which do not preclude in any way the use of more stringent controls later on, then why not utilize such economical control methods now; c) if other sulfur compounds are important, there should be standards developed based upon adequate research; and d) reducing the allowable sulfur emitted to unnecessarily low values everywhere will waste scarce natural resources (*i.e.*, low-sulfur coal, low-sulfur oil and natural gas) that are needed in regions where ambient air quality standards are most threatened.

Assessment of the effectiveness of SCS in lieu of constant emission controls (permanent use of low-sulfur fuels, flue gas desulfurization systems, or permanent production or generation curtailment) for reducing potential epidemiological or toxicological effects of atmospheric sulfates is beyond the scope of this chapter. However, the contents of this chapter are directed toward the following questions:

1. What constitutes the state of the art of SCS today for controlling SO_2 emissions?
2. What effectiveness and reliability can be expected from an SCS on a major source?
3. On what bases may a source be considered as a viable candidate for use of SCS?
4. How can air monitoring systems be used to gather meaningful information on sulfate concentrations in the vicinity of SCS controlled plants to provide a background of data for use in formulating a sulfate control plan?

GENERAL THEORY OF SUPPLEMENTARY CONTROL SYSTEMS

Ambient concentrations of air pollutants vary appreciably with the meteorological variables affecting atmospheric dispersion. Regional air quality control and maintenance planning activities have historically focused on relieving local "hot spots" of air pollutant concentrations by influencing the *spatial* distribution of pollutant sources in response to knowledge of source distribution and meteorological conditions. Supplementary Control Systems, or more correctly Dynamic Controls (DEC), play a parallel role but are primarily concerned with influencing the *temporal* variation of emission rate in response to knowledge of meteorological conditions within a defined source array.

Fundamentally, Supplementary Control Systems in which stack emissions of pollutants are curtailed during periods when meteorological conditions conducive to ground-level concentrations in excess of the NAAQS exist or are anticipated to exist. Conversely, when meteorological conditions conducive to low ground level concentrations exist, emission rates are increased. A block diagram of a typical operational SCS is shown in Figure 20.1.

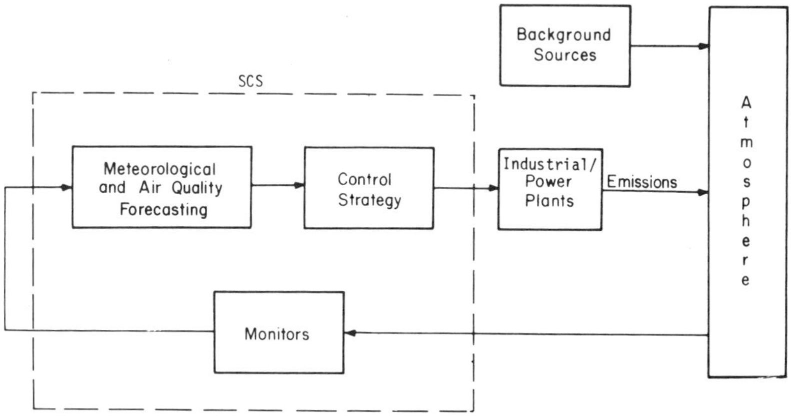

Figure 20.1. Supplementary Control System components.

Figure 20.2 represents a representative cumulative frequency distribution of short averaging time concentrations observed at a monitoring location influenced by an isolated source. The value C_s has been designated on the abscissa to indicate the value of an air quality standard. On the ordinate axis, the value F_s denotes the allowed frequency of occurrence of concentration values exceeding the standard C_s (*i.e.*, once per year as required by Federal regulations). In Figure 20.2a, the cumulative frequency of occurrence of concentrations for the "uncontrolled" source is above F_s at $C = C_s$ and an ambient air quality standard has therefore been transgressed. Direct application of a constant emission control which uniformly reduces emissions by 50% (by changing from 2% to 1% sulfur content fuel, for example) would move every value of F from the abscissa value C to a value C/2 yielding the curve labeled "control F." This curve shows compliance with the standard.

An alternative to constant emission reductions to meet standards is SCS, which focuses on reducing emissions during expected occurrences of high ground level concentrations. Figure 20.2b shows the frequency distribution resulting from such a control system. Note that the shape of the curve for occurrences of high concentrations is altered to achieve the goal of compliance with standards. The prime advantages of the use of SCS for achieving short averaging time standards are: a) the lower quantities of valuable low-sulfur fuels required, and b) greater range of control options for compliance under the most adverse meteorological conditions.

The above discussion relates to the use of SCS to achieve compliance with air quality standards having short averaging times such as the 3-hour and 24-hour SO_2 standard. In general, these are the standards whose maintenance will be required by a regulatory agency as certification for SCS candidacy. However, SCS is also applicable to the problems of achieving longer averaging time standards and of reducing long-term average concentrations, which are of major concern in regional air quality maintenance planning.

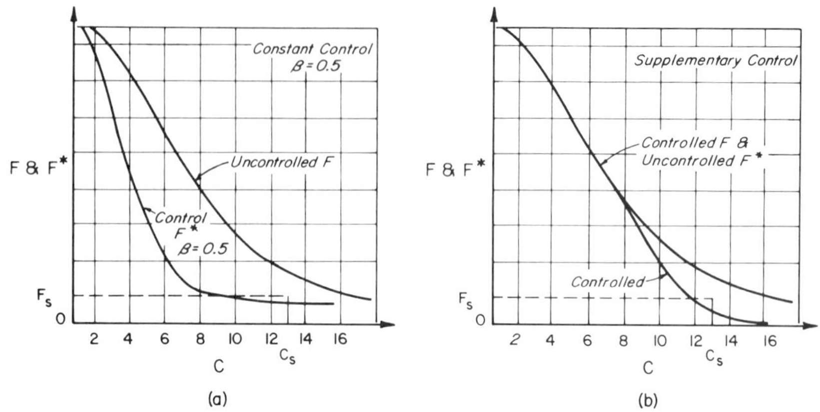

Figure 20.2. Representative cumulative frequency distributions for maximum concentrations under conditions of: a) no emission controls and constant emission reduction, and b) supplementary emission controls.

Assume that concentrations at any ground level location in the influence of a controllable source may be described at any time, t, by the relation

$$C(t) = Q(t) \cdot M(t) \tag{1}$$

where Q(t) is the source emission rate and M(t) is the function relating meteorological dispersion conditions and emission rate to ground level concentration.

We can define the variable Q(t) and M(t) in terms of time averaged values (barred quantities) and time dependent variations from the average (primed quantities) as

$$M(t) = M + M'(t) \text{ and} \tag{2a}$$

$$Q(t) = Q + Q'(t) \tag{2b}$$

where $\overline{M'(t)}$ and $\overline{Q'(t)}$ are, by definition equal to zero.

The average value of C(t) over an averaging time can then be shown to be given by

$$\overline{C} = \overline{Q} \cdot \overline{M} + \overline{Q'(t)\, M'(t)} \tag{3}$$

The first term on the right-hand side is the product of the average emission rate and the average meteorological function. The second term arises from the covariance of the departure of M and Q from the mean values. If M and Q were totally uncorrelated (blended fuels or random use of fuels), this term would be zero. An intelligently designed SCS, however, will have small emissions associated with large values of M and vice-versa such that the covariance term would be negative. In this way the value of $\overline{C}$ can be reduced below the value $\overline{Q} \cdot \overline{M}$. With this in mind, we can identify two quite different types of applications for SCS:

1. Given a set of standards as goals for $\overline{C}$, make maximum use of available higher-sulfur content fuels (in Equation 3 constrain $\overline{C}$ and maximize $\overline{Q}$, or
2. Given constraints on the total sulfur allowed to be emitted, minimize the average concentration (in Equation 3, constrain $\overline{Q}$ and minimize $\overline{C}$).

The first application is the most commonly stressed form of SCS as it offers a methodology for meeting air quality standards at least economic costs. The second application is, however, equally valid and points to potential use of SCS in minimizing air pollution levels in conjunction with future constant emission control systems.

Equation (3) also points to the air quality advantage of keeping fuels of differing sulfur contents apart and available for use in SCS over blending fuels to a homogeneous sulfur content, provided fuel-switching capabilities exist at a given plant.

SCS in the broadest sense offers a scientific methodology which, given constraints on low-sulfur fuel availability, the ability to shift or reduce load, or the availability and reliability of fuel cleaning techniques, can be utilized to minimize not only the occurrence of very high peak concentrations, but also long-term average concentrations. It is for this reason that SCS should not be considered to be only an interim tool to be phased out when "permanent" controls are fully available. SCS as a methodology deserves a long-term, permanent place alongside of fuel or stack gas cleaning methods and regional planning measures as a viable means of maintaining air quality.

PHYSICAL ASPECTS OF AN OPERATIONAL SCS PROGRAM

There are many possible modes of operation for the meteorological and air quality prediction function of Figure 20.1. Their differences lie in the types of information available and whether a real time model is used (a model here is meant to be a set of equations relating weather and emission data to ambient SO_2 concentrations). Four possible modes are: a) air quality data only: no real time model; b) meteorological data only: no real time model; c) meteorological data only: real time model; d) meteorological and air quality data: real time model.

In the first two modes, nonreal-time (off-line) studies using models are done to provide "rules" which are used in real-time operation (on-line). The best SCS performance has been shown to result from using the fourth mode which combines meteorological data, air quality data, and a real time model.

There are various ways the fourth model can be implemented. Two of them are: a) meteorological data only as input to a real-time model: air quality data used only as a real time "check;" b) both meteorological and air quality data are used as inputs to real-time model. Best performance results from the second case where all available data is used directly in one model in a self-consistent fashion.

It is conceptually possible to rely on the U.S. Weather Bureau for meteorological predictions. However, this can result in unsatisfactory SCS performance because the Weather Bureau does not predict (forecast) the meteorological parameters of most importance to SCS operations. In general, SCS operation requires specially trained meteorologists.

The Supplementary Control System which provides best operational performance includes a real-time air quality and meteorological network, telemetry equipment and computer software programs developed to analyze and predict ambient air quality over the region of impact of the plant. This basic system is then combined with data related to plant emissions characteristics and operating plans to provide a fuel or load management program for dynamic emissions control. The system requires the identification, implementation, where necessary, and coordination of the following elements:

1. A control center, with telecommunications and computer facilities for the receipt, processing and analysis of all data, the generation of all predictions and control decisions, and the monitoring of all SCS subsystems. This center may be located on the plant site or be remotely located with telecommunication tie-in to the plant operations center.
2. Input facilities, including: a) monitoring stations in the plant area to measure, record and transmit SO_2 and meteorological data; b) the plant control center to measure, record, and transmit parameters describing actual and scheduled plant load and emissions and to plan and transmit a preferred control strategy.
3. Air quality predictive models of demonstrated reliability to assimilate the data and produce the predicted air quality.
4. A reliable, objective decision model for controlling plant operations.
5. Qualified personnel to prepare accurate air pollution meteorological forecasts, as input to the air quality prediction models;

to operate and manage the SCS hardware and software, and to ensure the consistency and reliability of the inputs and outputs.

6. Thorough operations and maintenance procedures to ensure the orderly functioning of the system, and to provide for regular checks on performance.
7. A program for qualitative and quantitative evaluation of the system which will lead to improvement of system performance as a history of operational experience is developed.

RELIABILITY OF SCS SYSTEMS

Of key concern regarding SCS is the question of reliability of such systems in meeting specified air quality standards or goals. Stated in terms of probability of achieving standards, overall reliability becomes a function of: a) air quality forecasting accuracy *combined* with the design of the operational control strategy, and b) reliability of other system components. Thus, deficiencies in air quality forecasting abilities can be offset by the adoption of an overly conservative control strategy to obtain a reliable SCS. The operational costs of such an overly conservative system, however, would be higher.

It is desirable in practice to establish as accurate an air quality forecasting system as possible to support any given SCS. This requires optimizing the accuracy of the three components to an air quality forecasting system. These components are: a) the meteorological forecast as it relates to the prediction of meteorological parameters influencing dispersion, b) the air quality models that use the meteorological inputs to predict concentrations, and c) the emission models used to predict effluent emission rates.

A methodology for quantifying the relationships between system reliability, control strategies and fuel usage for a given power plant has recently been developed for EPA by Gaertner *et al.*[1]

In this study, a simulation model of SCS was developed involving the interaction of an air quality forecast model (made up of the three components discussed above) and a control strategy logic. Uncertainties in the air quality forecast can be assumed, and alternative control strategies can be tested with respect to their reliability to meet standards or goals and, therefore, the effectiveness of each control strategy can be quantified. Example analyses suggest tradeoffs among various SCS component accuracies and optimal methods of achieving any given system reliability.

A key element in assessing system reliability is the identification of the uncertainty of each component of the air quality forecasting system. Specifically, the expected uncertainty in projecting future emission rates, future meteorological conditions and the accuracy of air quality dispersion models needs to be identified both initially and then as part of a periodic updating procedure for improving the reliability of an operational SCS with time. With proper data logging procedures absolutely objective analysis techniques can be employed to assess these component uncertainties and to use experience gained to suggest methods for improving system reliability.

RESULTS FROM OPERATING SYSTEMS

Tennessee Valley Authority (TVA)

Noncontinuous SO_2 emission limitation programs have been operated by TVA since September, 1969, when the first program was initiated for the Paradise Steam Plant in west-central Kentucky.[2] Two others are in the process of being implemented, one at TVA's Widows Creek Steam Plant in northern Alabama and another at the Cumberland Steam Plant, located about 50 miles west of Nashville, Tennessee. Analysis of SO_2 data before and during implementation of the SO_2 emission limitation program at Paradise Steam Plant indicates the program has been effective in improving the air quality in the region surrounding the plant. A brief summary of each of the TVA programs follows.

Paradise Steam Plant

The generating plant consists of three units—an 1150-MW unit and two 704-MW units each with a 600-foot stack in a rural river valley of west-central Kentucky. Stack emissions are reduced when on-site meteorological measurements indicate plume dispersion conditions are such that they would adversely affect local air quality. During these periods, plant generating load is lowered as required to prevent ambient ground level SO_2 concentrations from exceeding a 30-minute average concentration of 0.8 ppm* as determined by the results from a computer program that uses the meteorological measurements as input to a limited mixing plume dispersion model.

During the SO_2 limitation implementation period, the on-site meteorologist continuously monitors the meteorological and air quality information and informs the plant operator if conditions have improved to the point where normal plant operations may be resumed. From September, 1969, through November, 1972, there were 106 days when meteorological conditions called for some SO_2 emission limitations. However, there were only 41 days when SO_2 emissions were actually limited because on the other days the unit outages of the plant were already below full-load operation.

The SO_2 data shown in Table 20.1 demonstrates the effectiveness

*This 30-minute concentration threshold limit is nearly equal to peak-to-mean relationship to the national and Kentucky ambient air quality standards for SO_2 of 0.5 and 0.14 ppm for 3-hour and 24-hour averaging periods, respectively.[3]

of the program. For example, there were two instances of exceeding the 0.5 ppm 3-hour average SO_2 standard during a 21-month operational period of the program while there were ten such incidents during the 21-month period prior to the program operation. For the 24-hour federal primary SO_2 standard (0.14 ppm) the number of occurrences exceeding the standard were eight before and zero after the SCS program was implemented.

Table 20.1. Frequency Distribution of Measured SO_2 Concentrations Before and After Implementation of SO_2 Emission Limitation Program—Paradise Steam Plant.[4]

3-Hour Average SO_2 Concentration Distribution

3-hour Average SO_2 Conc. Range (ppm)	*Frequency* Before (January 1, 1968–September 19, 1969)	*Frequency* After (September 19, 1969–June 25, 1971)
0	51,097	55,961
.01–.10	10,574	9,363
.11–.20	443	275
.21–.30	102	59
.31–.40	37	28
.41–.50	16	10
.51–.60	7	2
.61–.70	2	0
.71–.80	1	0

24-Hour Average SO_2 Concentration Distribution

24-hour Average SO_2 Conc. Range (ppm)	*Frequency* Before (January 1, 1968–September 19, 1969)	*Frequency* After (September 19, 1969–June 25, 1971)
0	3,064	3,752
.001–.02	4,248	4,043
.021–.04	400	305
.041–.06	92	73
.061–.08	24	19
.081–.10	14	9
.101–.12	6	3
.121–.14	4	2
.141–.16	7	0
.161–.18	1	0

Widows Creek Steam Plant

The program was started in September, 1970, and expanded in August, 1972, when a meteorologist was assigned to the program full-time on-site. The plant is in a region with complex terrain (ridge-valley) features in northern Alabama. There are a total of eight units—six units of 125-MW each with 170-ft stacks and two units of 500-MW each with 492-ft stacks. The program was more limited than the Paradise Steam Plant program in that it was designed to prevent high SO_2 concentrations on an adjacent plateau across the river from the plant and about 250 m above the plant site in the valley.

The SO_2 emission limitation program was implemented only in mid-morning through early afternoon when critical meteorological conditions that affect plant dispersion are most likely. The emissions reduction procedure is to burn low-sulfur coal in six of the small units and reduce load in all units. The program was upgraded in the fall of 1973 to provide 24-hr/day coverage for all areas around the plant. Results were not available for this augmented program as of this writing.

Cumberland Steam Plant Program

This program is essentially similar to the program at Paradise because of similar plant size and surrounding terrain. The plant has two 1300-MW units—each with a 1000-foot stack. No results are available as yet.

American Smelting and Refining Company (ASARCO)—Tacoma, Washington

An SCS program has been operational since 1969 at Tacoma, Washington. The plant has a 565-foot stack and a limited plume rise (15 ft/sec exit velocity). Terrain is somewhat hilly with maximum elevations nearby rising to within 250 ft of stack top.[5]

Meteorological forecasts which are continuously updated provide the basis for reduction of plant operations.

Sulfur dioxide data from five ASARCO monitoring stations show continuous and significant improvement over the years. National Ambient Air Quality Standards have not been exceeded since March, 1971, at any monitoring stations in the area. Local standards are more stringent and have been exceeded. Average load reduction due to SCS is $\backsim$16% of full output.

ASARCO—El Paso, Texas

SCS reliability was evaluated for a 36-day period in 1971 at the ASARCO El Paso Smelter.[1] It is in a remote location in a mountainous area. Stack heights are 828 ft and 611 ft for the copper and lead smelting operations, respectively. The top of the stacks are at sea level elevations of approximately 4500 and 4300 ft. To the northwest of the plant site the Franklin Mountain Range has peaks of 6000-6600 ft with the range extending in a north-south direction.

Results from the case study indicate:

1. SCS measures were effective in lowering high SO_2 concentrations east of the smelter but were not as effective in reducing SO_2 concentrations northwest of the plant.
2. Percentage reductions in average SO_2 concentrations at a 17-station monitoring network were 28–65%.
3. The 24-hour ambient SO_2 standard of 0.14 ppm was exceeded at two stations and the 3-hour standard of 0.5 ppm was exceeded at one station during the 36-day period. Experience previous to implementation of the SCS system indicated standards would have been exceeded far more often during the analysis period without the SCS operating.

The study also recommended means for improving reliability—primarily through improved prediction and modeling of flow over the mountainous terrain.

Dow Chemical

After receiving approval by the Michigan Air Pollution Control Commission, an SCS system was designed and implemented for Dow Chemical during 1974 and it is presently being operated routinely around-the-clock.[6] Dow provides desired operating modes for their plants and a sequence of preferred steps they will take if it becomes necessary to reduce SO_2 emissions to maintain standards. Dow is provided (three times daily at 8-hour intervals) with recommended plant operating conditions for the next 30 hours based on output from SO_2 prediction models and an operational computer decision model. This latter model specifies, for given meteorological conditions, the plant operations required to have SO_2 concentrations stay below ambient standards.

The program is quite sophisticated in its capability to specify rather subtle changes in operation of each of 12 boilers exhausted through nine stacks at the Dow facility. Some of these boilers are on coal and some of these coal boilers are capable of being switched to oil. Other boilers are on oil or gas and load may be reduced at these boilers.

Experience with the program thus far indicates a significant reduction in average SO_2 levels and only one instance in six months of the daily SO_2 standard being exceeded at the monitoring locations (Table 20.2). In 1972, before the SCS was operational, the 24-hour federal primary SO_2 standard was exceeded at the monitoring sites a total of 33 times and there were 23 instances of the secondary 3-hour standard of 0.5 ppm being exceeded (Table 20.3).

Table 20.2. Number of Times Standards were Violated Near Dow *With* SCS.

	Month (1974)			
	September	*October*	*November*	*December*
Number of Days[a] with Control Action Required/Days in Month	12/30	25/31	18/30	23/31
Number of AAQS Violations	1[b]	1[c]	0	0

[a]On approximately one-third of these days only limited control steps were recommended to Dow.
[b]No action taken in response to control recommendations.
[c]Control action taken was insufficient.

Table 20.3. Number of Times and Percentage Frequency[a] Standards were Violated Near Dow *Without* SCS.

	Monitoring Site							
	2		3		4		5	
Standard	*Number*	*%*	*Number*	*%*	*Number*	*%*	*Number*	*%*
24-hour: 0.14 ppm	20	5.5	10	2.7	2	0.5	1	0.3
24-hour: 0.14 ppm (Midnight to Midnight only)	17	4.6	5	1.3	2	0.5	1	0.3
3-hour: 0.50 ppm	20	0.7	2	0.07	1	0.03	0	0

[a]Based on 366 possible independent 24-hour periods and on 2928 independent 3-hour periods in 1972.

Greater Boston

Forecasts have been prepared in a program for the North Shore of eastern Massachusetts just outside the City of Boston since January, 1972, for use as the basis for selective fuel burning.[7] This program is being sponsored by New England Power Company (NEPC) and Cambridge Electric (CE) as a demonstration for the technical feasibility of fuel management. Verification of the 24-hour SO_2 forecasts indicates that predictions are within 0.01 ppm over 90% of the time. Until the present time, the fuel switching forecast program has been an exercise, without actual switching taking place. It has proved, however, that it is possible to predict SO_2 concentrations (in the Boston area, at least) to a greater degree of accuracy than is probably required for any viable fuel management program. Results for two years of operation are shown in Figure 20.3.

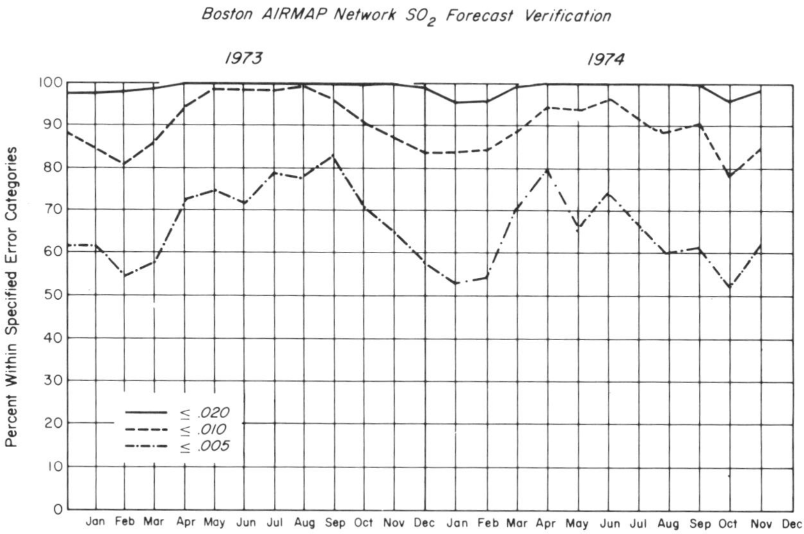

Figure 20.3. Forecasting accuracy for SO_2 concentrations in the Greater Boston area.

Ontario Hydro

Some form of dynamic emissions control system has been operated by Ontario Hydro since the late 1960s. In earlier years, the control strategies used were less sophisticated than present methods. At the present time DEC is being operated for three plants in Ontario.[8]

At the R. L. Hearn plant in Central Toronto, there are eight generating units with a total capacity of 1400 MW and one 700-foot stack. Four of the units operate on natural gas and the other four are on coal or gas. The control strategy is to switch mostly to gas when winds are onshore from Lake Ontario blowing over Toronto. With a forecast of offshore winds (out over the lake), coal is burned on the units that can burn either coal or gas.

Because the plant configuration has been changed from all coal and short stacks in the mid 1960s to its present configuration, it is difficult to evaluate the impact of the DEC system on air quality.

At the Lakeview plant, which has four 500-foot stacks and a capacity of ~2200 MW, a DEC system was operating on a trial basis during 1974 and is mandatory this year. The plant is located about one mile from the western boundary of the City of Toronto. It burns only coal with an average sulfur content of 2.2%. The control strategy is to predict meteorological conditions and reduce load under certain conditions. There is a prespecified schedule of plant operating conditions (required to stay within standards) associated with several different meteorological conditions (*e.g.*, onshore and offshore winds, inversions). Ambient levels of SO_2 are thereby reduced under certain prespecified meteorological conditions.

A third plant where SCS has been implemented by Ontario Hydro is their Lampton plant near Sarnia, Ontario. It is located in relatively flat terrain, has two stacks of ~500 ft each and burns coal. Rated capacity is estimated near 2000 MW. A Pollution Index is forecast and load switching is accomplished based on this index. The frequency of occurrence when the one-hour Ontario standard of 0.30 ppm has been exceeded has dropped noticeably since implementation of the SCS.

Long Island Lighting Company (LILCO)

For the past year LILCO has been fuel switching at two of its power plants—Northport and Port Jefferson. Northport has three 600-foot stacks fed by boilers of 360 MW capacity each. The Port Jefferson plant has two base loading units with boilers rated at 175 MW each feeding a stack 425 ft high and two peaking units of 95 MW each feeding one stack 300 ft high.

The control strategy used by LILCO is based on observed monitoring data only. When the SO_2 concentration averages 0.20 ppm or higher for 2 hours, or when the 4-hour average is 0.10 ppm or higher at any of the 8 monitors, LILCO switches from high-sulfur oil (2.8% sulfur content) to low-sulfur oil (1.0% sulfur content).

In 1974, at Port Jefferson, switching was needed in four instances. At Northport, it was required only once. There were no violations of national or state SO_2 standards in the past year at any monitor.

Pennsylvania Electric Co. (Penelec)

A demonstration project for SCS systems is being performed for the AEC by the Massachusetts Institute of Technology (MIT) Energy Laboratory and Environmental Research & Technology, Inc. (ERT) in cooperation with the Pennsylvania Electric Co. (Penelec).[9] Unique aspects of the project include extension of the state-of-the-art for SCS through the design, implementation and testing of a dynamic control system on the four interconnected, mine-mouth coal-fired plants operated by Pennsylvania Electric Company near Johnstown, Pennsylvania. In particular, four goals have been defined:

1. To demonstrate that an SCS can be designed and operated reliably and economically for multiple interconnected sources in complex terrain.
2. To provide data on the environmental costs or benefits and the economic and reliability impacts on the power system for comparison with other control schemes.

3. To identify the problems and to suggest their possible solutions in the transfer of SCS technology to other plants and locations.
4. To identify the stochastic aspects of SCS and to develop means for dealing with them.

Wind direction, wind speed, atmospheric stability and mixing depth are predicted operationally three times per day for an area in Pennsylvania in the vicinity of the Chestnut Ridge monitoring network. These real-time predictions of meteorological parameters are input to a sophisticated ERT numerical diffusion model that includes terrain information and produces forecasts of pollutant concentrations. Actual emission rates and fuel type are used as input. To demonstrate its feasibility for those conditions that are calculated by the model to produce violation of standards, comparison with observed pollutant measurements and simulation runs of different fuel-type (lower sulfur) inputs to the model are made.

GENERAL CRITERIA FOR SCS CANDIDACY

The establishment of rational criteria for SCS candidacy is necessarily couched in the definition of requirements for technical feasibility and reliability at a given plant, economic reasonableness of alternative control strategies, and successful enforcement by regulatory agencies.

Requirements for technical feasibility and reliability include:

1. A demonstration that reasonable forecasting accuracy is possible provided that sufficient meteorological, emissions and ambient air quality data in the form of an adequate monitoring system can be implemented. Predictive dispersion models should be able to account for complex terrain, mesometeorological (lake or sea breeze fumigation and valley drainage) and micrometeorological (building effects) conditions to give a reliable prediction of ground-level concentrations. This should include a demonstration that background or multi-source interactions can be readily taken into account in the plant control strategy.
2. A demonstration that the possible degree and response time of plant emission reduction options are compatible with the predicted requirements of a dynamic control system.
3. A demonstration that system components will be adequate and reliable to assure required real-time data capture, predictive results, and, ultimately, control decisions to prevent standard violations.

Candidacy is also very much of an economic concern. Even if a plant can satisfy the technical requirements for SCS, the economic costs of achieving acceptable level of control sophistication may preclude the use of SCS. Of major concern is the availability and cost of low-sulfur fuels, the cost of plant or system retrofit to permit fuel switching, and the availability of adequate utility system reserve capacity at an acceptable cost in light of diurnal or seasonal demand. Economic criteria must be developed on the plant or system basis and cannot be defined generically.

Starting with the principles guiding present sulfur oxide control strategies, criteria can be set forth, which would provide a framework around which successful enforcement of any SCS can be tied. These criteria have been developed over the last several years in response to the special scrutiny enforcement of SCS has received within EPA. These criteria are the following:

1. The systems must be objective; *i.e.*, the same control procedure must result from similar meteorological conditions.
2. Air quality control agencies must retain ultimate control over the daily performance of the systems.
3. The systems must be reliable (as opposed to accurate).
4. The systems must be flexible enough to allow other sources (and therefore differing background concentrations) to be easily taken into account.
5. There must be unequivocal evidence of compliance to emission limitations.

Together, the technical, economic and regulatory criteria addressed above, if properly met, should allow SCS to be implemented as an effective, reliable, cost-effective, and enforceable method for maintaining the NAAQS.

SULFATE: MONITORING AS A COMPLEMENT TO SCS

Clearly, the most pressing need in the entire sulfur oxide control strategy controversy is for more information on the origin, transformation, transport and scavenging of sulfates. The possible serious toxicological influence of atmospheric particulate matter for which water soluble sulfates are a proxy has been raised by the EPA as the result of the analysis of data collected in the Community Health and Environmental Surveillance System (CHESS) studies. Data on atmospheric sulfate concentration are also available from the EPA National Air Surveillance Network (NASN) and from several state-run air quality monitoring programs. Very little data on atmospheric sulfate levels, however, has been collected by industry and utilities whose operation of sulfur oxide emitting processes may be severely influenced by this new sulfates issue. When the regulations are established for sulfates, processes now in compliance with sulfur dioxide and particulate standards could be forced to exercise additional controls on SO_x emissions to comply with the standards on sulfates. It is, therefore, important to gather information on the atmospheric sulfate concentrations present in the vicinity of industrial and utility plants to provide a background of data for use in formulating an intelligent control plan.

Atmospheric sulfate measurements which utilize high volume samplers do not lend themselves to short response data retrieval normally required in SCS. However, ambient sulfate monitoring, when coupled with the data generated by an operational SCS, can provide a direct means for establishing a cause-effect relationship between the SO_2 emissions from a given source and the observation of sulfates at a given receptor or monitoring site. A procedure for defining this relationship at an SCS controlled plant is described below.

Establishing this relationship for SO_2 is often possible since the range of significant SO_2 impact from a source has been shown to be relatively short range—approximately 10 miles. The rate of oxidation is very slow in the atmosphere: on the order of 1–3% of the SO_2 reacting per hour. This means that the range of impact of sulfates formed from SO_x emissions from a given source is long range, up to 1000 miles or more. Resolving the impact of a given SO_x source on observed atmospheric sulfate levels is therefore difficult. Control strategies for SO_x emissions based on sulfate occurrence may, therefore, be more regional- than individual-source oriented. However, information on sulfate aerosol concentrations in the vicinity of a given source is useful in determining the requirement for local versus regional control.

Standard high-volume samplers currently in use as part of SCS monitoring networks can be used for this purpose. A typical sulfate monitoring program should include the collection of 24-hour high-volume filters over at least a year from two or more sampling stations in each network. If two stations are used, one should be in the prevailing downwind direction from the plant, and the other in a direction with low-frequency (upwind) from the plant.

Filters should be analyzed for water soluble sulfate. The pH and conductivity of the aqueous extract from each filter should also be determined to serve as a future documentation of data quality. Output data should include atmospheric sulfate concentration as well as conventionally reported total atmospheric particulate loading.

These sulfate data when properly coupled with an analysis of meteorological and SO_2 ground level concentrations gathered during annual SCS operations can be used to develop the following significant products:

1. Documented annual average sulfate concentrations in the vicinity of the plant with analysis of seasonal trends.

2. Analysis of the persistence of high and low concentration episodes as related to synoptic, meso- and microscale meterological conditions.
3. Frequency distributions of sulfate concentrations by season with analysis of meteorological conditions associated with peak concentrations.
4. Statistical analysis of the variability of sulfate concentrations for upwind and downwind cases.
5. Stratification of sulfate concentrations as a function of types of air mass passing over the region by season.

The type of data generated by the operational SCS can greatly facilitate the development of the above analyses to identify the relationship between regional and local sulfate concentrations. Such an interface also permits the possible prediction of sulfate levels as trends analyses are developed and causal relationships are better understood.

CONCLUSIONS

It is evident that supplementary control systems are a feasible and reliable means of achieving and maintaining short- and long-term air quality standards. Several continuing operational case studies show that a properly designed and implemented SCS offers a scientific methodology for meeting air quality goals while minimizing use of costly low-sulfur fuels or permanent emissions control measures. As constant emission control systems become more technically and economically viable, SCS will still afford a complementary control option to reduce capital and operating cost and maintain air quality during system outages.

The decision to use SCS in industrial or utility operations must consider its technical feasibility, operational reliability, economic reasonableness, and regulatory enforceability. Because of its potential flexibility in treating multiple sources, the scale of SCS applications can readily be expanded from the single isolated plant concept to multi-source or regional use.

Operational SCS also affords the plant operator the opportunity to gather site-specific sulfate data as a direct means for establishing a cause-effect relationship between SO_2 emissions from a given source and sulfate measurements made in the surrounding monitoring network. This data can result in a better understanding of the relationship between regional and local sulfate levels for potential use in formulating an intelligent sulfate control plan.

REFERENCES

1. Gaertner, J. P., B. A. Egan, N. E. Gaut, J. S. Lague, T. F. Lavery, H. S. Rosenblum and F. C. Schweppe. "Analysis of the Reliability of a Supplemen-

tary Control System for SO_2 Emissions from a Point Source," ERT Project Report P-669, Prepared for U.S. Environmental Protection Agency, Environmental Research & Technology, Inc., Concord, Massachusetts, June, 1974.
2. Montgomery, T. L. and J. W. Frey. "Tall Stacks and Intermittent Control of SO_2 Emissions—TVA Experience and Plans," American Mining Congress, Las Vegas, Nevada, October 7-10, 1974.
3. Montgomery, T. L., S. B. Carpenter and H. E. Lindley. "The Relationship between Peak and Mean SO_2 Concentrations," American Meteorological Society Conference of Air Pollution Meteorology, Raleigh, North Carolina, April, 1971.
4. Montgomery, T. L. *et al.* "Control of Ambient SO_2 Concentrations by Noncontinuous Emission Limitation on Large Coal-Fired Power Plants," 102nd Annual Meeting of AIM, M & Petroleum Engineers, Chicago, February 25-March 1, 1973.
5. Nelson, D. W., M. A. Yeager and C. K. Guptill. "Closed-Loop Control System for Sulfur Dioxide Emissions from Non-Ferrous Smelters," (New York, N.Y.: American Smelting and Refining Company, 1973).
6. Spiegler, D. B. and E. Newman. "Evaluation of Various Operating Modes at the Dow-Midland Boiler Facilities Relative to Compliance with Ambient Air Quality Standards," *Analysis of Air Quality Measurements,* Vol. 2 (Concord, Massachusetts: Environmental Research & Technology, Inc., 1973).
7. Spiegler, D. B. (Ed.). *ERT AIRMAP Newsletter,* **3**, Nos. 1 and 2 (Concord, Massachusetts: Environmental Research & Technology, Inc., 1974).
8. Boyer, A., personal communication (January, 1975).
9. Ruane, M. R., D. M. Fyock and B. A. Egan. "Supplementary Control Systems —A Demonstration." Submitted to the 1975 Summer Power Meeting IEEE Power Engineering Society.

ADDITIONAL REFERENCES

Dunlap, R. W. "Tall Stacks and Intermittent Control Systems." Discussion document prepared for the National Academy of Sciences/National Academy of Engineering in Their Power Plant Emissions—Air Quality Study. Carnegie-Mellon University, Pittsburgh, Pennsylvania, December 10, 1974.

Environmental Research & Technology, Inc. "The Impact on Air Quality of the Pioneer Valley AIRMAP Group and the Feasibility of a Fuel Use Management Program," ERT Project Report P-500 (Concord, Massachusetts: Environmental Research & Technology, Inc., 1972).

Federal Energy Administration. "The Clean Fuels Deficit — A Clean Air Act Problem," (Washington, D.C.: Office of Environmental Assessment and Impact, 1974).

Finklea, J. F., D. B. Turner, G. G. Akland, R. F. Larsen, V. Hasselblad and S. D. Shearer. "Briefing Notes — A Status Report on Sulfur Oxides," (Research Triangle Park, North Carolina: U.S. Environmental Protection Agency, Office of Research and Development, 1974).

"Nevada SO_2 Control Strategy," *Fed. Reg.* **39**, No. 209 (1974).

"Requirements for Preparation, Adoption and Submittal of Implementation Plans — Prevention of Air Pollution Emergency Episodes," *Fed. Reg.* **36**, No. 206 (1971).

Train, R. E. "Implications of Alternative Policies for the Use of Permanent Controls and Supplemental Control Systems (SCS)," Report prepared by EPA for the Secretary of the Interior, November 14, 1974.

"Use of Supplementary Control Systems and Implementation of Secondary Standards," *Fed. Reg.* **38**, No. 178 (1973).

21

Tall Stacks and Intermittent Control of SO_2 Emissions: TVA Experience*

T. L. Montgomery and **J. W. Frey**
Tennessee Valley Authority, Muscle Shoals, Alabama

INTRODUCTION

The Tennessee Valley Authority, in producing about 7% of the United States' electrical energy, operates 12 large coal-fired power plants that comprise 72% of TVA's installed electric generating capacity. They burn about 38 million tons of coal annually, or about 6% of the coal used in the United States. The locations of these twelve plants and basic facts about them are shown in Figure 21.1 and Table 21.1.

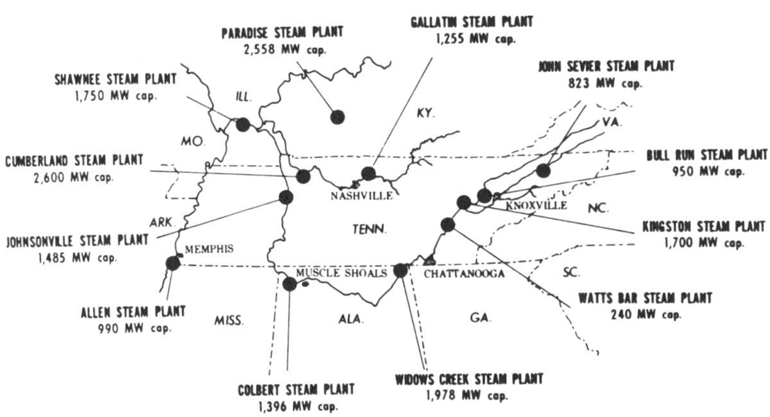

Figure 21.1. TVA coal-fired power plants.

*This chapter was originally published in *Mining Congress Journal* (January, 1975) and is reprinted by permission of the American Mining Congress.

Table 21.1. Facts About Coal-fired Power Plants.

Plant	*No. of Units*	*Construction Started*	*Units Placed in Service*	*Capacity (kW)*	*24-Hour Coal Use at Full Load (tons)*	*Cost ($ millions)*
Watts Bar	4	1940	1942–1945	240,000	3,040	$ 22
Johnsonville	10	1949	1951–1959	1,485,200	13,266	196
Widows Creek	8	1950	1952–1954 1961–1965	1,977,985	16,230	280
Shawnee	10	1951	1953–1957	1,750,000	14,040	223
Kingston	9	1951	1954–1955	1,700,000	14,256	198
Colbert	5	1951	1955, 1965	1,396,500[b]	11,832	175
John Sevier	4	1952	1955–1957	846,500	7,392	117
Gallatin	4	1953	1956–1959	1,255,200	9,636	143
Allen[a]	3	1956	1959	990,000[b]	7,200	[a]
Paradise	3	1959	1963, 1970	2,558,200	21,016	366
Bull Run	1	1962	1967	950,000	7,560	143
Cumberland	2	1968	1972–1973	2,600,000	22,500	392
Totals	63			17,749,585	147,968	$2,255

[a]Allen plant built by Memphis Light, Gas & Water Division, leased by TVA in 1965. Cost not included in total.
[b]In addition, gas turbine peaking units have been installed at Allen (620,800 kW) and Colbert (476,000 kW).

Table 21.2. National Ambient Air Quality Standards for SO_2

	Standard[a]	
Averaging Time	*Primary (ppm)*	*Secondary (ppm)*
3 hours[b]	—	0.5
24 hours[b]	0.14	—
1 year	0.03	—

[a]The primary standards are designed to protect public health, and the secondary standard is designed to protect public welfare.
[b]The 3-hour and 24-hour standards are not to be exceeded more than once per year, and the annual standard is not to be exceeded.

The Clean Air Act Amendments of 1970 required the Environmental Protection Agency to set ambient air quality standards for various air pollutants.[1] Sulfur dioxide is an air pollutant of great concern in the operation of coal-fired power plants, and the National Ambient Air Quality Standards that EPA set for SO_2 are shown in Table 21.2. As a part of their Clean Air Act requirements,[2] most states have adopted SO_2 emission standards as the control strategy designed to ensure that ambient SO_2 standards will be met and maintained. The SO_2 emission standards applicable to TVA coal-fired power plants are shown in Table 21.3.

TVA was aware that some positive action would have to be taken concerning SO_2 emissions from its coal-fired power plants. All generating units could not be operated to meet SO_2 emission standards by the mid-1975 ambient standard deadlines. Early in 1973, TVA adopted a strategy designed to ensure that ambient SO_2 standards

Table 21.3. SO_2 Emission Standards Applicable to TVA Coal-fired Power Plants.[b]

State	*Plant*	*SO_2 Emission Standard*
Alabama	Colbert	1.8 lb/10^6 Btu
	Widows Creek	1.2 lb/10^6 Btu
Kentucky	Paradise	2.0 lb/10^6 Btu
	Shawnee	2.0 lb/10^6 Btu
Tennessee	Allen	4.0 lb/10^6 Btu
	Bull Run	4.0 lb/10^6 Btu
	Cumberland	4.0 lb/10^6 Btu
	Gallatin	4.0 lb/10^6 Btu
	John Sevier	4.0 lb/10^6 Btu
	Johnsonville	620 ppm[a]
	Kingston	620 ppm[a]
	Watts Bar	4.0 lb/10^6 Btu

[a]The Tennessee standard of 620 ppm is approximately equivalent to 1.2 lb/10^6 Btu, which is the national new source emission standard.

[b]The above standards are to be met no later than 1975 in Alabama and Tennessee, and by 1978 in Kentucky.

would be met in the vicinity of all of its coal-fired power plants by mid-1975. Ambient SO_2 standards were already being met around three of the smaller TVA plants, and no additional measures were needed to control ambient levels of SO_2. Also, as a result of a recent relaxation of the emission standards in parts of Tennessee, two of these plants, Bull Run and John Sevier, are now in compliance with SO_2 emission standards. For the other nine plants, ambient SO_2 standards will be met and maintained as follows: a) sulfur dioxide emission limitation (SDEL) programs will be designed for the nine plants and implemented. These SDEL programs are meteorologically based techniques[3,4] for the intermittent control of SO_2 emissions; b) taller stacks will be constructed and put into operation at the Kingston, Shawnee, and Widows Creek plants. The use of these taller stacks will reduce the complexity and improve the effectiveness of the SDEL programs at the three plants.

TVA is constructing a full-scale experimental flue gas desulfurization system (SO_2 scrubber) on a 550-MW coal-fired unit at its Widows Creek power plant. The scrubber will be a part of the overall ambient SO_2 control strategy.

TALL STACKS

The development of plume dispersion models for relating tall stack emissions to ambient SO_2 concentrations has led to the conclusion that tall stacks are an effective means for controlling ambient SO_2 concentrations in the vicinity of large coal-fired power plants.[5,6] As newer and larger TVA coal-fired plants were built, taller stacks were constructed to effectively disperse the SO_2 before it reached ground level (Table 21.4). At the Cumberland plant, 1000-foot stacks serve each of the two 1300-MW units (Figure 21.2).

Table 21.4. Stack Height Information—TVA Coal-fired Power Plants.

Plant	*Year Last Unit Placed in Service*	*Height of Tallest Stack (ft)*
Watts Bar	1945	177
Kingston	1955	300
Shawnee	1957	250
John Sevier	1957	350
Johnsonville	1959	400
Allen	1959	400
Gallatin	1959	500
Widows Creek	1965	500
Colbert	1965	500
Bull Run	1967	800
Paradise	1970	800
Cumberland	1973	1000

Figure 21.2 Cumberland coal-fired power plant.

An evaluation of potential SO_2 control strategies at a number of coal-fired power plants led to the decision to build taller stacks at three plants—Kingston, Shawnee, and Widows Creek. The comparative benefits of the proposed taller stacks over the existing stacks can be obtained from Table 21.5. With existing stacks, a calculated SO_2 emission level less than 1.0 lb/10^6 Btu would be needed to meet ambient standards at two of these three plants. To achieve this emission level would require using coal with only about 0.6% sulfur. However, with taller stacks, ambient SO_2 standards can be met with an emission level of about 4.0 lb/10^6 Btu. This emission level is about that which results from burning 2.3% sulfur coal. Since this coal is generally available in greater quantities than 0.6% sulfur coal, these new taller stacks will be of great benefit in controlling ambient SO_2 concentrations around these plants while at the same time permitting the continued use of regional coal supplies.

SULFUR DIOXIDE EMISSION LIMITATION PROGRAMS

The SDEL Concept

TVA has maintained a long-standing monitoring program at its steam plants, most of which are in relatively isolated rural areas.

Table 21.5. Characteristics of Existing and Planned Taller Stacks—TVA Coal-fired Power Plants.

Plant	*Units*	*Stack Height (ft)*	*SO_2 Emission Level Needed to Meet SO_2 Ambient Standards (lb $SO_2/10^6$ Btu)*
Kingston	1–4	250 }	0.9
	5–9	300 }	
	1–9	1000[a]	4.8
Shawnee	1–10	250	2.3
	1–10	800[a]	4.5
Widows Creek	1–6	270 }	0.7
	7–8	500 }	
	1–6	1000[a] }	3.7
	7–8	500 }	

[a]Planned new stacks—to be in operation in 1976 for Kingston and 1977 for Shawnee and Widows Creek.

Many instrument-years of recorded ambient SO_2 concentrations show that more than 95% of the time these concentrations are either not measurable or are only a fraction of the present primary or secondary ambient SO_2 standards. At infrequent intervals, periods of adverse dispersion occur. During these periods atmospheric dispersion processes are inhibited and SO_2 concentrations in excess of ambient standards are liable to occur. These concentrations are usually confined to particular locations within 2–6 miles of the plant, depending upon the actual meteorological conditions. The SDEL program is a technique for controlling ambient SO_2 concentrations by limiting SO_2 emissions during these adverse atmospheric dispersion periods.

From many years of study, considerable information has been obtained through which the relationships between emissions and ambient concentrations may be understood. This knowledge is coupled with monitoring results and with experience gained from intensive field studies and is applied to develop an SDEL program that is operated to determine objectively and routinely an emission limit that will ensure that ambient concentrations will remain below the level of ambient standards. These programs, which need to be implemented only about 5% of the time when adverse dispersion conditions occur, reduce SO_2 concentrations around a given plant.

Sufficient control can be obtained with these programs to meet existing ambient SO_2 standards continuously, for all wind directions, independent of SO_2 monitor locations. These programs operate as follows: a) meteorological and, in some cases, ambient SO_2 information is taken at or near the plant once a day or more often; b) this information is entered into a computer for detailed evaluations, which include calculations of atmospheric dispersion; c) if prevailing or projected meteorological conditions are such that adverse dispersion could produce significant ambient SO_2 concentrations, the computer results indicate the need to control plant SO_2 emissions to the level necessary to meet ambient SO_2 standards. The SO_2 emission level is controlled by reducing plant generation or by switching, on a short-term basis, to coal with a lower sulfur content.

The Paradise SDEL Experience

In September 1969, TVA began its first SDEL program at the Paradise power plant in Kentucky.[3,4] The program was initiated concurrently with the startup of a new third generating unit, which increased the capacity of the plant by 80%. For this particular program, meteorological measurements are made each morning to determine whether adverse dispersion conditions will exist that day. If so, steps are taken to reduce the SO_2 emission rate by reducing plant generation. The program has operated successfully five years to date. Levels of the primary ambient SO_2 standard are no longer being exceeded as they were before the program was initiated. Data for the 24-hour average health-related primary standard for SO_2 are shown in Table 21.6 for periods before and after the program began. These data show that primary standard levels were exceeded eight

Table 21.6. Frequency Distribution of Measured SO_2 Concentrations[a] Before and After Implementation of SO_2 Emission Limitation Program—Paradise Coal-fired Power Plant.

	Frequency	
24-Hour Average SO_2 Conc. Range (ppm)	*Before (April 29, 1963–September 18, 1969)*	*After (September 19, 1969–March 31, 1974)*
0	8232	7611
.001–.02	6052	6752
.021–.04	549	496
.041–.06	121	110
.061–.08	29	27
.081–.10	14	11
.101–.12	7	5
.121–.14 [b]	4	4
.141–.16	7	0
.161–.18	1	0

[a]From four SO_2 recorders 1963, and 1964; increased to five SO_2 recorders on April 20, 1965; increased to fourteen SO_2 recorders on January 1, 1968; reduced to seven SO_2 recorders on July 1, 1971; increased to nine SO_2 recorders on January 1, 1974.

[b]National Primary Ambient Air Quality Standard for SO_2 is 0.14 ppm, not to be exceeded more than once per year.

times before the SDEL program was initiated but at no time afterward. Although these measurements were largely recorded in the plant's north-northeastern sector where extensive air monitoring has been conducted, generation reductions are always implemented independent of wind direction or monitor locations.

The TVA System of SDEL Programs

A detailed plant-by-plant analysis of ambient conditions around all TVA power plants was made using ambient monitoring information and dispersion modeling. Figure 21.3 displays some of the results of this analysis and shows the 24-hour average ambient SO_2 concentrations around all 12 TVA power plants under three types of conditions: a) no SO_2 control, b) fixed SO_2 emission control as governed by emission standards now applicable, and c) SDEL program control. Figure 21.3c shows that ambient SO_2 standards are expected to be met when SDEL program control is used. By contrast, Figure 21.3b shows that compliance with present emission standards will not achieve corresponding compliance with the 24-hour average health-related primary ambient standard for SO_2 at two of the TVA plants—Widows Creek and Kingston—primarily because of elevated terrain near these plants. Results of a detail analysis of SO_2 concentration levels, that can be expected in the vicinity of all TVA plants, are presented in Table 21.7. This analysis shows that the large, rela-

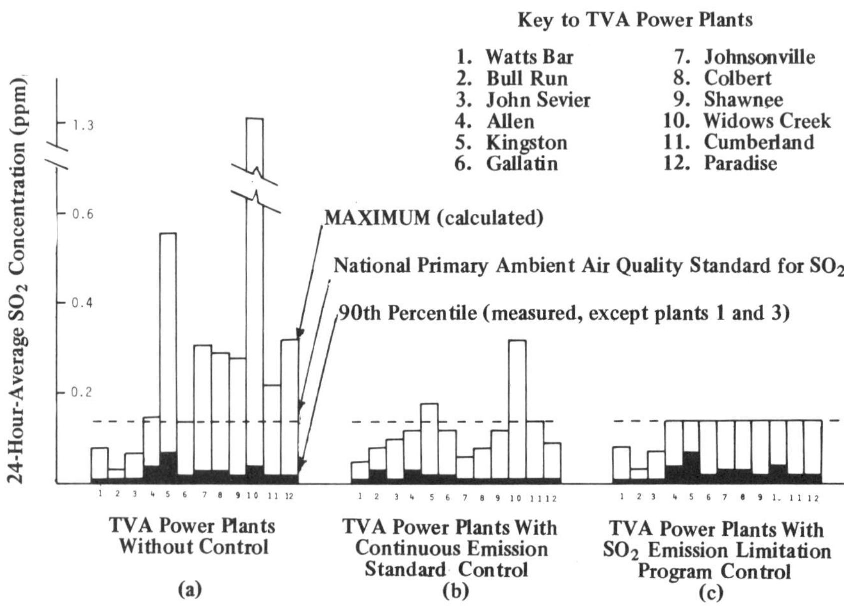

Figure 21.3. Ambient air quality conditions at TVA coal-fired power plants.

Table 21.7. Ambient SO_2 Concentrations—TVA Coal-fired Power Plants.

Plant	*Maximum Measured SO_2 Concentrations[a] (ppm)*			*Maximum Calculated SO_2 Concentrations[b] (ppm)*		
	3-hour Average	*24-hour Average*	*Annual Average*	*3-hour Average*	*24-hour Average*	*Annual Average*
Allen	0.35	0.08	0.011	0.65	0.15	0.007
Bull Run	0.21	0.05	0.005	0.14	0.03	0.002
Colbert	0.79	0.24	0.011	1.19	0.29	0.012
Cumberland	0.32	0.08	0.005	1.09	0.22	0.011
Gallatin	0.47	0.14	0.005	0.63	0.14	0.006
John Sevier	0.28	0.06	N A[c]	0.29	0.07	0.003
Johnsonville	0.61	0.16	0.013	1.31	0.31	0.014
Kingston	0.83	0.45	0.020	1.01	0.56	0.026
Paradise	0.79	0.17	0.011	1.54	0.32	0.016
Shawnee	0.67	0.24	0.020	1.08	0.28	0.011
Watts Bar	0.35	0.08	N A[c]	0.37	0.08	0.004
Widows Creek	3.03	0.52	0.019	4.85	1.31	0.018

[a]The actual maximum measurement recorded at each plant during the complete history of that plant's monitoring network through March 31, 1974.
[b]Based on FY 1973 average emissions.
[c]Value not available.

tively isolated power plants in the TVA system have problems with short-term ambient SO_2 standards only.

The SDEL programs have been divided into two types designated Class I and Class II. Class I programs, already in use at the Allen, Cumberland, Gallatin, and Paradise plants, involve meteorological measurements and predictions that are used for dispersion modeling (Figure 21.4) but do not include direct feedback from ambient SO_2 monitors. Class II programs, which will be implemented at the Colbert, Johnsonville, Kingston, Shawnee, and Widows Creek plants, will involve feedback from ambient SO_2 monitors in addition to the meteorological measurements, predictions, and dispersion modeling (Figure 21.5). Development of both Class I and Class II programs involves extensive field studies at each plant, analyses of the data collected, and design of the particular dispersion model or models to be applied in each case. Completion of these programs depends heavily upon the installation of all necessary equipment and the selection and training of staff to operate them.

Figure 21.6 shows the development schedule for all of the SDEL programs and the date when new tall stacks are scheduled to be put into operation for three plants.

Tall Stacks and SDEL Program Control

Applicability

Since SDEL programs involve the use of very little equipment,

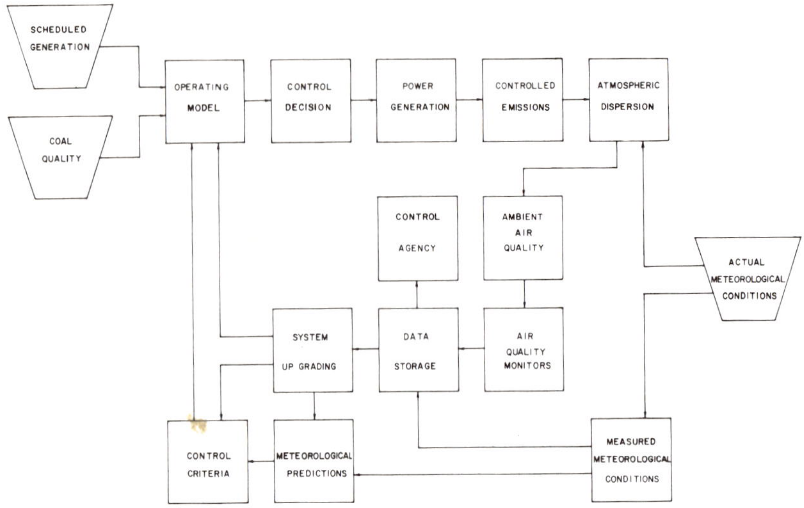

Figure 21.4. Elements of TVA's Class I SDEL programs—Allen, Cumberland, Gallatin, and Paradise power plants.

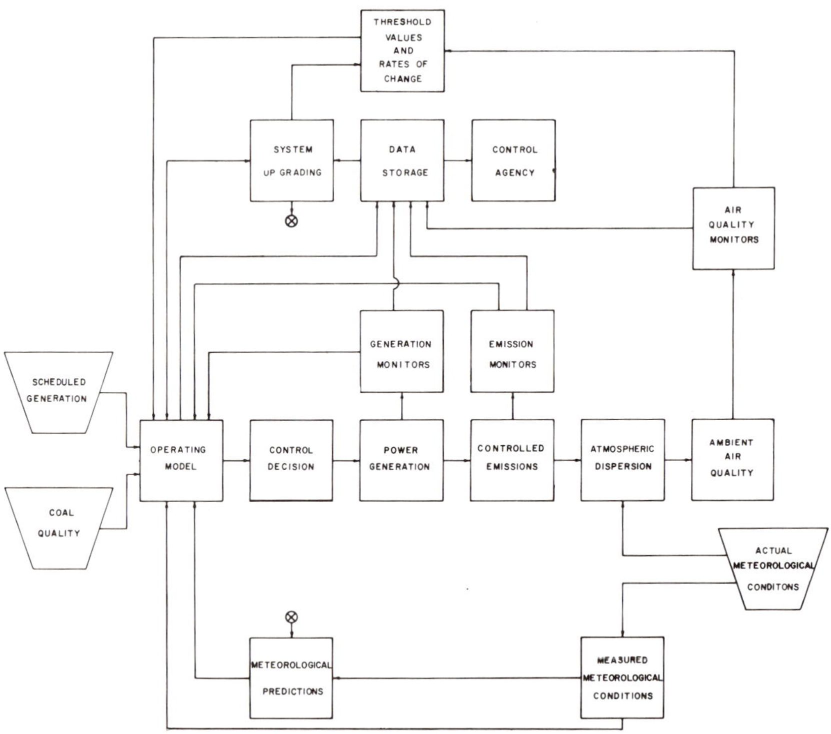

Figure 21.5. Elements of TVA's Class II SDEL programs—Colbert, Johnsonville, Kingston, Shawnee and Widows Creek power plants.

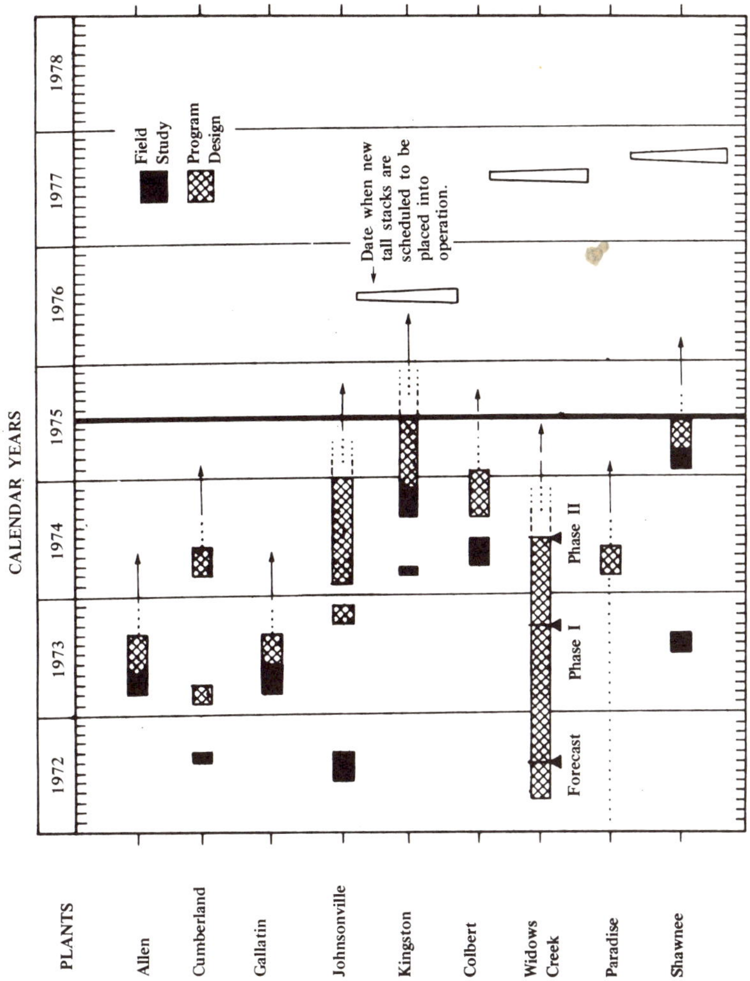

Figure 21.6. Development schedule—sulfur dioxide emission limitation (SDEL) programs—TVA coal-fired power plants.

compared to that required for flue gas desulfurization systems, SDEL program *availability* is excellent. The technology for tall stack construction is already available. Detailed cost estimates for the SDEL programs are given in Table 21.8. Average developmental and operational costs are \$0.581 and \$0.232 million. By contrast, based upon the estimated cost of the single experimental SO_2 scrubber for the Widows Creek unit 8, it might cost \$100 million to equip a large power plant with SO_2 scrubbers and another \$20 million per year to operate it. Therefore, SDEL programs are highly *cost-effective* compared to SO_2 scrubbers—the cost-effectiveness factor ranges from about 10–100. Paradise monitoring results, noted previously in Table 21.6, demonstrate that SDEL programs can be *reliable* in meeting ambient SO_2 standards. Furthermore, the programs are *enforceable*. The emission limits that are dictated by an SDEL program at regular intervals can be looked upon as variable emission standards that can be enforced in the usual manner. Finally, some secondary environ-

Table 21.8. Cost Estimate[a] SO_2 Emission Limitation Programs—TVA Coal-fired Power Plants.

	Capital Cost (Including First Year Development)				
Plant	*Existing Air Monitors*[b] *($1000)*	*New Air Monitors ($1000)*	*SDEL Program Development ($1000)*	*Total ($1000)*	*Annual Operating Cost ($1000)*
Allen	19	94	269	382	143
Colbert	19	170	269	458	188
Cumberland	170	227	352	749	329
Gallatin	0	94	270	364	132
Johnsonville	19	208	269	496	214
Kingston					
Existing stacks	38	189	352	579	218
Tall stacks			21	21	(218)
Paradise	132	227	270	629	304
Shawnee					
Existing stacks	57	245	269	571	252
Tall stacks			21	21	(252)
Widows Creek					
Existing stacks	113	302	526	941	343
Tall stacks			21	21	(343)
Total, all plants	567	1756	2909	5232	2124
Average cost per plant	63	195	323	581	232

[a]Includes only program development and operating costs, exclusive of power replacement or increased fuel costs.

[b]Installed, as of September 1973.

mental problems associated with SO_2 scrubbing, such as solid waste and water pollution, are avoided by using tall stacks and SDEL program control.

Suitability

Tall stacks and SDEL programs have been opposed as control techniques primarily because they rely on the assimilative and dispersive capacity of the atmosphere and, therefore, do not substantially reduce total emissions. It has also been suggested that SO_2 emissions of the present level may contribute, through secondary processes, to effects other than those now known.

From a local viewpoint, the concentration of sulfur compounds in the air, rather than the total quantity emitted, is the principal factor that determines the quality of air and its effects on health and welfare. As has been stated, sulfur dioxide emitted from sources within the Tennessee Valley region does not normally result in concentrations that approach the level of either the primary or secondary ambient standard. As long as the combination of emission rate and meteorological conditions maintains a ground level concentration below that which causes adverse effects, that is, within the National Primary and Secondary Ambient Air Quality Standards for SO_2, the rate or total quantity of emissions should be of lesser importance. SDEL program control and tall stacks at TVA's relatively isolated coal-fired power plants facilitate the practical attainment of these standards.

From a global point of view, it has been postulated that man-made additions to the sulfur cycle[7] have the potential to adversely affect aquatic and terrestrial ecosystems. As in most areas of the United States, the SO_2 discharged to the atmosphere in the Tennessee Valley region may undergo, in addition to and in conjunction with dilution and dispersion, a broad combination of secondary processes. As is the case with local SO_2 concentrations, the relative importance of these processes depends upon meteorological conditions. Very little is known about the detailed mechanism of the individual processes and much less is known about the combined ones. Therefore, one of the objectives of atmospheric chemistry research[8] is to explain the fate of sulfur oxides emitted to the atmosphere, their known lifetimes considering the removal mechanisms that affect them, the degree to which their form may be altered, and the ways in which they may contribute to possibly adverse secondary effects. The absence of necessary information from which to define existing levels of secondary substances and upon which to base criteria for standards for these substances precludes the promulgation of such standards at this time. Tall stacks and SDEL programs are effective control measures for attaining and maintaining ambient SO_2 standards until the nature of these processes and their potential effects can be fully ascertained.

REFERENCES

1. 77 Stat. 392 as amended, 42 U.S.C. §§ 1857-1857k (1970; Supp. II, 1972), 88 Stat. 246 (1974).
2. Approval and Promulgation of Implementation Plans, 40 C.F.R. Part 52.
3. Leavitt, J.M. *et al.* "Meterological Program for Limiting Power Plant Stack Emissions," *J. Air Poll. Cont. Assoc.* **21**, No. 7, 400-405 (1971).
4. Montgomery, T. L., *et al.* "Controlling Ambient SO_2," *J. Metals* (June, 1973).
5. Carpenter, S. B., *et al.* "Principal Plume Dispersion Models: TVA Power Plants," *J. Air Poll. Cont. Assoc.* **21**, No. 8, 491-495 (1971).
6. Thomas, F. W. *et al.* "Stacks—How High? *J. Air. Poll. Cont. Assoc.* **13**, No. 5, 198-204 (1963).
7. Kellogg, W. W., *et al.* "The Sulfur Cycle," *Science* **175**, No. 4022, 587-596 (1972).
8. Huang, Chao-Ming, *et al.* "Investigation of Atmospheric Chemical Interactions in Coal-Fired Power Plant Plumes," Chattanooga, Tennessee (May, 1974).

Section V
Removal of Sulfur Dioxide from Stack Gases

22

Status of Sulfur Dioxide Removal Systems for the Electric Utility Industry

Timothy W. Devitt and **Thomas C. Ponder, Jr.**
PEDCo-Environmental Specialists, Inc.
Cincinnati, Ohio

This chapter provides an initial evaluation of the status of flue gas desulfurization (FGD) technology in the United States. As shown in Table 22.1, by February 1, 1975 there were 21 operating FGD systems on utility size boilers, 19 units under construction and 61 units planned. The planned units include those for which there has been a contract awarded, a letter of intent signed or the utility is presently requesting or evaluating bids. Of the operating systems, fourteen are lime- or limestone-based (including limestone injection), two are magnesium oxide scrubbing, two are sodium carbonate, two are double alkali and one is catalytic oxidation.

Table 22.1. Number and Total of MW of FGD Systems.

Status	*No. of Units*	*MW*
Operational	21	4,471
Under construction	19	6,337
Planned		
Contract awarded	9	3,851
Letter of intent	5	910
Requesting/evaluating bids	13	7,022
Considering only FGD Systems	34	13,814
Total	101	36,405

Figure 22.1 illustrates the projected increase in the number of megawatts of power production equipped with FGD systems. By late 1975, approximately 7000 MW of capacity will be installed, of which 50% will be retrofit applications and 50% will be on new power plants. By 1980, approximately 35,000 MW of capacity will be installed, of which 30 and 70% will be retrofit and new installations, respectively.

Figure 22.2 illustrates the megawatts of capacity that will be controlled by FGD process. In 1974, approximately 80% was controlled by lime and limestone scrubbing processes; in 1977, over 90% will be lime or limestone scrubbing.

SO_2 removal efficiencies for the operating units range from approximately 70–90%, and particulate removal efficiencies generally are above 99% for those units designed for particulate removal. Installation size varies from about 30 MW to over 800 MW. FGD systems have been applied to both low (0.4–1.0%) and high (6.0%) sulfur coals. The nonregenerable systems have used a variety of sludge disposal methods including ponding, dewatering by vacuum filtration and landfilling, and fixation followed by landfilling. The regenerable systems have been limited to date to recovering sulfuric acid although a sulfur recovery unit is being installed.

RELIABILITY AND AVAILABILITY

The reliability and availability of FGD systems has been a major question. During plant inspections, it was apparent that most plants had overcome the "chemical" problems; most operators felt they could control the process chemistry and hence minimize or eliminate scaling and related problems. However many units were and still are faced with a number of mechanical problems (*e.g.*, frozen venturi throat drives, fan vibration). While most of these problems are not major in that they reoccur on an infrequent basis, they do lead to lower unit availability.

Over the past year, the availability of flue gas desulfurization systems has improved. The Cholla unit of Arizona Public Service has operated with availabilities in excess of 90% in 1974. The Will County Station of Commonwealth Edison has had availabilities of over 90% for the past six months. The LaCygne unit has been averaging approximately 80% availability. The Paddy's Run unit of Louisville Gas and Electric, which is on a seasonal boiler and is used to carry summer peak loads, has operated essentially whenever it was required. In addition to these systems, others have shown high availabilities. Southern California Edison's 160 MW horizontal test unit at their Mojave Station has operated with availabilities of approximately 90% over the past year. The operating history of these and other units is presented in Table 22.2.

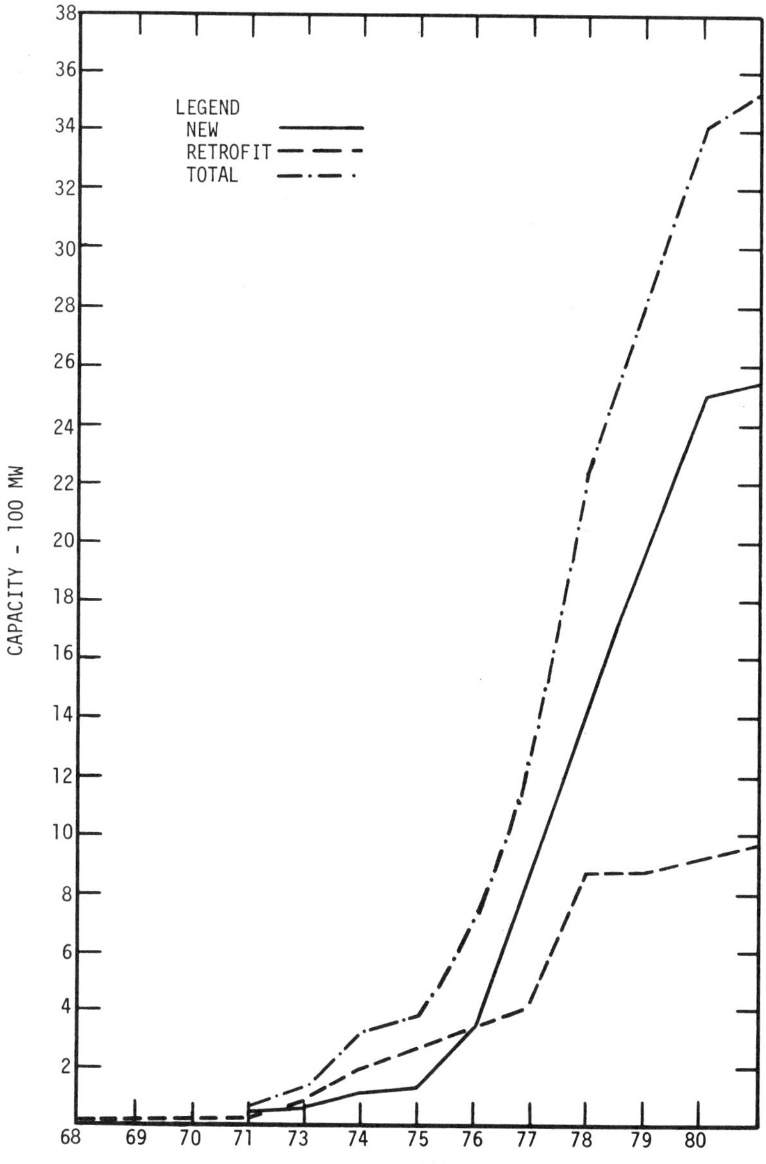

Figure 22.1. Present and projected FGD systems capacity new/retrofit/total vs. year.

Table 22.2. Operating Summary for FGD Systems.

Utility and Unit Name	*Status*
Arizona Public Service Cholla No. 1	Since start-up in October 1973 the system availability has averaged over 90%. There have been some mechanical problems, the most persistent being vibration in the reheater section, and liquid and gas leaks. More recent problems include significant corrosion attack on the vessel walls of module B, which does not have packing.
Boston Edison Mystic No. 6	Scrubber availability reported for 1973 ranged from 73% in August to 13% in December. The decreasing availability was due to deterioration and subsequent overhaul of process equipment. Availability figures for 1974 are: March-87%, April-81%, May-57% and June-80%. This system is down with no immediate plans for restarting since the EPA demonstration program is completed. Boston Edison is evaluating the data collected during the demonstration to determine the course of future action.
Commonwealth Edison Will County No. 1	Availability of Module A has averaged over 90% since April 1, 1974. It was 96% in July, 93% in August, 85% in September, 94% in October, 97% in November, 99% in December, 99% in January, and 99% in February. Module B is being modified to make this module as reliable as Module A. Leaks are still occurring in the reheat tube bundles.
Dairyland Power Coop Alma Station	This experimental wet lime furnace injection system has been operational since 1971 for periods up to 30 days. Recently the unit was tested using western low-sulfur coal. The two-day test showed an approximate 50% SO_2 removal efficiency, but precipitator performance was adversely affected. The FGD project is shut down indefinitely.
Duquesne Light Phillips Station	At present five of the six boilers at the station have been tied into the scrubber system. Four scrubber trains are installed. Each train includes a venturi scrubber for particulate control. Only one train (No. 1) includes a second stage venturi for SO_2 control. The plan attempts to operate trains No. 1 and No. 4 continuously. Availability was indicated to be 100% in January 1975. Fly ash overloads clarifiers so that all four units cannot be simultaneously operated at capacity.
General Motors Parma Plant	System has been operational since February 17 after shutdown for modifications to scrubber piping. Modules have operated on a relatively continuous basis, depending on boiler load. A. D. Little is currently running material balance tests and other operational tests for Federal EPA.
Gulf Power Scholz Plant	The Scholz Plant has three, 20-MW demonstration units. The first unit, the double alkali system, was placed in operation February 5, 1975. In the first five days of operation the system was down a total of 8 hours because of a vacuum-coupling failure. The second demonstration unit, the Chiyoda Thoroughbred 101 System, started operation in March, 1975. The third unit, the carbon adsorption system, was placed in operation in April, 1975.

Illinois Power Wood River	The FGD system has had significant problems since startup in 1972. The longest sustained operation was about 24 hours in July 1973; unit operated at a reduced rate for 140 hours in 1972, 160 hours in 1973, and 55 hours in 1974. SO_2 removal efficiency was 85%. Presently, the plant auxiliary equipment is being checked out. Burner modifications are complete and plant restart is anticipated in mid-April.
Kansas City Power and Light Hawthorn No. 3	The FGD unit has been out of service all year because system cannot be started without shutting down boiler, and gas curtailment precludes boiler restart. Boiler is now down, and oil-fired light-off burners are being installed. Because of continued operational problems since startup, the availability factor for the systems has been low (in the range of 30–40%). The most recent operation of the FGD system has shown improved availability of about 70 percent.
Kansas City Power and Light Hawthorn No. 4	This unit was converted from furnace injection to tail end injection. The plant experienced more problems with this unit than with Hawthorn 3. SO_2 removal efficiency is about 70%. The boiler was down most of Dec. to repair the turbine. FGD unit has been out of service all year because system cannot be started without shutting down boiler, and gas curtailment precludes boiler restart. Oil fired light-off burners are to be installed.
Kansas City Power and Light LaCygne	From February, 1973–February, 1974 boiler debugging problems caused 112 outages (9 were due to FGD modules). Operation of boiler and FGD system have since improved. System availability for the past 12 months is calculated to be 76%. Overall SO_2 efficiency is 70%. Boiler down from November 21 until late February for annual turbine overhaul.
Kansas Power and Light Lawrence No. 4	The system has undergone a number of process modifications. At present the system is capable of sustained operation although SO_2 removal efficiency is limited to only 75% and the demisters must be washed daily (automatic washer). Manual washing of the demisters is required every two weeks. The unit is badly corroded and will be replaced by a new electrostatic precipitator and FGD system in 1977.
Kansas Power and Light Lawrence No. 5	The FGD system availability is reported to be adequate for the operation of this particular station. Station load is reduced to 50% every night. Therefore, as many as four modules can be shut down for cleaning and repair each evening.
Key West Utility Board Key West	FGD system operation resumed January 7 and continued till January 28. Shutdown was caused by damage to baffles in the ductwork system. The FGD system was scheduled to restart in mid-March.

Table 22.2. Operating Summary for FGD Systems. (Continued)

Utility and Unit Name	*Status*
Louisville Gas and Electric Paddy's Run	Unit operated with 70% availability since startup in April to December 1973. Boiler was shut down in December 1973 because of low demand, and restarted in July 1974. Availability since restarting has been high with no forced shutdown (however, during short boiler runs the FGD system was bypassed). SO_2 removal efficiency 80%. Boiler down due to low demand but may restart for lime test runs if EPA funding request approved.
Nevada Power Reid Gardner No. 1	FGD unit performed satisfactorily since startup 12 months ago. Availability was over 90% and operation was only interrupted by lack of trona. SO_2 removal efficiency is 85%. During December the FGD system operated with 95% availability on trona supplemented with sodium carbonate brine. Availability for January was only about 65% because of frozen carbonate feed lines.
Nevada Power Reid Gardner No. 2	FGD unit performed satisfactorily since startup 12 months ago. Availability was over 90% and operation was only interrupted by lack of trona. SO_2 removal efficiency is 85%. During December the FGD system operated with 95% availability on trona supplemented with sodium carbonate brine. January availability was about 94%. Loss of availability was attributed mainly to frozen sodium carbonate feed lines.
Potomac Electric and Power Dickerson No. 3	The FGD unit operated intermittently during 1974 and was frequently down for process changes. December availability was 64%. SO_2 removal efficiency exceeds 90%. The unit was shut down January 28 and remained out of service for 8–12 weeks due to turbine overhaul.
Southern California Edison Mohave No. 1	This experimental vertical module started up October 31 and operated 500 hours to date with 56% availability. Major problems experienced included migration and loss of ping pong balls in TCA tower causing gas channeling. Also solid buildup on demister trap out tray which is located below the demister.
Southern California Edison Mohave No. 2	Since startup of formal test program on January 16, 1974 and up to December 20, 1974 this experimental horizontal module operated 5280 hours with an overall availability of 77.4%. SO_2 removal efficiency was 90 in flue gas to simulate conditions from combustion of high sulfur coal. This module is being dismantled so that it can be shipped to the Four Corners Plant.
Tennessee Valley Authority Shawnee No. 10	This is an experimental system funded by US EPA. There are three types of scrubbers (10 MW equivalent each) being evaluated under a variety of operational modes.

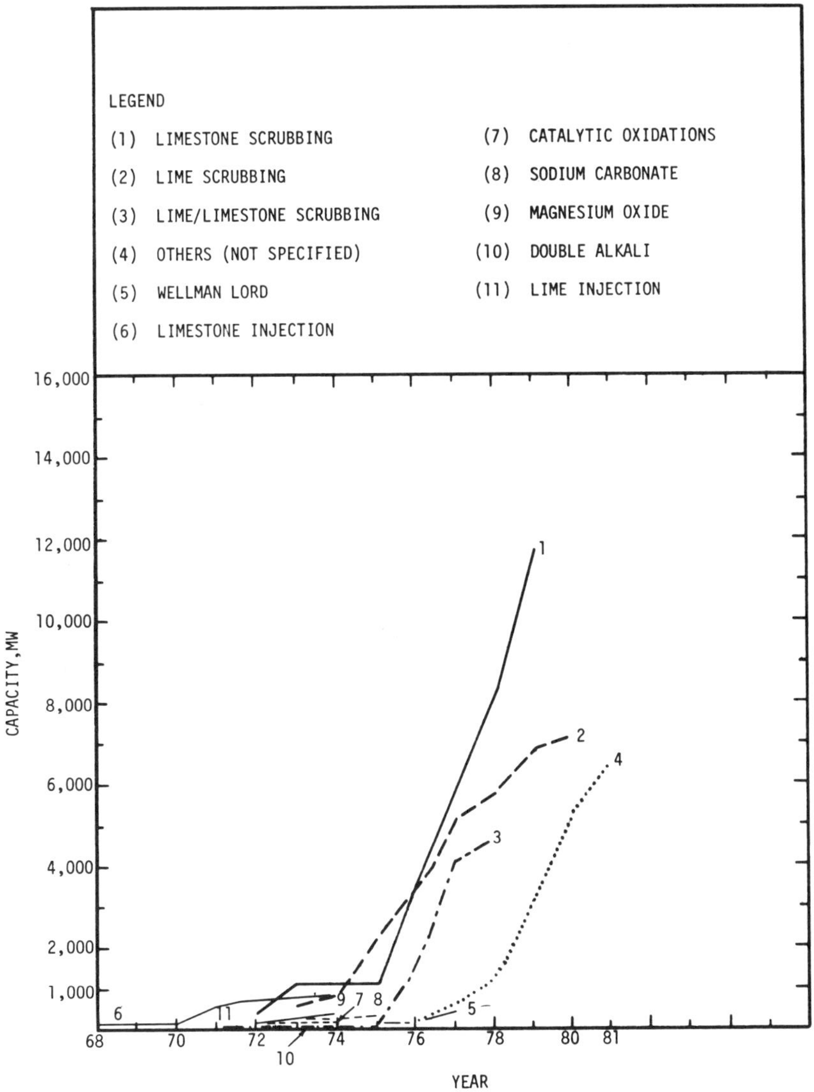

Figure 22.2. Present and projected capacities of FGD processes.

CONTROL COSTS FOR FLUE GAS DESULFURIZATION SYSTEMS

In addition to evaluating the status of flue gas desulfurization, a number of studies have evaluated the cost of controlling the particulate and sulfur dioxide emissions from utility boilers. Current work includes determining the feasibility and environmental impact of converting selected utility boilers to coal-firing to conserve the nation's gas and oil supplies.

Estimates are based on vendor quotes for different sizes of equipment. These sizes cover the total range which may be encountered in flue gas desulfurization. An equation is derived for each equipment item which covers the entire size range. These equations may be linear, exponential, or stepped. Once the total equipment cost is known, installation factors are used to obtain the direct installed cost. Indirect costs such as engineering, contractors fee, and freight are added to obtain the total installed cost.

Using these cost estimating programs, it is possible to generate costs for a base case power plant. These costs are shown in Table 22.3 All costs are expressed in $/kw of installed capacity. The base case is for a new, single boiler, 500-MW plant burning 1% sulfur coal or 3.5% sulfur coal. The Wellman-Lord and the Limestone scrubbing systems were designed to control sulfur dioxide emissions to meet the Federal New Source Performance Standards (1.2 lb/mm Btu).

Table 22.3. Base Case.

	Flue Gas Desulfurization System			
	Limestone		*Wellman-Lord*	
Application	*Low-Sulfur Coal*	*High-Sulfur Coal*	*Low-Sulfur Coal*	*High-Sulfur Coal*
SO_2 control	60	71	69	104
Total	60	71	69	104

The impact of sulfur content on capital cost is much less for the limestone system than the Wellman-Lord system. The regeneration equipment in the Wellman-Lord system is more cost-sensitive to sulfur content than are limestone preparation and sludge disposal facilities. Table 22.4 shows the effect of particle removal on the cost of the FGD system. About $20/kw are required to add a venturi scrubber to an absorber system.

Table 22.4. Base Case with Particulate Removal.

	Flue Gas Desulfurization System			
	Limestone		*Wellman-Lord*	
Application	*Low-Sulfur Coal*	*High-Sulfur Coal*	*Low-Sulfur Coal*	*High-Sulfur Coal*
SO_2 control	60	71	69	104
Particulate control	20	18	22	22
Total	80	89	91	126

In many cases, an FGD system is added to an existing unit. If the retrofit is difficult, it could add \$15–\$20 to the cost of FGD systems as illustrated in Table 22.5. In many cases retrofits are simple and add essentially nothing to the cost.

Table 22.5. Base Case with Particulate Removal and Retrofit.

	Flue Gas Desulfurization System			
	Limestone		*Wellman-Lord*	
Application	*Low-Sulfur Coal*	*High-Sulfur Coal*	*Low-Sulfur Coal*	*High-Sulfur Coal*
SO_2 control	60	71	69	104
Particulate control	20	18	22	22
Retrofit	18	19	25	26
Total	98	108	116	152

Table 22.6 illustrates several other items. These include overdesign of the system to allow for high excess air (30–40%) or an extra absorber train to gain added reliability. Such redundancy can add \$30/kw or more to the cost of the control system.

Table 22.6. Base Case with Particulate Removal, Retrofit and Redundancy.

	Flue Gas Desulfurization System			
	Limestone		*Wellman-Lord*	
Application	*Low-Sulfur Coal*	*High-Sulfur Coal*	*Low-Sulfur Coal*	*High-Sulfur Coal*
Base case	60	71	69	104
Particulate removal	20	18	22	22
Retrofit	18	19	25	26
Redundancy	42	42	43	42
Total	140	150	159	194

Thus, costs for FGD systems can range between \$60/kw and \$150/kw depending upon design philosophy and site specific factors.

WHY SO MUCH?

In 1968, scrubber system costs were estimated to be about \$20–\$30/kw. However, these early systems were comparatively simple processes and not very reliable. Costs have risen to increase process reliability. Scrubbers are now designed with more expensive corrosion/erosion resistant materials. Bypasses have been added to allow the boiler to operate when the scrubber is down. Most pumps are spared. Inflation has been another major factor. During 1974, as shown in Figure 22.3, chemical plant costs increased by 20%. Curves have been projected at 7.5% and 15% to show what the future may hold.

Now that sufficient scrubber reliabilities are being attained, emphasis is shifting to cost/effective process design. Scrubber internals

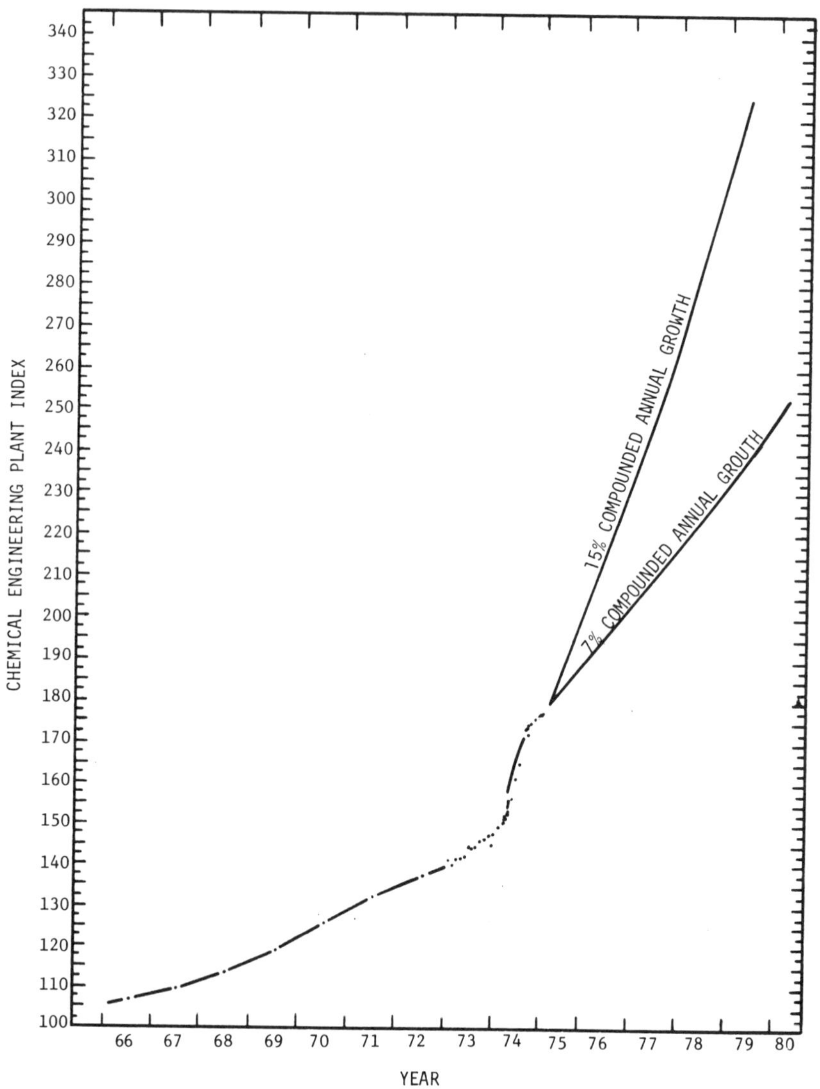

Figure 22.3. Escalation factor.

are being removed to reduce costs and minimize scaling and plugging potential. Coatings and rubber linings are being used for process vessels and piping to minimize corrosion and erosion problems. Reheaters are no longer being put in the flue gas stream, but cheaper, lower alloy reheaters are being used to heat ambient air for mixing with flue gas. Less spares (*e.g.*, pumps) are being required as successful designs are being recognized.

Thus in terms of January 1975 dollars, FGD costs for the majority of future systems will probably stabilize in the \$60–\$80/kw range.

23

Cost Comparisons of Flue Gas Desulfurization Systems

G. G. McGlamery and **R. L. Torstrick**
Tennessee Valley Authority, Muscle Shoals, Alabama

After several years of intensive process development, several power plant stack gas SO_2 removal systems are now advancing from the pilot plant stage to demonstration-scale applications. With the installation of these larger demonstration-size facilities on power units around the United States, it should now be possible to more accurately predict their costs so that utility executives can better choose between alternatives.

Under the Clean Air Act of 1967 and its amendments, the Federal Environmental Protection Agency has funded research and development on SO_2 removal processes including several conceptual design and cost studies. In these earlier efforts, many design assumptions were necessary and cost estimate accuracy was questionable since technology was in a state of flux and available design data were limited. Equipment costs were sketchy since most vendors had yet to fabricate and erect the large gas scrubbing devices required for full-scale systems. Furthermore, very little corrosion data were available to predict materials of construction for the services involved. In many cases, optimism of the process developers tended to maximize process potential and minimize problem areas such as erosion, scaling, solids disposal, sulfite oxidation, mist elimination, gas reheat, operational turndown, and pH control.

Finally, after many pilot plant tests and both encouraging and disappointing experiences, five processes have emerged from the many proposed as the initial systems for installation. These are the limestone slurry process and the lime slurry process, both of which are throwaway systems (no salable by-product), and the magnesia slurry — regeneration to sulfuric acid, sodium solution scrubbing — SO_2 reduction to sulfur, and catalytic oxidation — 80% sulfuric acid processes. In cooperation with participating utilities and process developers, EPA is currently funding large-scale test and demonstration projects on each of these processes. The processes and the associated projects are shown in Table 23.1.

Now that many of the unknowns for these systems have surfaced, remedies been prescribed, and large-scale projects started, more accurate assessment of process costs should be possible. The objective of an EPA-TVA study completed in early 1974 to prepare a set of detailed capital and operating cost estimates for comparison of the five leading processes on a common, uniform basis. Unfortunately, the recent rapid construction cost escalation has once again made it difficult to accurately project total costs; however, comparative costs should be relatively assessable.

PROCESS DEFINITION

A brief description of the five processes and the organizations supplying representative systems data are given below.

1. *Limestone Slurry Scrubbing.* Stack gas is washed with a recirculating slurry (pH of 5.8–6.4) of limestone and reacted calcium salts in water using a two-stage (venturi and mobile bed) scrubber system for particulate and SO_2 removal. Limestone feed is wet ground prior to addition to the scrubber effluent hold tank. Calcium sulfite and sulfate salts are withdrawn to a disposal area for discard. Reheat of stack gas to 175°F is provided. Design is based on data taken from EPA-TVA-Bechtel Shawnee test program. A flow diagram is shown in Figure 23.1.

2. *Lime Slurry Scrubbing.* Stack gas is washed with a recirculating slurry (pH of 6.0–8.0) of calcined limestone (lime) and reacted calcium salts in water using two stages of venturi scrubbing. Lime is purchased from "across the fence" calcination operation, slaked, and added to both circulation streams. Calcium sulfite and sulfate is withdrawn to a disposal area for discard. Reheat of stack gas to 175°F is provided. Design is based on data provided by Chemical Construction Corporation (Chemico). The representative flow diagram is shown in Figure 23.2.

3. *Magnesia Slurry Scrubbing — Regeneration to H_2SO_4.* Stack gas is washed using two separate stages of venturi scrubbing—the first utilizing water for removal of particulates, and the second utilizing a recirculating slurry (pH 7.5–8.5) of magnesia (MgO) and reacted magnesium-sulfur salts in water for removal of SO_2. Makeup magnesia is slaked and added to cover only handling losses since sulfates

Table 23.1. EPA-Sponsored Stack Gas Desulfurization Demonstration Systems.

EPA-Sponsored Process (by-product)	*Cooperating Utility*	*Process Developer*	*Location*	*Unit Size and Type*	*Expected Startup*
Limestone slurry scrubbing (sludge)	TVA	Bechtel and others	Shawnee unit 10 Paducah, Ky.	10 MW coal	Underway
Lime slurry scrubbing (sludge)	TVA	Chemico, Bechtel, and others	Shawnee unit 10 Paducah, Ky.	10 MW coal	Underway
Magnesia slurry scrubbing - regeneration (98% sulfuric acid)	Boston Edison	Chemico-Basic	Mystic Station 6 Boston, Mass.	150 MW oil	Completed
Catalytic oxidation (80% sulfuric acid)	Illinois Power	Monsanto	Wood River Station 4 East Alton, Ill.	110 MW coal	Under review
Sodium scrubbing - regeneration (sulfur)	Northern Indiana Public Service Co.	Davy Powergas Allied Chemical	D. H. Mitchell Station 11 Gary, Ind.	115 MW coal	1976

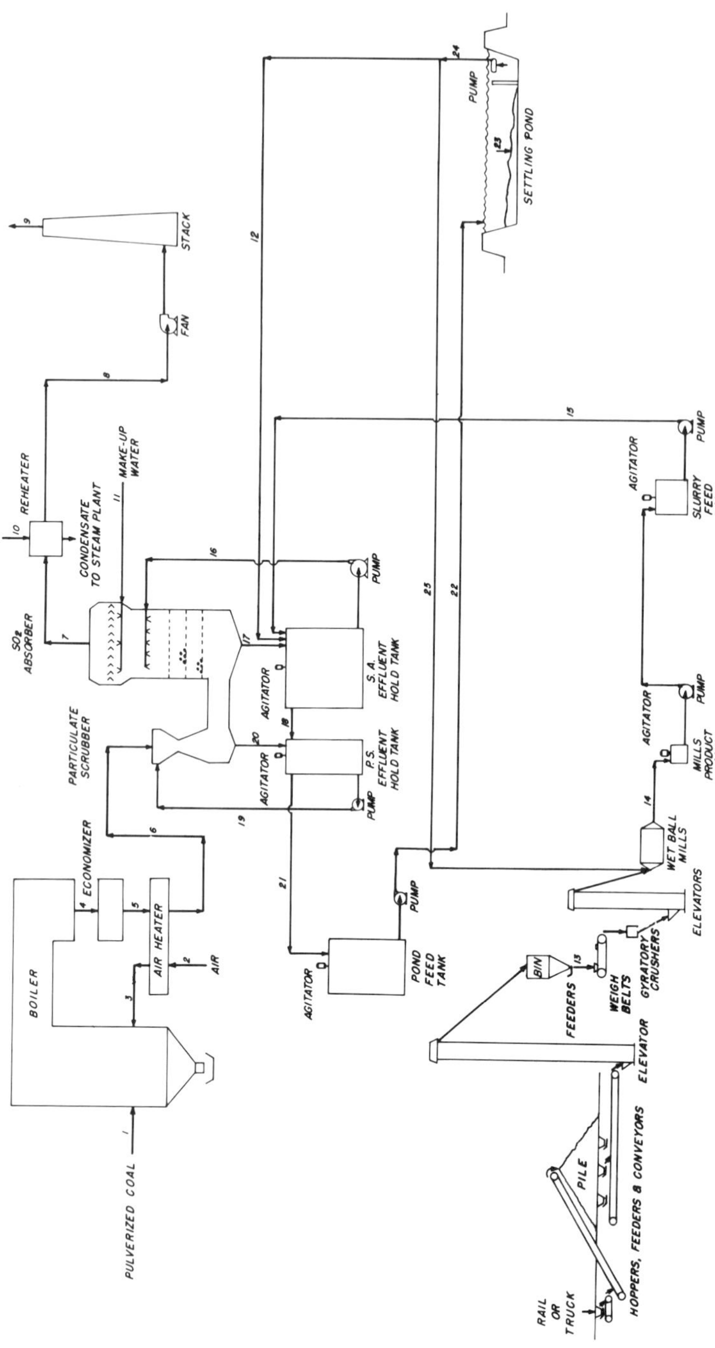

Figure 23.1. Limestone slurry process. Flow diagram.

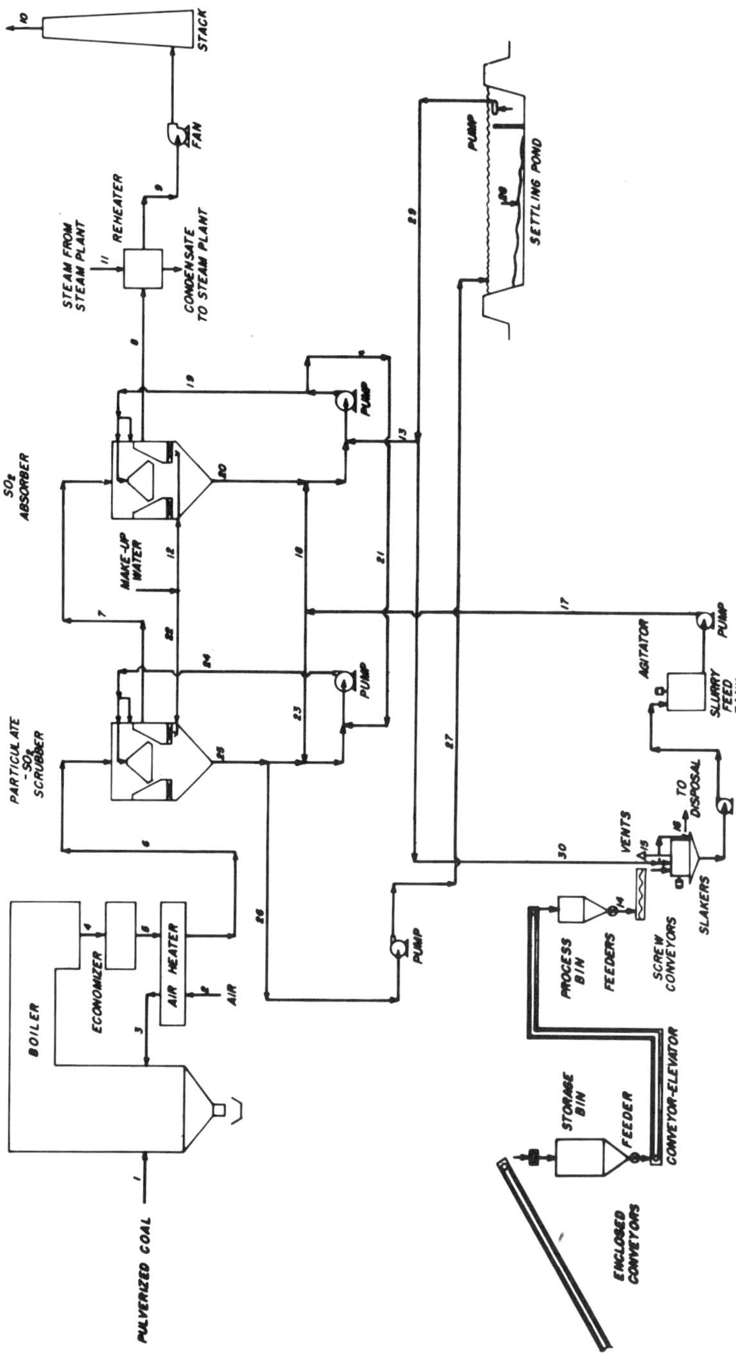

Figure 23.2. Lime slurry process. Flow diagram.

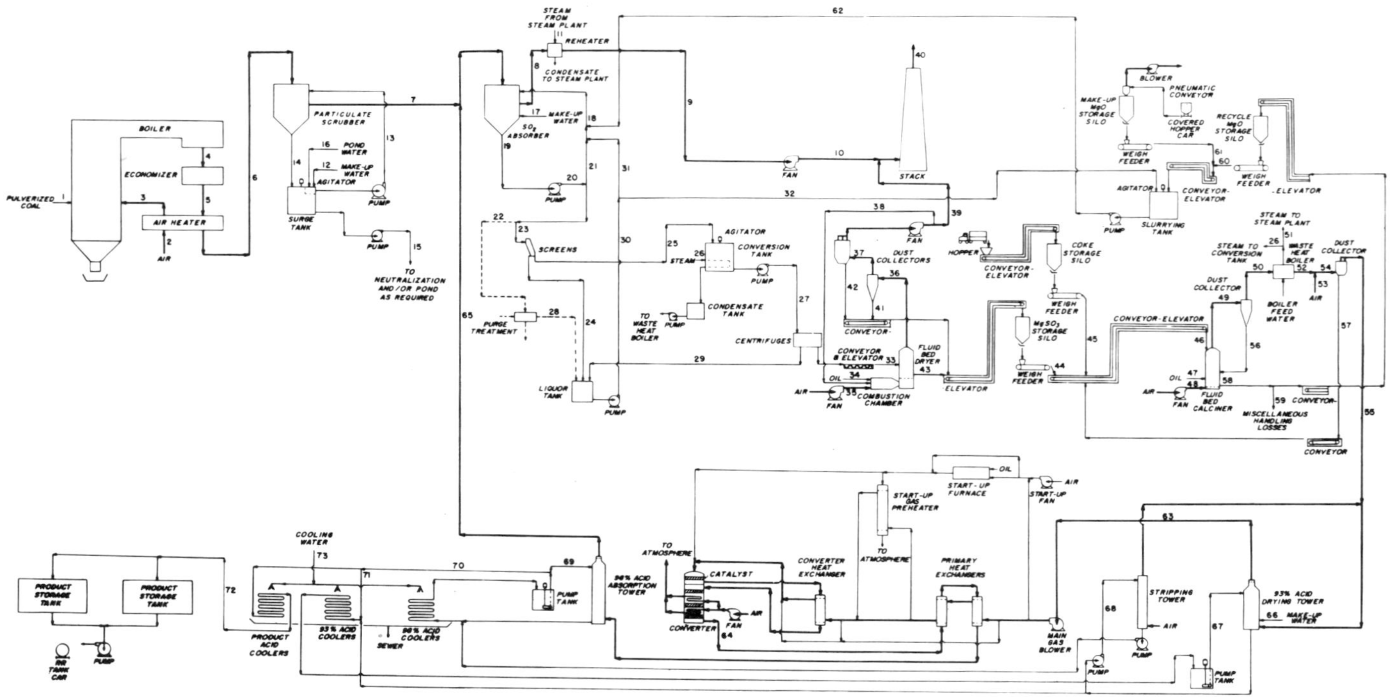

Figure 23.3. Magnesia slurry-regeneration process. Flow diagram.

formed are reduced during regeneration. Slurry from the SO_2 scrubber is dewatered, dried, calcined, and recycled during which concentrated SO_2 is evolved to a contact sulfuric acid plant producing 98% acid. Design is based on data supplied by Chemico-Basic Corporation. A flow diagram is shown in Figure 23.3.

4. *Sodium Solution Scrubbing—SO_2 Regeneration and Reduction to Sulfur*. Stack gas is washed with water in a venturi scrubber for removal of particulates and then washed in a valve tray scrubber with a recirculating solution of sodium salts in water for SO_2 removal. Makeup sodium carbonate is added to cover losses due to handling and oxidation of sodium sulfite to sulfate. Sodium sulfate crystals are purged from the system, dried, and sold. Water is evaporated from the scrubbing solution using a single-effect evaporator to crystallize and thermally decompose sodium bisulfite, driving off concentrated SO_2. The resulting sodium sulfite is recycled to the scrubber and the SO_2 is reacted with methane for reduction to elemental sulfur. The regeneration and reduction areas are designed for 100% of power unit load. Design for the scrubbing-evaporator-crystallizer system is provided by Davy Powergas Inc. (Wellman-Lord process) and data for the SO_2 reduction unit are provided by Allied Chemical Corporation. Figure 23.4 describes the processing steps.

5. *Catalytic Oxidation*. Stack gas is first cleaned of particulates by a high-temperature electrostatic precipitator; then, the SO_2 is catalytically converted to SO_3 and available excess heat is recovered. The SO_3 reacts with moisture in the stack gas to form H_2SO_4 mist which is scrubbed in a packed tower using a recirculating acid stream to yield 80% acid. Mist is removed by a Brink mist eliminator and the clean 254°F gas is exhausted to the stack. Design is based on data supplied by Monsanto Company, developers of the Cat-Ox process. Shown in Figure 23.5 is a flowsheet.

Representative flow diagrams, material balances, control diagrams, plant layouts, and equipment arrangements were prepared for the base case (new 500-MW coal-fired unit, 3.5% S in fuel, 90% SO_2 removal) of each process. Together with detailed equipment descriptions, this background defines the systems estimated.

MAJOR COST FACTORS

From previous economic studies, the following factors are considered to be the most important:

1. *Project schedule and location*. Project assumed to start in mid-1972 with three-year construction period ending mid-1975. Midpoint of construction costs mid-1974; *Chemical Engineering* Cost Index — 160.2. Startup — mid-1975. A midwestern plant location is assumed.

2. *Power unit size*. Costs for three unit sizes—200, 500, and 1000 MW—are projected.

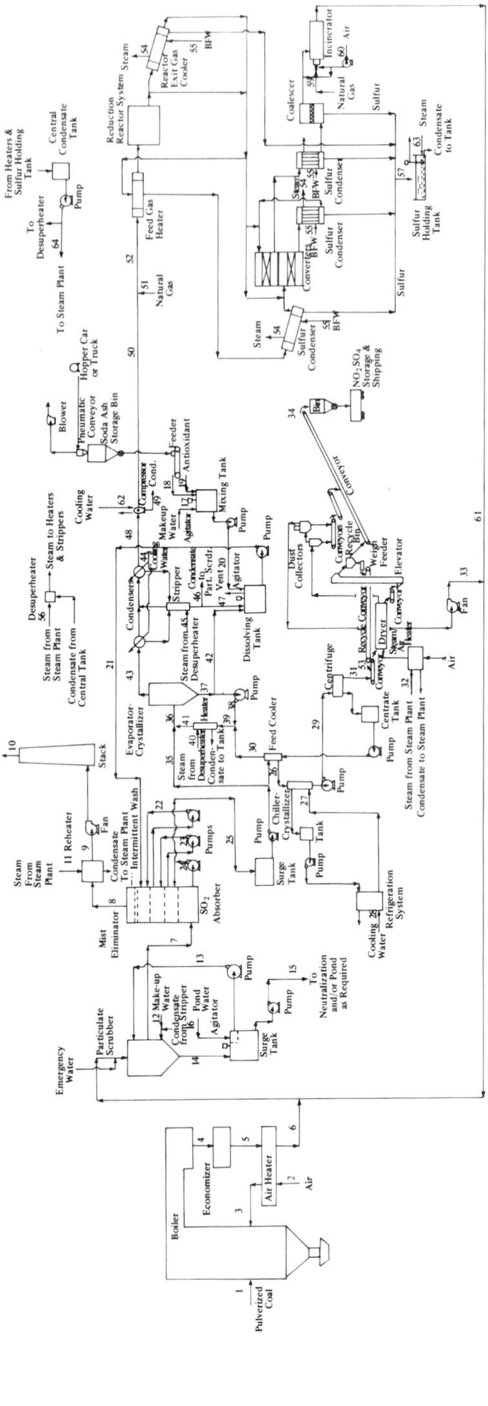

Figure 23.4. Sodium solution-SO_2 reduction process. Flow diagram.

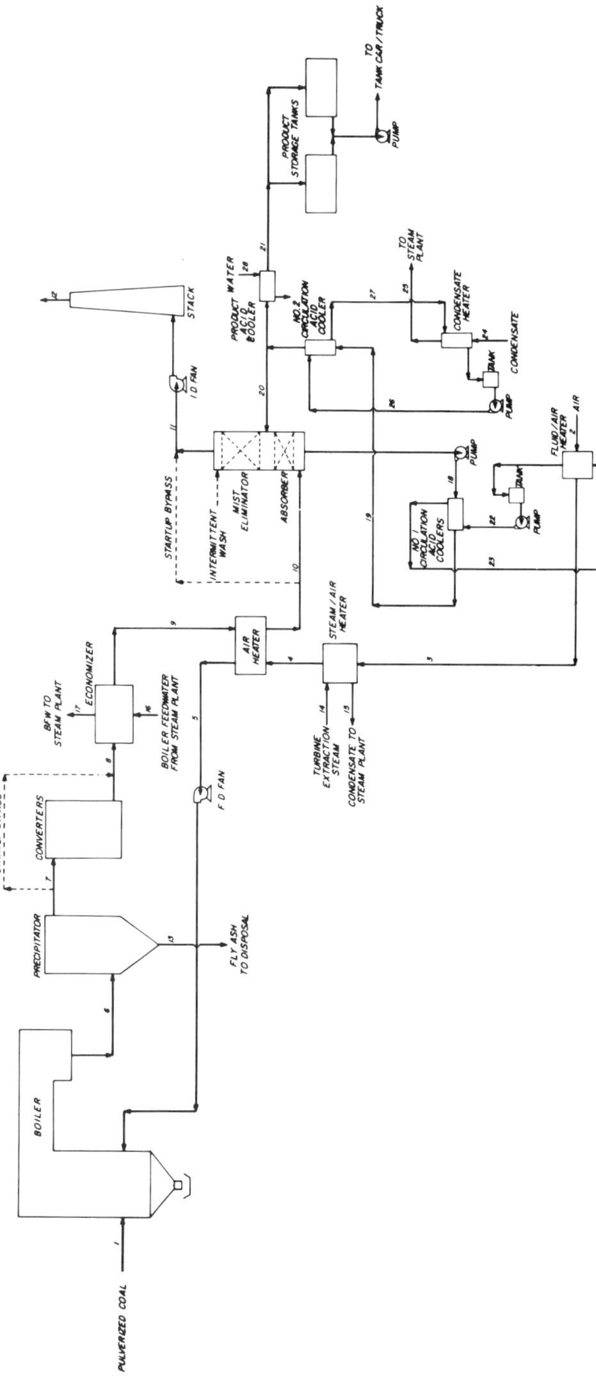

Figure 23.5. Catalytic Oxidation process. Flow diagram.

3. *Fuel type.* Systems for both coal- and oil-fired units are costed — coal 12,000 Btu/lb, 12% ash, oil — 18,500 Btu/lb, 0.1% ash.

4. *Sulfur content of fuel.* Costs for three sulfur levels are evaluated for each fuel—2.0, 3.5, and 5.0% for coal; 1.0, 2.5, and 4.0% for oil.

5. *Plant status.* Although systems for both new and existing power units are evaluated, only a simple, moderately difficult (scrubbing system installed on vacant space beyond the stack) retrofit is estimated since such systems can vary over such a wide range of configurations and restrictions. New units are designed for a 30-year life, 127,500 hours of operation. Costs for new and existing systems are not directly comparable.

6. *SO_2 removal.* Since all five processes are capable of 90% SO_2 removal and future demands for emission control may exceed present standards, 90% removal is specified as the base value. For those processes in which cost-effective design changes could be identified, 80% removal is also projected.

7. *Particulate Removal.* Costs are included for 98.7% particulate removal (to meet EPA standard of 0.1 lb/million Btu heat input) on new coal-fired systems except Cat-Ox which requires 99.9% removal for process reasons (restricted to 0.005 gr/scf prior to entering converter). Existing units are assumed to be already equipped with 98.7% electrostatic participators. Because of this provision, the investment and operating cost results for existing coal-fired systems appear lower than for new units. To cover this disparity, a special case is examined where full particulate removal must be added to an existing unit. Oil-fired units do not require dust removal facilities. Existing Cat-Ox installations require incremental additional participator to reduce loading to 0.005 gr/scf.

8. *Raw materials and catalysts.* Assuming startup in 1975, midwestern 1975 delivered prices are projected and sensitivity analysis is used to evaluate variance.

9. *Labor.* 1975 midwestern operating labor rates are projected. Sensitivity analysis is used to evaluate overall operating labor cost variance. Operating labor is escalated over the life of the project for only a few special cases to show effect.

10. *Utilities.* Recent energy cost escalation is recognized and 1975 values are projected. Values used in operating cost estimates for utilities supplied by power plant cover all costs for generation including return on investment, depreciation, and income taxes.

11. *Maintenance.* Various levels are analyzed by sensitivity analysis.

12. *Capital charges.* Regulated (profit and taxes included) economic basis is used. Annual operating cost estimates utilize a base

value of 14.9% of fixed investment (10% cost of money). Level is varied by sensitivity analysis.

13. *On-stream time*. Annual operating costs were projected for operating times of 7000, 5000, 3500, and 1500 hours per year. Later, these values are used to project a lifetime cost over a predefined 30-year declining operating schedule.

14. *Solids disposal*. Either on-site ponding or off-site disposal costs is included for limestone and lime processes. On-site ponding includes prorated costs for calcium solids to cover pumping and piping to and from the pond, plus a 40-foot-deep, clay-lined pond, sized to meet requirements over the remaining life of the power unit. Off-site disposal includes proration of a thickener, filter, piping, pumps, cake conveyors, and loader at the power unit. Off-site charges are levied on a fee per ton of wet solids basis assumed to cover all contractor expenses for hauling, treatment, and final disposal. Sensitivity anlysis is used to evaluate variance.

15. *Net sales revenue*. Base values—\$8/ton 100% H_2SO_4 as 98% H_2SO_4, \$6/ton 100% H_2SO_4 as 80% H_2SO_4, \$25/short ton for sulfur, \$20/ton for sodium sulfate are used. Variances are covered by sensitivity analysis.

Other important design and cost assumptions defined for consistent evaluation are:

1. Fly ash disposal facilities proration (ponds, pipes, pumps) not included. Water balance is based on closed-loop operation.
2. System design assumed not "first of a kind;" no redundancy is included; only pumps are spared; experienced design and construction team is assumed utilized.
3. Stack gas reheated to 175°F except Cat-Ox process.
4. Product storage — 30 days except for Na_2SO_4 — 7 days.
5. Equipment, material, and construction labor shortages with accompanying overtime pay incentive not considered.

PRESENTATION OF RESULTS

Capital Investment

In the detailed cost study, several methods are used to present the results. For investment data, the following types of information are presented for each process: (a) Equipment list, cost, size-scale factor, and data source—presented for base case (new 500-MW coal-fired unit burning coal with 3.5% S); (b) Summarized process area equipment and installation cost breakdown—presented for new 500-MW limestone system (See table 23.2); (c) Case variations—area investment summaries presented for each of the 16 case variations. See Tables 23.3–23.7 for examples.

Table 23.2. Limestone Slurry Process—Total Capital Investment Requirements

	Raw materials handling	*Food preparation*	*Particulate scrubbing*	*SO_2 scrubbing*	*Reheat*
Direct Costs					
Equipment					
Material	103	379	1,126	2,254	410
Labor	29	57	114	335	80
Piping and insulation					
Material	1	14	102	341	11
Labor	1	28	95	313	20
Ductwork, chutes, and supports					
Material	15	6	733	377	—
Labor	8	6	582	418	—
Concrete foundations					
Material	14	7	12	24	—
Labor	81	38	48	81	—
Excavation, site preparation, railroads, roads, and pond	—	—	—	—	—
Structural					
Material	11	—	84	85	—
Labor	10	—	38	41	—
Electrical					
Material	39	58	61	101	1
Labor	83	92	73	112	1
Instruments					
Material	8	49	77	162	21
Labor	4	25	38	81	11
Paint and miscellaneous					
Material	1	1	3	3	—
Labor	5	6	15	15	1
Buildings					
Material	—	65	—	—	—
Labor	—	67	—	—	—
Land	6	1	2	2	—
Construction facilities	—	—	—	—	—
Subtotal direct investment	419	899	3,203	4,745	556
Indirect Costs					
Engineering design and supervision	38	81	288	427	50
Construction field expense	46	99	352	522	61
Contractor fees	21	45	160	237	28
Contingency	42	90	320	474	56
Subtotal fixed investment	566	1,214	4,323	6,405	751
Allowance for startup and modifications	45	97	346	513	60
Interest during construction	45	97	346	513	60
Total capital investment	656	1,408	5,015	7,431	871
Percent of total capital investment	2.6	5.6	19.9	29.5	3.5

[a]Basis:

500-MW new coal-fired power unit, 3.5% S in fuel; 90% SO_2 removal; on-site disposal.
Stack gas reheat to 175°F by indirect steam reheat.
Disposal pond located 1 mile from power plant.
Midwest plan location represents project beginning mid-1972, ending mid-1975. Average cost basis for scaling, mid-1974.

Base Case[a] Summary—Process Equipment and Installation Analysis.[b]

Fans	*Solids disposal*	*Utilities*	*Services*	*Construction facilities 5%*	*Total*	*Percent of direct investment*	*Percent of total capital investment*
285	60	—	107	—	4,724	29.4	18.8
34	3	—	25	—	677	4.2	2.7
—	86	6	—	—	561	3.5	2.2
—	78	11	—	—	546	3.4	2.2
146	—	—	—	—	1,277	7.9	5.1
53	—	—	—	—	1,067	6.6	4.2
4	1	—	—	—	62	0.4	0.3
21	5	—	—	—	274	1.7	1.1
—	3,028	—	339	—	3,367	21.0	13.4
—	—	—	—	—	180	1.1	0.7
—	2	—	—	—	91	0.6	0.4
133	50	10	—	—	453	2.8	1.8
155	184	10	—	—	710	4.4	2.8
15	11	16	—	—	359	2.2	1.4
7	5	8	—	—	179	1.1	0.7
—	2	3	—	—	13	0.1	0.1
1	13	3	—	—	59	0.4	0.2
—	—	—	113	—	178	1.1	0.7
—	—	—	40	—	107	0.7	0.4
—	395	—	14	—	420	2.6	1.7
—	—	—	—	765	765	4.8	3.0
854	3,923	67	638	765	16,069	100.0	63.9
77	353	6	57	69	1,446	9.0	5.7
94	432	8	70	84	1,768	11.0	7.0
43	196	3	32	38	803	5.0	3.2
85	392	7	64	77	1,607	10.0	6.4
1,153	5,296	91	861	1,033	21,693	135.0	86.2
92	424	7	69	82	1,735	10.8	6.9
92	424	7	69	82	1,735	10.8	6.9
1,337	6,144	105	999	1,197	25,163	156.6	
5.3	24.4	0.4	0.4	4.8			100.0

Minimum in-process storage; only pumps are spared.
Remaining life of power unit, 25 yr.
Investment requirements for removal and disposal of fly ash excluded.
Construction labor shortages with accompanying overtime pay incentive not considered.
[b]In thousands of dollars.

Table 23.3. Limestone Slurry Process Summary of Estimated Fixed Investment[a] (500-MW *new* coal-fired power unit, 3.5% S in fuel; 90% SO_2 removal; onsite solids disposal).

	Investment ($)	*Percent of subtotal direct investment*
Limestone receiving and storage (hoppers, feeders, conveyors, elevators, and bins)	419,000	2.6
Feed preparation (feeders, crushers, elevators, ball mills, tanks, and pumps)	899,000	5.6
Particulate scrubbers and inlet ducts (4 scrubbers including common feed plenum, effluent hold tanks, agitators, and pumps)	3,203,000	19.9
Sulfur dioxide scrubbers and ducts (4 scrubbers including mist eliminators, effluent hold tanks, agitators, pumps, and exhaust gas ducts to inlet of fan)	4,745,000	29.5
Stack gas reheat (4 indirect steam reheaters)	556,000	3.5
Fans (4 fans including exhaust gas ducts and dampers between fan and stack gas plenum)	854,000	5.3
Calcium solids disposal (on-site disposal facilities including feed tank, agitator, slurry disposal pumps, pond, liner, and pond water return pumps)	3,923,000	24.4
Utilities (instrument air generation and supply system, plus distribution systems for obtaining process steam, water, and electricity from the power plant)	67,000	0.4
Service facilities (buildings, shops, stores, site development, roads, railroads, and walkways)	765,000	4.0
Construction facilities	638,000	4.8
Subtotal direct investment	16,069,000	100.0
Engineering design and supervision	1,446,000	9.0
Construction field expense	1,768,000	11.0
Contractor fees	803,000	5.0
Contingency	1,607,000	10.0
Subtotal fixed investment	21,693,000	135.0
Allowance for startup and modifications	1,735,000	10.8
Interest during construction (8%/annum rate)	1,735,000	10.8
Total capital investment	25,163,000	156.6

[a]Basis:

Stack gas reheat to 175°F by indirect steam reheat.

Disposal pond located 1 mile from power plant.

Midwest plant location represents project beginning mid-1972, ending mid-1975. Average cost basis for scaling, mid-1974.

Minimum in process storage; only pumps are spared.

Investment requirements for disposal of fly ash excluded.

Construction labor shortages with accompanying overtime pay incentive not considered.

Table 23.4. Lime Slurry Process. Venturi-Venturi Scrubbing Scheme. Summary of Estimated Fixed Investment[a] (500-MW *new* coal-fired power unit, 3.5% S in fuel; 90% SO_2 removal; onsite solids disposal).

	Investment (\$)	*Percent of subtotal direct investment*
Lime receiving and storage (bins, feeders, conveyors, and elevators)	795,000	5.5
Feed preparation (conveyors, slakers, tanks, agitators, and pumps)	387,000	2.7
Particulate—sulfur dioxide scrubbers and inlet ducts (4 scrubbers including common feed plenum and pumps)	4,017,000	28.1
Sulfur dioxide scrubbers and ducts (4 scrubbers including mist eliminators, pumps, and exhaust gas ducts to inlet of fans)	3,153,000	22.0
Stack gas reheat (4 indirect steam reheaters)	542,000	3.8
Fans (4 fans including exhaust gas ducts and dampers between fan and stack gas plenum)	767,000	5.4
Calcium solids disposal (on-site disposal facilities including slurry disposal pumps, pond, liner, and pond water return pumps)	3,356,000	23.4
Utilities (instrument air generation and supply system, plus distribution systems for obtaining process steam, water, and electricity from the power plant)	67,000	0.5
Service facilities (buildings, shops, stores, site development, roads, railroads, and walkways)	552,000	3.8
Construction facilities	682,000	4.8
Subtotal direct investment	14,318,000	100.0
Engineering design and supervision	1,289,000	9.0
Construction field expense	1,575,000	11.0
Contractor fees	716,000	5.0
Contingency	1,432,000	10.0
Subtotal fixed investment	19,330,000	135.0
Allowance for startup and modifications	1,546,000	10.8
Interest during construction (8%/annum rate)	1,546,000	10.8
Total capital investment	22,422,000	156.6

[a]Basis:

Stack gas reheat to 175°F by indirect steam reheat.

Disposal pond located 1 mile from power plant.

Midwest plant location represents project beginning mid-1972, ending mid-1975. Average cost basis for scaling, mid-1974.

Minimum in process storage; only pumps are spared.

Investment requirements for disposal of fly ash excluded.

Construction labor shortages with accompanying overtime pay incentive not considered.

Table 23.5. Magnesia Slurry-Regeneration Process. Summary of Estimated Fixed Investments[a] (500-MW *new* coal-fired power unit, 3.5% S in fuel; 90% SO_2 removal; 15.8 tons/hr 100% H_2SO_4).

	Investment ($)	*Percent of subtotal direct investment*
Magnesium oxide and coke receiving and storage (pneumatic conveyor and blower, hoppers, conveyors, elevators, and storage silos)	192,000	1.2
Feed preparation (weigh feeders, conveyors, elevators, slurrying tank, agitator, and pumps)	238,000	1.5
Particulate scrubbers and inlet ducts (4 scrubbers including common feed plenum, effluent hold tanks, agitators, pumps, and fly ash neutralization facilities)	3,966,000	24.3
Sulfur dioxide scrubbers and ducts (4 scrubbers including mist eliminators, pumps, and exhaust gas ducts to inlet of fan)	2,592,000	15.9
Stack gas reheat (4 indirect steam reheaters)	509,000	3.1
Fans (4 fans including exhaust gas ducts and dampers between fan and stack gas plenum)	741,000	4.5
Slurry processing (screens, tanks, pumps, agitators and heating coils, centrifuges, conveyors, and elevators)	711,000	4.3
Drying (fluid bed dryer, fans, combustion chamber, dust collectors, conveyors, elevators, and $MgSO_3$ storage silo)	972,000	5.9
Calcining (fluid bed calciner, fans, weigh feeders, conveyors, elevators, waste heat boiler, dust collectors, and recycle MgO storage silo)	1,108,000	6.8
Sulfuric acid plant (complete contact unit for sulfuric acid production including dry gas purification system)	3,197,000	19.6
Sulfuric acid storage (storage and shipping facilities for 30 days production of H_2SO_4)	278,000	1.7
Utilities (instrument air generation and supply system, fuel oil storage and supply system, and distribution systems for obtaining process steam, water, and electricity from power plant)	269,000	1.6
Service facilities (buildings, shops, stores, site development, roads, railroads, and walkways)	783,000	4.8
Construction facilities	778,000	4.8
Subtotal direct investment	16,334,000	100.0
Engineering design and supervision	1,797,000	11.0
Construction field expense	1,797,000	11.0
Contractor fees	817,000	5.0
Contingency	1,633,000	10.0
Subtotal fixed investment	22,378,000	137.0
Allowance for startup and modifications	2,238,000	13.7
Interest during construction) 8%/annum rate)	1,790,000	11.0
Total capital investment	26,406,000	161.7

[a]Basis:

Stack gas reheat to 175°F by indirect steam reheat.

Midwest plant location represents projects beginning mid-1972, ending mid-1975. Average cost basis for scaling, mid-1974.

Minimum in process storage; only pumps are spared.

Fly ash slurry neutralized before disposal; closed loop water utilization for first stage.

Investment requirements for disposal of fly ash excluded.

Construction labor shortages with accompanying overtime pay incentive not considered.

Table 23.6. Sodium Solution-SO_2 Reduction Process. Summary of Estimated Fixed Investment[a] (500-MW *new* coal-fired power unit, 3.5% S in fuel; 90% SO_2 removal; 4.7 tons/hr sulfur).

	Investment ($)	*Percent of subtotal direct investment*
Soda ash and antioxidant receiving, storage, and preparation (pneumatic conveyor and blower, feeders, mixing tank, agitator, and pumps)	225,000	1.2
Particulate scrubbers and inlet ducts (4 scrubbers including common feed plenum, effluent hold tanks, agitators, pumps, and fly ash neutralization facilities)	3,846,000	20.4
Sulfur dioxide scrubbers and ducts (4 scrubbers including mist eliminators, pumps, and exhaust gas ducts to inlet of fan)	4,269,000	22.6
Stack gas reheat (4 indirect steam reheaters)	539,000	2.9
Fans (4 fans including exhaust gas ducts and dampers between fans and stack gas plenum)	889,000	4.7
Purge treatment (refrigeration system, chiller-crystallizer, feed coolers, centrifuge, rotary dryer, steam/air heater, fan, dust collectors, feeders, tanks, agitators, pumps, conveyors, elevator, and bins)	1,473,000	7.8
Sulfur dioxide regeneration (evaporator-crystallizers, heaters, condensers, strippers, compressers, desuperheater, tanks, agitators, and pumps)	2,717,000	14.4
Sulfur dioxide reduction unit	2,921,000	15.5
Sulfur storage (storage and shipping facilities for 30 days production of molten sulfur)	227,000	1.2
Utilities (instrument air generation and supply system, and distribution systems for obtaining process steam, water, and electricity from power plant)	195,000	1.0
Service facilities (buildings, shops, stores, site development, roads, railroads, and walkways)	662,000	3.5
Construction facilities	898,000	4.8
Subtotal direct investment	18,861,000	100.0
Engineering design and supervision	2,075,C00	11.0
Construction field expense	2,075,000	11.0
Contractor fees	943,000	5.0
Contingency	1,886,000	10.0
Subtotal fixed investment	25,840,000	137.0
Allowance for startup and modifications	2,584,000	13.7
Interest during construction (8%/annum rate)	2,067,000	11.0
Total capital investment	30,491,000	161.7

[a]Basis:

Stack gas reheat to 175°F by indirect steam reheat.

Midwest plant location represents project beginning mid-1972, ending mid-1975. Average cost basis for scaling, mid-1974.

Minimum in process storage; only pumps are spared.

Fly ash slurry neutralized before disposal; closed loop water utilization for first stage.

Investment requirements for disposal of fly ash excluded.

Construction labor shortages with accompanying overtime pay incentive not considered.

Table 23.7. Catalytic Oxidation Process. Summary of Estimated Fixed Investment[a] (500-MW *new* coal-fired power unit, 3.5% S in fuel; 90% SO_2 removal; 15.7 tons/hr 100% H_2SO_4).

	Investment ($)	*Percent of subtotal direct investment*
Converter and absorber startup bypass ducts and dampers	491,000	1.9
Electrostatic precipitators and inlet ducts (4 high temperature electrostatic precipitators including common feed plenum)	8,736,000	34.4
Sulfur dioxide converters and ducts (4 converters including catalyst sifter, hopper, storage bin, conveyors, and elevators)	2,145,000	8.5
Heat recovery and ducts (4 steam/air heaters and 4 fluid/air heaters including ducts between economizers and air heaters, and combustion air ducts and dampers between powerhouse and air heaters; investment credit for use of smaller air heaters included)	1,475,000	5.8
Fans (4 ID fans including exhaust gas ducts and dampers between ID fans and stack gas plenum)	1,412,000	5.6
Sulfuric acid absorbers and coolers (2 absorbers including mist eliminators, coolers, tanks, pumps, and ducts and dampers between air heaters and ID fans)	8,917,000	35.2
Sulfuric acid storage (storage and shipping facilities for 30 days production of H_2SO_4)	409,000	1.6
Utilities (instrument air generation and supply system, and distribution systems for obtaining process steam, water, and electricity from power plant)	57,000	0.2
Service facilities (buildings, shops, stores, site development, roads, railroads, and walkways)	518,000	2.0
Construction facilities	1,208,000	4.8
Subtotal direct investment	25,368,000	100.0
Engineering design and supervision	2,790,000	11.0
Construction field expense	2,790,000	11.0
Contractor fees	1,268,000	5.0
Contingency	2,537,000	10.0
Subtotal fixed investment	34,753,000	137.0
Allowance for startup and modifications	3,475,000	13.7
Interest during construction (8%/annum rate)	2,780,000	11.0
Total capital investment excluding catalyst	41,008,000	161.7
Catalyst	1,728,000	6.8
Total capital investment	42,736,000	168.5

[a]Basis:

Midwest plant location represents projects beginning mid-1972, ending mid-1975. Average cost basis for scaling, mid-1974.

Only pumps are spared.

Investment requirements for disposal of fly ash excluded.

Construction labor shortages with accompanying overtime pay incentive not considered.

Shown in Table 23.8 is a summary of total investment for the 16 case variations of the five leading processes. The relative ranking of the processes for new coal-fired units (3.5% S) in order of increasing investment is as follows:

1. Lime slurry
2. Limestone slurry
3. Magnesia slurry-regeneration
4. Sodium solution-SO_2 reduction
5. Catalytic oxidation

Shown in Figures 23.6 and 23.7 are the effects of unit size and sulfur content of coal on investment for installations on new units. In comparing these rankings, it should be pointed out that the lime slurry process does not provide facilities for calcining limestone, and includes a minimum of facilities for storage of the calcined material.

An important result derived by comparing the investment projections in this study is the relatively minor investment savings realized (3.2–4.5%) in designing for 80% SO_2 removal as compared to 90%.

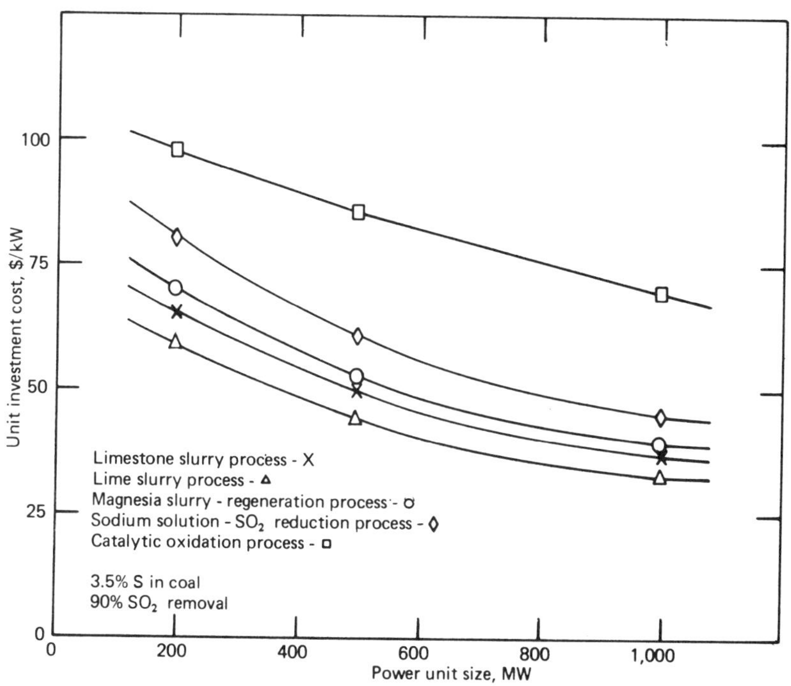

Figure 23.6. New coal-fired units—the effect of power unit size on unit investment cost, dollars per kilowatt.

Table 23.8. Summary of Total Capital Investment Requirements.[a]

Case	Years life	Limestone process ($)	Limestone process ($/kW)	Lime process ($)	Lime process ($/kW)	Magnesia process ($)	Magnesia process ($/kW)	Sodium process ($)	Sodium process ($/kW)	Cat-Ox process[b] ($)	Cat-Ox process[b] ($/kW)
Coal-fired power unit											
90% SO_2 removal; on-site solids disposal											
200 MW new, 3.5% sulfur	30	13,031,000	65.2	11,749,000	58.7	14,139,000	70.7	16,198,000	81.0	19,537,000	97.7
200 MW existing, 3.5% sulfur	20	11,344,000	56.7	13,036,000	65.2	14,372,000	71.9	17,149,000	85.7	17,735,000	88.7
500 MW existing, 3.5% sulfur	25	23,088,000	46.2	26,027,000	52.1	26,026,000	52.1	31,208,000	62.4	37,907,000	75.8
500 MW new, 2.0% sulfur	30	22,600,000	45.2	20,232,000	40.5	22,958,000	45.9	26,706,000	53.4	42,520,000	85.0
500 MW new, 3.5% sulfur	30	25,163,000	50.3	22,422,000	44.8	26,406,000	52.8	30,491,000	61.0	42,736,000	85.5
500 MW new, 5.0% sulfur	30	27,343,000	54.7	24,272,000	48.5	29,355,000	58.7	33,709,000	67.4	42,928,000	85.9
1000 MW existing, 3.5% sulfur	25	35,133,000	35.1	38,133,000	38.1	38,717,000	38.7	47,721,000	47.7	62,913,000	62.9
1000 MW new, 3.5% sulfur	30	37,725,000	37.7	32,765,000	32.8	38,865,000	38.9	45,832,000	45.8	69,889,000	69.9
80% SO_2 removal; on-site solids disposal											
500 MW new, 3.5% sulfur	30	24,267,000	48.5	21,586,000	43.2	25,568,000	51.1	29,127,000	58.3	—	—
90% SO_2 removal; off-site solids disposal											
500 MW new, 3.5% sulfur	30	20,532,000	41.1	18,323,000	36.6	—	—	—	—	—	—
90% SO_2 removal; on-site solids disposal (existing unit without existing particulate collection facilities)											
500 MW existing, 3.5% sulfur	25	29,996,000	60.0	26,090,000	52.2	32,213,000	64.4	37,957,000	75.9	43,816,000	87.6
Oil-fired power unit											
90% SO_2 removal; on-site solids disposal											
200 MW new, 2.5% sulfur	30	8,263,000	41.3	9,482,000	47.4	8,861,000	44.3	10,324,000	51.6	13,069,000	65.3
500 MW new, 1.0% sulfur	30	12,935,000	25.9	15,961,000	31.9	12,695,000	25.4	15,198,000	30.4	28,067,000	56.1
500 MW new, 2.5% sulfur	30	15,473,000	30.9	18,148,000	36.3	16,080,000	32.2	18,949,000	37.9	28,277,000	56.6
500 MW new, 4.0% sulfur	30	17,481,000	35.0	19,861,000	39.7	18,765,000	37.5	21,893,000	43.8	28,449,000	56.9
500 MW existing, 2.5% sulfur	25	18,657,000	37.3	21,817,000	43.6	20,376,000	40.8	24,445,000	48.9	32,824,000	65.6
1000 MW new, 2.5% sulfur	30	23,384,000	23.4	26,341,000	26.3	23,656,000	23.7	28,765,000	28.8	46,356,000	46.4

[a]Midwest plant location represents project beginning mid-1972, ending mid-1975. Average cost basis for scaling, mid-1974. Minimum in process storage; only pumps are spared. Investment requirements for disposal of fly ash excluded. Construction labor shortages with accompanying overtime pay incentive not considered.

[b]All Cat-Ox installations require particulate removal to 0.005 gr/scf prior to entering converter. Because existing units are assumed to already meet EPA standards (0.1 lb particulate/mm Btu of heat input). Only incremental additional precipitator is required.

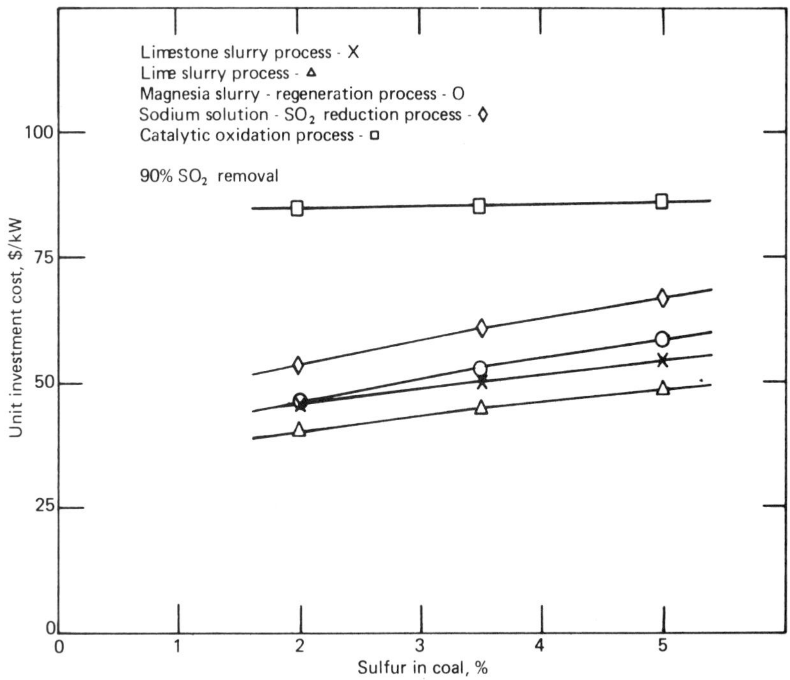

Figure 23.7. 500-MW new coal-fired units—effect of sulfur content of coal on unit investment cost.

The projected investments for limestone and lime slurry processes designed for off-site solids disposal represent savings equivalent to approximately 18% of the projected on-site investment. However, these projections do not include the capital for off-site waste treatment facilities.

Accuracy

In reviewing the capital investment estimates for these five processes, it should be understood that the base case process definitions and estimates represent a generalized cost. As an indication of how the project scope and corresponding investment could vary, the effect of "add ons" to the limestone slurry process base case estimate is shown in Table 23.9. Such "add ons" as delays in equipment delivery, physical space limitations, additional redundancy requirements, taller stacks, inclusion of a closed-loop fly ash disposal pond, and overtime to accelerate the project completion schedule could more than double the investment for the limestone slurry process. Excluding these additional costs, but considering only the data available for this appraisal and the depth of investigation, the estimated investment accuracy for each of the processes is given in Table 23.10.

Table 23.9. Limestone Slurry Process Investment with Modified Project Scope.

	Investment, (\$/kW)
BASE INVESTMENT - LIMESTONE SLURRY PROCESS (Including Fly Ash Removal But Not Disposal) 500-MW new coal-fired unit buring coal with 3.5% S, 12% ash, 90% SO_2 removal, 30-year life 127,500 hours operation, on-site solids disposal, proved system, only pumps spared, no bypass ducts, experienced design and construction team, no overtime, 3-year program, 5%/year escalation, mid-1974 cost basis for scaling	50.30
A. Overtime to accelerate project or cover local demand requirements (50% of construction labor requirements)	3.20
B. Research and development costs for first-of-a-kind process technology (as allowed by FPC accounting practice)	5.00
C. Power generation capital for lost capacity (normally covered by appropriate operating costs for power used in process)	4.50
D. Reliability provisions with added redundancy of scrubbers, other equipment, ducts and dampers, instrumentation for change over (assumes no permission to run power plant without meeting SO_2 removal emission standards at all times)	7.00
E. Additional bypass ducts and dampers	1.00
F. Retrofit difficulty—moderate, space available beyond stack, less than three shutdowns required for tie-ins, field fabrication feasible	10.00
G. Fly ash pond including closed-loop provisions	5.50
H. 500-foot stack added to project cost	6.00
I. Air quality monitoring system, 2–15 mile radius, 10 stations	0.70
J. Cost escalation of 10% per year instead of 5%	4.80
K. Possible delay of up to 2 years in equipment and material deliveries (1977 completion instead of 1975)	15.00
Total	113.00

Annual Operating Costs

For the annual operating costs covering 7000 hours per year operation, the following types of displays are used for each process: (a) Summarized effect of individual process areas on total costs—presented for base case and existing 500-MW limestone case (Table 23.11); (b) Case variations—operating cost summaries presented for each of the 16 case variations. See Tables 23.12–23.16.

Table 23.10. Projected Investment Estimates. Accuracy Based on Available Data and Depth of Investigation.

Process	*% range*
Limestone slurry	+20 to −5
Lime slurry	+20 to −10
Magnesia slurry - regeneration	+25 to −15
Sodium solution - regeneration	+25 to −10
Catalytic oxidation (Cat-Ox)	+20 to −10

Table 23.17 presents the total average annual operating costs for the 16 case variations of all five processes excluding credit for any by-products produced. Corresponding to new 3.5% S coal-fired units, the relative ranking of average annual operating costs for the processes (base case) is as follows:

1. Limestone slurry
2. Lime slurry
3. Catalytic oxidation
4. Magnesia slurry - regeneration
5. Sodium solution - SO_2 reduction

Some important effects are not readily seen in comparing the ranking of processes for only one plant size and sulfur level. As determined in this study, the relative ranking of operating costs of the various processes is sensitive to changes in the assumed sulfur content of the fuel. As an illustration, a 500-MW, 1% oil-fired power unit using the catalytic oxidation process has the highest operating cost of the five processes; but for fuel oils with sulfur contents greater than 4.0%, this process has the lowest cost. Although the sodium process is not the lowest cost process for any of the cases presented in this study, it becomes more competitive for low-sulfur fuel oil-fired units. Furthermore, at the expense of small additional investment, its operating cost could be lowered by designing the process with multiple-effect evaporators, thereby reducing the overall energy requirements about 18%.

In all of the projected annual operating cost estimates, capital charges are the most significant components followed by maintenance, energy, and raw material costs in varying orders of importance. For the magnesia and sodium processes, which require thermal energy for regeneration of absorbent, steam or fuel oil are the second most significant components. Lime cost (which requires thermal energy in preparation) is the second most predominant item for the lime slurry process, whereas maintenance is second in predominance for the limestone slurry and catalytic oxidation processes.

The effect on annual operating costs of designing for 80% SO_2 removal as compared to 90% is relatively small. Operating cost savings range from 3.6–4.2% of the projected operating costs at 90% removal for the limestone and lime slurry processes.

The effect of excluding product credits from the annual operating cost estimates is illustrated in Table 23.18 which compares the projected annual base case operating costs for the five processes with the revenue from sale of by-products.

Lifetime Operating Cost

Assumed credits for salable by-products are reflected in the lifetime operating costs. The lifetime operating costs are based on a

Table 23.11. Limestone Slurry Process—Total Average Annual Operating

	Total		*Raw materials handling*	*Food preparation*	*Partculate scrubbing*	*SO_2 scrubbing*
Direct capital investment, $	16,069,000		419,000	899,000	3,203,000	4,745,000
Total capital investment, $	25,163,000		656,000	1,408,000	5,016,000	7,430,000
Direct costs	*Unit cost, $*	*Raw materials*				
Delivered raw material						
Limestone	4.00/ton					
Annual quantity, tons		175,000				
Annual cost, $		700,000				
Sub-total raw material cost		700,000				
Conversion costs						
Operating labor and supervision	8.00/man-hr					
Annual quantity, man-hr			4,170	6,470	4,900	4,900
Annual cost, $			33,400	51,700	39,200	39,200
Utilities						
Steam	0.70/M lb					
Annual quantity, M lb						
Annual cost, $						
Process water	0.08/M gal					
Annual quantity, M gal						174,700
Annual cost, $						14,000
Electricity	0.010/kWh					
Annual quantity, kWh			390,000	5,150,000	6,440,000	22,530,000
Annual cost, $			3,900	51,500	64,400	225,300
Maintenance (labor and material)						
Percent of direct investment			6	8	10	11
Annual cost, $			25,000	71,000	320,000	522,000
Analyses	12.00/hr					
Annual quantity, hr		1,520			760	1,140
Annual cost, $		18,200			9,100	13,700
Subtotal conversion costs		18,200	62,300	174,200	432,700	814,200
Subtotal direct costs		718,200	62,300	174,200	432,700	814,200
Indirect costs						
Average capital charges at 14.9% of capital investment			97,700	209,800	747,400	1,107,000
Overhead						
Plant, 20% of conversion costs		3,600	12,500	34,800	86,500	162,800
Administrative, 10% of operating labor			3,300	5,200	3,900	3,900
Subtotal indirect costs		3,600	113,500	349,800	837,800	1,273,800
Total normal operating cost		721,800	175,800	424,000	1,270,500	2,088,000
Percent of total annual operating cost		9.37	2.28	5.50	16.49	27.11
	Dollars/ton coal burned	*Mills/kWh*	*Cents/million Btu heat input*		*Dollars/ton sulfur removed*	
Equivalent total unit operating cost	5.87	2.30	24.45		214.68	

[a]Basis:

500-MW new coal-fired power unit, 3.5% in fuel; 90% SO_2 removed; on-site solids disposal.
Remaining life of power plant, 30 yr.
Coal burned, 1,312,500 tons/yr, 9,000 Btu/kWh.
Stack gas reheat to 175°F.

Costs — *Base Case*[a] Summary—Area Contribution Analysis.

Reheat	*Fans*	*Calcium solids disposal*	*Utilities*	*Services*	*Construction facilities*	*Total annual quantities*	*Total annual dollars*	*Percent of total annual operating cost*
556,000	854,000	3,923,000	67,000	638,000	765,000			
871,000	1,337,000	6,143,000	105,000	999,000	1,198,000			
						175,000		
							700,000	9.09
							700,000	9.09
1,250	1,250	3,340				26,280		2.73
10,000	10,000	26,700					210,000	
492,800						492,800		
345,000							345,800	4.48
		75,600				250,300		
		6,000					20,000	0.26
	43,080,000	700,000	230,000	220,000		78,740,000		
	430,800	7,000	2,300	2,200			787,400	10.22
8	8	6				8		
44,500	68,000	235,000					1,285,500	16.69
		380				3,800		
		4,600					45,600	0.59
399,500	508,800	279,300	2,300	2,200			2,693,700	34.97
399,500	508,800	279,300	2,300	2,200			3,393,700	44.06
129,800	199,200	915,300	15,600	148,500	178,500		3,749,300	48.68
79,900	101,800	55,900	500	400			538,700	6.99
1,000	1,000	2,700						0.27
210,700	302,000	973,900	16,100	149,300	178,500		4,309,000	55.94
610,200	810,800	1,253,200	18,400	151,500	178,500		7,702,700	
7.92	10.53	16.27	0.24	1.97	2.32			100.00

Power unit on-stream time, 7,000 hr/yr.
Midwest plant location, 1975 operating costs.
Sulfur removed, 35,880 tons/yr; solids disposal, 206,000 tons/yr calcium solids including only hydrate water.
Investment and operating cost for disposal of fly ash excluded.

Table 23.12. Limestone Slurry Process. Total Average Annual Operating Costs - Regulated Utility Economics[a] (500-MW *new* coal-fired power unit, 3.5% S in fuel; 90% SO_2 removal; on-site solids disposal).

	Annual quantity	*Unit cost ($)*	*Total annual cost ($)*	*Percent of total annual operating cost*
Direct Costs				
Delivered raw material				
Limestone	175.0 M tons	4.00/ton	700,000	9.09
Subtotal raw material			700,000	9.09
Conversion costs				
Operating labor and supervision	26,280 man-hr	8.00/man-hr	210,200	2.73
Utilities				
Steam	492,800 M lb	0.70/M lb	345,000	4.48
Process Water	250,300 M gal	0.08/M gal	20,000	0.26
Electricity	78,740,000 kWh	0.010/kWh	787,400	10.22
Maintenance				
Labor and material, .08 x 16,069,000			1,285,500	16.69
Analyses			45,600	0.59
Subtotal conversion costs			2,693,700	34.97
Subtotal direct costs			3,393,700	44.06
Indirect Costs				
Average capital charges at 14.9% of total capital investment			3,749,300	48.68
Overhead				
Plant, 20% of conversion costs			538,700	6.99
Administrative, 10% of operating labor			21,000	0.27
Subtotal indirect costs			4,309,000	55.94
Total annual operating costs			7,702,700	100.00
	Dollars/ton coal burned	*Mills/kWh*	*Cents/million Btu heat input*	*Dollars/ton sulfur removed*
Equivalent unit operating cost	5.87	2.20	24.45	214.68

[a]Basis:
- Remaining life of power plant, 30 yr.
- Coal burned, 1,312,500 tons/yr, 9,000 Btu/kWh.
- Stack gas reheat to 175°F.
- Power unit on-stream time, 7,000 hr/yr.
- Midwest plant location, 1975 operating costs.
- Total capital investment, $25,163,000; subtotal direct investment, $16,069,000.
- Working capital, $572,600.
- Investment and operating cost for disposal of fly ash excluded.

Table 23.13. Lime Slurry Process. Venturi - Venturi Scrubbing Scheme. Total Average Annual Operating Costs - Regulated Utility Economics[a] (500-MW *new* coal-fired power unit, 3.5% S in fuel; 90% SO_2 removal; on-site solids disposal).

	Annual quantity	*Unit cost ($)*	*Total annual cost ($)*	*Percent of total annual operating cost*
Direct Costs				
Delivered raw material				
Lime	81.2 M tons	22.00/ton	1,786,400	22.05
Subtotal raw material			1,786,400	22.05
Conversion costs				
Operating labor and supervision	22,320 man-hr	8.00/man-hr	178,600	2.20
Utilities				
Steam	490,000 M lb	0.70/M lb	343,000	4.23
Process water	241,900 M gal	0.08/M gal	19,400	0.24
Electricity	74,100,000 kWh	0.010/kWh	741,000	9.15
Maintenance				
Labor and material, .08 x 14,318,000			1,145,400	14.14
Analyses			36,500	0.45
Subtotal conversion costs			2,463,900	30.41
Subtotal direct costs			4,250,300	52.46
Indirect Costs				
Average capital charges at 14.9% of total capital investment			3,340,900	41.24
Overhead				
Plant, 20% of conversion costs			492,800	6.08
Administrative, 10% of operating labor			17,900	0.22
Subtotal indirect costs			3,851,600	47.54
Total annual operating costs			8,101,900	100.00
	Dollars/ton coal burned	*Mills/kWh*	*Cents/million Btu heat input*	*Dollars/ton sulfur removed*
Equivalent unit operating cost	6.17	2.31	25.72	225.81

[a]Basis:

Remaining life of power plant, 30 yr.
Coal burned, 1,312,500 tons/yr, 9,000 Btu/kWh.
Stack gas reheat to 175°F.
Power unit on-stream time, 7,000 hr/yr.
Midwest plant location, 1975 operating costs.
Total capital investment, $22,422,000; subtotal direct investment, $14,318,000.
Working capital, $744,000.
Investment and operating cost for disposal of fly ash excluded.

Table 23.14. Magnesia Slurry - Regeneration Process. Total Average Annual Operating Costs - Regulated Utility Economics[a] (500-MW *new* coal-fired power unit, 3.5% S in fuel; 90% SO_2 removal; 110,400 tons/yr 100% H_2SO_4).

	Annual quantity	*Unit cost ($)*	*Total annual cost ($)*	*Percent of total annual operating cost*
Direct Costs				
Delivered raw material				
Lime (1st stage neutralization)	134 tons	26.00/ton	3,500	0.04
Magnesium oxide (98%)	1,086 tons	155.00/ton	168,300	1.83
Coke	763 tons	15.00/ton	11,400	0.12
Catalyst	1,800 liters	1.65/liter	3,000	0.03
Subtotal raw material			186,200	2.02
Conversion costs				
Operating labor and supervision	39,200 man-hr	8.00/man-hr	313,600	3.40
Utilities				
Fuel oil (No. 6)	5,356,000 gal	0.23/gal	1,231,900	13.39
Steam	440,000 M lb	0.70/M lb	308,000	3.34
Heat credit	20,300 MM Btu	−0.60/MM Btu	(12,200)	(0.13)
Process water	2,207,500 M gal	0.04/M gal	88,300	0.96
Electricity	71,060,000 kWh	0.010/kWh	710,600	7.71
Maintenance				
Labor and material, .07 x 16,334,000			1,143,400	12.41
Analyses			102,000	1.11
Subtotal conversion costs			3,885,600	42.19
Subtotal direct costs			4,071,800	44.21
Indirect Costs				
Average capital charges at 14.9% of total capital investment			3,934,500	42.71
Overhead				
Plant, 20% of conversion costs			777,100	8.44
Administrative and marketing, 11% of conversion costs			427,400	4.64
Subtotal indirect costs			5,139,000	55.79
Total annual operating costs			9,210,800	100.00

	Dollars/ton 100% H_2SO_4	*Dollars/ton coal burned*	*Mills/kWh*	*Cents/million Btu heat input*	*Dollars/ton sulfur removed*
Equivalent unit operating cost	83.43	7.02	2.63	29.24	255.43

[a]Basis:

Remaining life of power plant, 30 yr.
Coal burned, 1,312,500 tons/yr, 9,000 Btu/kWh.
Stack gas reheat to 175°F.
Power unit on-stream time, 7,000 hr/yr.
Midwest plant location, 1975 operating costs.
Total capital investment, $26,406,000; subtotal direct investment, $16,334,000.
Working capital, $721,000.
Investment and operating cost for disposal of fly ash excluded.

Table 23.15. Sodium Solution - SO_2 Reduction Process. Total Average Annual Operating Costs - Regulated Utility Economics[a] (500-MW *new* coal-fired power unit, 3.5% S in fuel; 90% SO_2 removal; 32,700 tons/yr sulfur).

	Annual quantity	*Unit cost ($)*	*Total annual cost ($)*	*Percent of total annual operating cost*
Direct Costs				
Delivered raw materials				
Lime (1st stage neutralization)	134.1 tons	26.00/ton	3,500	0.03
Soda ash	9,300 tons	52.00/ton	483,600	4.17
Antioxidant	317,100 lb	2.00/lb	634,200	5.47
Catalyst			12,000	0.10
Subtotal raw materials			1,133,300	9.77
Conversion costs				
Operating labor and supervision	46,500 man-hr	8.00/man-hr	372,000	3.21
Utilities				
Natural gas	509,500 mcf	1.00/mcf	509,500	4.39
Steam	2,137,800 M lb	0.70/M lb	1,496,500	12.90
Heat credit	62,900 MM Btu	−0.60/MM Btu	(37,700)	(0.33)
Process water	9,953,400 M gal	0.02/M gal	199,100	1.72
Electricity	74,190,000 kWh	0.010/kWh	741,900	6.39
Maintenance				
Labor and material, .06 x 18,861,000			1,131,700	9.75
Analyses			109,900	0.95
Subtotal conversion costs			4,522,900	38.98
Subtotal direct costs			5,656,200	48.75
Indirect Costs				
Average capital charges at 14.9% of total capital investment			4,543,200	39.16
Overhead				
Plant, 20% of conversion costs			904,600	7.80
Administrative and marketing			497,500	4.29
Subtotal indirect costs			5,945,300	51.25
Total annual operating costs			11,601,500	100.00

	Dollars/ton product sulfur	*Dollars/ton coal burned*	*Mills/kWh*	*Cents/million Btu heat input*	*Dollars/ton sulfur removed*
Equivalent unit operating cost	354.79	8.84	3.31	36.83	323.34

[a]Basis:
Remaining life of power plant, 30 yr.
Coal burned, 1,312,500 tons/yr, 9,000 Btu/kWh.
Stack gas reheat to 175°F.
Power unit on-stream time, 7,000 hr/yr.
Midwest plant location, 1975 operating costs.
Total capital investment, $30,491,000; subtotal direct investment, $18,861,000.
Working capital, $1,015,500.
Investment and operating cost for disposal of fly ash excluded.

Table 23.16. Catalytic Oxidation Process. Total Average Annual Operating Costs - Regulated Utility Economics[a] (500-MW *new*, coal-fired power unit, 3.5% S in fuel; 90% SO_2 removal; 109,900 tons/yr 100% H_2SO_4).

	Annual quantity	*Unit cost ($)*	*Total annual cost ($)*	*Percent of total annual operating cost*
Direct Costs				
Delivered raw material				
Catalyst	104,700 liters	1.65/liter	172,800	1.95
Subtotal raw material			172,800	1.95
Conversion costs				
Operating labor and supervision	7,890 man-hr	8.00/man-hr	63,100	0.71
Utilities				
Steam	179,000 M lb	0.70/M lb	125,300	1.41
Heat credit	987,000 MM Btu	−0.60/MM Btu	(592,200)	(6.67)
Process water	312,000 M gal	0.08/M gal	25,000	0.28
Electricity	90,440,000 kWh	0.010/kWh	904,400	10.19
Maintenance				
Labor and material, .04 x 25,368,000			1,014,700	11.43
Analyses			48,000	0.54
Subtotal conversion costs			1,588,300	17.89
Subtotal direct costs			1,761,100	19.84
Indirect Costs				
Average capital charges at 14.9% of total capital investment			6,367,700	71.76
Overhead				
Plant, 20% of conversion costs			317,700	3.58
Administrative and marketing			427,400	4.82
Subtotal indirect costs			7,112,800	80.16
Total annual operating costs			8,873,900	100.00

	Dollars/ton 100%H_2SO_4	*Dollars/ton coal burned*	*Mills/kWh*	*Cents/million Btu heat input*	*Dollars/ton sulfur removed*
Equivalent unit operating cost	80.75	6.76	2.54	28.17	247.32

[a]Basis:
Remaining life of power plant, 30 yr.
Coal burned, 1,312,000 tons/yr, 9,000 Btu/kWh.
Stack gas heat to 175°F.
Power unit on-stream time, 7,000 hr/yr.
Midwest plant location, 1975 operating costs.
Total capital investment, $42,736,000; subtotal direct investment, $25,368,000.
Working capital, $365,700.
Investment and operating cost for disposal of fly ash excluded.

Table 23.17. Summary - Total Average Annual Operating Costs[a] (Excluding Credit for By-products).

Case	Years life	Limestone process		Lime process		Magnesia process		Sodium process		Cat-Ox process	
		Total annual operating cost ($)	*Mills/kWh*	*Total annual operating cost ($)*	*Mills/kWh*	*Total annual operating cost ($)*	*Mills/kWh*	*Total annual operating cost ($)*	*Mills/kWh*	*Total annual operating cost ($)*	*Mills/kWh*
Coal-fired power unit											
90% SO_2 removal; on-site solids disposal											
200 MW new, 3.5% sulfur	30	3,921,500	2.80	4,163,900	2.97	4,776,800	3.41	5,971,700	4.27	4,232,700	3.02
200 MW existing, 3.5% sulfur	20	3,867,100	2.76	4,822,000	3.44	5,091,200	3.64	7,377,700	5.27	5,849,400	4.18
500 MW existing, 3.5% sulfur	25	7,892,600	2.26	9,612,400	2.75	9,607,900	2.75	14,658,000	4.19	12,399,600	3.54
500 MW new, 2.0% sulfur	30	6,774,700	1.94	6,915,100	1.98	7,523,400	2.15	9,101,700	2.60	8,801,200	2.51
500 MW new, 3.5% sulfur	30	7,702,700	2.20	8,101,900	2.31	9,210,800	2.63	11,601,500	3.31	8,873,900	2.54
500 MW new, 5.0% sulfur	30	8,522,200	2.43	9,170,100	2.62	10,768,500	3.08	13,983,300	4.00	8,940,500	2.55
1000 MW existing, 3.5% sulfur	25	12,752,900	1.82	15,301,400	2.19	15,481,900	2.21	25,118,500	3.59	21,460,800	3.07
1000 MW new, 3.5% sulfur	30	11,874,100	1.70	12,553,100	1.79	14,347,000	2.05	18,391,300	2.63	13,957,600	1.99
80% SO_2 removal; on-site solids disposal											
500 MW new; 3.5% sulfur	30	7,378,000	2.11	7,806,900	2.23	8,789,700	2.51	10,834,300	3.10	—	—
90% SO_2 removal; off-site solids disposal											
500 MW new; 3.5% sulfur	30	8,376,500	2.39	8,641,000	2.47	—	—	—	—	—	—
90% SO_2 removal; (existing unit without existing particulate collection facilities)											
500 MW existing, 3.5% sulfur	25	9,573,400	2.74	9,728,300	2.78	11,227,300	3.21	16,389,200	4.68	13,598,300	3.89
Oil-fired power unit											
90% SO_2 removal; on-site solids disposal											
200 MW new, 2.5% sulfur	30	2,842,000	2.03	3,413,500	2.44	3,204,400	2.29	4,269,200	3.05	2,750,100	1.96
500 MW new, 1.0% sulfur	30	4.732,500	1.35	5,748,600	1.64	4,633,100	1.32	5,854,700	1.67	5,743,600	1.64
500 MW new, 2.5% sulfur	30	5,564,400	1.59	6,852,800	1.96	6,092,700	1.74	8,305,100	2.37	5,677,500	1.62
500 MW new, 2.5% sulfur	30	6,281,800	1.79	7,742,300	2.21	7,393,500	2.11	10,640,500	3.04	5,565,100	1.59
500 MW existing, 2.5% sulfur	25	6,587,300	1.88	8,001,500	2.29	7,308,700	2.09	10,261,600	2.93	11,126,100	3.18
1000 MW new,2.5% sulfur	30	8,987,400	1.28	10,795,200	1.54	9,715,900	1.39	13,686,200	1.96	8,911,900	1.27

[a]Power unit on-stream time, 7000 hr/yr. Midwest plant location, 1975 operating costs. Investment and operating cost for disposal of fly ash exclued.

Table 23.18. Comparison of Projected Annual Operating Costs and Product Credits for Base Case Estimates at 7000 Hours Annual Operation.

	Total average annual operating cost	*Annual credit for by-products*	*Net annual operating cost*
Process	*($)*	*($)*[a]	*($)*
Limestone	7,702,700	—	7,702,700
Lime	8,101,900	—	8,101,900
Magnesia	9,210,800	883,200	8,327,600
Sodium	11,601,500	1,077,500	10,524,000
Cat-Ox	8,873,900	659,400	8,214,500

[a]Corresponds to credit of \$8/ton 100% H_2SO_4 as 98% H_2SO_4 for magnesia slurry process; \$25/short ton sulfur, \$20/ton Na_2SO_4 for the sodium solution process; and \$6/ ton 100% H_2SO_4 as 80% H_2SO_4 for the catalytic oxidation process.

prescribed 30-year operating profile—7000 hours the first 10 years; 5000 hours the next 5 years; 3500 hours the next 5 years; 1500 hours the last 10 years. Table 23.19 presents a summary of the cumulative discounted process costs and equivalent levelized unit increase (decrease) in the cost of power over the life of the power unit for the five processes. Table 23.20 shows the revenue assumed for each product and the actual and discounted cumulative credit for products which are included in the cumulative discounted process costs for the base case. Figures 23.8 and 23.9 show the effect of power unit size and sulfur content of coal on the levelized unit operating cost. With some exceptions, the relative ranking and trends in levelized unit operating costs are somewhat similiar to those projected and discussed for annual operating costs. Lifetime levelized unit operating costs are slightly higher than corresponding average annual unit operating costs because of the declining operating profile of the power unit.

Table 23.20. Lifetime By-product Production and Credit. Base Case, 500-MW New, Coal-Fired, 3.5% S, 30-year Remaining Life.

	Equivalent lifetime production		*Net revenue, \$/short ton*	*Cumulative revenue*	
Process	*Product*	*Short tons*		*Actual \$*	*Discounted \$*
Magnesia slurry - regeneration	100% H_2SO_4[a]	2,011,500	8.00	16,092,000	6,923,300
Sodium solution -	Sulfur	595,000	25.00		
SO_2 reduction	Na_2SO_4	237,000	20.00	19,627,500	8,446,300
Catalytic oxidation	100% H_2SO_4[b]	2,002,500	6.00	12,015,000	5,168,900

[a]As 98% H_2SO_4.
[b]As 80% H_2SO_4.

Table 23.19. Summary - Cumulative Discounted Process Costs and Equivalent Levelized Unit Increase (Decrease) in Cost of Power over the Life of the Power Unit[a] (including Credit for By-products).

Case	Years life	Limestone process: Cumulative present worth net increase (decrease) in cost of power[b] ($)	Limestone process: Levelized increase (decrease) in unit operating cost, mills/kWh[c]	Lime process: Cumulative present worth net increase (decrease) in cost of power[b] ($)	Lime process: Levelized increase (decrease) in unit operating cost, mills/kWh[c]	Magnesia process: Cumulative present worth net increase (decrease) in cost of power[b] ($)	Magnesia process: Levelized increase (decrease) in unit operating cost, mills/kWh[c]	Sodium process: Cumulative present worth net increase (decrease) in cost of power[b] ($)	Sodium process: Levelized increase (decrease) in unit operating cost, mills/kWh[c]	Cat-Ox process: Cumulative present worth net increase (decrease) in cost of power[b] ($)	Cat-Ox process: Levelized increase (decrease) in unit operating cost, mills/kWh[c]
Coal-fired power unit											
90% SO_2 removal; on-site solids disposal											
200 MW new, 3.5% sulfur	30	40,142,800	3.66	41,112,500	3.75	44,860,300	4.09	55,045,000	5.02	44,823,500	4.08
200 MW existing, 3.5% sulfur	20	29,067,800	4.73	34,979,000	5.69	36,106,200	5.87	48,568,300	7.90	42,423,000	6.90
500 MW existing, 3.5% sulfur	25	70,550,000	3.09	84,117,600	3.69	78,292,200	3.43	113,985,500	5.00	106,607,800	4.67
500 MW new, 2.0% sulfur	30	69,314,200	2.53	68,709,000	2.50	71,503,600	2.61	85,604,900	3.12	95,780,300	3.49
500 MW new, 3.5% sulfur	30	78,439,900	2.86	79,593,300	2.90	84,249,500	3.07	104,292,300	3.80	94,320,900	3.44
500 MW new, 5.0% sulfur	30	86,426,800	3.15	89,293,900	3.25	95,621,900	3.49	121,660,300	4.43	92,805,300	3.38
1000 MW existing, 3.5% sulfur	25	111,985,400	2.45	130,977,300	2.87	121,156,200	2.66	188,464,400	4.13	181,012,700	3.97
1000 MW new, 3.5% sulfur	30	120,015,500	2.19	121,789,900	2.22	126,808,000	2.31	160,375,200	2.92	148,117,600	2.70
80% SO_2 removal; on-site solids disposal											
500 MW new, 3.5% sulfur	30	75,259,300	2.74	76,687,900	2.80	81,119,800	2.96	98,245,000	3.58	—	—
90% SO_2 removal; off-site solids disposal											
500 MW new, 3.5% sulfur	30	80,426,200	2.93	80,903,300	2.95	—	—	—	—	—	—
90% SO_2 removal (existing unit without existing particulate collection facilities)											
500 MW existing, 3.5% sulfur	25	87,143,300	3.82	84,924,200	3.72	93,875,800	4.12	130,713,900	5.73	119,124,800	5.22
Oil-fired power unit											
90% SO_2 removal; on-site solids disposal											
200 MW new, 2.5% sulfur	30	28,281,000	2.58	33,612,500	3.06	30,089,900	2.74	39,147,900	3.57	29,653,800	2.70
500 MW new, 1.0% sulfur	30	46,404,800	1.69	56,505,800	2.06	44,030,300	1.61	55,025,400	2.01	63,574,000	2.32
500 MW new, 2.5% sulfur	30	54,743,900	2.00	66,727,200	2.43	55,673,400	2.03	74,204,600	2.70	61,591,500	2.25
500 MW, new, 4.0% sulfur	30	61,808,400	2.25	74,929,500	2.73	65,572,800	2.39	91,887,900	3.35	59,249,500	2.16
500 MW existing, 2.5% sulfur	25	58,353,800	2.56	70,126,200	3.07	61,393,300	2.69	82,852,700	3.63	96,305,200	4.22
1000 MW new, 2.5% sulfur	30	87,171,700	1.59	103,411,900	1.88	85,962,400	1.57	118,705,700	2.16	96,899,300	1.77

[a]Basis:
Over previously defined power unit operating profile. 30 yr life;
7000 hr - 10 yr, 5000 hr - 5 yr, 3500 hr - 5 yr, 1500 hr - 10 yr.
Midwest plant location, 1975 operating costs.
Investment and operating cost for disposal of fly ash excluded.
Constant labor cost assumed over life of project.

[b]Discounted at 10% to initial year.

[c]Equivalent to discounted process cost over life of power unit.

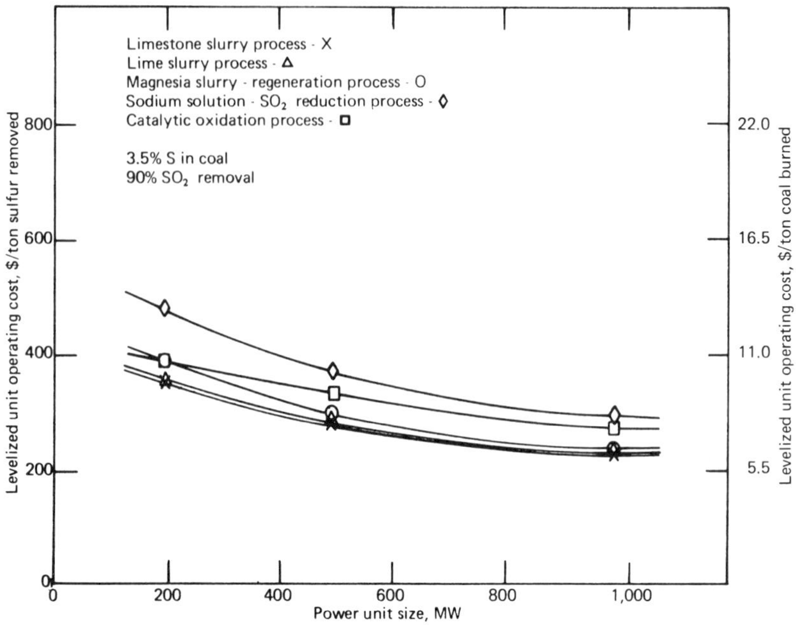

Figure 23.8. New coal-fired units - the effect of power unit size on levelized unit operating cost-regulated economics.

TVA IN-HOUSE ESTIMATES

In addition to the more detailed estimates prepared for EPA, a series of in-house studies have also been undertaken by TVA to determine the attractiveness of several process options. These include limestone scrubbing using benzoic acid as a slurry dissolution promoter; lime scrubbing using a venturi - mobile bed contactor rather than two venturi stages; lime scrubbing including a limestone calcination unit; sodium scrubbing with sulfuric acid production rather than sulfur; and two dilute mode double-alkali systems—one using ammonia scrubbing, the other sodium scrubbing. Because the data available for the double-alkali process evaluations were limited, the accuracy of these estimates is less than the others.

Seven TVA case estimates were prepared. When evaluating lime scrubbing with a mobile bed contactor, both a new and a retrofit case (500 MW, 3.5% S in coal) were estimated to determine the merit of using such a single-stage device for SO_2 removal. For the other five process options, only new systems were evaluated. Except for specific process requirements, the same criteria used to prepare the EPA estimates were used for these evaluations. The summarized investment and operating cost results are given in Table 23.21.

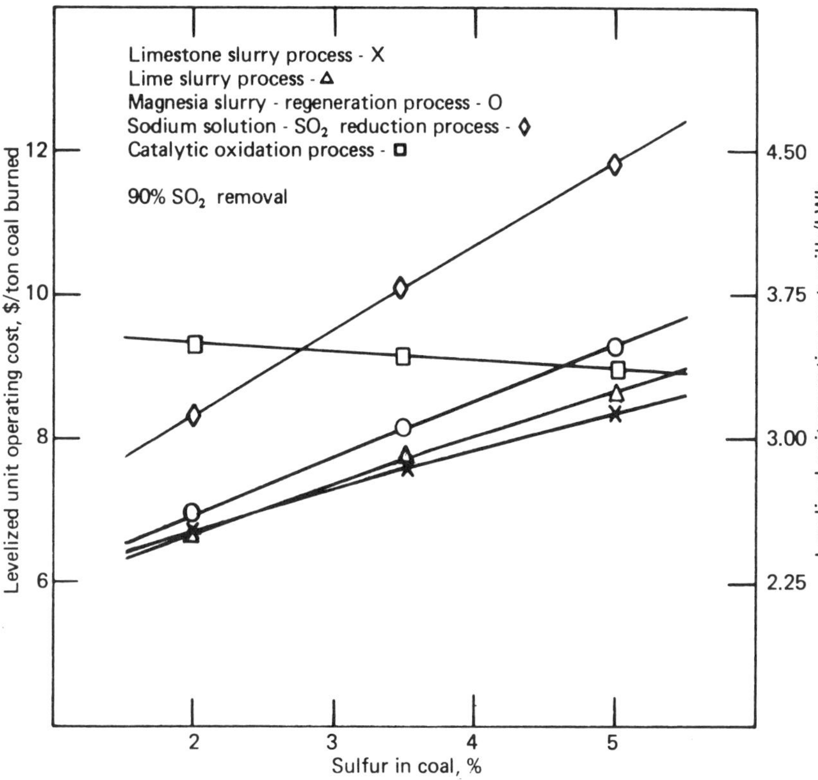

Figure 23.9. 500-MW new coal-fired units - effect of sulfur content on coal levelized unit operating cost-regulated economics.

CONCLUSIONS

Conclusions derived from the EPA-TVA detailed study results are as follows:

Investment

1. For coal-fired systems, the lime scrubbing process has the lowest investment and catalytic oxidation has the highest. Because lime process definition specifies venturi scrubbers and two stages are required for SO_2 removal, the limestone and magnesia processes become comparatively less expensive on oil-fired units; limestone investment is the lowest and magnesia investment is very close. All four wet scrubbing processes are within 21% of each other.

2. As sulfur content of fuel increases, the relative investment ranking of the wet scrubbing processes changes; lime is still lowest for coal-fired units, but our oil-fired units, limestone or magnesia investments are the lowest. Although these investments are very close

Table 23.21. Investment and Operating Cost Results from TVA In-house Studies.[a]

Process	Years life	*Capital investment* ($)	*Capital investment* ($/kW)	*Average annual operating costs* ($)	*Average annual operating costs* (mills/kWh)
Limestone-benzoic acid slurry process[b]					
500 MW *new* 3.5% sulfur	30	23,473,000	46.9	7,439,000	2.13
Lime slurry process; venturi-mobile bed scrubbing scheme[b]					
500 MW *new* 3.5% sulfur	30	23,363,000	46.7	8,146,000	2.33
Lime slurry process; mobile bed scrubbing scheme[b]					
500 MW *existing,*[c] 3.5% sulfur	25	20,327,000	40.7	8,270,600	2.36
Lime slurry process; venturi - venturi scrubbing scheme with onsite calcination[b]					
500 MW *new* 3.5% sulfur	30	27,452,000	54.9	8,101,900[d]	2.31[d]
Sodium solution - H_2SO_4 production process					
500 MW *new* 3.5% sulfur	30	30,138,000	60.3	11,054,800	3.16
Ammonia-limestone/lime double-alkali process[b]					
500 MW *new* 3.5% sulfur	30	31,087,000	62.2	9,669;300	2.76
Sodium-limestone double-alkali process[b]					
500 MW *new* 3.5% sulfur	30	30,759,000	61.5	9,545,800	2.73

[a]Stack gas reheat to 175°F. Midwest plant location represents project beginning mid-1972, ending mid-1975. Average cost basis for scaling, mid-1974. Operating costs based on power unit on-stream time of 7000 hours; costs correspond to 1975. Minimum in process storage; only pumps are spared. Investment and operating cost for disposal of fly ash excluded. Construction labor shortages with accompanying overtime pay incentive not considered.
[b]Limestone raw material cost, $4/ton; lime raw material cost, $22.00/ton. On-site disposal pond located one mile from power plant.
[c]Assumes 98.7% effective electrostatic precipitator already installed.
[d]Same as operating cost for lime slurry process in EPA study, since EPA study utilized estimated production cost to establish purchased cost of lime.

throughout the range, magnesia investment is slightly lower for the low-sulfur oil, and limestone investment is slightly lower for the medium- and high-sulfur oils.

3. The Cat-Ox process has the poorest investment economy of scale (unit size) of all five systems; however, for sulfur content of fuel, Cat-Ox has the best scale factor. A reheat (existing) Cat-Ox system requiring full particulate removal facilities to .005 gr/scf was found to be only 2.5% higher than a new integrated unit.

4. Plant age is an important factor only in the limestone and lime

processes where pond size depends on remaining plant life; older units can best use limestone or lime processes.

5. Removal of only 80% of the SO_2 (which would meet Federal new source emission standards for 3.5% S) instead of 90% decreases investment by only 3 to 5%.

Operating Cost

1. For both 3.5% S coal-fired and 2.5% oil-fired power units under the premises used, limestone has the lowest annual operating cost and sodium the highest. Even after credits for recovery products are applied, the lifetime operating costs are higher for the regenerable systems than for the throw-away systems. Cat-Ox becomes competitive with limestone on new oil-fired units, especially for smaller units.

2. As sulfur content of fuel increases, the relative operating cost for all five systems changes. Cat-Ox lifetime operating costs go from highest at low-sulfur levels to third on coal-fired units and first on oil-fired units with high-sulfur fuel levels. These are some of the most interesting results of the study. The heat credit for Cat-Ox becomes significant at high-sulfur levels.

3. The lime and sodium processes rank high in raw material cost and magnesia and Cat-Ox are low. Limestone has the largest range in cost of all the raw materials.

4. Sodium scrubbing has the highest total labor cost and Cat-Ox the lowest; however, labor is one of the smallest components (1–3% of total) for all five processes.

5. Energy costs are significant for all systems; applications of sodium scrubbing on existing units require the greatest amount of energy (35% of total operating cost). If a double-effect evaporator-crystallizer is utilized in the sodium system, overall energy costs could be reduced as much as 18%. The magnesia process is also energy intensive; the Cat-Ox system uses the lowest amount (5%).

6. Although expense for antioxidant (sodium system) is high, it is justified to keep sulfate formation down and save Na_2CO_3 makeup.

7. Maintenance is quite significant ranging from 7% of total operating cost for a Cat-Ox case to 17% for a limestone case.

8. Capital charges are the largest individual component of operating costs for all five processes. For new Cat-Ox systems capital costs are 72% of total. For the other processes, capital charges are in the range of 39–50%. A change in depreciation rate or cost of money will obviously affect Cat-Ox the most.

9. Only about 4–5% of total operating cost is saved when 80% SO_2 removal is provided instead of 90%.

10. A Cat-Ox process on an existing plant has a 40% greater oper-

ating cost than on a new system. This is caused by the high energy required for stack gas reheat from 310–890°F prior to conversion of SO_2 to SO_3.

11. As would be expected, the scrubbing steps of each process are the highest cost operations.

12. For the energy-intensive processes (magnesia, sodium), oil-fired systems are at a disadvantage because of the high cost of fuel oil relative to coal ($1.53/MM Btu compared to $0.54/MM Btu).

13. Sale of by-products at the values assumed in the study would reduce the base case lifetime operating cost 7.2% for the magnesia, 7.1% for the sodium, and 4.9% for the Cat-Ox process.

14. Because of product revenues, the relative ranking of the magnesia scrubbing process on oil-fired units improves under lifetime operating costs until it is the lowest cost system above 800-MW size.

15. Labor cost escalation (7.5% per year) over the project life would add about 7–10% to total project cost.

16. Regardless of which process is utilized, the increase in the cost of power to consumers for the base case is projected to range from 2.86–3.80 mills/kWh. For all case variations, projected costs could range from 1.57–7.90 mills/kWh.

Although there are differences in the depth and quality of the EPA-sponsored evaluations and the TVA in-house estimates, some additional conclusions can be derived from comparing all the base case results as shown in Table 23.22.

1. For limestone scrubbing, process savings from the addition of benzoic acid to reduce limestone stoichiometry (and therefore raw material and solids disposal costs) are relatively small.

2. Costs for a two-stage venturi lime system and a venturi - mobile bed lime system on a new coal-fired unit are reasonably close; however, for an existing coal-fired power plant or oil-fired unit not requiring additional particulate removal facilities, the mobile bed scrubber option requires about 22% less capital and 14% less operating cost.

3. Production of sulfuric acid is slightly less expensive than production of sulfur when sodium scrubbing - regeneration is used. Depending on prevailing market value of by-product acid and sulfur, the expected revenues could determine which option is best (2.8 times as much acid as sulfur).

4. Based on the "state of the art" at this time, the costs of the two dilute mode double-alkali options are about the same, but at least 20–25% more costly than lime or limestone scrubbing.

Table 23.22. Comparison of Results of EPA-Sponsored Evaluations and TVA In-house Estimates. New 500-MW, 3.5% Sulfur Coal-fired Units, 30 Years Life (Exceptions Noted).

	Results			
	EPA-sponsored evaluations		*TVA in-house estimates*	
Process	*Capital investment ($)*	*Average annual operating cost ($)*	*Capital investment ($)*	*Average annual operating cost ($)*
Limestone				
Limestone slurry process	25,163,000	7,702,700	—	—
Limestone-benzoic acid slurry process	—	—	23,473,000	7,439,000
Lime				
Lime slurry process (venturi-venturi scheme)	22,422,000	8,101,900	—	—
Lime slurry process (venturi-mobile bed scheme)	—	—	23,363,000	8,146,000
Lime slurry process (venturi-venturi scheme)[a]	26,027,000[a]	9,612,400[a]	—	—
Lime slurry process (mobile bed scheme)[a]	—	—	20,327,000[a]	8,270,600[a]
Lime slurry process with on-site lime production (venturi-venturi scheme)	—	—	27,452,000	8,101,900
Magnesia				
Magnesia slurry - regeneration process	26,406,000	9,210,800	—	—
Sodium				
Sodium solution - SO_2 reduction process	30,491,000	11,601,500	—	—
Sodium solution - H_2SO_4 production process	—	—	30,138,000	11,054,800
Cat-Ox				
Catalytic oxidation process	42,736,000	8,873,900	—	—
Double alkali				
Ammonia-limestone/lime double-alkali process	—	—	31,087,000	9,669,300
Sodium-limestone double-alkali process	—	—	30,759,000	9,545,800

[a]Existing 500-MW, 3.5% sulfur coal-fired unit, 25 years remaining life.

24

Pilot Plant Study of the Ammonia Absorption—Ammonium Bisulfate Regeneration Process

Claude E. Breed
Tennessee Valley Authority, Muscle Shoals, Alabama

Gerald A. Hollinden
Tennessee Valley Authority, Chattanooga, Tennessee

Absorption of sulfur dioxide from plant flue gases using ammoniacal solution and subsequent regeneration of the absorbing solution has been proposed or demonstrated in several processes. Lepsoe and Kirkpatrick[1] report that one of the earliest of these is the Cominco process in which SO_2 from a smelter operation is absorbed in an ammoniacal solution, the solution is acidulated to produce ammonium sulfate for disposal as a fertilizer, and the evolved SO_2 is sent to an acid plant. TVA piloted a similar process on power plant stack gases in the 1950s.[2] Other processes that produce fertilizers as the end product have been developed and are in full-scale operation on sulfuric acid plant tail gases in Czechoslovakia and Romania.[3,4] These processes use nitric or phosphoric acid to release the SO_2 and produce marketable fertilizers. Limited long-term markets for ammonium sulfate and the constraints on the location of power plants to fertilizer manufacturing centers limit their usage in the U.S.

Regeneration processes that do not rely on fertilizer marketing have been studied extensively. H. F. Johnstone[5] developed a process in the 1930s that produced only SO_2. Steam stripping was employed to recover SO_2 and to regenerate the ammoniacal solution for reuse in the scrubber. Since concentrated SO_2 is the major product, either sulfuric acid, liquid SO_2, or elemental sulfur can be the final product, depending on the need of the user. Although the Johnstone process has been operated successfully on a small pilot scale and presupposes no link to a fertilizer plant, it possesses some undesirable characteristics. Energy requirements for the stripping step are relatively high, 10–15 pounds of steam per pound SO_2. Oxidation products are difficult to purge from the system without loss of active species as well. The occurrence of undesirable disproportionation reactions in the steam stripper further aggravates the oxidation problem.

Development of the Johnstone process has been vigorously pursued in the USSR. Several pilot plant studies led to installation of a large-scale (about 70 MW) system on a coal-fired power plant in Moscow in 1952 and the system was operated until the power plant was converted to burn gas in 1961. Regeneration was based on steam stripping. A wet electrostatic precipitator was reported effective in controlling fume emission. During the period 1968–1972, an ammonia scrubbing process was used on a 15-MW oil-fired power plant located at Ufa in the Ural Mountains. Both direct oxidation to produce ammonium sulfate and regeneration by the autoclave process were tested. Corrosion was a serious problem.

Design work has been started on an ammonia scrubbing system for a full-scale (about 250 MW) oil-fired boiler near Moscow. Completion in 1977 is planned. Thermal regeneration will be used for recovery of SO_2, but the TVA-EPA ammonium bisulfate process is also being considered. Ammonia scrubbing is included in the agreement for formal exchange of technology between the U.S. and the USSR.

Other regenerative schemes not dependent on the fertilizer market structure are mentioned below.

The Institut Francais de Petrole (IFP) in France is developing a process based on ammonia scrubbing. The various steps of the process have been piloted separately at various locations. Two integrated systems are now being installed in Japan; one on a Claus plant and one at a refinery.

In March 1974, Nippon Kokan KK began operating a full-scale ammonia/lime double-alkali unit for treating 150,000 Nm^3/h waste gas from an iron ore sintering plant (300,000 Nm^3/h capacity) at the Keihin Works, Kawasaki, Japan. An ammonia-based fume from the absorber is considered a major problem. The calcium salts in the system are converted to gypsum and discarded. No additional units are currently being installed by Nippon Kokan KK using the ammonia double-alkali scrubbing process.

In September 1973, TVA modified its limestone/wet-scrubbing pilot plant to obtain preliminary data on an ammonia-limestone/lime (double alkali) scrubbing process. Good SO_2 removal was obtained (90% +). Two major problems identified during the initial period of this pilot plant operation were (a) excessive losses of soluble ammonia-sulfur salts with the discarded solids (calcium sulfite, fly ash, and unreacted limestone) from the filtration step, and (b) formation of a dense, persistent plume by the scrubbed gases exhausted to the atmosphere.

THE TVA-EPA AMMONIA ABSORPTION – AMMONIUM BISULFATE REGENERATION PROCESS

In 1968, EPA contracted with TVA to begin a pilot plant program to study in depth the ammonia absorption process as applied to power plants. The pilot plant was located at TVA's Colbert Power Plant, a pulverized coal-fired installation in northwest Alabama. The initial phase of study covered a number of variables in the absorption step. The results led to an amended program to include a process for regenerating the liquor from the absorber operation. The regeneration process selected for the second phase of pilot plant study was the ammonium bisulfate process. Two articles[6,7] cover the early absorption study and work on the regeneration process.

The ammonium bisulfate regeneration process (ABS process) uses a version of the Hixon-Miller[8] scheme for release of SO_2 from ammoniacal solutions by acidification and subsequent decomposition of the ammonium sulfate product to ammonia for recycling to the absorber and ammonium bisulfate for recycling to the acidulator. The net absorption reactions are

$$(1)\ NH_3 + SO_2 + H_2O \rightarrow NH_4HSO_3$$
$$(2)\ NH_4HSO_3 + NH_3 \rightarrow (NH_4)_2SO_3$$
$$(3)\ (NH_4)_2SO_3 + SO_2 + H_2O \rightarrow 2NH_4HSO_3$$

The net regeneration reactions are

$$(4)\ NH_4HSO_3 + NH_4HSO_4 \rightarrow (NH_4)_2SO_4 + H_2O + SO_2\uparrow$$
$$(5)\ (NH_4)_2SO_3 + 2NH_4HSO_4 \rightarrow 2(NH_4)_2SO_4 + H_2O + SO_2\uparrow$$
$$(6)\ (NH_4)_2SO_4 \xrightarrow{700^\circ F} NH_4HSO_4 + NH_3\uparrow$$

As in the Johnstone process, the only product from the ABS process is a gaseous stream of SO_2. The ABS process has advantages over the sulfite steam stripping process in that disproportionation reactions are avoided and sulfites are not purged from the system. Also, energy requirements for the ABS process are estimated to be approximately one-half to two-thirds that of the Johnstone process. For simplicity, the pilot plant is divided into the absorption and regeneration sections.

Absorption Section

A flowsheet of the absorption section of the pilot plant is shown in Figure 24.1. The pilot plant is designed to treat approximately 3000 cfm of flue gas at absorber conditions (saturated at about 125°F).

The gas flows through a prewash section containing a venturi-type element. The conditioned gas is then routed to the absorption tower. The absorber consists of as many as four independent absorbing stages (valve tray elements).

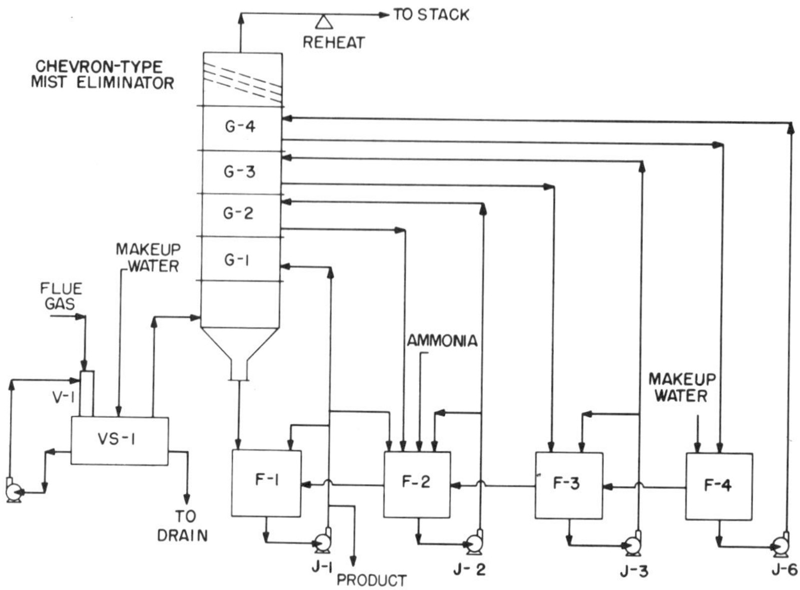

Figure 24.1. Absorption section.

The makeup water to the absorber is added to the top recirculating state (G-4) to maintain a low salt concentration at the top of the tower to control the ammonia loss from the system. Ammonia is added to the second absorption stage (G-2). A portion of the liquor from the bottom tray is pumped to G-2 to decrease the pH of the liquor on the second stage. Decreasing the pH reduces the vapor pressure of ammonia which decreases the possibility of gas phase reactions of ammonia, SO_2, and water to form a fume in the absorber. Product liquor is withdrawn from the first stage (G-1) and is stored in surge tanks for use in the regeneration section. The scrubbed flue gas is exhausted to the atmosphere either with or without reheating.

A discussion of the individual steps follows:

Gas Pretreatment

The inlet flue gas enters the pretreatment section (Figure 24.2) through a 1 by 1-foot duct, which contains a venturi element similar to Environeering Inc.'s Ventri-Rod element. The element consists of a number of ¾-inch rods mounted perpendicular to the gas flow. The number of rods can be varied to increase or decrease the pressure drop (ΔP) across the element. Recirculating wash water from a sump beneath the venturi element is sprayed concurrently with the gas stream immediately upstream from the element.

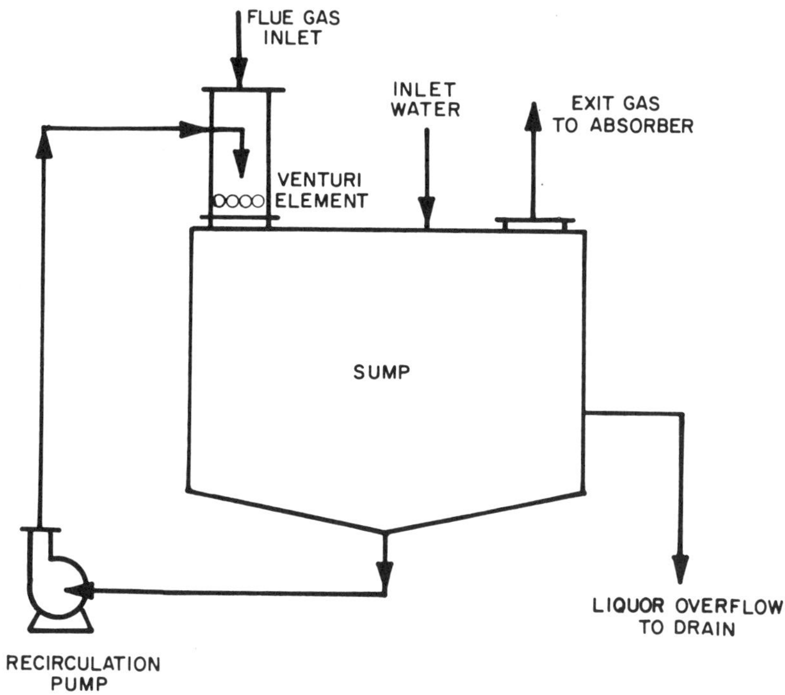

Figure 24.2. Prewash section.

The gas and spray water passes through the venturi element and into the sump. The velocity of the flue gas decreases in the sump to 8 ft/sec allowing some of the spray water entrained in the venturi to separate. The gas travels the length of the sump and is routed through a 14-inch-diameter duct to the absorber.

The purpose of the gas pretreatment is to humidify and cool the flue gas and to remove fly ash and chlorides from the gas stream. Humidification, cooling, and chloride removal are considered necessary for reducing the plume emission from the absorber. Fly ash is removed before the absorption section to decrease the amount of solids in the scrubber liquor. These solids could cause major problems in the absorption and regeneration sections.

In the test program, the pressure drop across the venturi and the liquid-to-gas ratio (L/G, gal/1000 ft^3) of wash water to the element were varied to determine their effects on the particulate, sulfur dioxide, and chloride removal, and on mist carryover. The pressure drop ranged from 0.5–15.0 inches of water and the L/G from 10–30. The flue gas used was taken both upstream and downstream from an electrostatic precipitator. The results of these tests are summarized below.

1. The inlet fly ash loading was reduced an average of 93% when the inlet dust loading was 4–6 gr/dscf. No reduction in dust loading was obtained when the inlet loading was low (0.01–0.1 gr/dscf).

2. No direct relationship was established between the SO_2 removal and the L/G, pressure drop, or pH of the wash water in the prewash section. Between 7 and 23% of the inlet SO_2 (2600 ppm) was removed in the section.

3. An increase in the pressure drop across the venturi element and the L/G to the element increased the mist carryover into the absorber.

4. Essentially all of the chloride in the flue gas (35 ppm average) was removed in the venturi section at the anticipated normal operating conditions.

To minimize mist carryover during absorber tests, the venturi prewash section was operated at as low a pressure drop and L/G as possible consistant with good humidification. Based on actual operation, these conditions are a pressure drop of 5–6 inches of water and an L/G of about 10.

There were three major problem areas identified during the operation of the prewash section: (a) low pH of sump liquor (corrosion), (b) disposal of low pH purge water from prewash sump, and (c) mist carryover from the prewash section to the absorber section.

During initial operation, the quantity of makeup water added to the prewash section was that consumed in the process for humidification and absorber product liquor bleedoff, approximately 1.2 gpm. At this flow rate, the pH of the wash liquor was 1.0 or less. The makeup water throughput was increased to 15–30 gpm to increase the pH of the wash water and to reduce its corrosion potential. This raised the pH of the wash water to approximately 2.5, but corrosion continued.

An inspection of the gas pretreatment section was made after 2000 hours of operation. The entire pretreatment section, constructed of Type 316L stainless steel, was badly damaged by corrosion. The rods across the throat of the venturi were pitted badly, especially on the ends, and several needed replacing. The liquor sump walls were corroded with the most severe attack above the normal gas-liquor interface. In this area, pits 40–60 mils deep were common. The gas duct from the sump to the absorber and the sample probes inside the duct (316 S.S.) were also severely corroded.

The gas prewash section was operated "open loop" to minimize corrosion of the equipment. The problem of disposing and/or utilizing low pH water will be common to most SO_2 removal processes that require a prewash. It is recognized that in a commercial operation, this loop would have to be closed to meet environmental regulations. Attempts to close the loop around the prewash section will be studied in future operations.

Mist carryover from the venturi section to the absorber dilutes the absorber product liquor and transports dissolved and undissolved solids into the absorber section. Both of these problems are discussed in more detail in the absorber section of this chapter.

Currently, the prewash section is being redesigned to resolve the corrosion and mist carryover problems. To protect against corrosion, the sump and exit gas ducts will be constructed of FRP; the venturi throat will probably be lined with neoprene. Several types of corrosion-resistant materials will be tested for use as rods in the venturi. The new prewash section will include a mist eliminator (probably a chevron in the horizontal position) to reduce the amount of carryover to the absorber.

SO_2 Absorber. The absorber shell consists of 32-inch by 32-inch by 4-foot sections. At normal operating conditions, the gas velocity in the tower is approximately 7 ft/sec. Two types of absorption elements, marble beds and valve trays, have been tested.

A two-pass chevron mist eliminator is located in the vertical position above the top absorption stage.

Two series of tests have been completed. During the first test, the absorber consisted of one valve tray followed by two marble beds, Figure 24.3. Tests were made to determine the highest liquor concentration that could be obtained using this configuration while limiting the outlet SO_2 concentration to $\leqslant$ 250 ppm and the ammonia concentration to $\leqslant$ 50 ppm. Product liquors with high concentrations require less energy for regeneration and smaller regeneration equipment.

The maximum product liquor concentration obtained using this configuration was a C_A (mols of active ammonia present as sulfite and bisulfite per 100 mols of total water) value of 14. Factors limiting the C_A in the initial series were (a) the diluting effect of the mist carryover from the gas pretreatment section, and (b) the SO_2 and NH_3 vapor pressure over the liquor on the top stage. An increase in the concentration of the product increases the liquor concentrations throughout the tower. An additional absorption stage was required to reduce the ammonia and SO_2 vapor pressure over the liquor on the top stage to meet the arbitrarily set limits.

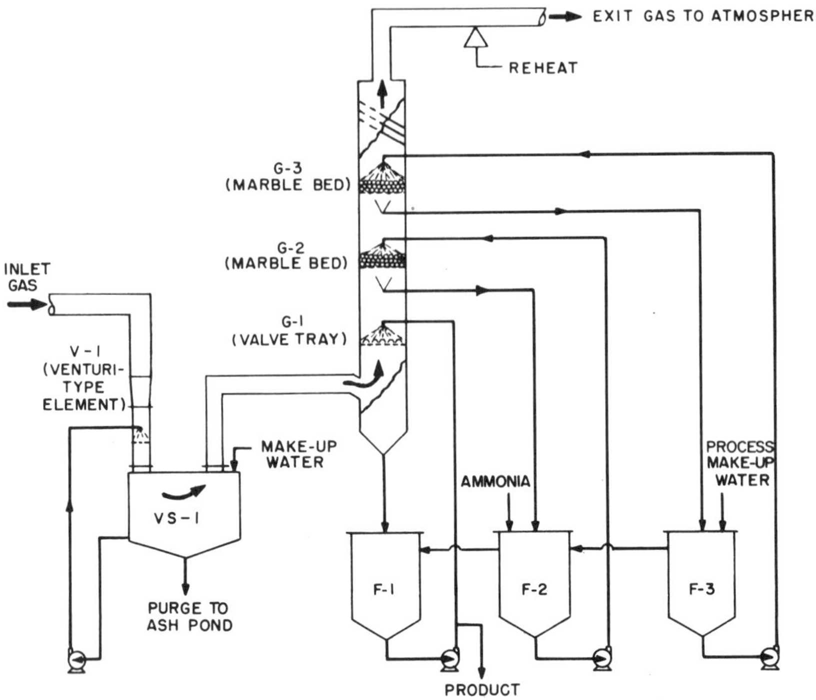

Figure 24.3. Three-stage absorption.

The second series of tests was made with a fourth valve tray stage added to the absorber. The two marble-bed absorption elements were replaced with valve trays during the modification because of occasional excessive stage-to-stage weepage which upset the liquor concentrations throughout the absorption tower. Experience showed that this problem does not occur with the valve tray when the gas flow rate remains relatively constant although weepage from the valve trays is a problem when the gas flow varies.

Tests using this absorber arrangement indicated that product liquor could be produced that had C_A's of 20 and above. Table 24.1 shows data from the absorber operation while producing liquor with a C_A of 21. No absorber operational problems were encountered during operation at the high concentration.

After determining that high liquor concentrations could be produced in the modified absorber, it was operated to produce liquor with C_A's of 12 to 15 for use in the regeneration section. Table 24.2 shows typical data from this operation. Solutions with concentrations in this range are acceptable for regeneration in the ammonium bisulfate process and are much easier to achieve within the limits of

Table 24.1. Typical Absorber Loop Test—Ammonium Bisulfate Pilot Plant.

Test Conditions	
Gas to absorber	
Flow rate, scfm at 32°F	2350
Temperature, °F	295
SO_2, ppm	2800
Gas leaving venturi	
Temperature, °F	120
SO_2, ppm	2720
Gas leaving first stage	
Temperature, °F	123
SO_2, ppm	1160
Gas leaving second stage	
Temperature, °F	124
SO_2, ppm	430
Gas leaving third stage	
Temperature, °F	115
SO_2 ppm	510
Gas leaving fourth stage	
Temperature, °F	114
SO_2 ppm	390
SO_2 removal, %	84
Fresh water feed to absorber, gpm	0.2
Ammonia feed to absorber, lb/hr	22.5
Liquor flow from first stage to second stage, gpm	1.6
Liquor to first stage (product)	
C_A	21.1
S/C_A	0.73
pH	6.2
Specific gravity	1.286
Liquor to second stage	
C_A	21.5
S/C_A	0.67
pH	6.3
Specific gravity	1.282
Liquor to third stage	
C_A	6.6
S/C_A	0.78
pH	5.9
Specific gravity	1.138
Liquor to fourth stage	
C_A	2.5
S/C_A	0.86
pH	5.7
Specific gravity	1.086

SO_2 and ammonia loss and plume emission than are the more concentrated solutions.

The most apparent problem in the absorption step is that of fume formation, Figure 24.4. From results with a single-stage marble-bed absorber, it was concluded that an acceptable plume (defined as

Table 24.2. Typical Absorber Loop Test—Ammonium Bisulfate Pilot Plant.

Test Conditions	
Gas to absorber	
Flow rate, scfm at 32°F	2350
Temperature, °F	305
SO_2 ppm	2680
Gas leaving venturi	
Temperature, °F	123
SO_2, ppm	2520
Gas leaving first stage	
Temperature, °F	121
SO_2, ppm	1240
Gas leaving second stage	
Temperature, °F	122
SO_2, ppm	330
Gas leaving third stage	
Temperature, °F	116
SO_2, ppm	320
Gas leaving fourth stage	
Temperature, °F	115
SO_2 ppm	230
NH_3, ppm	10
SO_2 removal, %	91
Fresh water feed to absorber, gpm	0.3
Ammonia feed to absorber, lb/hr	21.5
Liquid flow from first stage to second stage, gpm	0.74
Liquor to first stage (product)	
C_A	14.7
S/C_A	0.80
pH	5.9
Specific gravity	1.236
Liquor to second stage	
C_A	15.3
S/C_A	0.66
pH	6.4
Specific gravity	1.220
Liquor to third stage	
C_A	4.0
S/C_A	0.84
pH	5.9
Specific gravity	1.093
Liquor to fourth stage	
C_A	1.6
S/C_A	0.94
pH	5.6
Specific gravity	1.042

$<$ 5% opacity for this pilot plant) could be obtained when a) a water wash is used ahead of the absorption stage, b) the absorber and all ducts are insulated, and c) reheat is applied as required to dissipate the steam plume.

Additional tests were run using a single-stage valve tray absorber

Figure 24.4. Unacceptable plume during routine operation; product from absorber: $C_A = 10.0$, $S/C_A = 0.8$; no reheat.

to obtain data on the plume emitted from an uninsulated absorber as compared with an insulated absorber. The absorption stage was sandwiched between two water-wash stages. Either or both of the water stages were turned off to determine the effect on plume opacity. The C_A values of the absorption liquor tested were 2, 7, and 12. The S/C_A ratio (mols of SO_2 present as sulfite and bisulfite per mol of active ammonia present as sulfite and bisulfite) was controlled near 0.8. The exit gas was reheated to 140° and 175°F in each test.

The water wash from the bottom decreased the plume opacity in all cases. Without water wash, the plume opacity was 5–10%. With water wash, the opacity decreased to 0%. The same amount of reheat (200°F) was applied in both cases to eliminate the steam plume. The water wash from above lowered the opacity only at a C_A of 12. The combination of a water wash on the bottom tray and reheat resulted in an opacity of 5% or less for all three levels of C_A's.

The plume was also observed during multistage absorber operation. The insulated absorber shown in Figure 24.3 contained three absorption stages (1 valve tray and 2 marble beds) preceded by a prewash section. A steam and particulate plume was emitted from the absorber in these multi-stage absorber tests. In most cases, a reheated exit gas temperature of 10°–20°F above the calculated value at which the steam plume should dissipate was required to reduce the plume to less than or equal to 5% opacity. A previous test series[7] indicated that a single-stage absorber could be operated with an acceptable plume while reheating to or just above the calculated temperatures. The only operational changes made in comparison to the previous series were three stages of absorption versus one and higher liquor concentration on all stages. Mist carryover from the more concentrated liquor on the top stage probably accounts for the increase in reheat temperature required to maintain the same opacity. During reheat, the mist is evaporated, thus increasing the vapor pressure of SO_2 and NH_3 in the exit gases. A plume then results from the gas phase reaction of SO_2 and ammonia in the presence of water in the gas duct and/or in the atmosphere.

When the fourth absorption stage was added to the tower, the insulation was removed and not replaced. As in the 3-stage insulated absorber operation, the plume from the uninsulated 4-stage absorber was controlled at 5% opacity in most cases by reheating the exit gas 10°–20°F above the temperature required to dissipate the steam plume.

The use of the prewash, the reheat, and dilute solutions at the top of the tower are considered necessary in controlling the plume from the absorber. The value of insulating the absorber has not been fully evaluated and no recommendation can be made at this time.

In all cases, the ambient weather conditions have an overriding effect on plume emission. On cold wet days, the plume cannot be reheated to a high enough temperature to eliminate the plume. On warm days with high humidities, the plume reforms outside the absorber 10–20 ft downwind of the stack.

In two sampling and analysis series, EPA measured the levels of ammonia and SO_2 in the exit gas as 1.48 ± 4.17 ppm and 234 ± 88 ppm, respectively. Particulate mass emissions and particulate size distribution of the effluent particulate were also determined by EPA. The measured effluent particulate emission concentration was 0.057 ± 0.015 gr/dscf and 0.127 ± 0.015 gr/dscf. Calculations indicate that the measured particulate concentration in the exit gas could not totally be accounted for via the gas phase reaction between NH_3 and SO_2 but was primarily due to mist carryover from the absorber. Size distribution analysis indicated that 88.1% ± 2.6 of the particulate was finer than about 0.8 μ. Analysis of the emitted particulate by x-ray diffraction technique and wet chemical methods indicated that the particulate was pure ammonium sulfate.

The precipitation of solids in the absorption section may present another major problem. A yellow solid, tentatively identified petrographically as a homogeneous iron-ammonia-sulfur compound was present in all of the absorber liquors. Fly ash is the probable source of iron in the solid. Although most of the fly ash is removed in the prewash section, some passes through the venturi and is removed in the absorber. The mist carryover from the prewash section also contains dissolved iron along with other metals.

A small quantity of the solids in the product liquor may be sufficient to adversely affect the production of crystalline ammonium sulfate in the regeneration section. Also the solids contain iron which may catalize the decomposition of ammonia during thermal decomposition of ammonium sulfate. Attempts to remove the solids by filtration were unsuccessful because the precipitated solids and fly ash form a gelatinous, thixotropic material (Figure 24.5) that blinded the filter media. The material is removed from the product liquor in the pilot plant by settling in the product storage tanks and then decanting the clear supernatant liquor for use in the regeneration section. This method of solids separation is acceptable for the pilot plant operation but may prove impractical for a commercial operation.

Oxidation of sulfite to sulfate occurred in the absorption section. In the range of liquor concentrations tested, the oxidation ranged from 10–15% of the absorbed SO_2. An average value of 13% was obtained over an extended operating period using the 4-stage absorber. This

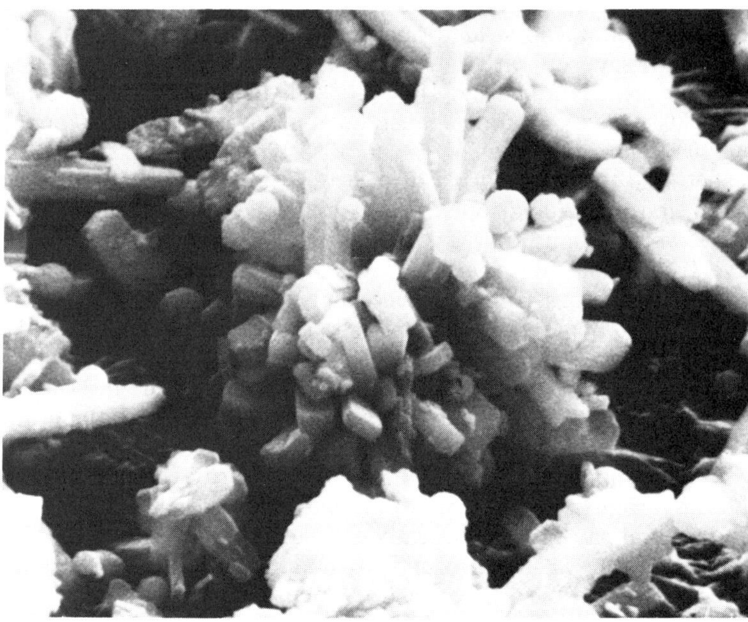

Figure 24.5. Photomicrograph of solids precipitated is the absorber loop mixed with fly ash.

oxidation by-product would have to be purged from a closed-loop system and probably sold as a fertilizer.

An equipment inspection of the absorption tower was made to locate the areas in which corrosion is a problem.

The absorber shell is a combination of six 4-foot sections salvaged from various absorbers. Data pertinent to each section are:

Section	*Description*	*Exposure (hr)*	*Material of construction, Type stainless steel*
1	Transition piece for gas inlet and liquor oulet	14,000	316
2	Section between first and second absorption stages	10,000	304
3	Section between second and third absorption stages	4,000	304
4	Section between third and fourth absorption stages	4,000	304
5	Section between fourth stage and demister	14,000	316
6	Demister housing	10,000	304

The first section, which has been in service 14,000 hours, had only a few small pits. The second and sixth sections, exposed for 10,000 hours, were severely pitted. The third and fourth sections, exposed for 4000 hours, contained very few pits. The fifth section was used for about 13,000 hours as an exhaust duct after the demister and for about 1000 hours as a housing for an absorbing element. The walls of this section were severely pitted; most of the pitting occurred in the location above the demister.

The tower contains four Type 316 stainless steel absorption trays (valve trays) spaced 4 feet apart. The bottom tray (G-1) was used for about 3100 hours, and the three upper trays (G-2,-3, and-4) were used about 700 hours. The bottom side of G-1 was pitted, but the top side showed little corrosion. The support angles located below the tray (Type 316 or 304) were severely pitted. Some of the internal piping (probably Type 304) to the tray was severely pitted while other sections (Type 316) were only slightly affected. The temperature of the liquor normally in contact with the bottom tray was 125°F (52°C) with an average pH of 5.8 and a C_A value of 8–20.

The bottom tray was used for about 1200 hours of its total exposure time, as a water wash stage and for humidification of the inlet flue gas. During this time, the operating conditions were similar to the operating conditions of the venturi. Most of the corrosion probably occurred during this time.

The second, third, and fourth trays (G-2,-3, and-4), supports, and internal piping showed no signs of corrosion after the 700 hours of operation. The temperature of the absorbing liquor on these stages was approximately 120°F; the pH value ranged from 5.4 to 7.0.

The chevron mist eliminator is located in the uppermost 4-foot section of the absorber. This section of the absorber had been in service about 10,000 hours and was corroded severely. Severe pitting, 30 mils deep had occurred where drainage from the mist eliminator flowed down the walls. The top and bottom of the mist eliminator blades were badly pitted.

The transition piece (Type 316L sainless steel) that joins the top of the absorber to the 14-inch-diameter flue gas exit duct was in service for about 700 hours with no apparent corrosion.

Conclusions drawn from the equipment inspection are:

1. Type 316 stainless steel is not suitable for use in the prewash (venturi) section.

2. Type 316 stainless steel is suitable for service in the absorber shell and in the internal piping.

3. Use of Type 304 stainless steel, which pitted badly in the absorber shell and piping when exposed for 10,000 hours, is marginal.

4. The demist section is subject to severe corrosion, and use of both Types 304 and 316 stainless steel in this area is marginal.

Regeneration Section

A flowsheet of the regeneration section of the pilot plant is shown in Figure 24.6. The absorber product liquor is fed to an acidulator. Here the liquor is mixed with an acid ion to chemically release the SO_2. A portion of the released SO_2 is evolved in the acidulator. In past operation, the acid ion was supplied by plant grade sulfuric acid. In an intergrated ABS system, the acid ion will be supplied by ammonium bisulfate generated by thermal decomposition of ammonium sulfate.

The liquor from the acidulator overflows into a stripper where the remaining SO_2 is stripped from the liquor with a countercurrent flow of air.

The stripped liquor (ammonium sulfate solution) is metered into the evaporator-crystallizer where water is evaporated to produce a slurry of ammonium sulfate crystals. The crystals are separated from the mother liquor in a continuous belt filter and dried. The crystals will be fed to the ammonium sulfate decomposer when it is available.

A discussion of the individual steps follows.

Acidulation and Stripping

The acidulation and stripping vessels are shown in Figure 24.7. The acidulator is constructed from a 6-foot section of a 12-inch stainless

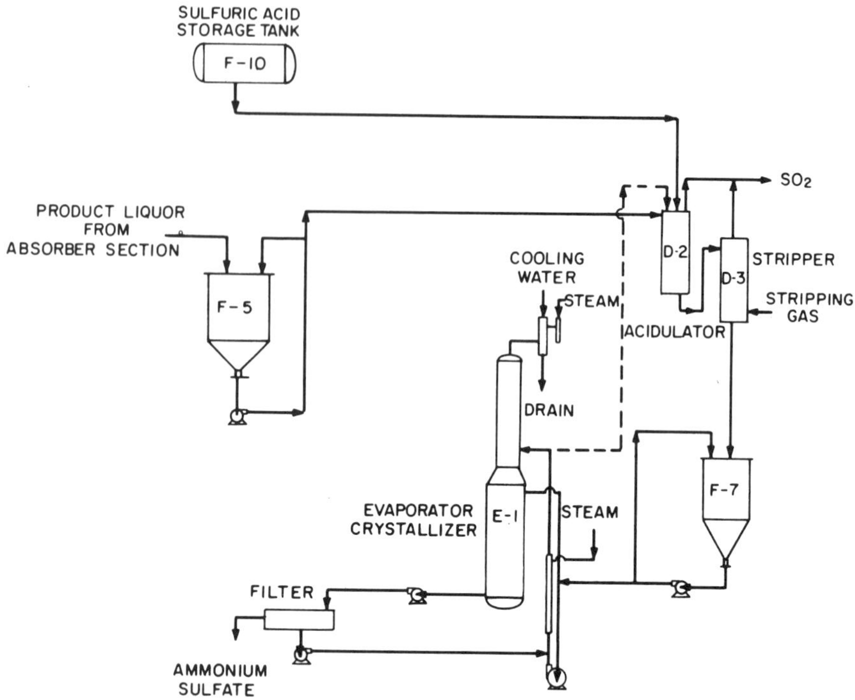

Figure 24.6. Regeneration section.

steel pipe and is coated internally with DuPont's TFE Teflon for corrosion protection. A cone mixer, located at the top of the acidulator, is used for mixing the absorber product liquor with sulfuric acid. The acidulator is mounted so that the point of gravity overflow to the stripper can be raised or lowered to vary the retention time of the material in the acidulator.

The stripping vessel was constructed of a 6-foot length of 12-inch stainless steel pipe and the inside was coated with Teflon. The stripper contains a 4-foot section of packing (polypropylene Pall rings). Acidulated liquor enters the top of the vessel and flows countercurrent to a stream of stripping gas (air) entering the vessel near the bottom. The SO_2 stripped from the liquor joins the SO_2 released in the acidulator and is routed back to the power plant ductwork. The stripped liquor (ammonium sulfate solution) is then fed to the evaporator crystallizer.

Typical data from a series of acidulation and stripping operations are shown in Table 24.3. In these tests, liquor from the absorption section having a C_A value of 14.7 was metered to the acidulator. Sulfuric acid (93%) was also metered to the acidulator at a rate to give an acid ion to ammonia ion ratio of 1.5. The liquor and acid were at ambient temperature (about 80°F). Stripping gas (air) also at ambient temperature was used to remove the chemically released SO_2 from the solution. The air flow rate was 10 cfm (30 cfm/gal of solution). Under these conditions, 96% of the SO_2 in the ammonium sulfite - bisulfite liquor to the acidulator was removed during the acidulation and stripping operation.

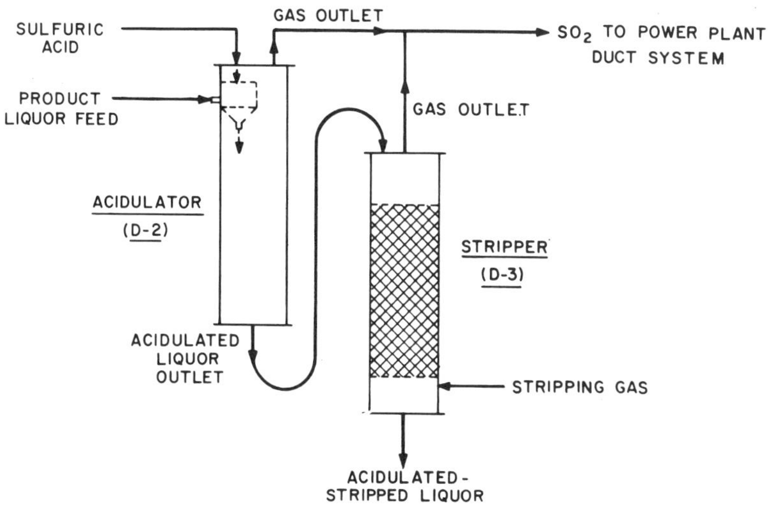

Figure 24.7. Acidulator-stripper.

The SO_2 remaining in the acidulated and stripped liquor is released in the evaporation-crystallization step and leaves the process with the steam from this equipment. Because this is a potential pollution problem additional efforts are being made to further decrease the SO_2 remaining in the acidulated and stripped liquor. New designs for both the acidulator and stripper are being considered to improve the SO_2 release from this operation.

The acidulator and stripper were inspected after 1800 hours of operation. The Teflon coating in the acidulator had failed and was peeling from the cone mixer and acidulator walls. The exposed stainless steel bottom of the cone mixer were severely damaged. The bare stainless steel acidulator walls were coated with a rust-colored deposit.

The stripper was found to be in good condition. No peeling of the Teflon coating was noted and the Pall rings showed no signs of wear.

Evaporator-Crystallizer

The evaporator-crystallizer used to produce ammonium sulfate from the acidulated and stripped liquor is ≊ 30 feet high, 2 feet in

Table 24.3. Typical Regeneration Loop Operation. Ammonium Bisulfate Pilot Plant.

Test Conditions	
Acidulator	
Liquor feed in	
C_A	14.7
S/C_A	0.81
pH	6.2
Specific gravity	1.244
Flow rate, gpm	0.3
Sulfuric acid	
Flow rate, gpm	0.07
Percent sulfuric acid	93
Stoichiometry[a]	1.5
Liquor flow out	
pH	1.8
Specific gravity	1.244
Percent SO_2 release	90
Stripper	
Stripping gas	
Type gas	Air
Flow rate, cfm at 70°F	10
Liquor flow out	
pH	1.9
Specific gravity	1.240
Percent SO_2 release	59
Overall % SO_2 release	96

[a]Stoichiometry is the ratio of acid ion to mols NH_3 as ammonium bisulfite and ammonium sulfite.

diameter at the bottom, and has a rated capacity of 350 pounds of ammonium sulfate crystals per hour. The entire unit is constructed of Type 316L stainless steel.

The crystallizer was received and installed near the end of the fiscal year and most of the operating time was spent in getting the equipment to perform according to specifications. Limited tests after operation was established indicated that the evaporator-crystallizer was adequately sized to remove water from the liquor at the highest flow rate anticipated during the test program. Crystals of ammonium sulfate produced in the crystallizer were sent to a belt filter.

Ammonium sulfate removal was accomplished by use of a belt filter and a product dryer which removed the approximately 150 lb/hr of ammonium sulfate crystals produced in the crystallizer. The belt filter was borrowed from another pilot plant and proved to be greatly oversized for the ABS study. The belt has an effective filtration area of 10 ft^2 and was operated at its lowest operating speed of about 2 ft/min. Intermittent operation was required even at the lowest belt speed in order to build a cake thick enough to maintain a vacuum in the system. Hinderance from ferrous ammonium sulfite, the subject of earlier concern, did not materialize and satisfactory filtration rates were obtained.

The crystals were dewatered from about 80% to about 5% water on the belt and were then fed to a propane gas-fired dryer (1 ft diameter by 12 ft long). Product from the dryer contained less than 1% water and was free flowing.

Figure 24.8 is a photograph of a typical group of crystals. Approximately 70% of the crystals were retained on a 35-mesh screen.

Ammonium Sulfate Decomposer

The ammonium sulfate decomposer will be used to decompose the sulfate crystals to ammonia and ammonium bisulfate. The ammonia will be returned to the absorption section and the ammonium bisulfate to the acidulator.

The preliminary project plans for a completed ABS process included a thermal decomposer. However, the piloting and scaleup of that decomposer which uses direct combustion of petroleum products as the heat source has not been completed. The plans have now been modified to include an electrically heated decomposer. A preliminary design for the pilot unit has been completed using the following design criteria:

Capacity	≊ 2 MW
Ammonium sulfate feed rate	200–250 lb/hr
Melt temperature	650°–750°F
Melt volume	12–14 ft^3 (1500 lb)
Flow rate of melt	2 ft^3/hr (0.25 gpm)
Steam feed rate	40–75 lb/hr at process temp.
Off-gas rate	≊ 3000-5000 ft^3/hr
Net electrical power input to process	45 kW (≊ 350 kWh/ton A/S decomposed)

Future Studies

Results of past operations have shown that the ammonium bisulfate process is a promising candidate for second generation SO_2 removal systems. Continuing pilot plant activities will be directed toward long-term operation of an integrated ABS system and demonstrated performance of 90% + SO_2 removal. The approach to this objective will include:

1. Closing the water balance around the gas prewash section and reducing carryover into the absorber.
2. Further reducing the plume emissions from the absorber.
3. Improving the acidificaton and stripping steps to obtain a complete release of SO_2 using ABS melt for acidulation.
4. Improving the solids separation from the absorber product liquor.

Figure 24.8. Crystals of ammonium sulfate produced in the ABS process; one scale division = 1 millimeter.

5. Studying ammonium sulfate crystallization (energy required, crystal growth, crystal size, etc.).

REFERENCES

1. Lepsoe, R. and W. S. Kirkpatrick. "SO_2 Recovery at Trail," *Trans. Can. Inst. Mining Met.* **XL**,399-404 (1937).
2. Hein, L. B., A. B. Phillips, and R. D. Young. "Recovery of SO_2 from Coal Combustion Stack Gases." In *Problems and Control of Air Pollution* Frederick S. Mallatte, Ed. (New York, N.Y.: Van Nostrand Reinhold 1955), pp 155–69.
3. *Sulfur* **80** (1), 36-37 (1969).
4. Rumanian Ministry of Petroleum Industry and Chemistry. "Ammonium Sulfate" Brit. Pat. 1,097,257 (January 3, 1968).
5. Johnstone, H. F. "Recovery of SO_2 from Waste Gases." *Ind. Eng. Chem.* **29** (12), 1396-98 (1937).
6. Tennessee Valley Authority. "Pilot-Plant Study of an Ammonia Absorption - Ammonium Bisulfate Regeneration Process, Topical Report Phases I and II." Report Y-83, Prepared for U.S. Environmental Protection Agency (EPA), Environmental Protection Technology Series, EPA-650/2-74—049-a (June, 1974).
7. Hollinden, G. A., N. D. Moore, P. C. Williamson; and D. A. Denny. "Removal of Sulfur Dioxide from Stack Gases by Scrubbing with Ammoniacal Solutions: Pilot-Scale Studies at TVA." *Proceedings: Flue Gas Desulfurization Symposium* (1973) pp 961-96, Environmental Protection Technology Series, EPA-650/2-73-038 (December, 1973).
8. Hixon, A. W. and R. Miller. "Recovery of Acidic Gases." U.S. Pat. 2,405,747 August 13, 1946).

25

Flue Gas Desulfurization by Fly Ash

Harold I. Zeliger
Chemical and Environmental Consultant
Spring Valley, New York 10977

INTRODUCTION

The lignites and subbituminous coals of the western United States are generally low in sulfur content, yet many cannot meet the new federal SO_2 emission standards of 1.2 lb SO_2 per MM kw of power produced without some SO_2 removal. Flue gas desulfurization (FGD) can be accomplished by utilizing several processes. The following processes are either currently in use or have been advanced sufficiently to where they will shortly be in commercial operation.

1. Limestone
2. Lime
3. Lime/Limestone
4. Fly Ash/Lime
5. Wellman-Lord
6. Magnesium Oxide
7. Catalytic Oxidation
8. Aqueous Sodium
9. Double Alkali
10. Sulfuric Acid
11. Dry adsorption

In a study resulting in the recommendation of an FGD process for the Milton R. Young Station, Center Unit 2, Center, North Dakota, operated by the Square Butte Electric Cooperative, available FGD processes were analyzed. This chapter a) presents a summary of the chemistries of 11 FGD processes listed above, b) describes the reasons for choosing a fly ash/lime system for the Milton R. Young station, and c) discusses the design parameters for this particular unit.

FGD CHEMISTRY

Following are the reactions that summarize the chemical processes occurring in flue gas desulfurization. There are many variables associated with scrubber chemistry. Time is extremely important since heterogeneous reactions are taking place. Design of equipment and mode of mixing are variables that influence contact between liquids, solids and gases. Depending upon the process, temperature and particle size of solids are also important factors. Since scrubber chemistry is very complicated exact mechanisms are difficult to elucidate and the equations that follow must be considered as working hypotheses of what is really occurring. In some cases they may only be approximations.

Limestone[1]

This is a wet nonregenerative process. The overall reaction in limestone scrubbing is

$$SO_2\ (g) + CaCO_3\ (s) + \tfrac{1}{2}H_2O \rightarrow CaSO_3 \cdot \tfrac{1}{2}H_2O\ (s) + CO_2\ (g) \quad (1)$$

The sequential steps are believed to be as follows:
The SO_2 dissolves and the Sulfurous acid ionizes as shown by equations (2)–(5).

$$SO_2\ (g) \rightleftharpoons SO_2\ (aq) \quad (2)$$

$$SO_2\ (aq) + H_2O \rightleftharpoons H_2SO_3 \quad (3)$$

$$H_2SO_3 \rightleftharpoons H^+ + HSO_3^- \quad (4)$$

$$HSO_3^- \rightleftharpoons H^+ + SO_3^= \quad (5)$$

Simultaneously, limestone dissolves into the scrubbing liquor as shown by Equations (6) and (7).

$$CaCO_3\ (s) \rightleftharpoons CaCO_3\ (aq) \quad (6)$$

$$CaCO_3\ (aq) + H^+ \rightleftharpoons Ca^{++} + HCO_3^- \quad (7)$$

Precipitation takes place when calcium ions combine with sulfite ions to form the hemihydrate as shown in Equation (8).

$$Ca^{++} + SO_3^= + \tfrac{1}{2}H_2O \rightarrow CaSO_3 \cdot \tfrac{1}{2}H_2O\ (s) \quad (8)$$

The sulfite ion may be oxidized to sulfate ion and precipitated as gypsum as shown in Equations (9) and (10).

$$SO_3^= + \tfrac{1}{2}O_2 \rightarrow SO_4^= \quad (9)$$

$$Ca^{++} + SO_4^= + 2H_2O \rightarrow CaSO_4 \cdot 2H_2O\ (s) \quad (10)$$

SO_2 may also be oxidized to SO_3 dissolved, ionized and precipitated in a manner analagous to SO_2, as shown by Equations (11)–(15).

$$SO_2\ (g) + \tfrac{1}{2}O_2\ (g) \rightarrow SO_3\ (g) \quad (11)$$

$$SO_3\ (g) \rightleftharpoons SO_3\ (aq) \quad (12)$$

$$SO_3\ (aq) + H_2O \rightleftharpoons H_2SO_4 \tag{13}$$

$$H_2SO_4 \rightleftharpoons H^+ + HSO_4^- \tag{14}$$

$$HSO_4^- \rightleftharpoons H^+ + SO_4^= \tag{15}$$

$$Ca^{++} + SO_4^= + 2H_2O \rightarrow CaSO_4 \cdot 2H_2O\ (s) \tag{16}$$

The system is acidic and CO_2, either from the flue gas or from $CaCO_3$ reacts with water as shown in Equations (17), (18) and (19).

$$CO_2\ (g) + H_2O \rightleftharpoons H_2CO_3 \tag{17}$$

$$H_2CO_3 \rightleftharpoons H^+ + HCO_3^- \tag{18}$$

$$HCO_3^- \rightleftharpoons H^+ + CO_3^= \tag{19}$$

Lime[2]

This is a wet, nonregenerative process. The overall reaction of lime scrubbing is given by Equation (10).

$$SO_2\ (g) + CaO\ (s) + \tfrac{1}{2}H_2O \rightarrow CaSO_3 \cdot \tfrac{1}{2}H_2O \tag{20}$$

The conversion of SO_2 into sulfite and sulfate is the same as that shown above for limestone, Equations (2)–(5), (8) and (16).

Equations (21)–(23) summarize the reaction in which lime is dissolved and ionized.

$$CaO + H_2O \rightleftharpoons Ca(OH)_2\ (s) \tag{21}$$

$$Ca(OH)_2\ (s) \rightleftharpoons Ca(OH)_2\ (aq) \tag{22}$$

$$Ca(OH)_2\ (aq) \rightleftharpoons Ca^{++} + 2OH^- \tag{23}$$

The precipitation of calcium sulfite and calcium sulfate is identical with that shown for limestone above—Equations (8), (10) and (16).

Carbonate introduced into the system from reactions summarized by Equations (17)–(19) also reacts with calcium ion as follows.

$$Ca(OH)_2\ (aq) + CO_3^= \rightleftharpoons CaCO_3 + 2OH^- \tag{24}$$

The acid produced by ionization of sulfite/sulfate and the base produced from the $Ca(OH)_2$ react to neutralize each other.

$$H^+ + OH^- \rightleftharpoons H_2O \tag{25}$$

Limestone/Lime

The chemistry is simply a combination of the equations shown for limestone and lime individually, Equations (1-25).

Lime/Fly Ash

The chemistry of the lime scrubbing has been presented. Fly ash contains quantities of lime as well as other alkalis which vary with coal composition. This lime may be obtained by extraction with water, as schematically indicated by Equation (26)

$$CaO \cdot \text{Inerts (s)} + H_2O \rightarrow Ca(OH)_2 \text{ (s)} + \text{Inerts} \quad (26)$$
(fly ash)

The $Ca(OH)_2$ is ionized as shown by Equation (23).

Wellman-Lord[3]

The Wellman Lord process is a wet system with by product recovery. The process is based on a sodium sulfite/bisulfite regenerative cycle. The following equations summarize the process.
Adsorption:

$$Na_2SO_3 + H_2O + SO_2 \text{ (g)} \rightleftharpoons 2NaHSO_3 \quad (27)$$

Recovery of Reagent:

$$2NaHSO_3 \rightleftharpoons Na_2SO_3 + SO_2 + H_2O \text{ (g)} \quad (28)$$

$$NaHSO_3 + NaOH \rightleftharpoons Na_2SO_3 + H_2O \quad (29)$$

Condenser:

$$SO_2 \text{ (g)} + H_2O \text{ (g)} \rightarrow H_2O \text{ (1)} + SO_2 \text{ (g) (pure)} \quad (30)$$

Overall:

$$SO_2 \text{ (flue gas)} \rightarrow SO_2 \text{ (pure gas)} \quad (31)$$

There is some loss of reagent, due to oxidation of sodium sulfite to nonregenerable sodium sulfate as shown by the following equations:

$$Na_2SO_3 + \tfrac{1}{2} O_2 \rightarrow Na_2SO_4 \quad (32)$$

$$2Na_2SO_3 + SO_3 + H_2O \rightarrow Na_2SO_4 + 2NaHSO_3 \quad (33)$$

Magnesium Oxide[4,5]

The chemistry of the magnesium oxide process is somewhat similar to that for calcium oxide, although unlike the case for calcium where only one hydrated sulfite is formed, here three are formed. Also, due to the high cost of the reagent, magnesium oxide must be recycled. Equation (33)–(44) show the absorption and drying process, and Equations (45)–(50) show the regeneration process.
Absorption:

$$MgO \text{ (s)} + H_2O \text{ (1)} \rightleftharpoons Mg(OH)_2 \text{ (s)} \quad (34)$$

$$Mg(OH)_2 \text{ (s)} \rightleftharpoons Mg(OH)_2 \text{ (aq)} \quad (35)$$

$$Mg(OH)_2 \text{ (aq)} \rightleftharpoons Mg^{++} + 2OH^- \quad (36)$$

$$SO_2 \text{ (g)} + H_2O \rightleftharpoons SO_2 \text{ (aq)} \quad (37)$$

$$SO_2 \text{ (aq)} + H_2O \rightleftharpoons H_2SO_3 \quad (38)$$

$$H_2SO_3 \rightleftharpoons H^+ + HSO_3^- \quad (39)$$

$$HSO_3^- \rightleftharpoons H^+ + SO_3^= \quad (40)$$

$$Mg^{++} + SO_3^= + 3H_2O \rightarrow MgSO_3 \cdot 3H_2O \quad (41)$$

$$Mg^{++} + SO_3^= + 6H_2O \rightarrow MgSO_3 \cdot 6H_2O \quad (42)$$

$$Mg^{++} + SO_3^{=} + 7H_2O \rightarrow MgSO_3 \cdot 7H_2O \quad (43)$$

$$MgSO_3 \cdot 6H_2O \rightarrow MgSO_3 \cdot 3H_2O + 3H_2O \quad (44)$$

$$MgSO_3 \cdot \tfrac{1}{2}O_2 + 7H_2O \rightarrow MgSO_4 \cdot 7H_2O \quad (45)$$

The Drying Reaction:

$$MgSO_3 \cdot 3H_2O \xrightarrow{\text{heat}} MgSO_3 + 3H_2O \quad (46)$$

$$MgSO_3 \cdot 6H_2O \xrightarrow{\text{heat}} MgSO_3 + 6H_2O \quad (47)$$

$$MgSO_4 \cdot 7H_2O \xrightarrow{\text{heat}} MgSO_3 + 7H_2O \quad (48)$$

$$MgSO_4 \cdot 7H_2O \xrightarrow{\text{heat}} MgSO_4 + 7H_2O \quad (49)$$

At the recycling plant, magnesium oxide is regenerated. Coke is used to reduce the sulfate.

$$MgSO_3 \xrightarrow{\text{heat}} MgO + SO_2 \quad (50)$$

$$MgSO_4 + \tfrac{1}{2}C \xrightarrow{\text{heat}} MgO + SO_2 + \tfrac{1}{2}CO_2 \quad (51)$$

At the recovery plant the sulfur dioxide is oxidized and converted to sulfuric acid.

$$SO_2\,(g) + \tfrac{1}{2}O_2\,(g) \rightarrow SO_3\,(g) \quad (52)$$

$$SO_3\,(g) + H_2O\,(1) \rightarrow H_2SO_4 \quad (53)$$

Catalytic Oxidation[6,7]

The chemistry of catalytic oxidation is relatively simple. The SO_2 containing flue gas is passed over a catalytic bed, oxidized to SO_3 and reacted with water to produce usuable sulfuric acid, as shown by Equations (54) and (55).

$$SO_2\,(g) + \tfrac{1}{2}O_2 \xrightarrow[\text{Catalyst}]{\text{hot}} SO_3\,(g) \quad (54)$$

$$SO_3\,(g) + H_2O\,(1) \rightarrow H_2SO_4 \quad (55)$$

Aqueous Sodium[8]

Nonregenerable sodium scrubbing is the simplest and most reliable process for removing SO_2 from flue gas.[3] The process involves simply scrubbing with a solution of sodium hydroxide or sodium carbonate to produce a solution of dissolved sodium/sulfur salts, as shown by Equations (56) and (57).

$$2NaOH + SO_2 \rightarrow Na_2SO_3 + H_2O \quad (56)$$

$$Na_2CO_3 + SO_2 \rightarrow Na_2SO_3 + CO_2 \quad (57)$$

Following scrubbing, the solution is oxidized to produce a neutral solution of sodium sulfate (with no oxygen demand) which is then discarded.

$$Na_2SO_3 + \frac{1}{2}O_2 \rightarrow Na_2SO_4 \quad (58)$$

Double Alkali[8]

SO_2 absorption takes place in sodium sulfite solution to produce a bisulfite scrubber effluent solution as follows:

$$2NaOH + SO_2 \rightarrow Na_2SO_3 + H_2O \quad (59)$$

$$Na_2CO_3 + H_2O + SO_2 \rightarrow Na_2SO_3 + H_2O \quad (60)$$

$$Na_2SO_3 + H_2O + SO_2 \rightarrow 2NaHSO_3 \quad (61)$$

In the regenerative process, lime or limestone is used to recover soluble alkali.

$$CaO + H_2O \rightleftharpoons Ca(OH)_2 \quad (62)$$

$$2NaHSO_3 + Ca(OH)_2 \rightarrow Na_2SO_3 + CaSO_3 \cdot \frac{1}{2}H_2O + 3/2H_2O \quad (63)$$

$$2NaHSO_3 + CaCO_3 \rightleftharpoons Na_2SO_3 + CO_2 + H_2O \quad (64)$$

In theory, the lime neutralization reaction should go to completion, whereas complete neutralization of bisulfite with limestone is not possible. Using lime, the regeneration can be carried beyond neutralization to generate sodium hydroxide:

$$Na_2SO_3 + Ca(OH)_2 \rightleftharpoons 2NaOH + CaSO_3 \cdot \frac{1}{2}H_2O + 3/2H_2O \quad (65)$$

As with limestone or lime scrubbing, formation of sulfate may occur in the scrubber liquor as follows:

$$2NaOH + SO_3 \rightarrow NaSO_4 + H_2O \quad (66)$$

$$Na_2CO_3 + SO_3 \rightarrow Na_2SO_4 + CO_2\ (g) \quad (67)$$

$$Na_2SO_3 + \frac{1}{2}O_2 \rightarrow Na_2SO_4 \quad (68)$$

Equations (66) and (67) represent reactions that may occur due to the presence of SO_3 in the flue gas. Equation (68) represents oxidation in the regenerative process and is the principle source of sulfate. Sulfate that is formed may also be accommodated in a dual alkali system, since it may be reacted with lime to produce soluble alkali as shown by Equation (69).

$$Na_2SO_4 + Ca(OH)_2 + 2H_2O \rightleftharpoons 2NaOH + CaSO_4 \cdot 2H_2O \quad (69)$$

Sulfuric Acid[9]

The Chiyoda Process is a lime scrubbing process but is different in that the absorbent liquor in the Chiyoda Process is an acid one rather than a basic one as in other processes. The reactions are summarized by the following equations:
In the absorber:

$$H_2O\ (1) + SO_2\ (g) \rightleftharpoons H_2SO_3\ (aq) \quad (70)$$

$$H_2O\ (1) + SO_3\ (g) \rightleftharpoons H_2SO_4\ (aq) \tag{71}$$

In the oxidizer, using an iron catalyst:

$$H_2SO_3 + \tfrac{1}{2}O_2 \xrightarrow[\text{Catalyst}]{\text{Fe}} H_2SO_4 \tag{72}$$

In the reactor:

$$H_2SO_4 + CaO + H_2O \rightarrow CaSO_4 \cdot 2H_2O \tag{73}$$

The gypsum is produced as a recoverable by-product.

Dry Adsorption[10]

SO_2 removal from the flue gas stream is accomplished by using activated carbon. SO_2 is oxidized to SO_3 by oxygen in the flue gas. In a secondary reaction, SO_3 is converted to sulfuric acid by water vapor in the flue gas. The acid is retained in the pores of the carbon where it is firmly adsorbed.

The amount of SO_2 that can be absorbed by the carbon ranges between 5 and 15% and is a function of the nature of the activated carbon. Once saturated, the carbon must be regenerated at temperatures of 750–1500°F depending upon the nature of the carbon used. Upon heating, the sulfuric acid is reduced to SO_2 by the carbon. CO_2 and H_2O are also liberated.

Adsorption:

$$SO_2 + \tfrac{1}{2}O_2 \xrightarrow{300^\circ F} SO_3 \text{ (a surface reaction)} \tag{74}$$

$$SO_3 + H_2O \rightarrow H_2SO_4 \tag{75}$$

Desorption:

$$H_2SO_4 + \tfrac{1}{2}C \xrightarrow{750\text{-}1500^\circ F} H_2O + SO_2 + \tfrac{1}{2}CO_2 \tag{76}$$

Equations (74) and (75) are exothermic. In addition, equation (74) requires that the reaction temperature can be about 275°F to maximize rate of reaction and adsorption efficiency.

FLY ASH/LIME SYSTEM

When a fly ash/lime system was selected for the Milton R. Young Station, several parameters were taken into consideration. These are:

1. The Milton R. Young Station units 1 and 2 have precipitators to remove particulates.
2. Both units have cyclone boilers which produce high quality, small particle sized fly ash.
3. Both units are fueled by low-sulfur (0.5–1.3% sulfur range, average of 0.55% sulfur) lignite which is site-specific.
4. The fly ash produced has no commercial value.

5. The fly ash is high in alkalinity,[11] consisting of 25% calcium oxide, 40% of which (or 10% of the total fly ash) is estimated to be available for SO_2 removal. With this high alkalinity content, no additional lime will be necessary when average sulfur content coal is being burned and only minimal additions of lime will be necessary when high-sulfur coal will be burned for the plant to meet SO_2 levels in flue gas required by the New Source Performance Standards.
6. The site is remote, inland and has difficult access. This would make reagent transport for another type of system difficult.
7. The site is in an arid region, with low water availability.
8. The site has limited space, making the storing of reagents difficult.
9. The topography of the region makes large scale ponding physically and economically unacceptable.
10. The site is a mine mouth and the plant is between one and four miles of a surface strip mine. The unique feature may be taken advantage of by the disposal of sludge in an economically and environmentally acceptable manner. The sludge may be spread to dry and then covered with top soil.

DESIGN PARAMETERS

The parameters described above dictated that effluent discharges be minimized and that water use be kept as low as possible. A closed loop system was designed. Sludge will be dewatered by a clarifier, a thickener and finally a vacuum filter. This will result in a 60% solid filter cake that will be trucked to the mine for disposal.

Since the station already has a precipitator and because only about 60% of the SO_2 in the flue gas must be removed to comply with regulations covering stack emissions, a spray tower absorber was recommended so as to minimize the pressure drop penalty. Two absorbers, each capable of scrubbing 60% of the flue gas, have been specified.

Reheat, which is necessary to prevent condensation in the stack, will be accomplished through bypassing of a fraction of the incoming flue gas and mixing of the bypassed fraction with the absorber outlet fraction (See Figure 25.1.). The relatively low SO_2 removal requirement enables this design feature to be incorporated.

The following systems comprise this fly ash/lime process. The entire process is represented by Figure 25.1.

Absorber Gas System

The gas enters the system after passing through the precipitators and exit fans in two parallel streams. Each stream has a bypass for

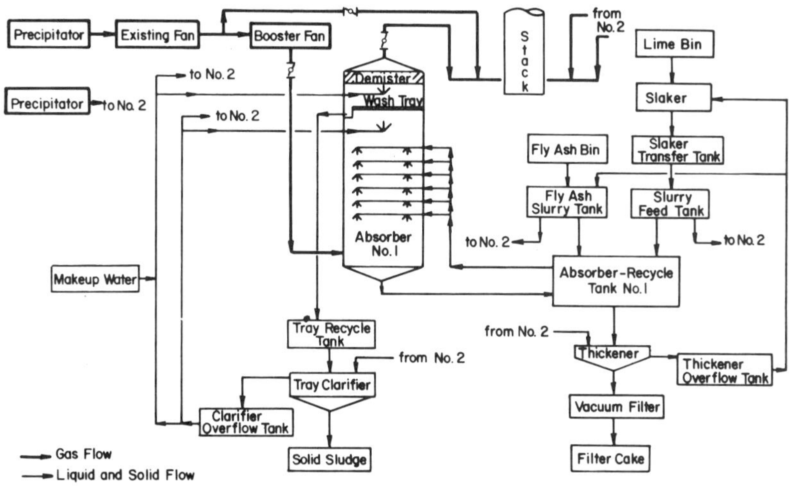

Figure 25.1. Fly Ash/Lime FGD Process flow diagram (no scale).

the fraction of the gas used to provide reheat. Dampers are incorporated so that each of the absorbers may be completely bypassed and isolated, in the event of failure, without necessitating total shut down.

Absorber-Recycle System

This is a closed loop system. The absorbers are spray towers in which the gas is passed through liquor streams composed of 12% solids. The wash tray is an important feature of this design, since it minimizes scaling and plugging of demisters. The liquid to gas ratio in the absorber towers is 80/1,000 CFM. The absorber recycle tanks are sized for eight minutes retention time to allow chemical reaction to be carried to completion.

Reagent System

Fly ash is conveyed to the fly ash slurry tank where most of the lime extraction takes place. The fly ash slurry is then pumped to the absorber-recycle tanks for use in the spray towers. Lime is slaked and transferred through the slaker transfer tank (a small tank used only for transfer purposes) to the slurry feed tank where it is stored until needed in the absorber-recycle tanks. The lime is used when needed as a supplementary reagent. In addition, the lime is also used to control pH in the absorber-recycle tanks. pH must be maintained between 5.0 and 6.0 to achieve maximum utilization of the alkalinity contained in the fly ash. When the sulfur content of the coal being burned is 0.84% or less, lime is not needed for scrubbing purposes. At these levels of SO_2, fly ash is able to accomplish the entire task.

Disposal Recycle System

The slurry bled to the thickener from the absorber-recycle tanks is concentrated to 30% solids. Additional water is removed by the vacuum filter to produce a filter cake that contains 60% solids. The thickener overflow water is transferred via the thickener overflow tank to the slaker and fly ash slurry tank for reuse.

Clarifier System

The stream that exits from the wash tray is collected in the tray recycle tanks and transferred to the tray clarifiers where solid sludge (30% solids) is removed and the overflow is recycled. Although not shown in Figure 25.1, the solids sludge may be transferred to the vacuum filter for further concentration and removal of water. All makeup water coming into the system enters above the wash tray.

Overall Material Balance

Figure 25.2 shows material balance in pounds per minute for the process. The upper numbers in the boxes are for scrubbing SO_2 from gas streams that result from the combustion of 1.3% sulfur coal. The lower numbers represent 0.82% sulfur coal data. These data show that for the 0.82% sulfur coal, all the alkali required comes from the fly ash. The solids formed are chemical reaction products. Since the calcium oxide content of the fly ash is 10% of the weight of the fly ash, the total solids passing through the system is considerably higher than shown in Figure 25.2. For the high sulfur case, the fly ash inerts would be 750.6 lb/min and the total solids coming out of the system would be 1,183.6 lb/min.

Water use for this process is very low. For the 1.3% sulfur coal, 550 gpm is the total water use.

In conclusion, the process described above was designed specifically for the site on which it is to be erected.

ACKNOWLEDGEMENT

We are indebted to Messers Sam Davidman, Richard W. Goodwin and Joseph P. Olszewski of Sanderson and Porter for their valuable assistance. We also wish to thank the management of Square Butte Electric Cooperative for permission to incorporate information about their installation.

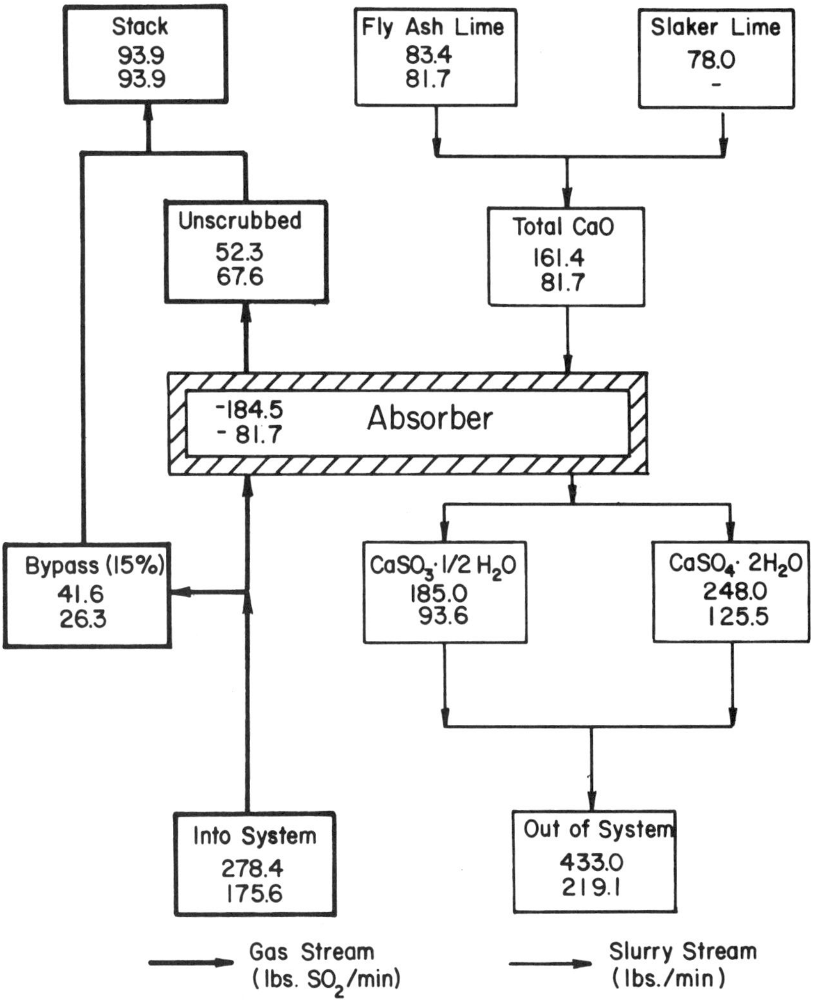

Figure 25.2. Material Balance (lb/min) for a Fly Ash/Lime FGD Process. Upper numbers are for scrubbing of SO_2 from gas streams resulting from combustion of 1.30% sulfur coal. Lower numbers represent 0.82% sulfur coal data.

REFERENCES

1. Gogineni, M. R., W. C. Taylor, A. L. Plumley and J. Jonakin. "Wet Scrubbing of Sulfur Oxides From Flue Gases," Presented at A.C.S. Meeting, New York 1972.
2. Devitt, T. W. and F. K. Zada. "Status of Flue Gas Desulfurization Systems

in the United States," Presented at E.P.A. FGD Symposium, Atlanta, Georgia, 1974.
3. Slack, A. V. "Removing SO_2 from Stack Gases," *Environ. Sci. Technol.* 7, No. 2 (1973).
4. Quigley, C. P. "Sulfur in Utility Fuels: The Growing Dilemma," Progress Report, Magnesium Oxide Scrubbing System at Boston Edison's Mystic Station, Presented at the Electrical World Tech. Conf., Chicago, 1972.
5. Koehler, G. R. "New England SO_2 Recovery Project System Performance," Presented at the Amer. Inst. Chem. Eng. Meeting, Philadelphia, 1973.
6. Horlacher, W. R. "Review of SO_2 Recovery Processes," Presented at Oklahoma State Meeting of the Amer. Inst. of Chem. Eng., Bartlesville, Oklahoma, 1972.
7. Miller, W. E. "The Cat - OX Project at Illinois Power," Presented at Amer. Inst. of Chem. Eng. Meeting, Philadelphia, 1973.
8. La Mantia, C. R., R. R. Lunt, J. E. Oberhaltzer, E. L. Field and N. Kaplan. "EPA-ADL Duel Alkali Program - Interim Results," Presented at EPA Flue Gas Desulfurization Symposium, Atlanta, Georgia, 1974.
9. Idemura, H. "New Flue Gas Desulfurization Process," Presented at ACHEMA Congress, Frankfurt, Germany, 1973.
10. Bischoff, W. F. "Dry Adsorption System for Flue Gas Clean Up," Presented at E.P.A. Flue Gas Desulfurization Symposium, New Orleans, Louisiana, 1973.
11. U.S. Bureau of Mines, unpublished Data.

26
The Centripetal Vortex Contactor for SO_2 Absorption and Fine Particulate Removal

Richard P. Tennyson and **Ernest J. Bouffard**
Entoleter, Inc., New Haven, Connecticut.

To achieve the emissions standards of federal and local regulatory agencies in an era of energy shortage, careful study must be made of the energy efficiency of the emission control device selected.

The method addressed here is a wet scrubber that depends on spray contact for particle interception. Various mechanical arrangements have been used for atomization and acceleration of the scrubbing liquid into the contacting zone.

Although the venturi or kinetic orifice had been the industry standard, the centripetal vortex scrubber is a new high-performance contactor which has achieved acceptance in a wide variety of particulate and gaseous emissions over the past few years.

Empirical investigations have shown that for conventional spray contactors, there is a relationship between energy consumption per unit of gas treated and collection efficiency. This relationship has become generally known as the "contacting power" rule and is usually expressed in the form $N_t = (\frac{1}{1\text{-}n}) = aP$ here $N_t = \log_e (\frac{1}{1\text{- collection efficiency}})$. P equals external power input per unit of gas treated. For a given dispersoid, an improvement in the gas scrubber design results in an increase in "a." The centripetal vortex principle increases "a" by 25% over conventional contactors (Figure 26.1).

Since population density and specific surface areas vary inversely with droplet diameter, it is reasonable to expect that the centripetal vortex with its small droplet diameter would exhibit both superior

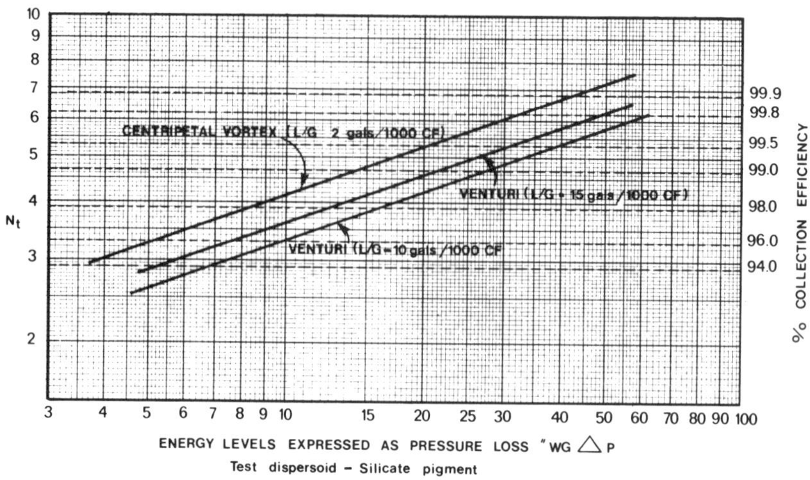

Figure 26.1. Comparative performance curve: Centripetal vortex vs. venturi scrubber expresses the equation $N_t = (\frac{1}{1\text{-}n}) = aP$ described in text.

mass transfer capabilities and particulate removal. Gas enters the contactor section tangentially, establishing a vortex pattern around a circular arrangement of vanes (vane cage). The stationary vanes in this contactor element direct and distribute the gas to establish a vortex. Scrubbing liquid is introduced without sprays or restrictive orifices in one of three ways: a) pumped to the periphery of the vane cage in a tangential direction, b) fed to the inside of the cage through a central feed pipe, or c) a liquid level is established in the inlet or contactor section of the device. For the third method, a length of pipe extending from the bottom of the vane cage contactor below the liquid level acts as a suction tube to draw liquid a distance equal to the draft loss of differential pressure across the contactor. In this mode, the scrubber is considered self-pumping. A pump need only supply the liquid necessary to maintain the level diminished by evaporative loss and solids blowdown. Regardless of the method by which the liquid is delivered, once introduced into the vortex zone of the contactor, it is broken up into very fine droplets, which approach the velocity of the gas.

A major percentage of the fine droplets formed remain in a stabilized orbit maintaining a balance between centrifugal and centripetal forces. Those droplets that have achieved a larger mass through agglomeration or interception of contaminant leave the contact zone through large relief ports in each vane and serve to totally flush and precondition the inlet section and incoming gas stream.

The large droplets that escape the vortex tangentially against the incoming contaminated gas stream are eventually collected and discharged through a slurry drain. Meanwhile, scrubbed gas exits through the vortex eye and travels vertically to the entrainment separator which is close-coupled to the contactor section. The average droplet size in the rotating centripetal vortex cloud thus established will depend on the velocity or pressure loss through the contactor. Higher velocities will establish a finer average droplet, thereby increasing the fine particulate removal rate.

In the contact zone there are three basic droplet size ranges. The average droplet that is rotating in the centripetal vortex cloud is maintained in equilibrium between its centrifugal force tending to throw it out and the centripetal force of the gas stream driving it toward the eye of the vortex.

The second basic droplet size is a much larger droplet, formed through agglomeration or nucleation. These droplets, by virtue of their mass, have become unstable in the high-speed vortex. They move to the periphery of the vortex and on a random basis find their way through the slots in each vane of the contactor. Those droplets that do not find an opening will either impact against the solid portion of the vane or be swept in and torn up by the shearing action of the high-velocity gas stream and rejoin the orbiting field as equilibrium droplets. The droplets that do not escape act as a prequench and low-energy spray contactor for removal of large particulates in the incoming gas stream.

The third sized droplet in this contact zone is the finer droplet, the mass of which is insufficient to resist the centripetal force of the gases moving to the eye of the vortex. Theses droplets are carried up to the entrainment separator where they are impinged and inertially separated from the gas stream (Figure 26.2).

The centripetal vortex principle can be adapted to variable pressure drop design, which increases its flexibility in gas cleaning applications where the gas volume changes substantially as a result of process variations. A recent refinement of the centripetal vortex contactor places a second contactor vane cage inside the first, creating a two-stage countercurrent liquid/gas contactor. The first or outer vane cage is fed independently of the inner cage by drawing scrubbing liquor up through the annular suction tube from a large internal sump reservoir. This liquid forms the first or outer scrubbing cloud and uses already spent scrubbing liquid. Through a center feed tube to the inner vane cage, cooler, fresh liquid is pumped, and a second or inner scrubbing cloud is formed, which contacts the already scrubbed gases from the outer vane cage. Final contact of liquid and gas occurs between the cleanest, coolest liquid and the cleanest gas.

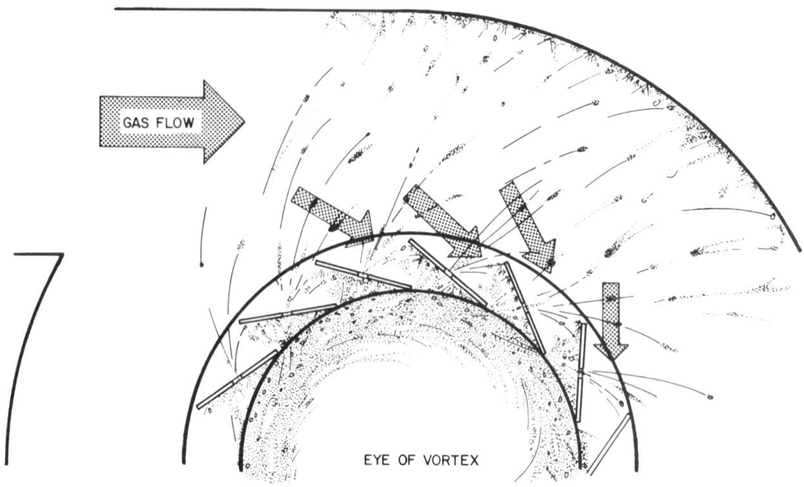

Figure 26.2. Centripetal Vortex droplet dispersion.

The result is a countercurrent flow providing maximum absorption driving force, minimum slurry concentration within the entrainment separator section, and a maximum solids concentration in the slurry discharge.

The centripetal vortex scrubbers can also employ an adjustable contactor vane cage. The vane cage is raised or lowered within a blinding cylinder. Raising it reduces the area through which the gas must pass, thereby creating a greater resistance. Dropping the vane cage creates the opposite effect. In fluctuating gas volumes such as in a steam boiler, the vane cage can be raised and lowered manually or automatically to maintain a constant pressure drop at varying gas flow rates.

The ability to attain extended residence time between the gas and the scrubbing liquid permits substantial reduction in the amount of liquids required per unit volume of gas. Some units operate on particulate applications as low as 2 gal/1000 CFM. For the purpose of absorption or chemical reaction, units operate as high as 20 gal/1000 CFM. The operating pressure drop range can be as low as 3″ w.c. to as high as 100″ w.c., thus spanning the range of low-energy wet centrifugal scrubbers to high-energy venturis.

Some examples of performance are as follows:

1. A phosphate calciner emission of 40,000 ACFM required 14″ w.c. pressure drop with the centripetal vortex to achieve process weight codes versus 35″ w.c. pressure drop required by venturi scrubbers.

2. An asphalt batch plant aggregate dryer emission of 42,000 ACFM required 7″ w.c. pressure drop through a centripetal vortex to achieve less than .02 gr/SCF outlet loading with 6–10 gr/SCF inlet loading. This unit replaced an existing low-energy scrubber with the same pressure drop which was achieving a .19 gr/SCF outlet emission.
3. A pilot centripetal vortex contactor was used on a coal dryer at a coal preparation plant. To achieve a .02 gr/SCF emission, a venturi operating on this plant required 38″–40″ w.c. pressure drop. The centripetal vortex unit required 14″ w.c. to achieve an emission lower than .02 gr/SCF. Two 200,000-CFM units have been purchased and will be installed on new coal dryers in Pennsylvania.

Having established basic operating principles and the performance advantages of the centripetal vortex contactor for particle removal, the next logical progression would be control of fly ash and SO_2 emissions from power boilers.

Early laboratory and field pilot unit tests using either lime or sodium hydroxide scrubbing liquors showed promising performance for SO_2 and flyash removal. Process data generated from the system discussed here were derived from a pilot test program conducted at a boiler installation during the fall and winter of 1972–73.

A pilot slip stream of 1500–2000 CFM was taken from one of two stacks serving two spreader stoker-fed 60,000 lb/hr steam power boilers. Following the pilot work, a system was built and installed.

The system consists of two scrubbers, auxiliary equipment, and control instrumentation. Each scrubber is designed for an inlet of 45,000 ACFM at 350–400°F (including 5% moisture) with constant pressure to a turndown rate of 15,000 ACFM. Below 15,000 AFCM down to 5000 ACFM, ambient air is drawn into each system's fan inlet plenum.

A constant pressure is maintained across each scrubber by the use of a pressure drop sensor and control loop to operate the scrubber's adjustable vane cage contactor. The pressure drop desired can be dialed in at the control panel from 4″ w.c. to 12″ w.c. This provides for the performance needs dictated by the ash content of the coal. The pressure drop selected to meet guarantees and at which preliminary performance tests have been run is 7″ w.c. across each scrubber.

The system uses an aqueous solution of sodium hydroxide which, from an operating standpoint, consists of a sodium sulfite-bisulfite recycle. The liquor is maintained at pH7. The liquid to gas ratio is 6 gal/1000 CFM. An average system purge of 20 gpm of liquor is maintained.

When using less than 1% sulfur coal, the average removal is

greater than 90%. This equates to an inlet concentration range of 250–350 ppm with outlet emissions of 25–35 ppm (6 lb/hr SO_2). When using 2% sulfur coal, the boilers emit 700–800 ppm of SO_2. The removal efficiency of the scrubbers with all other conditions maintained is 75% or 200 ppm of SO_2.

The system has been operational since September, 1974, with 98–99% scrubber operating time. The average reagent cost when burning 2–2.5% sulfur coal is about $2.00 per ton of coal burned.

The plan was to install scrubbers that would achieve particulate emission standards and, by the addition of a caustic feed system, be able to maintain a pH that would pretreat the effluent and reduce system corrosion. Since this customer was burning a low-sulfur coal (.7–.9%) the reagent use could be maintained at a very low rate. The system has the ability to burn higher-sulfur coal and comply with emissions standards. Since some industry reports show as much as $20.00 per ton cost differential between low-sulfur coal (less than 1%) and moderately high-sulfur coal (2–3%), the dollar savings become obvious.

SUGGESTED READING

1. Walling, J. C. "New Centripetal Vortex Contactors," *J. Air Poll. Cont. Assoc.*, (November, 1972).
2. Sparrs, L. E. and M. J. Pilat. "Diffusion Forces and Particulate Scrubber Efficiencies," PACA paper #69-82.
3. Walton, W. H. and A. Woolcock. "The Suppression of Airborne Dust by Water Spray," *Internat. J. Air Poll.* **III**, 129 (1960).
4. Semrau, K. T. "Dust Scrubber Design - A Critique on the State of the Art," *J. Air Poll. Cont. Assoc.* **13** (12) 587-594 (December, 1963).
5. Nukiyama, S. and Y. Tanasawa. "An Experiment on the Atomization of Liquid by Means of an Air Stream," *Trans. Soc. Mech. Engr. (Japan)* **IV** (14) 86, (1938).
6. Calvert, S. "Source Control by Liquid Scrubbing," In *Air Pollution*, A. C. Stern, Editor, 2nd ed., Vol. III (New York, N.Y.: Academic Press, 1968).
7. Bakke, E. and S. H. Reiter. "An Economic Optimization of a Venturi Scrubber with a Cyclonic Water Separator," APCA paper #70–7.
8. Hesketh, H. E., A. J. Engle and Calvert. "Atomization and Acceleration of Liquid in a Venturi Scrubber," APCA paper #69–83.
9. Walker, A. B. and R. M. Hall. "Operating Experience with a Flooded Disc Scrubber - a New Variable Throat Orifice Contactor," *J. Air Poll. Cont. Assoc.* **18** (5) 319–323 (1968).

27

Removal of Sulfur Dioxide from Stack Gases by Scrubbing with Limestone Slurry: Effect of an Organic Additive

Gerald A. Hollinden and **Neal D. Moore**
Power Research Staff, Tennessee Valley Authority,
Chattanooga, Tennessee

James J. Schultz
Division of Chemical Development, Tennessee Valley Authority,
Muscle Shoals, Alabama

INTRODUCTION

The dissolution of limestone is probably the major rate limiting step in the removal of sulfur dioxide from stack gases by scrubbing with a limestone slurry. There have been numerous efforts devoted to increasing the rate of limestone dissolution, such as: (a) decreasing the limestone particle size and hence increasing the effective surface area, (b) pH control and (c) addition of an acid to solubilize limestone. Earlier work[1] showed that reducing the limestone particle size increased sulfur dioxide removal efficiency, but only marginally. Using the Radian equilibrium program,[2] one can show that the solubility of limestone increases rapidly with decrease in pH as shown in Table 27.1. However, overall system control at low pH is difficult and rules out this approach. A similar increase in solubility can be achieved by adding another acid provided that the calcium salt of the added acid is soluble, and the ionization constant of the added acid is greater than that of carbonic acid H_2CO_3 ($K_1 = 4.4 \times 10^{-7}$) but less than that of sulfurous acid, H_2SO_3 ($K_1 = 1.3 \times 10^{-2}$).

Hatfield and Potts[3] suggested that the ideal additive would be soluble in water and limestone scrubbing slurry; nonvolatile and stable under scrubbing conditions; soluble with Ca, Mg, or other metals

Table 27.1. Solubility of $CaCO_3$ in Water at 50°C (122°F)[a]

Ca (*ppm*)	P_{CO_2} (*atm*)	*pH*
9.5	0.000136	8.5
16.8	0.000803	8.0
30.7	0.00466	7.5
57.7	0.0271	7.0
111.1	0.159	6.5
221.7	0.950	6.0
1020.3	36.3	5.0

[a]Calculated values using Radian's equilibrium program (PB 193029) for CaO-CO_2-H_2O system.

likely to be found in the solution; and commercially available at reasonable cost.

Further small-scale tests by Hatfield and Potts indicated that the addition of benzoic acid not only improves the effectiveness of limestone slurry in removing sulfur dioxide from stack gas but also has the further promise of increasing limestone utilization. Two earlier patents[4,5] issued by the United States Patent Office discuss somewhat similar approaches to that of benzoic acid addition.

As a further development of the benzoic acid system, TVA performed a pilot plant study of the process in 1973. A description of the pilot plant and results obtained are presented in this paper.

PILOT PLANT OPERATION

The evaluation of the effect of the benzoic additive on the operating characteristics of the limestone — wet-scrubbing process was made in a pilot plant at TVA's Colbert Power Plant in northwest Alabama near Muscle Shoals. The pilot plant was operated a total of 1400 hours using the benzoic acid additive. During the first period of operation (500 hr), several short-term tests were made to determine (a) the effect of the benzoic acid additive on the operating characteristics of the scrubbing system at various design conditions and to compare these data with similar data obtained during previous operation without the additive; and (b) operating problems attributed to the presence of benzoic acid.

After the short-term tests were completed, an extended (900 hr) test was made. This test was designed to evaluate the long-term effect of the benzoic acid additive while operating at conditions identified during the short-term tests that were considered to be optimum. A discussion of the tests follows.

Description of Process and Pilot Plant System

The pilot plant scrubber is a scale model of the system planned for the 550-MW Widows Creek unit. It is designed for a gas flow of 2400 acfm at 120°F. This flow is equivalent to a generating capacity of one megawatt. The model is used to determine the operating characteristics of the Widows Creek design and to identify changes that may be required. The pilot plant scrubber system is constructed of Type 316L stainless steel and is suitable for testing with alkalies other than limestone. A perspective view of the pilot plant is shown in Figure 27.1. A process flow diagram is shown in Figure 27.2.

Flue gas for the pilot plant contains about 2800 ppm SO_2 and can be obtained from either of two pulverized coal-fired boilers. The flue gas from one boiler can be withdrawn from (a) a point between the mechanical precipitator and the electrostatic precipitator (fly ash loading of about 2 gr/scf), or (b) a point downstream from the electrostatic precipitator (fly ash loading of about 0.1 gr/scf). Flue gas from the other boiler is withdrawn from a point upstream from the electrostatic precipitator. This gas contains the full loading of fly ash (4–5 gr/scf) because there are no mechanical precipitators upstream from the electrostatic precipitator.

The pilot plant is operated with a forced-draft gas flow configuration with the blowers located on the dry side of the scrubber. This is the same configuration planned for the Widows Creek unit. The blower system has a capacity of 3600 acfm (120°F) at 30 inches of H_2O pressure differential although the normal gas flow is 2400 acfm.

The flue gas flows downward through a venturi and discharges into a hopper-bottom below. The elbow connects the venturi to the absorber. The venturi is irrigated with scrubbing slurry at a liquid-to-gas ratio (L/G, gal/1000 ft^3 gas) of 10. The slurry drains from the elbow into a retention tank. The retention time of the scrubbing liquor duplicates the Widows Creek design and is 14 minutes based on the pumping rate (24 gpm) to the venturi. The pressure differential across the venturi is set at about 5 inches H_2O. However, it can be adjusted by varying the opening between two movable throat plates.

The gas, after passing through the venturi and elbow, flows upward through the absorber. The velocity of the gas through the absorber was normally maintained at 12 ft/sec (2400 acfm at outlet conditions) to duplicate the Widows Creek design. At maximum blower capacity, the velocity of the gas through the absorber can be increased to about 18 ft/sec. The absorber contains five grids (65% open area) spaced at 5-foot intervals. It can be converted to a Turbulent Contact Absorber by adding mobile packing spheres. Scrubbing slurry from the absorber retention tank is fed to the top of the absorber (using a rubber-lined centrifugal pump) at an L/G of 50. The scrubbing slurry is dispersed across the top of the absorber by using an open pipe distributor. The retention time in the tank is 7 minutes based on a pumping rate of 120 gpm.

The gas from the absorber passes through a mist elimination sys-

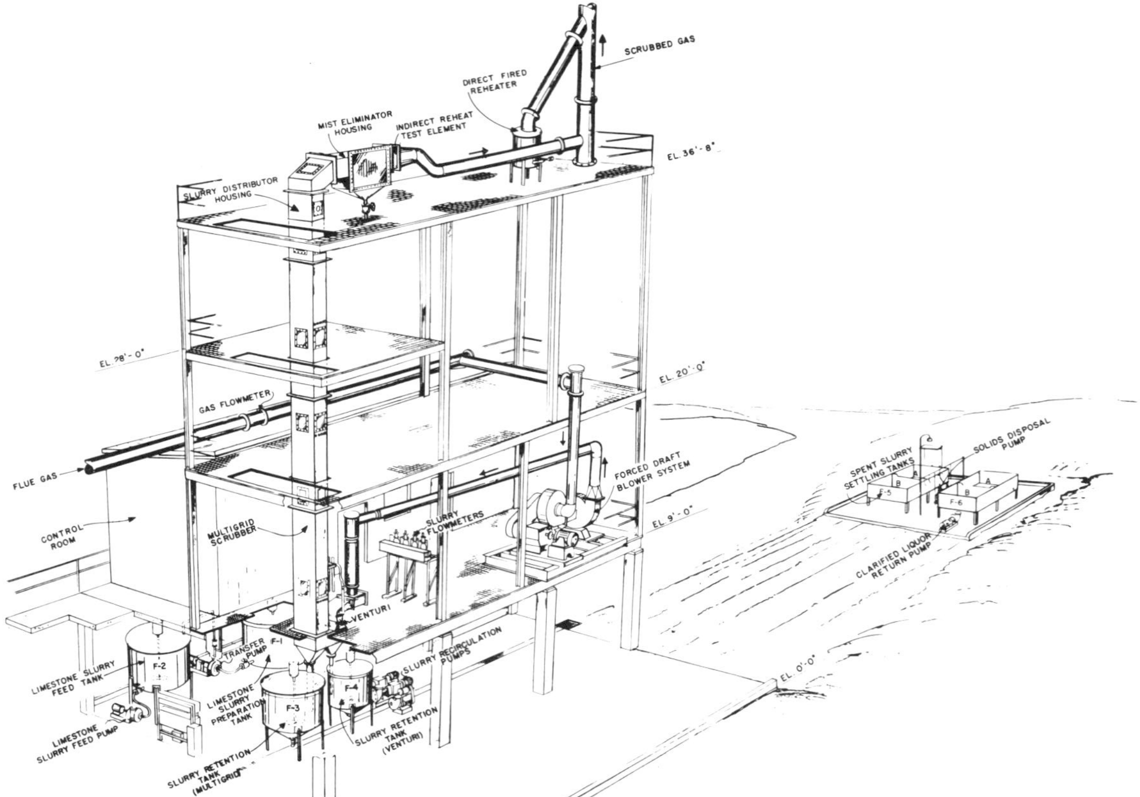

Figure 27.1. Perspective view—pilot plant model of Widows Creek unit 8. Limestone-wet-scrubbing system.

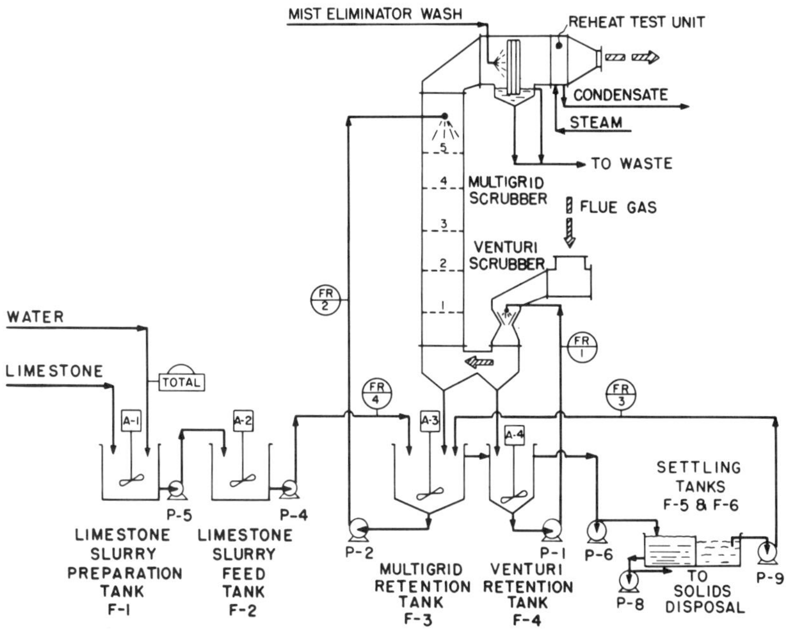

Figure 27.2. Flow diagram—pilot plant model of Widows Creek unit 8. Limestone-wet-scrubbing system.

tem before it is discharged to the atmosphere. The mist elimination system closely duplicates the configuration planned for Widows Creek. It consists of a horizontal housing equipped with a four-pass chevron vane element. The superficial velocity of the gas through the housing is 9 ft/sec when the multigrid scrubber is operated at 12 ft/sec. The vanes of the chevron element are made of Type 316L stainless steel and are spaced on 1½-inch centers. The element, which is 13½ inches deep, is washed continuously from the upstream side with fresh water on a once-through basis.

An indirect heating system will be used to reheat the scrubbed flue gas in the Widows Creek installation. Various heat exchange tube materials have been tested in the pilot plant for this application. Tubes made of Inconel 625, Incoloy 825, and Hastelloy C-276 have shown the most resistance to attack. The temperature of the flue is not increased significantly as it passes through the pilot plant indirect reheat test element because of the small heat transfer surface. A direct-fired reheat system (propane fuel) is used to heat the gas before it is exhausted to the atmosphere when tests are made to evaluate the plume appearance at various reheat temperatures and weather conditions.

Pulverized limestone (75%-200 mesh, purchased in bags) is slurried and fed to the absorber retention tank to maintain the desired Ca:SO_2 mole ratio based on the concentration of SO_2 in the inlet flue gas. When benzoic acid (flake material purchased in bags) is used, it is fed to the system along with the fresh limestone slurry.

The gas flow to the scrubber system is measured with a venturi flow element installed in the hot gas inlet duct. The indicated gas flow is checked routinely with pitot tube measurements.

The SO_2 concentration of the flue gas before and after the scrubber is monitored continuously with a Du Pont ultraviolet analyzer. The accuracy of the ultraviolet analyzer is checked routinely by using a wet chemical method.

Magnetic flowmeters equipped with electronic recorders are used to measure and record the slurry flows. The pH of the slurry in the absorber retention tank is monitored continuously with a pH measuring system manufactured by Universal Interloc, Inc. This system consists of a waterproof sensor which is submerged directly in the retention tank. The solids content of the scrubbing slurry is monitored continuously with a differential pressure bubble tube device.

Short-Term Tests

A series of short-term tests were made to determine the effect of the following variables on the SO_2 removal efficiency and the operating characteristics of the system while using the benzoic acid additive.

- Ca:SO_2 mole ratio
- Undissolved solids content in the scrubbing slurry
- Liquid-to-gas ratio (L/G) in the absorber
- Type of packing used in the absorber

These variables were selected for evaluation because they can significantly affect the operating cost of a commercial-size SO_2 removal system. Similar data on the effects of these variables were obtained previously while operating without benzoic acid.

The pilot plant was operated in a closed-loop configuration (no overflow from disposal pond) as shown in Figure 27.3. Limestone (76% -200 mesh, 65% -325 mesh) was fed to the absorber retention tank to maintain Ca:SO_2 mole ratios of 1.5, 1.2, and 1.0 based on the concentration of SO_2 in the inlet flue gas. Benzoic acid was added to the system continuously along with the fresh limestone slurry to make up for losses and to maintain a concentration of 0.10% by weight in the scrubbing slurry. This concentration of benzoic acid was selected on the basis of bench-scale tests and is sufficient to feed (at an L/G of 50) one mole of benzoic acid to the absorber for each mole of SO_2 in the inlet flue gas. The temperature of the scrub-

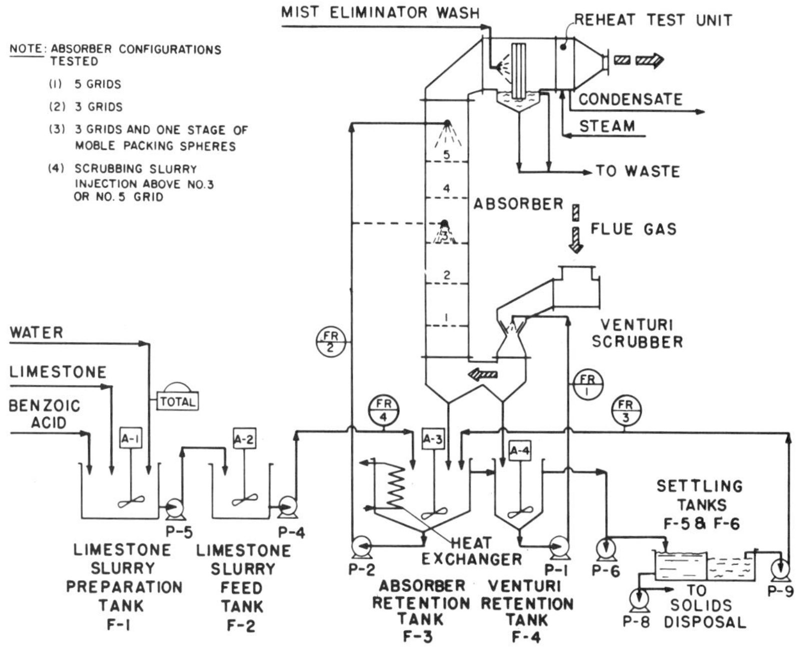

Figure 27.3. Flow diagram—short-term tests using benzoic acid in limestone-wet-scrubbing pilot plant.

bing slurry was maintained at 120°F by using a heat exchanger submerged in the absorber retention tank.

The SO_2 removal efficiency averaged 75% when the absorber was operated with five grids (65% open area) compared with 68% when the benzoic acid additive was not used. When benzoic acid was present and five grids were used, the efficiency was not affected when the Ca:SO_2 mole ratio and the undissolved solids content were reduced from 1.5 to 1.0 and 15 to 6%, respectively, and the L/G was increased from 50 to 60. During previous operation without benzoic acid, the SO_2 removal efficiency declined to less than 60% and control of the pH of the scrubbing slurry could not be maintained when the Ca:SO_2 mole ratio and the undissolved solids content of the scrubbing slurry were reduced to less than 1.3 and 10%, respectively. At these conditions, increasing the L/G was not effective in stabilizing the SO_2 removal efficiency or pH of the scrubbing slurry. The SO_2 removal efficiency increased 10 percentage points (from 75–85%) when one stage (12 in. deep) of mobile packing spheres was added to the absorber between the bottom two grids. During previous operation without benzoic acid, a similar increase (10 percentage points, from 68–78%) in SO_2 removal efficiency was obtained by adding two stages of packing spheres to the absorber. A comparison of these data with data obtained while operating without benzoic acid is shown in Table 27.2.

The spent solids from the limestone - wet-scrubbing process do not compact well. The settled solids normally contain about 60% water by weight even after extended periods of settling. The settling and compaction characteristics of the spent solids (final volume of settled solids) did not improve when benzoic acid was used. However, the settled solids with the benzoic acid additive contained less water. The reduced amount of water contained in the settled solids would reduce the purge from the system and less makeup water would be required. This may cause an operating problem because the normal water makeup due to pump seals and soot blower condensate may exceed the system requirement. A comparison of the chemical composition and the settling and compaction characteristics of the spent slurry while operating with and without the benzoic acid additive is shown in Table 27.3.

Operation of the pilot plant was routine during the short-term test series. However, the following problems were identified:

1. Scaling of the mist eliminator,
2. Coloration of the clarified liquor in the solids disposal pond and,
3. The makeup rate of benzoic acid had to be increased to about 30% above the calculated value to maintain a concentration of 0.10% in the scrubbing slurry.

The absorber wall and grid surfaces were generally clean but a uniform deposit of scale coated the two downstream passes of the four-pass chevron mist eliminator. The scale was about ¼ inch thick and was a mixture of about 90% gypsum and 10% fly ash. Petrographic analysis of the scale showed that the gypsum crystals were well formed and apparently grew on the blades slowly by reaction of dissolved calcium in finely divided entrained mist and residual SO_2 in the flue gas. The mist eliminator was washed continuously from the upstream side with river water on a once-through basis at a rate of 1 gpm/ft^2. This wash was effective in keeping the first two passes clean, but apparently it did not penetrate the chevron sufficiently to rinse the two back passes.

The clarified liquor returned to the scrubbing system from the spent slurry settling tank was deep red in color. The color disappeared when the pH of the clarified liquor was reduced from the normal 7.4 to below 5.8. The coloring agent has not been isolated but is related to the presence of benzoic acid and iron. The color of the clarified liquor did not cause any operating problems, but it is objectionable for aesthetic reasons.

Benzoic acid was added continuously along with the fresh limestone slurry to replace the acid lost with the settled solids. A make-

Table 27.2. Short-Term Test Data With and Without Benzoic Acid Addition. Limestone - Wet-Scrubbing Pilot Plant.[a]

	Absorber packing		*Scrubbing slurry*					
					L/G			
Test No.	*Number of grids*[b]	*Spheres, stages*[c]	*Ca:SO_2 mole ratio*	*Benzoic acid (%)*	*Venturi*	*Absorber*	*Solids (%)*	*SO_2 removal efficiency (%)*
1 — (1)	5	0	1.5	0	10	50	15	68
1 — (2)	5	0	1.5	0.12	10	50	15	75
1 — (3)	5	0	1.0	0.12	10	50	15	77
2 — (1)	5	0	1.2	0	10	50	15	64
2 — (2)	5	0	1.2	0.10	10	50	15	75
3 — (1)	5	1	1.2	0	10	50	15	75
3 — (2)	5	1	1.2	0.12	10	50	15	85
3 — (3)	5	1	1.2	0.12	10	50	6	85
3 — (4)	5	1	1.0	0.12	10	60	6	88
4 — (1)	3	1	1.0	0.14	10	50	6	79
5 — (1)	3	0	1.0	0.10	10	55	7	62

[a]Inlet flue gas contained 3 to 5 gr/std ft^3 of fly ash and 2200–2800 ppm SO_2.
[b]65% open grids spaced at 5-ft intervals.
[c]Polypropylene (TCA) spheres, 12 in. deep.

Table 27.3. Chemical and Settling Data—Spent Scrubbing Slurry. Limestone - Wet-Scrubbing Pilot Plant Operation With and Without Benzoic Acid.

Sample No.	*Percent solids in slurry at start of test*	*Volume percent after* 2 days *Solids*	*Volume percent after* 2 days *Liquid*	*Volume percent after* 20 days *Solids*	*Volume percent after* 20 days *Liquid*	*Percent water in settled solids after* 2 days	*Percent water in settled solids after* 20 days
Settling and compaction data							
1 (without benzoic acid, P-38)	15	38	62	33	67	70	67
2 (with benzoic acid, M-6)	22	38	62	36	64	52	50

	Benzoic acid	*Total sulfur*	*Sulfur as Sulfate*	*Sulfur as Sulfite*	*Ca*	*Mg*	*Cl*	*Total dissolved solids*	*pH*
Chemical analysis, % by weight									
1 (dry solids)	0	10.5	2.0	8.5	25.0	0.90	—[a]	—	—
(liquid)	0	0.07	0.07	Nil	0.10	0.05	0.17	1.00	7.7
2 (dry solids)	Nil	12.0	3.8	8.2	21.2	0.44	—[a]	—	—
(liquid)	0.11	0.06	0.06	Nil	0.20	0.05	0.32	1.24	7.6

[a]Not analyzed.

up rate of 14 pounds per day was required in the 1-MW pilot plant to maintain a concentration of 0.10% benzoic acid in the scrubbing slurry. Only about 75% of the benzoic acid added to the system could be accounted for by chemical analysis. The reason for the discrepancy in the material balance is not known, but it may be caused by degradation of the benzoic acid.

Long-Term Test

After completion of the series of short-term tests, a long-term test was made using the following operating conditions which were considered to be optimum.

• Gas velocity through absorber	12 ft/s at 120°F
• Type of packing in absorber	3 grids and 1 stage of TCA spheres
• Pressure drop across venturi	5–7 in H_2O
• L/G to venturi	10 gal/1000 ft^3
• L/G to absorber	50–60 gal/1000 ft^3
• $Ca:SO_2$ mole ratio	1.0
• Undissolved solids content of scrubbing slurry in absorber circuit	5%
• Benzoic acid content of scrubbing slurry	0.10%

Selection of these conditions was based on the short-term test data and the proposed design criteria for the Widows Creek system.

The pilot plant was operated in a closed-loop configuration according to the flow diagram shown in Figure 27.4. The top two grids (Nos. 4 and 5) were removed from the absorber. This was done to provide more freeboard between the top grid (where atomization of the scrubbing slurry occurs) and the inlet to the mist eliminator, thus reducing the potential for scaling caused by entrained liquor. One stage (12-in. static depth) of TCA spheres (standard polypropylene) was installed between the No. 2 and No. 3 wire mesh grids. Scrubbing slurry containing 5% undissolved solids and 0.10% benzoic acid was fed to the absorber through an open-pipe distributor located 15 feet above the No. 3 (top) grid. The temperature of the scrubbing slurry was maintained at 120°F.

During the first 4 days of operation, there were large variations in the fly ash loading in the flue gas because of erratic operation of the boiler. During periods of extremely high fly ash loading, it was difficult to maintain control of the solids level in the scrubbing slurry and a major portion of the solids in the slurry was fly ash because of the low level of excess limestone. As a result, the pH of the scrubbing slurry declined from the normal 5.6 to about 4.8, and the SO_2 removal efficiency declined from 85 to 60%. During previous opera-

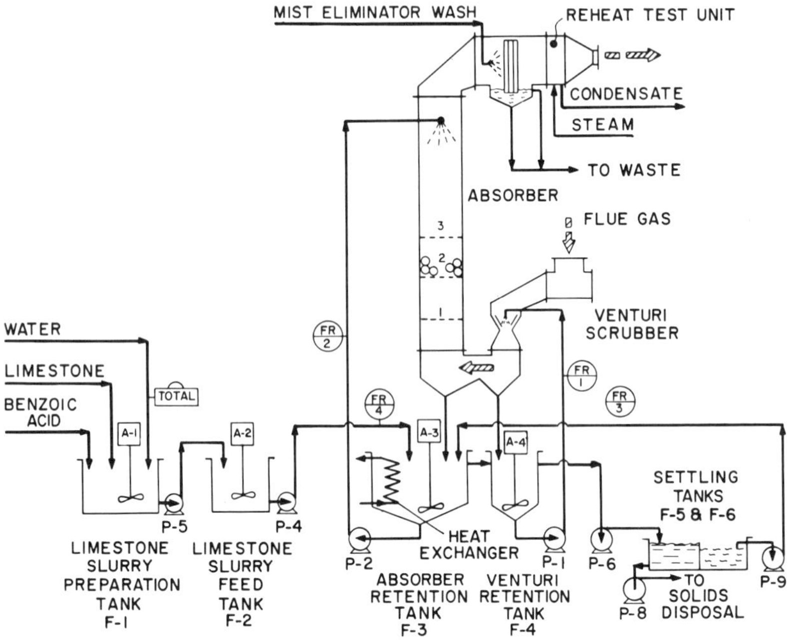

Figure 27.4. Flow diagram—long-term test using benzoic acid in limestone-wet-scrubbing pilot plant.

tion, when a large excess of limestone (Ca:SO_2 mole ratio of 1.5) was maintained in the scrubbing slurry, fluctuations in the ratio of fly ash to limestone were not critical.

After the operation of the boiler was stabilized, operation of the pilot plant was routine. The SO_2 and fly ash content of the inlet flue gas averaged 2600 ppm and 1.4 grains per standard cubic foot, respectively. The pressure drop across the system remained steady at 13 inches of H_2O (7-in. for the venturi and 6-in. for the absorber). The pH of the scrubbing slurry gradually declined from 5.5 to about 5.0 during the test. However, the SO_2 removal efficiency remained relatively steady at 85%. Typical trends in the operating data are shown in Figure 27.5.

During this period of operation, from 25–55% of the reacted SO_2 formed calcium sulfate. However, the settling and compaction characteristics of the reacted slurry were essentially the same as those obtained with a lower level of oxidation (10–20%) when benzoic acid was not used.

The color of the clarified liquor returned to the scrubbing system from the spent slurry settling pond continued to be dependent upon the pH which varied from 5.9 to 7.3. The liquor was deep red in color when the pH was above 7. The intensity of the color of the liquor decreased as the pH decreased and the color disappeared when the pH declined to 6 or less.

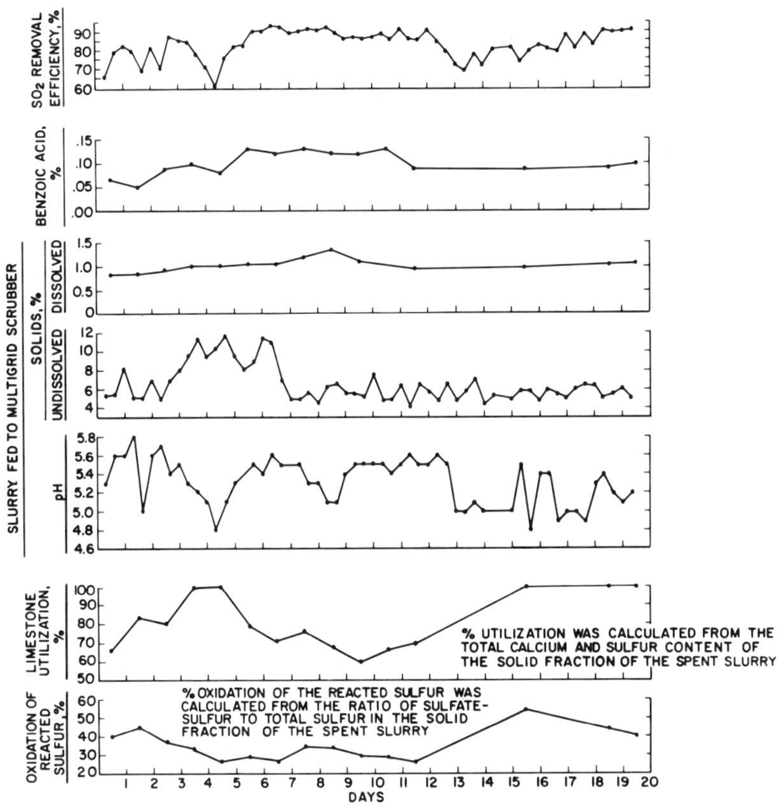

Figure 27.5. Typical operating trends—benzoic acid addition in limestone-wet-scrubbing pilot plant.

During this test, the makeup of benzoic acid [14 lb/day (MW)] required to maintain a concentration of 0.10% in the scrubbing slurry continued to be about 30% in excess of the calculated value.

The pilot plant was shut down and inspected after 20 days of continuous operation. The scrubber system was generally clean. Only minor deposits of hard scale (gypsum) had formed on about 10% of the scrubber wall and grid surfaces. The scale was about 1/16–1/8 inch thick and tended to bulge away from the scrubber walls and fall onto the grids. This did not cause plugging of the grids in the pilot plant, but it could be a problem in a large unit if the sections of scale were thicker and would not break up upon impact on the grids.

During the test, the four-pass chevron mist eliminator was washed continuously from the upstream side with river water (1 gpm/ft^2) on a once-through basis. The first two passes remained clean. The last two passes were partially coated with a deposit of scale about 1/16–1/8 inch thick. The deposit was identified petrographically as predominately gypsum. The amount of the deposit was much less than that which had formed during the previous short-term test period. No increase in pressure drop across the mist eliminator was observed. The decrease in the amount of deposits on the mist eliminator compared with the short-term operation may have been caused by (a) more freeboard between the top grid (where atomization of the scrubbing slurry occurs) and the mist eliminator and (b) the lower undissolved solids content (5 vs. 15%) of the scrubbing slurry used during this test.

The erosive characteristics of the scrubbing slurry containing the benzoic acid additive were evaluated. Scrubbing slurry containing 5–7% solids was fed to the venturi through three spiral-tip spray nozzles (Type 303 stainless steel) manufactured by Bete Fog Nozzle, Inc. The velocity of the slurry through the nozzles was about 30 ft/sec, and the pressure drop was 10 lb/in^2. Severe erosion occurred, and the nozzles had to be replaced after 10 days of operation. Bete nozzles were frequently used during previous operation of the pilot plant and gave good service for about 800 hours of operation using a 15% solids slurry.

The polypropylene packing spheres lost 16% of their original weight during 480 hours of operation with scrubbing slurry containing 5% undissolved solids. This rate of wear was identical to the rate that occurred with a 15% solids slurry and a Ca:S mole ratio of 1.5. The severe erosiveness of the slurry with the benzoic acid additive is attributed to the relatively large proportion of calcium-sulfur reaction products as a result of the low (1.0) Ca:S mole ratio. The calcium-sulfur reaction products form sharp-edged crystals that are more abrasive than fly ash or unreacted limestone. When the Ca:S mole ratio is maintained at 1.5, the slurry contains a large amount of the less abrasive limestone.

Economic Considerations

Although improved SO_2 removal efficiency can be obtained by using the benzoic acid additive, the primary advantage obtained from the additive is improved utilization of limestone. Improved utilization reduces those operating costs of the limestone—wet-scrubbing process which are attributed to (1) land and facilities for storage and final utilization of disposal of the spent slurry solids and (2) purchase, storage, and grinding of limestone.

By using the benzoic additive in the 550-MW Widows Creek installation, the Ca:SO_2 mole ratio (based on the SO_2 concentration in the inlet flue gas) can be reduced from 1.5 to 1.1 without reducing the SO_2 removal efficiency. This improved utilization of limestone will be effective in increasing the useful life of the spent slurry disposal pond 15%. This would be equivalent to a 25–30% increase in the pond life if fly ash were not included in the spent slurry.

The cost of the benzoic acid makeup for the Widows Creek system when operating at 550 megawatts will be about $1600 per day (based on the current price of $400 per ton for benzoic acid). The cost of benzoic acid will be essentially offset by a savings of about $1300 per day (based on $5.00 per ton, including grinding) resulting from the reduced consumption of limestone.

Conclusion

The addition of a small quantity of benzoic acid to the limestone–wet-scrubbing system is an effective way to improve the utilization of limestone and significantly reduce the amount of spent slurry solids discharged from the process. The long-term effects of the benzoic additive have not been completely identified or evaluated. Chemical scaling of the scrubber components is the most likely long-term operating problem attributed to the additive because of the increased amount of calcium-sulfur salts in solution.

REFERENCES

1. Potts, J. M., A. V. Slack, and J. D. Hatfield. "Removal of Sulfur Dioxide from Stack Gases by Scrubbing with Limestone Slurry: Small-Scale Studies at TVA," Second International/Limestone Wet-Scrubbing Symposium, New Orleans, Louisiana, November 8-12, 1971.
2. National Technical Information Service, 5285 Port Royal Road, Springfield, Virginia 22151.
3. Hatfield, J. D. and J. M. Potts. "Use of Weak Acids to Improve Sulfur Oxide Absorption by Limestone Slurries," American Institute of Chemical Engineers, Seventy-Second National Meeting, St. Louis, Missouri, May 21-24, 1972.
4. U.S. Patent 2,031,802, February 25, 1936, Daniel Tyrer, assigned to Imperial Chemical Industries Limited, a corporation of Great Britain.
5. U.S. Patent 3,632,306, January 4, 1969, John F. Villiers-Fisher, assigned to Chemical Construction Corporation, New York, New York.

Section VI
Control of Particles from Power Generation

28

Electrostatic Precipitator Design for Western Coals

Walter A. Baxter
Manager, Project Engineering
Environmental Elements Corporation, Baltimore, Maryland

The need for our country to lessen its dependence upon foreign oil coupled with our national objectives of cleaning up air pollution has focused the attention of the utility industry on the vast reserves of strippable low-sulfur coal contained in the 17 western states.

Although they vary in every property, we can make some generalizations about these coals as compared with eastern and midwestern fuels. Western coals will typically: have a low Btu content, many times have a moisture content over 20%, be low in sulfur content, often less than 0.5%, produce ash with high alkali content, and produce high-resistivity ash at normal air heater exit temperatures. A typical Wyoming coal and ash analysis is shown in Figure 28.1.

WYOMING COAL ANALYSIS	
BTU	8500
Volatile	30%
Carbon	42
Sulfur	0.5
Ash	6
Moisture	26

MINERAL ANALYSIS OF ASH	
SiO_2	30%
Al_2O_3	17
Fe_2O_3	6
CaO	24
MgO	4
P_2O_6	0.5
Na_2O	1.2
SO_3	1.1

Figure 28.1.

Starting at the mine pit through the coal preparation plant, conveyors, storage silos, pulverizers, furnace, boiler, air heater, precipitator, fans and ending at the stack, suppliers have found they must make special provisions in their equipment to accommodate the different properties of this fuel as compared with eastern or midwestern coals. This paper attempts an overview of the techniques in use today to adapt electrostatic precipitator design for the reliable high-efficiency collection of the fly ash emanating from the burning of these western coals.

PRECIPITATION PROCESS

To understand these techniques, we need some understanding of precipitation. Industrial electrostatic precipitation is basically a three-stage process:

1. Particle charging
2. Particle collection
3. Ash removal

Figure 28.2 shows the dominant relationships involved.

Let's limit ourselves to a simplified approach. Particle charge is dependent upon peak electrical field strength and particle surface

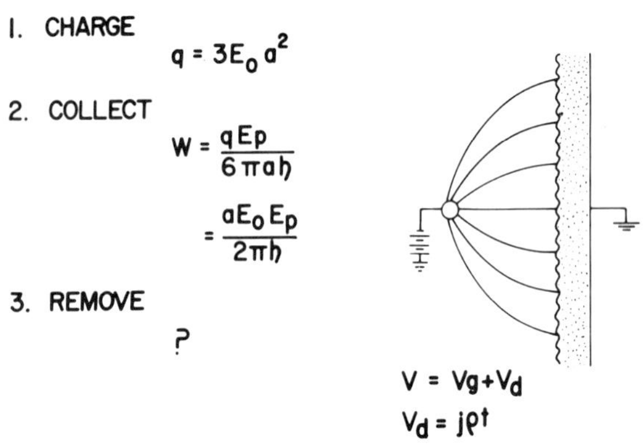

Figure 28.2. Precipitation process.

Where q = particle charge
E_o = charging field
a = particle radius
W = particle migration velocity
E_p = collecting field
η = gas viscosity
V = applied voltage
V_2 = voltage drop across gas space
V_d = voltage drop across dust layer
j = current density
ρ = ash resistivity
t = dust layer thickness

area. In turn, collection rate depends upon particle charge, collecting field strength and the gas viscous drag force. Charging and collecting western fly ash is not a difficult task at face value. However, ash removal from the collecting surface to the hopper below, and the interference to charging and collection that inadequate removal imposes, are the major engineering challenges in the precipitation of western fly ash.

The reason for this follows from the electrical nature of the process and is also illustrated in Figure 28.2. All current emitted from the discharge electrode must flow to the ground through the collecting surface and the dust layer adherent to it. Applied voltage divides between the gas space where it works to charge and collect the dispersoid and the dust layer where it works to hold ash to the plate. Voltage drop across the dust layer, if too high, not only steals charging and collecting voltage, but more importantly can give rise to localized electric fields at the dust surface that will positively ionize gas and produce back corona which, in its worst form, will bring precipitation to a virtual halt.

This voltage drop across the dust layer is dependent upon current flow, layer thickness, and ash conductivity. Reduction in current flow is somewhat self-defeating in that the process itself is dependent upon high fields and a copious supply of ions. Therefore, the major approaches to the precipitation of high resistivity ash involve either techniques to keep the dust layer thickness to a minimum or techniques to lower ash resistivity to acceptable values. We will discuss each in turn, but first a word about ash resistivity. Figure 28.3 shows the classic resistivity—temperature relationship. It can be seen that resistivity peaks close to normal air heater outlet temperature. Significant change in gas temperature in either direction works to increase the precipitability of a high-resistivity ash. Another method of changing resistivity is to shift the whole curve downward toward the lower curve in Figure 28.3 by the injection of small amounts of so-called "conditioning agents" into the process.

However done, the methods have the same objective, namely to keep dust layer voltage drop low enough so as not to impair the overall precipitation process. Now lets review the principal methods by which this problem is attacked. One of these methods attacks the problem directly, the others can be termed resistivity avoidance techniques.

HIGH ACCELERATION RAPPING

This concept, which is often referred to as the European approach, involves the use of rather massive rapping hammers, usually mechanically driven, intended to impart accelerations normal to the collection surface of from 50 to 200 gs for the purpose of limiting dust

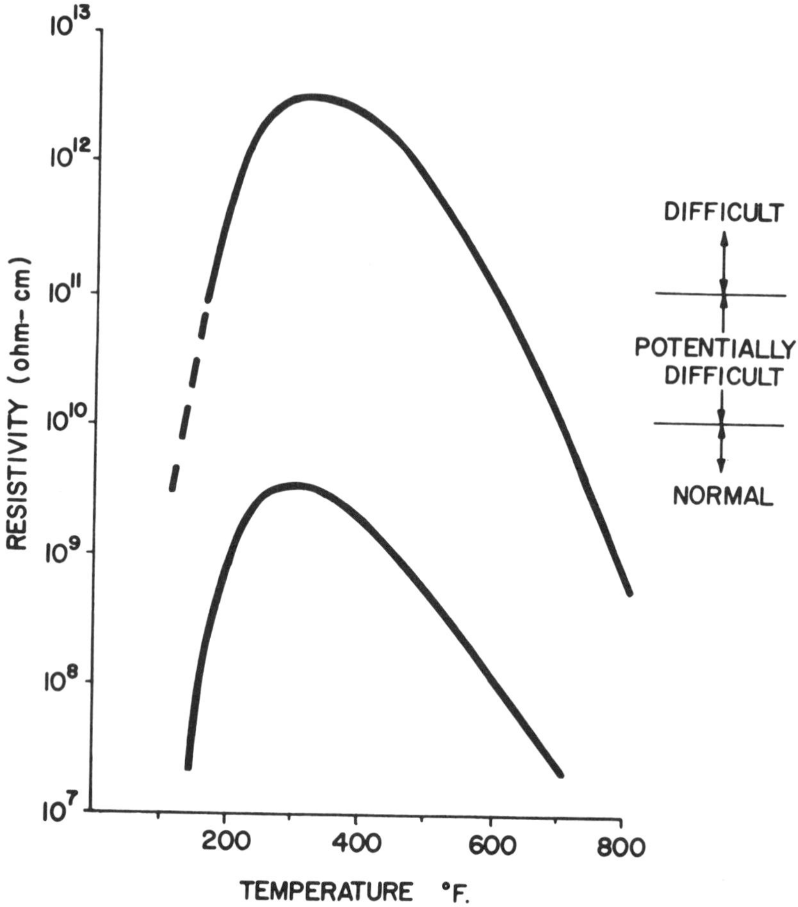

Figure 28.3. Ash resistivity.

build-up to a few millimeters, while allowing operation at normal air heater outlet temperatures. The precipitator operates at the peak of the upper resistivity curve of Figure 28.3 but with a plate ash layer thin enough to hold layer voltage drop within acceptable limits. Naturally, the entire structure and the internal components must be quite sturdy to withstand such forces. Ash layer conduction at these temperatures is principally a surface phenomenon and the designer looks to coal sulfur, its form, and its reaction with other ash constituents in determining the required precipitator size for any particular application.

Rapping is highly sectionalized, perhaps 800–2000 ft^2 of collecting surface per hammer. Figure 28.4 illustrates one tumbling hammer embodiment of this concept. These hammers are in the gas stream and require internal access for repair and physical change of the hammer for adjustment of rapper intensity. Mechanical rapping is also applied to the discharge electrode system, which is usually a heavy, rigid frame structure.

Figure 28.4. Tumbling hammer rapper.

General practice for these designs is to use rather large electrical bus sections operated at less than maximum current density to further minimize dust layer voltage drop. Wide plate to plate spacing (12″ and up) and low treatment zone gas velocity are also typical of this approach.

The result of this philosophy is a heavy, rather expensive precipitator which attacks the resistivity problem "head on." Correctly applied, it is a completely workable solution.

HIGH TEMPERATURE OPERATION

This is the most popular of the resistivity avoidance techniques and is often referred to as the American approach. As shown in Figure 28.5, the precipitator is merely moved to the inlet side of the air heater. This move shifts operation to the right hand, downside of the resistivity curve shown in Figure 28.3, but at the expense of 25–

40% increase in gas volume due to thermal expansion. Dust layer voltage drop is intended to be kept manageable by virtue of the resistivity reduction and layer thicknesses of ¼″ are acceptable. Ash layer current conduction is a volume rather than a surface phenomenon and the designer looks to temperature and ash chemistry and essentially ignores coal sulfur content in determining required precipitator size.

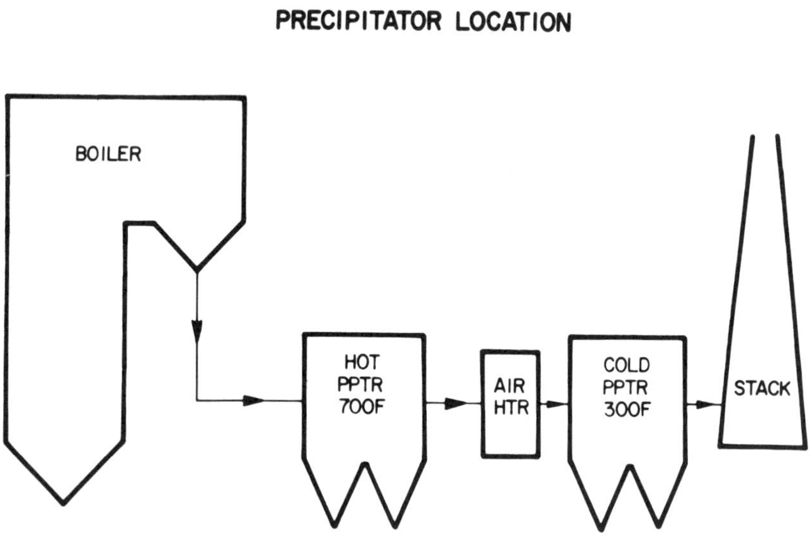

Figure 28.5. Precipitator location.

Functionally, the design is unchanged from American practice over the years and is shown in Figure 28.6. The plates and wires are rapped by externally mounted vibrators or hammers that are either electrically or pneumatically controlled and driven. Repair or intensity change can be done with the precipitator on-stream. Weighted wires are used instead of rigid frame electrodes. Because the design premise is that resistivity will not be a problem, rapping forces of 5 to 40 gs are sufficient and the entire structure does not need to be as massively constructed as in the European case. Rapping therefore, is not as sectionalized, and normal practice is to operate at high current density with a high degree of electrical sectionalization.

The principal advantages of this approach lie in the high current density, high precipitation rate and lower rapping requirements allowed by the shift to lower resistivity at 600–800°F. Disadvantages are the extra ductwork complexity often required, the additional gas

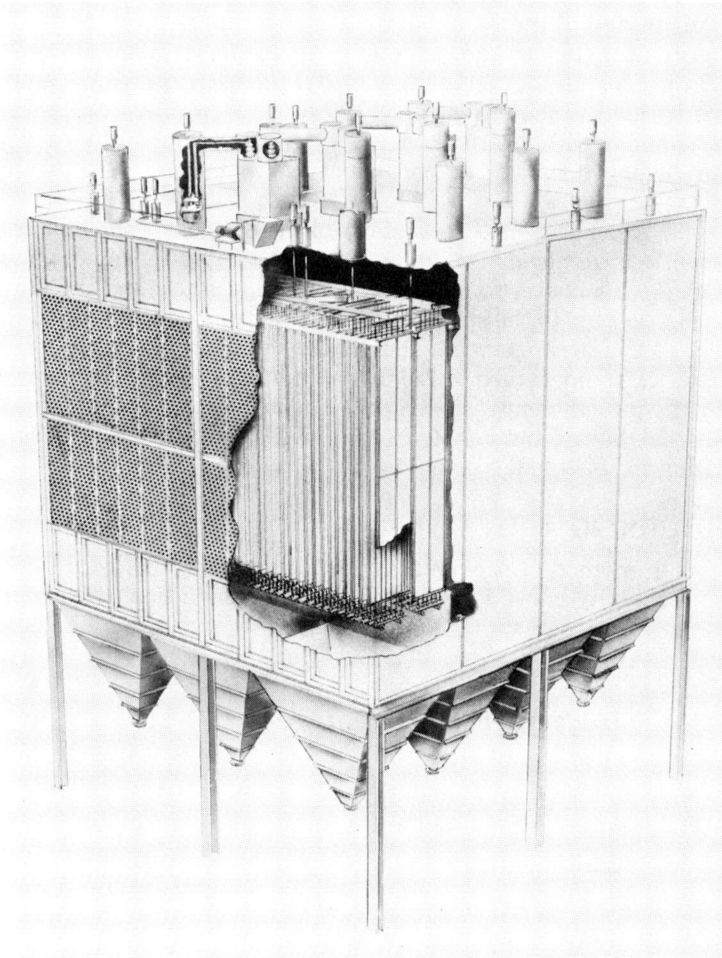

Figure 28.6. American precipitator design.

volume that must be handled, insulation and structural requirements for these high temperatures, and the penalty that any temperature drop upstream of the air heater imposes on overall plant thermal efficiency. Correctly applied, this concept has been proved by many successful operating installations.

LOW TEMPERATURE OPERATION

This concept merely involves operating a conventional cold side precipitator at a low enough temperature (say less than 230°F) to take advantage of the resistivity reduction apparent on the left side of Figure 28.3. It has had some success but is handicapped by the uncertainty of the exact shape of the resistivity curve and the temperature at which it peaks and, therefore, how low a temperature must be achieved for satisfactory operation. Of course, the potential for air heater pluggage or corrosion at low temperature is also a problem.

Berube[1] reported on one western utility's experience in the late 1960s with a rather novel approach involving increasing air flow through the air heater to lower flue gas temperature and then dumping the excess air to the stack downstream of the precipitator. Figure 28.7 shows this arrangement. This scheme enabled the utility to lower precipitator temperature to 220°F and meet emission codes but at a loss of boiler efficiency of about 1%. Thus, this scheme has some serious dawbacks, but as of a few months ago, the plant was still operating in this mode.

CONDITIONING

A fourth method of precipitating high resistivity ash is to shift the whole resistivity curve to a lower regime by conditioning. Condition-

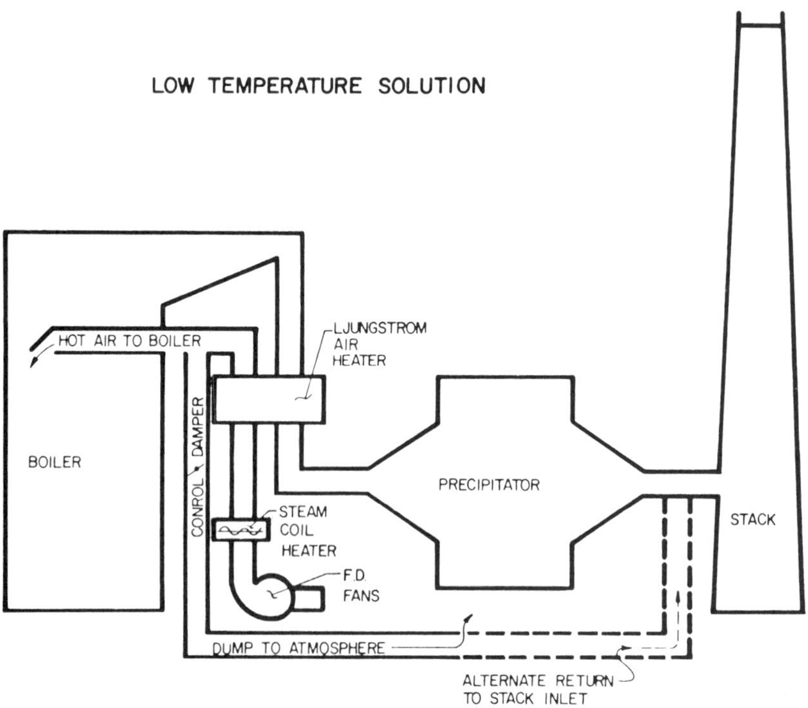

Figure 28.7. Low-temperature solution (after Berube).

ing is the act of adding trace amounts of a chemical to the process for the purpose of improving precipitability and is as old as the art of precipitation itself. White,[2] as early as 1912, reports the discovery of the efficacy of sulfur trioxide as a conditioner. Over the years, many industries and almost every precipitator manufacturer have experimented with conditioning, usually for performance upgrading, but occasionally at the design stage, to reduce the size of the original installation.

Almost every chemical compound imaginable has been tried for this purpose, but to date the most extensive use has been of sulfur trioxide, water, and ammonia compounds. Watson[3] and others have reported on the extensive use of gas conditioning in Australia. White[4] mentions some very interesting work using sodium compounds added to the coal as a conditioner. Recently, Tassicker,[5] has reported success with triethylamine in Australia. In the United States, there are firms that will design, furnish and install conditioning systems for power stations. One utility has purchased 11 of these systems for installation on its boilers.

The theories of how a conditioner works are as varied as their number. Sulfur trioxide appears to combine with water vapor and enhance conduction by surface deposition of sulfuric acid on the particulate. Work by Baxter[6] and others indicates that ammonia does not affect resistivity *per se*, but is effective as an agglomerative agent. Sodium enhances volume conductivity.

Although good correlations have been achieved between conditioning rate and precipitability at a single plant, it is difficult, if not impossible, to make such predictions in the design stage. Of course, conditioning burdens the operator with continuing chemical purchase and plant costs.

WET PRECIPITATION

One other method of combating resistivity problems is worthy of mention. Although, to my knowledge, never applied to utility scale precipitators, sprayed or irrigated plate precipitators have been on the scene for decades, blast furnace precipitators being a notable example. Wet precipitators, of course, burden the operator with all the problems attendant to handling and disposing of the resultant wet sludge. However, because the collected ash is continually washed off the plates, wetted precipitators effectively avoid the ash layer resistivity problem we have been discussing here. In combination with wet methods for sulfur dioxide removal, this concept may be worthy of reconsideration.

PRECIPITATOR SPECIFICATION

A concept—"Be Realistic"—is appropriate here concerning the interplay between the purchaser and the precipitator designer. In certain limited suitations it may be necessary to specify a very broad range of fuels to be burned over the 40-year life of a boiler. These broad "any coal" specifications will produce large expensive precipitator bids as the responsible designer tries to protect himself against almost any combination of ash chemical and physical characteristics. However, in some cases, these "any coal" requirements may merely represent an attempt to armor-plate the specification. This then results in an over-sized, over-priced offering that is not really in the best interests of either purchaser or supplier. This concept applies equally to all aspects of specifications. If 10% spare capacity is good, isn't 20% or 30% better? If 3/16″ mild steel is good, isn't ¼″ Corten® better? This is not to advise against conservatism. We have all worked too hard to get away from the inadequate designs of the '50s and '60s. Redundancy, conservative gas volume specifications, looking towards next year's emission codes etc., are all prudent. A clean stack is now necessary as a license to operate. However, it is also necessary to look at the businessman's performance criterion, "Return on Investment."

The best advice I can give here is dialogue. Talk with the vendors regarding your requirements and their relative importance. In today's environment, the precipitator manufacturer is as deeply committed to a clean stack as is the purchaser. Exchange of information on design alternatives and their cost vs. benefit is the best path I know towards a mutually satisfactory installation.

SUMMARY

The problem of electrostatic precipitation of high-resistivity western fly ash principally reduces to the prevention of localized high electric fields at the surface of the dust layer on the collecting electrode. The major solutions applied to date are the cold-side massive rapping European approach and the hot side resistivity avoidance American technique. Both have drawbacks and advantages. Correctly applied, either will work. There is absolutely no basis for statements that hot or cold "will not work." Rather, the question is which concept is the more economical design for a specific boiler intended for specific fuels interacting with the other constraints of total plant design.

To the extent that the utility can narrow these design inputs, particularly the fuel and ash characteristics which must be handled, precipitator costs can be minimized. If the purchaser must specify "any coal" or a very broad range of chemical and physical properties for the coal and ash, the responsible designer will provide a "worst case" design and the cost and size will be commensurately high.

REFERENCES

1. Berube, D. T. "Low Temperature Solution to High Resistivity Ash Problems," *Proceedings of the Electrostatic Precipitator Symposium,* Birmingham, Alabama (1971).
2. White, H. J. *Industrial Electrostatic Precipitation* (Reading, Massachusetts: Addison-Wesley Publishing Company, 1963).
3. Watson, K. S. and K. J. Blecher. "Further Investigation of Electrostatic Precipitators for Large Pulverised Fuel Fired Boilers," *Proceedings, Clean Air Conference,* Sydney, Australia, August, 1965.
4. White, H. J. "Resistivity Problems in Electrostatic Precipitation," *J. Air Poll. Cont. Assoc.* **24**, No. 4 (1974).
5. Tassicker, O. J. "Some Aspects of Electrostatic Precipitator Research in Australia," *J. Air Poll. Cont. Assoc.* **25**, No. 2 (1975).
6. Baxter, W. A. "Recent Electrostatic Precipitator Experience with Ammonia Conditioning of Power Boiler Flue Gases," *J. Air Poll. Cont. Assoc.* **18**, No. 12 (1968).

29

Significance of Particle Size and Resistivity Testing for Precipitator Sizing

Grady B. Nichols
Southern Research Institute
Birmingham, Alabama

INTRODUCTION

The behavior of an electrostatic precipitator is governed by the electrical conditions existing within the device and the properties of the particulate material to be collected. The limitations on electrical conditions within the unit are established by one of the two factors—either the mechanical conditions or the particulate resistivity depending upon the relative importance of each. The mechanical conditions dominate for resistivity values below the mid 10^{10} ohm-cm range while the particulate resistivity dominates for higher values of resistivity. The particle size distribution of the suspended material also determines the performance in that the collection efficiency varies with particle size. A detailed determination of these parameters is required if a meaningful analysis of the expected behavior of the unit is contemplated.

For the purposes of this chapter, the following assumptions are made: the mechanical condition of the precipitator is good, the gas velocity distribution is uniform, no gas sneakage exists either through the hopper area or the superstructure, and reentrainment due to rapping is negligible.

INFLUENCE OF PARTICULATE RESISTIVITY

The electrical conditions in an electrostatic precipitator directly influence the collection characteristics of the device. The current density (free ion density) in the interelectrode space influences the charging rate of all particles and the value of the electric field directly influences the total charge on particles charged by what is termed field charging. The collection force that acts to remove the previously charged particles is directly proportional to the value of the electric field adjacent to the collection electrodes. The electric field in the interelectrode region and the free ion density are governed by the electrical conditions of the precipitator.

Consider the electrical behavior of an electrostatic precipitator for the idealized case where no particulate matter is suspended in the effluent gas stream. As the applied voltage is increased from zero, as shown in Figure 29.1, no current flows until a sufficiently high field exists at the wire to cause corona initiation. Further increases in applied voltage causes an increase in current as shown by curve 1. The current will increase until the applied voltage is sufficient to

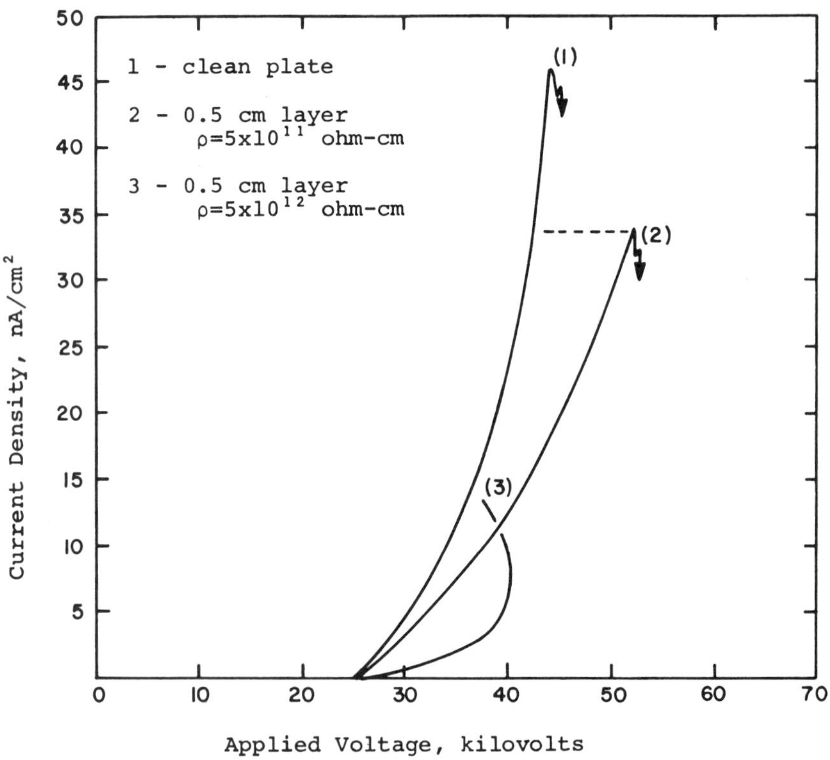

Figure 29.1. Simplified volt-ampere characteristic for an electrostatic precipitator. Plate spacing 10 inches wire diameter 0.109″ for various dust resistivities. A is collection electrode area in cm^2.

cause an electrical sparkover between the electrodes. The operating characteristics are limited in the situation by the corona wire size, electrode spacing, and effluent gas composition.

The addition of particulate matter in the gas stream and a dust layer with a resistivity less than 10^{10} ohm-cm causes no appreciable change in the voltage versus current characteristics; the dust may provide additional space charge with increased electric field at the plate but no change in the general shape of the V-I curve. The electric field in the dust layer is approximately the same as that in the interelectrode space and the V-I curves will be essentially identical.

Increasing the resistivity of the dust layer to about 5x10 ohm-cm causes an increase in the electric field in the dust layer. If we consider a 0.5 cm thick dust layer, the V-I curve will be shifted as shown by curve 2. The overall behavior of the precipitator will be unaffected until the electric field in the dust layer reaches the electrical breakdown strength of the gas in the intersticial region of the dust layer. When this occurs, the dust layer will break down and a spark through the dust layer will occur. This spark causes an immediate reduction in the voltage drop across the dust layer. Since the applied voltage is greater than that required for a clean plate sparkover, a spark will propagate across the interelectrode space. This acts to limit the operating current density to about 35 Na/cm^2, since sparking will occur there. Thus we see that the resistivity of the dust layer acts to limit the operating characteristics of the precipitator. An increase in the resistivity to 5 x 10 ohm-cm further limits the precipitator behavior. The electric field in the dust layer will approach the breakdown field at applied voltage below that necessary for sparkover. Under these conditions, the dust layer exhibits a general breakdown with electrical corona (back ionization) spread over large areas. This back ionization in the dust layer generates ions electrically opposite to that formed at the corona wire, tending to neutralize the charge on the dust particles and reduce the space charge component of the electric field adjacent to the collection electrode. This reduction in the electric field at the plate leads to a proportional reduction in the collecting force at the collection electrode.

This simplified discussion of the electrical behavior of an electrostatic precipitator indicates the need to determine the electrical resistivity of the particulate that is expected to be collected in the precipitator. The low-to-intermediate resistivity dusts do not limit the behavior of a collector but the high resistivity values will tend to either sparkover at reduced current densities or lead to a back ionization phenomenon depending on the actual value of the resistivity. Either of these conditions leads to a reduced collection efficiency in the precipitator.

INFLUENCE OF PARTICLE SIZE DISTRIBUTION ON PRECIPITATOR PERFORMANCE

The necessity for determining the particle size distribution is related to the fact that the collection efficiency as a function of particle size is not constant in an electrostatic precipitator. The electrical migration velocity is proportional to the particle radius for large particles that are charged predominantly by field charging. The migration velocity is related to the various precipitators as shown by

$$w \cong \frac{2a\epsilon_0 E_0 E_p}{\eta}$$

where: w = electrical migration velocity, cm/sec
a = particle radius, cm
ϵ_0 = permitivity of free space
ρ = electrical resistivity, ohm-cm
E = electric field, volt/cm
η = efficiency, %

Thus we find that the large particles are more easily collected than small ones for field charging. For particles smaller than about 0.5 μm, diffusion charging becomes the dominant mode of charging. However, the viscous drag force of the gas stream is reduced because of the reduced number of collisions with neutral gas molecules as the size approaches the mean free path of the molecules in the gas medium. In this region, the drag force is reduced by a slip correction factor to Stokes Law flow (Figure 29.2). This reduction in drag force causes an increase in the migration velocity for small particles. Figure 29.3 shows the variation in migration velocity with particle size for conditions that exist in a well-designed full-scale precipitator operating at an electric power station boiler.

The collection efficiency for each particle size is described by the familiar Deutsch Equation

$$\eta = 1 - \exp - (Aw/V)$$

If we utilize the migration velocities for Figure 29.3 together with the electrical conditions that exist at the collection electrode, to compute the collection efficiency as a function of particle size, we generate the relationship shown in Figure 29.4, the efficiency as a function of particle size. Thus we see a significant variation in collection efficiency as a function of particle size.

This points to the need to determine the particle size distribution of the material that is to be collected. For illustrative purposes, consider two fly ash particle size distributions as shown in Figure 29.5, each with a mass median diameter of 10 μm but with different slopes as shown. If we utilize the efficiency as a function of particle size to

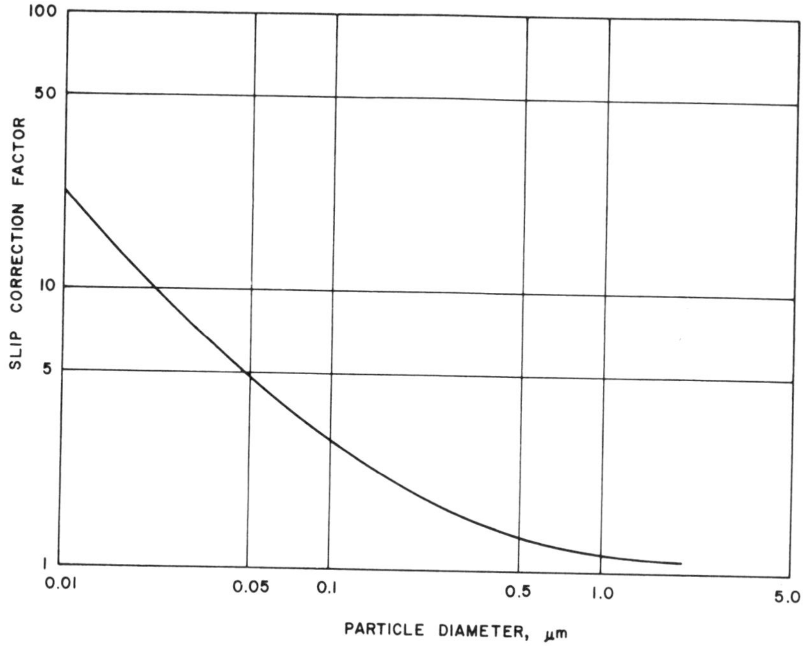

Figure 29.2. Slip correction factor to Stokes Law. Settling as a function of particle size.

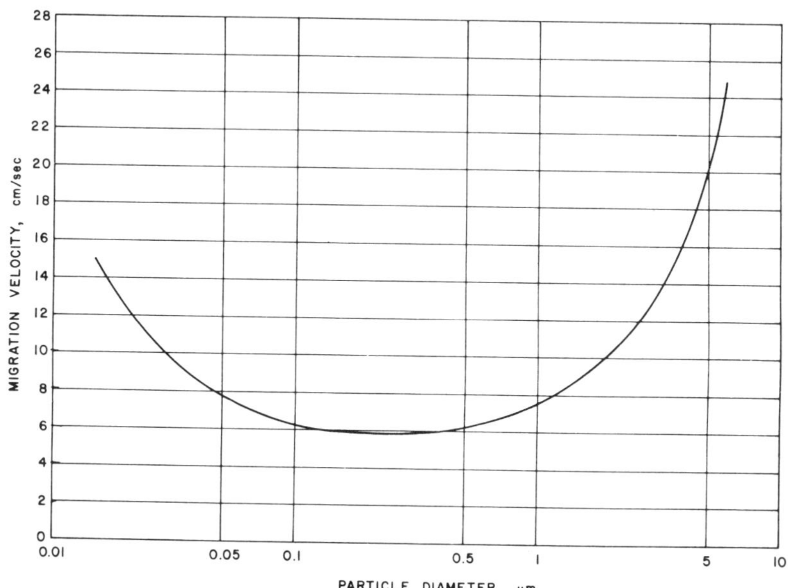

Figure 29.3. Theoretical migration velocity as a function of particle size for typical precipitator conditions.

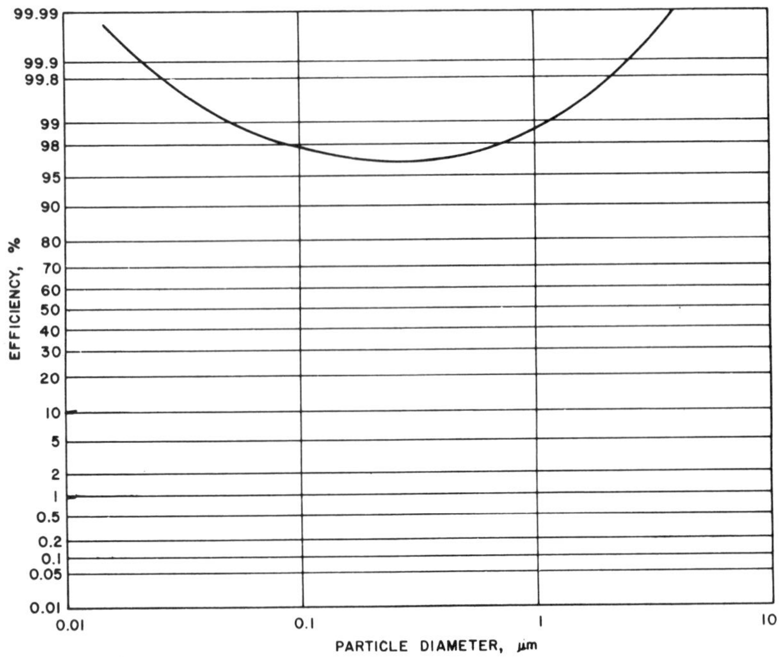

Figure 29.4. Theoretical collection efficiency as a function of particle size.

compute the overall collection efficiency for these two cases, we find the efficiency for the fine particle size ~98% and that for the coarse is 99.2% with a difference in emissions of a factor of 2.5.

CONCLUSION

The behavior of an eletrostatic precipitator is dependent upon the characteristics of the material to be collected and the electrical conditions that exist in the unit. A determination of both the particle size distribution and the electrical resistivity of the particulate is necessary if a reasonable analysis of the unit is expected.

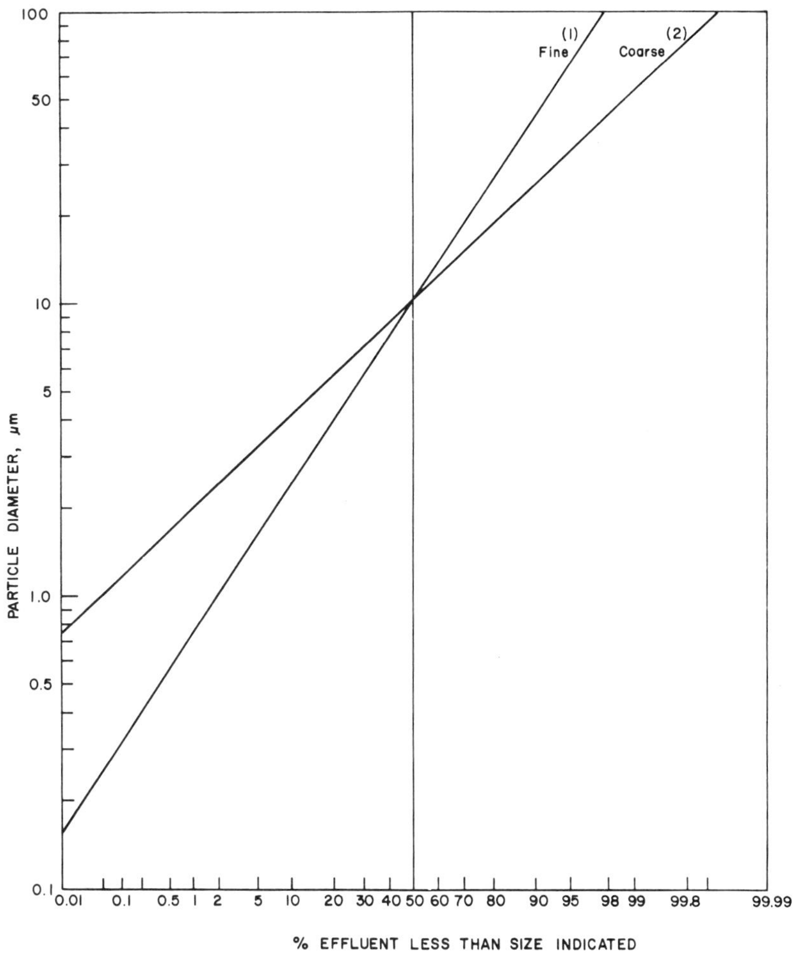

Figure 29.5. Particle size distributions used for illustration.

30

Electrostatic Precipitation: Predicted vs Actual Efficiencies

Kenneth R. Stamper and **Kenneth E. Noll**
The University of Tennessee
Knoxville, Tennessee

INTRODUCTION

The design of an electrostatic precipitator to meet a guaranteed collection efficiency is complex. This is due to the fact that a) the migration velocity is difficult to determine, and b) most design equations lack a term to describe reentrainment, and a polydispersed aerosol. The flow distribution of the gases also influences the collection efficiency as well as discharge electrode configuration, collecting plate geometry, and type of electrical control circuitry. Performance also depends on moisture content of the gas and on the temperature and chemical composition of the exhaust gas and particles.

As an aid to design evaluation and specification writing, this chapter will show the relationship between actual and predicted effective migration velocities (EMV). Test data from existing precipiator installations will be used to determine the actual EMV. Predicted EMV relationships obtained by using current literature will be compared to the test data.

METHODS AVAILABLE TO PREDICT COLLECTION EFFICIENCY

There are numerous methods available for predicting collection efficiency to compare to the actual performance data. The following methods will be utilized in this study:

1. Deutsch-Anderson[1]
2. Central Electric Generating Board (CEGB) empirical equation[2,3]

3. Selzler-Watson empirical equation[4]
4. Theodore Model[5]

A short description of each model follows and an example calculation is presented in the Appendix.

Deutsch-Anderson Formula With Efficiency Prediction Accounting for Size Distribution

The basic Deutsch-Anderson equation is

$$\text{Eff} = 1\text{-exp}\,(A\ W/Q)$$

where: A = collecting plate area
W = migration velocity
Q = flow volume

The Deutsch equation is for use on a monodispersed aerosol. However, one can obtain an efficiency given the Deutsch exponent and the polydispersed aerosol, assuming that it is log normally distributed.[6] The Omega term in the Deutsch exponent is determined by the equation

$$w_g = \frac{.0058E_1\ E_2 r}{N}$$

where: w_g = migration velocity, ft/sec
N = viscosity of the gas (air) at temp, poise
E_1 = peak discharge electrode field, $\frac{KV}{cm}$
E_2 = average collecting electrode field, $\frac{KV}{cm}$
r = half of mass median diameter, cm

CEGB Empirical Migration Velocity

The Central Electricity Generating Board in England developed an equation to describe the EMV of the particulate matter in their precipitator installations. The equation was the result of 74 efficiency tests at 19 power plants. The data from these tests were put into a computer linear multiple regression analysis program to derive equations relating various operating parameters and the effective migration velocity.

The final equation was

$$EMV = 31.2 - 4.8 1n(F) - 8x10^{-4}P - 4x10^{-8}g^2 - 8.1 \exp\,(-0.5S) - \frac{43.3}{A}$$

where: EMV = effective migration velocity, cm/sec
F = specific collecting area; (total plate area)/(total gas flow), $ft^2/ft^{3/sec}$
P = (total plate area)/(installed power), ft^2/KVA
g = specific surface area of inlet dust, cm^2/g
S = total sulphur content of the coal, %
A = ash content of coal, %

The "g" term can be computed from Table 30.1 if a Bahco particle analysis is available.

Table 30.1. Method of Calculating Specific Surface Area of Dust (based on a dust density of 2.0 gm/cc).[a]

1	*2*	*3*	*4*
Micron Size	*Percent In Range*	*Factor*	*2 x 3*
−5		7500	
5–10		4000	
10–20		2200	
+20		950	
		Total =	______

$$\text{Specific surface "g"} = \frac{\text{Total of (4)}}{100}\ \text{cm}^2/\text{g}$$

[a]See Nomenclature for Tables, pp. 449-450.

The Selzler-Watson Equation

The original empirically derived equation is given as follows:

$$\text{Efficiency} = 100\ [(1\text{-exp}(-.57 \times 203 \times (A/V)^{1.4} \times (kW/V)^{.6} \times (S/Ac)^{.22})]$$

where:
A = collecting plate area (1000s ft^2)
V = flue gas volumetric flow rate (1000s of ACFM)
kW = power input to discharge electrode kW
S/Ac = sulfur-to-ash ratio, by weight percent of coal combusted in the associated boiler

The authors used the least squares regression technique on data collected from 37 full-scale precipitator operations to develop the equation.

The number "203" in the exponent must be exchanged for "157.6" when either the precipitator is preceded by mechanical collectors or if the furnace used to burn the coal is a cyclone type. This smaller number is used to account for the difficulty involved in trying to collect the smaller particles.

N. W. Frisch and D. W. Coy expressed criticisms of the Selzler-Watson equation[7] and showed that power density expressed as kW/1000 ACFM is limited as a beneficial input into collection efficiency to account for such phenomena as sparkover and back corona.

W. D. Watson replied to this commentary with the coefficients of the equation modified.[8] The revised equation is as follows:

$$\text{Efficiency} = 100\ [(1\text{-exp}(-.57 \times 8.3 \times (A/V)^{.32} \times (kW/V)^{.16} \times (S)^{.16})]$$

where: A = collecting plate area in 1000s of ft^2
kW = corona power in kilowatts
V = flue gas flow rate 1000s of ACFM
S = sulphur content of coal (wt %)
8.3 = a constant for pulverized coal-fired units

Theodore Model

The Theodore model[5] suggests that for an electrostatic precipitator collecting fly ash with a 9-inch plate to plate spacing and a 14-foot plate length, efficiency varies with field strength according to the following equation:

$$\eta = 1 - c_1 e^{-c_2 E^2}$$

where: $c_1 = 0.7052$
$c_2 = 0.2004$
E = average actual field strength (KV/cm)

It should be noted that this model (used in the above manner) requires the input of actual field strength data. This quantity is influenced by the resistivity of the ash. Therefore, if this model is used to size a precipitator unit, assumptions would be required about actual field strength and ash resistivity.[9]

The authors give specific examples of how efficiency varies with plate length, plate spacing, and field strength. There are limitations imposed on each variable and base data implying the ranges of applicability of the model.

The actual tests taken on electrostatic precipitator A were within the suggested ranges and the physical design of the precipitator was such that the values in Theodore's Table 30.4 could be used to compare predicted and actual efficiencies.

ACTUAL PERFORMANCE DATA

The data on actual precipitator performance was obtained from TVA test and special study reports.[10] This data is summarized in Tables 30.2, 30.3, 30.4, and 30.5. The nomenclature used in the tables is identified at the end of the chapter.

Table 30.2 provides the bid specifications and physical design characteristics for the six electrostatic precipitators which were tested. Plate area of the precipitator varied from 24,750 to 158,540 ft^2. Specified flow varied from 245,000 acfm to 906,000 acfm with face velocities from 4.78 ft/sec to 7.95 ft/sec.

Table 30.3 and 30.3a show the number of tests conducted on each of the six precipitators. A total of 56 tests were conducted. This ranged from 34 tests on precipitator A to one test on precipitator F.

Tables 30.4 and 30.5 show the coal and ash composition for each test and also the particle size distributions (MMD and σ_g and particle density), which were obtained by Bahco analysis.

Table 30.2. Electrostatic Precipitator Bid Specifications and Physical Design Characteristics.[a]

Precip	Eff	Flow rate (KACFM)	Temp (°F)	Sulphur (wt %)	SCA	IKVA[b]	Plate Area (ft^2)	Plate Spacing (in)	Length of Discharge (wire ft)	Cross Sectional Area (ft^2)	# of Buses per Collector	Field Dimensions (H x w (ft)) #1	#2	#3	#4
A	90	292	330	2.7	102	96	29,700	9	19,800	743	4	30 x 4.5	30 x 4.5	30 x 6	—
B	95	245	330	1.0	101	65.58	24,750	9	16,500	514	2	23.75 x 9	23.75 x 9	—	—
C	98.5	410	320	1.0	342.4	328.8	157,500	11	102,000	1,605	8	35 x 9	35 x 12	35 x 12	35 x 12
D	98.5	410	320	1.0	342.4	328.8	157,500	11	102,000	1,605	8	35 x 9	35 x 12	35 x 12	35 x 12
E	97.0	906	340	—	175	277.8	158,840	11	99,330	2,660	12	35.5 x 9	35.5 x 9	35.5 x 9	—
F	97.0	906	340	—	175	277.8	158,840	11	99,330	2,660	12	35.5 x 9	35.5 x 9	35.5 x 9	—

[a]See Nomenclature for Tables, pp. 449-450.
[b]Installed kw capacity to discharge wire.

Table 30.3. Test Conditions for the Electrostatic Precipitator Samples.[a]

			Test Conditions								
Precip	*Test #*	*Test Date*	*Eff (%)*	*Flow (10^3cfm)*	*Temp (°F)*	*Gr Ld (gr/ft³)*	*Gr Ex (gr/ft³)*	*SCA*	*CnTime*	*FcVel (ft/sec)*	*AMV (cm/sec)*
A	1	7–11–69	95.8	237	303	1.7098	0.0712	125	2.82	5.32	12.88
A	2	7–11–69	95.6	232	303	1.5711	0.0697	128	2.88	5.20	12.40
A	3	7–17-69	98.0	228	312	1.557	0.0309	130	2.94	5.11	15.29
A	4	7–25–69	94.4	299	304	2.286	0.1284	99	2.24	6.71	14.79
A	5	7–31–69	96.6	255	272	1.3189	0.0455	116	2.62	5.72	14.80
A	6	7–31–69	95.7	257	272	1.3107	0.0559	116	2.60	5.76	13.78
A	7	8–5–69	94.9	330	308	1.652	0.0837	90	2.03	7.40	16.80
A	8	6–10–70	73.4	286	309	1.692	0.4480	104	2.34	6.42	6.47
A	9	6–15–70	83.3	285	303	1.787	0.294	104	2.35	6.39	8.74
A	10	6–18–70	47.0	280	317	2.230	1.180	106	2.39	6.28	3.04
A	11	6–22–70	68.7	299	311	2.159	0.674	99	2.24	6.71	5.96
A	12	6–23–70	68.8	300	311	1.361	0.431	99	2.23	6.73	5.98
A	13	6–23–70	68.9	302	311	1.744	0.546	98	2.21	6.77	6.05
A	14	6–29–70	78.8	277	313	1.806	0.386	107	2.41	6.21	7.65
A	15	6–29–70	81.8	279	313	2.072	0.377	106	2.40	6.26	8.17
A	16	6–30–70	76.3	289	316	1.892	0.477	103	2.31	6.48	7.10
A	17	7–1–70	85.8	314	314	1.913	0.270	95	2.13	7.04	10.44
A	18	7–2–70	85.3	299	316	2.173	0.319	99	2.24	6.71	9.83
A	19	7–6–70	92.6	295	306	1.392	0.102	101	2.27	6.62	13.10
A	20	7–6–70	77.8	283	306	1.383	0.311	105	2.36	6.35	7.28
A	21	7–7–70	91.1	237	313	1.645	0.145	125	2.83	5.32	9.83
A	22	7–8–70	89.9	285	310	1.465	0.149	104	2.35	6.39	11.20
A	23	7–8–70	71.3	285	310	1.260	0.362	104	2.35	6.39	6.10
A	24	7–8–70	79.3	280	310	1.606	0.334	106	2.39	6.28	7.55
A	25	7–9–70	94.9	279	304	1.737	0.087	106	2.40	6.26	14.26
A	26	7–9–70	88.3	279	304	2.099	0.246	106	2.40	6.26	10.28

A	27	7–10–70	82.0	283	304	1.635	0.379	105	2.36	6.35	8.30
A	28	7–10–70	81.3	296	304	1.304	0.243	100	2.26	6.64	8.52
A	29	7–13–70	74.0	290	310	1.601	0.418	102	2.31	6.51	6.70
A	30	7–14–70	85.6	287	309	1.463	0.214	103	2.33	6.44	9.56
A	31	7–14–70	78.6	275	309	1.487	0.319	108	2.43	6.17	7.25
A	32	7–15–70	88.5	227	311	1.119	0.129	131	2.95	5.09	8.39
A	33	7–15–70	88.0	222	311	1.352	0.162	134	3.01	4.98	8.04
A	34	7–15–70	87.2	223	311	1.661	0.214	133	3.00	5.00	7.85
B	35	8–11–70	82.6	217	308	3.060	0.533	114	2.67	7.04	7.85
B	36	8–12–70	63.7	308.5	316	3.429	1.220	80	1.88	10.00	6.56
B	37	8–12–70	55.2	308	316	2.532	1.129	80	1.88	9.99	5.20
B	38	8–13–70	72.6	282	314	2.848	0.745	88	2.05	9.14	7.22
B	39	8–13–70	69.1	283	315	3.003	0.928	88	2.04	9.18	6.79
B	40	8–14–70	59.8	342	334	2.689	1.089	73	1.69	11.09	6.56
B	41	8–14–70	50.7	342	334	2.327	1.145	73	1.69	11.09	5.09
B	42	8–17–70	85.2	275	318	2.399	0.276	90	2.10	8.92	10.46
B	43	8–18–70	91.1	187	304	2.754	0.244	132	3.10	6.06	9.32
B	44	8–18–70	92.8	188	305	3.364	0.242	131	3.08	6.10	10.14
B	45	8–19–70	69.1	342	337	1.317	0.593	73	1.60	11.09	8.45
B	46	8–19–70	71.1	341	334	1.396	0.571	73	1.70	11.06	8.93
B	47	8–20–70	94.6	185	291	2.247	0.126	133	3.13	6.00	11.31
B	48	8–20–70	94.8	177	304	3.076	0.164	140	3.27	5.74	11.03
C	49	5–30–73	99.38	338	312	1.655	0.0125	466	12.7	3.55	5.61
D	50	5–30–73	99.19	354	320	1.240	0.0096	445	12.3	3.67	5.48
D	51	5–31–73	99.39	345	309	1.579	0.0112	456.5	12.1	3.71	5.87
C	52	5–31–73	99.17	359	312	1.571	0.01577	439.7	12.7	3.54	5.27
C	53	6–5–73	98.24	340	335	1.568	0.0276	463	10.1	4.47	5.60
C	54	6–5–73	97.42	340	335	1.545	0.0398	463	10.1	4.47	5.07
E	55	6–26–73	97.6	805	356	1.720	0.0413	196.9	5.41	5.04	9.62
F	56	7–18–73	99.2	791	355	3.440	0.026	200.4	5.50	4.96	12.24

[a]See Nomenclature for Tables, pp. 449-450.

Table 30.3a. Electrostatic Precipitator Rating Conditions During Sampling[a]

			Rating Conditions									
Precip	*Test #*	*Test Date*	*VLT f1 (KV)*	*VLT f2 (KV)*	*VLT f3 (KV)*	*VLT f4 (KV)*	*CRT f1 (mamp)*	*CRT f2 (mamp)*	*CRT f3 (mamp)*	*CRT f4 (mamp)*	*Spk rt (#/min)*	*KVA*
A	1	7–11–69	(44.7)	(47.7)	(45.7)		147	287	210		285	29.86
A	2	7–11–69	(46.8)	(48.6)	(40.9)		147	287	210		309	29.42
A	3	7–17–69	(49.4)	(50.7)	(48.6)		147	280	200		496	31.2
A	4	7–25–69	(45.4)	(51.6)	(41.6)		173	293	220		517	32.1
A	5	7–31–69	(42.0)	(45.9)	(41.9)		160	280	290		282	31.7
A	6	7–31–69	(45.6)	(44.7)	(34.3)		156	260	260		289	27.6
A	7	8–5–69	(40.5)	(43.8)	(40.0)		216	426	320		198	40.2
A	8	6–10–70	(44.9)	(34.7)	(55.4)		103	67	60		153	10.25
A	9	6–15–70	(55.4)	(35.8)	(50.0)		40	140	200		819	17.23
A	10	6–18–70	(23.3)	(14.7)	(3.0)		20	67	10		371	1.48
A	11	6–22–70	(45.9)	(32.1)	(29.5)		40	80	240		313	11.48
A	12	6–23–70	(30.2)	(25.7)	(21.6)		40	80	150		317	6.05
A	13	6–23–70	(31.5)	(25.7)	(16.9)		33	80	170		325	5.97
A	14	6–29–70	(36.8)	(35.0)	(38.4)		57	140	140		355	12.37
A	15	6–29–70	(35.4)	(31.1)	(33.2)		50	100	70		321	7.20
A	16	6–30–70	(43.8)	(35.8)	(41.1)		80	187	150		959	16.36
A	17	7–1–70	(40.6)	(34.4)	(36.1)		90	267	190		996	19.70
A	18	7–2–70	(41.1)	(42.0)	(34.2)		47	80	110		440	9.05
A	19	7–6–70	(39.2)	(37.3)	(36.8)		97	207	240		318	20.36
A	20	7–6–70	(37.2)	(34.5)	(42.7)		43	93	80		345	8.22
A	21	7–7–70	(36.8)	(33.3)	(38.0)		53	113	120		338	10.27
A	22	7–8–70	(42.0)	(43.5)	(34.5)		57	187	260		321	19.50
A	23	7–8–70	(36.0)	(30.2)	(35.9)		43	107	70		343	7.29
A	24	7–8–70	(38.0)	(30.2)	(41.0)		43	107	70		346	7.74
A	25	7–9–70	(45.4)	(38.9)	(41.3)		140	147	300		322	24.46
A	26	7–9–70	(36.4)	(29.4)	(28.6)		50	106	110		338	8.08

A	27	7–10–70	(35.6)	(32.9)	(31.9)		47	100	90		340	7.83
A	28	7–10–70	(50.8)	(36.3)	(34.7)		67	153	120		328	13.12
A	29	7–13–70	(34.5)	(31.9)	(31.0)		47	107	90		348	7.83
A	30	7–14–70	(47.5)	(37.0)	(40.6)		140	173	130		335	18.33
A	31	7–14–70	(40.5)	(29.9)	(41.0)		53	120	70		348	8.61
A	32	7–15–70	(50.1)	(36.3)	(42.1)		87	173	160		332	17.37
A	33	7–15–70	(41.7)	(34.3)	(36.6)		50	107	110		351	9.78
A	34	7–15–70	(47.8)	(34.3)	(39.5)		43	107	110		352	10.07
B	35	8–11–70	(31.6)	(29.2)			85	100			200	5.61
B	36	8–12–70	(32.4)	(30.9)			105	140			186	7.73
B	37	8–12–70	(32.2)	(30.3)			105	130			191	7.32
B	38	8–13–70	(32.2)	(31.6)			100	140			171	7.64
B	39	8–13–70	(32.2)	(30.0)			100	130			165	7.12
B	40	8–14–70	(32.1)	(31.2)			120	160			174	8.84
B	41	8–14–70	(32.0)	(30.6)			105	140			172	7.64
B	42	8–17–70	(33.6)	(33.9)			190	370			404	18.93
B	43	8–18–70	(33.1)	(33.1)			110	190			430	9.93
B	44	8–18–70	(33.1)	(32.7)			110	190			428	9.85
B	45	8–19–70	(33.0)	(32.0)			190	230			374	13.63
B	46	8–19–70	(29.9)	(30.6)			135	220			326	10.77
B	47	8–20–70	(36.0)	(34.7)			195	330			285	18.47
B	48	8–20–70	(35.1)	(35.0)			180	310			316	17.17
C	49	5–30–73	39.5	49.5	40.5	41	400	130	260	560		55.73
D	50	5–30–73	40	50	41.5	43.5	390	90	380	550		59.8
D	51	5–31–73	40.5	39	43	43.5	70	30	360	520		42.11
C	52	5–31–73	41	53.5	41	42	460	240	190	570		63.43
C	53	6–5–73	42	53	38.5	40	460	260	80	340		40.00
C	54	6–5–73	42.5	53.5	40	42	450	250	100	340		50.78
E	55	6–26–75	43	41.75	28.25		390	430	810			66.65
F	56	7–18–73	49.25	41.25	44.50		540	820	970			88.08

[a]See Nomenclature for Tables, pp. 449-450.

Table 30.4. Chemical Analysis of Coal for Test Conditions.[a]

			Coal/Analysis							
Precip	*Test #*	*Test Date*	*Ht val (Btu/lb)*	*Tot C (%)*	*H (%)*	*O (%)*	*N (%)*	*Sulfur (%)*	*Ash (%)*	*Moist (%)*
A	1	7–11–69	12,700	70.2	5.0	7.9	1.7	3.3	11.9	11.9
A	2	7–11–69	12,520	69.9	4.9	7.5	1.6	3.6	12.5	12.2
A	3	7–17–69	12,300	68.6	4.9	7.9	1.5	3.0	14.1	9.2
A	4	7–25–69	12,530	70.2	4.7	7.5	1.5	2.9	13.2	9.6
A	5	7–31–69	12,510	70.1	4.7	8.2	1.5	2.3	13.2	10.7
A	6	7–31–69	12,560	71.0	4.7	9.7	1.5	1.9	12.8	10.9
A	7	8–5–69	12,430	68.1	4.7	6.8	1.5	4.6	14.3	7.9
A	8	6–10–70		70.0	4.6	7.5	1.5	1.4	15.0	
A	9	6–15–70		69.3	4.7	8.1	1.5	2.2	14.2	
A	10	6–18–70		58.3	4.0	5.4	1.3	2.4	28.6	
A	11	6–22–70		67.3	4.6	6.7	1.4	3.4	16.6	
A	12	6–23–70		67.1	4.5	8.1	1.4	2.6	16.3	
A	13	6–23–70		68.2	4.7	8.1	1.5	2.4	15.1	
A	14	6–29–70		66.5	4.4	7.3	1.6	2.9	17.3	
A	15	6–29–70		66.0	4.3	6.6	1.6	3.0	18.5	
A	16	6–30–70		67.5	4.5	8.0	1.6	3.0	15.3	
A	17	7–1–70		65.4	4.4	7.7	1.5	2.7	18.3	
A	18	7–2–70		67.3	4.5	7.9	1.4	2.7	15.9	
A	19	7–6–70		66.8	4.5	7.6	1.3	2.7	17.1	
A	20	7–6–70		68.3	4.6	7.3	1.3	2.7	15.8	
A	21	7–7–70		69.1	4.7	8.0	1.4	2.8	14.0	
A	22	7–8–70		69.6	4.7	7.9	1.4	2.7	13.7	
A	23	7–8–70		69.7	4.7	7.8	1.4	2.8	13.6	
A	24	7–8–70		69.4	4.7	7.9	1.5	2.8	13.7	
A	25	7–9–70		69.0	4.7	8.1	1.4	2.8	14.0	
A	26	7–9–70		70.4	4.7	7.5	1.5	2.7	13.2	

A	27	7–10–70		70.2	4.9	8.0	1.5	2.5	12.9	
A	28	7–10–70		69.1	4.7	8.2	1.4	2.6	14.0	
A	29	7–13–70		69.7	4.8	7.8	1.5	2.6	13.6	
A	30	7–14–70		68.6	4.7	7.8	1.6	2.5	14.8	
A	31	7–14–70		76.6	4.7	7.4	1.6	2.4	16.3	
A	32	7–15–70		63.9	4.3	7.6	1.5	2.5	20.2	
A	33	7–15–70		69.3	4.7	8.1	1.5	2.4	14.0	
A	34	7–15–70		67.4	4.6	8.0	1.4	2.4	16.2	
B	35	8–11–70	10,638					1.9	16.5	10.3
B	36	8–12–70	10,467					2.2	18.8	8.9
B	37	8–12–70	10,220					2.5	19.3	9.8
B	38	8–13–70	10,741					2.4	18.3	8.2
B	39	8–13–70	10,541					2.1	18.7	8.5
B	40	8–14–70	10,952					2.5	17.2	7.5
B	41	8–14–70	10,916					2.5	17.2	7.8
B	42	8–17–70	10,184					2.6	23.7	5.7
B	43	8–18–70	10,074					2.5	24.0	6.2
B	44	8–18–70	9,849					3.1	25.2	6.2
B	45	8–19–70	10,543					1.9	21.3	6.2
B	46	8–19–70	10,914					2.5	18.9	6.4
B	47	8–20–70	10,270					2.6	22.3	6.3
B	48	8–20–70	9,836					2.3	24.9	6.5
C	49	5–30–73	11,544					.84	14.2	7.2
D	50	5–30–73	11,100					.84	14.0	6.8
D	51	5–31–73	11,659					.84	16.6	6.5
C	52	5–31–73	11,621					.85	15.1	5.9
C	53	6–5–73	11,746					.85	14.6	5.9
C	54	6–5–73	11,775					.95	14.1	5.9
E	55	6–26–73	10,800	67.1	4.7	7.0	1.5	3.7	15.0	10.4
F	56	7–18–73	10,530	65.3	4.6	7.0	1.5	3.4	17.2	9.4

[a]See Nomenclature for Tables, pp. 449-450.

Table 30.5. Chemical and Particle Size Analysis of Fly Ash for Test Conditions.[a]

			Chemical Fly Ash Analysis										Particle Size Analysis		
Precip	Test #	Test Date	SiO_2 (%)	CaO (%)	Fe_2O_3 (%)	MgO (%)	Al_2O_3 (%)	Na_2O (%)	K_2O (%)	SO_3 (%)	P_2O_5	TiO_2	dens (gm/cc)	MMD (micron)	Sig G
A	1	7–11–69	50.1	3.5	13.4	1.2	22.6	0.7	3.0	1.9			2.35	8.4	1.99
A	2	7–11–69	50.1	3.5	13.4	1.2	22.6	0.7	3.0	1.9			2.35	8.4	1.99
A	3	7–17–69	49.5	4.8	12.6	1.1	22.2	1.4	2.7				(2.25)	(9.1)	(1.97)
A	4	7–25–69	48.7	3.6	15.9	0.9	23.4	0.9	2.9	1.4			2.18	10.0	2.20
A	5	7–21–69	52.7	3.4	8.9	1.3	24.7	1.7	2.9	1.5			2.18	9.0	1.72
A	6	7–31–69	52.7	3.4	8.9	1.3	24.7	1.7	2.9	1.5	(limestone inj.		2.18	9.0	1.72
A	7	8–5–69	48.7	3.6	15.9	0.9	23.4	0.9	2.9	1.5	rt. lb/hr)		2.40	10.0	2.20
A	8	6–10–70	31.6	27.2	5.9	0.9	14.0	0.8	1.8	5.0		22,000	(2.27)	(8.0)	(1.63)
A	9	6–15–70	31.6	27.2	5.9	0.9	14.0	0.8	1.8	5.0		19,000	(2.27)	(8.0)	(1.63)
A	10	6–18–70	31.6	27.2	5.9	0.9	14.0	0.8	1.8	5.0		20,000	(2.27)	(8.0)	(1.63)
A	11	6–22–70	32.0	29.0	7.7	1.4	13.9	0.6	1.9	4.8		20,000	(2.27)	(8.0)	(1.63)
A	12	6–23–70	34.2	22.2	7.7	1.1	14.9	0.8	2.3	1.4		11,000	(2.27)	8.4	1.74
A	13	6–23–70	31.0	29.3	7.0	1.0	13.8	0.7	1.8	5.2		19,000	(2.27)	8.0	1.63
A	14	6–29–70	34.2	22.2	7.7	1.1	14.9	0.8	2.3	1.4		10,300	(2.27)	8.4	1.74
A	15	6–29–70	31.0	29.3	7.0	1.0	13.8	0.7	1.8	5.2		20,000	(2.27)	8.0	1.63
A	16	6–30–70	32.0	29.0	7.7	1.4	13.9	0.6	1.9	4.8		20,000	(2.27)	(8.0)	(1.63)
A	17	7–1–70	34.2	22.2	7.7	1.1	14.9	0.8	2.3	1.4		11,000	(2.27)	8.4	1.74
A	18	7–2–70	31.0	29.3	7.0	1.0	13.8	0.7	1.8	5.2		19,000	(2.27)	8.0	1.63
A	19	7–6–70	48.2	3.9	12.4	1.3	20.6	1.1	2.7	1.4		0	(2.27)	8.4	1.87
A	20	7–6–70	34.2	22.2	7.7	1.1	14.9	0.8	2.3	1.4		10,000	(2.27)	8.4	1.74
A	21	7–7–70	34.2	22.2	7.7	1.1	14.9	0.8	2.3	1.4		10,000	(2.27)	8.4	1.74
A	22	7–8–70	48.2	3.9	12.4	1.3	20.6	1.1	2.7	1.4		0	(2.27)	8.4	1.87
A	23	7–8–70	41.7	13.5	9.8	1.2	17.3	0.9	2.1	3.6		6,600	(2.27)	(8.4)	(1.74)
A	24	7–8–70	41.7	13.5	9.8	1.2	17.3	0.9	2.1	3.6		4,500	(2.27)	(8.4)	(1.74)
A	25	7–9–70	48.2	3.9	12.4	1.3	20.6	1.1	2.7	1.4		0	(2.27)	8.4	1.87
A	26	7–9–70	41.7	13.5	9.8	1.2	17.3	0.9	2.1	3.6		4,600	(2.27)	(8.4)	(1.74)

A	27	7–10–70	41.7	13.5	9.8	1.2	17.3	0.9	2.1	3.6		2,500	(2.27)	(8.4)	(1.74)
A	28	7–10–70	41.7	13.5	9.8	1.2	17.3	0.9	2.1	3.6		2,400	(2.27)	(8.4)	(1.74)
A	29	7–13–70	32.0	29.0	7.7	1.4	13.9	0.6	1.9	4.8		21,000	(2.27)	(8.0)	(1.63)
A	30	7–14–70	48.2	3.9	12.4	1.3	20.6	1.1	2.7	1.4		0	(2.27)	8.4	1.87
A	31	7–14–70	41.7	13.5	9.8	1.2	17.3	0.9	2.1	3.6		2,800	(2.27)	(8.4)	(1.74)
A	32	7–15–70	48.2	3.9	12.4	1.3	20.6	1.1	2.7	1.4		0	(2.27)	8.4	1.87
A	33	7–15–70	41.7	13.5	9.8	1.2	17.3	0.9	2.1	3.6		2,600	(2.27)	(8.4)	1.74
A	34	7–15–70	34.2	22.2	7.7	1.1	14.9	0.8	2.3	1.4		11,000	(2.27)	8.4	1.74
B	35	8–11–70	52.0	1.1	6.8	1.5	27.8	0.3	3.4	0.7			2.14	11.5	2.17
B	36	8–12–70	51.9	1.1	6.3	1.4	28.0	0.3	3.5	0.5			2.20	11.2	2.37
B	37	8–12–70	51.9	1.1	6.3	1.4	28.0	0.3	3.5	0.5			2.20	11.2	2.37
B	38	8–13–70	50.3	1.4	6.6	1.2	28.1	0.3	3.8	0.5			2.24	12.0	2.42
B	39	8–13–70	52.0	1.1	6.7	1.2	28.2	0.3	3.4	0.4			2.14	11.5	2.17
B	40	8–14–70	52.5	1.2	6.2	1.2	28.8	0.2	3.3	0.5			2.31	10.3	2.28
B	41	8–14–70	52.5	1.2	6.2	1.2	28.8	0.2	3.3	0.5			2.31	10.3	2.28
B	42	8–17–70	50.3	1.4	6.6	1.2	28.1	0.3	3.8	0.5			2.24	12.0	2.42
B	43	8–18–70	53.5	1.1	6.3	1.4	28.1	0.3	3.8	0.5			2.27	12.0	2.17
B	44	8–18–70	53.5	1.1	6.3	1.4	28.1	0.3	3.8	0.5			2.27	12.0	2.17
B	45	8–19–70	52.5	1.2	6.2	1.2	28.8	0.2	3.3	0.5			2.31	10.3	2.28
B	46	8–19–70	52.5	1.2	6.2	1.2	28.8	0.2	3.3	0.5			2.31	10.3	2.28
B	47	8–20–70	53.5	1.1	6.3	1.4	28.1	0.3	3.8	0.5			2.27	12.0	2.17
B	48	8–20–70	53.5	1.1	6.3	1.4	28.1	0.3	3.8	0.5			2.27	12.0	2.17
C	49	5–30–73	52.0	1.8	6.1	1.3	27.0	0.3	3.3	0.6	1.1	0.1	2.25	4.6	2.24
D	50	5–30–73	48.8	2.2	7.0	1.2	27.0	0.4	3.3	0.6	1.4	0.1	2.27	4.4	2.29
D	51	5–31–73	50.2	2.3	6.9	1.3	27.7	0.3	3.5	0.7	1.4	0.1	2.34	1.0	5.8
C	52	5–31–73	50.9	2.3	6.9	1.3	27.3	0.3	3.7	0.7	1.2	0.1	2.35	1.3	4.46
C	53	6–5–73	51.3	2.3	6.9	1.3	26.8	0.3	3.6	0.5	1.1	0.1	2.34	1.2	4.83
C	54	6–5–73	50.4	2.3	7.1	1.3	26.2	0.3	3.6	0.7	1.1	0.1	2.30	2.25	3.74
E	55	6–26–73	48.8	5.6	14.8	1.3	19.2	0.8	3.1	2.3	0.3	0.1	2.56	5.2	3.17
F	56	7–18–73	50.1	5.9	12.3	1.4	18.9	1.1	3.3	2.0	0.3	0.1	2.38	6.6	1.80

[a]See Nomenclature for Tables, pp. 449-450.

The data given in the tables came from actual performance tests taken by the testing group of TVA's Department of Power Production. There are some instances where certain quantities had to be estimated. The field voltage values given in parentheses were estimated by a knowledge of the primary voltage, primary amperage and the secondary amperage, and estimating the efficiency of the transformer rectifier sets. The estimate is quite good and checks well with values of other performance tests where all four readings were given.

In some cases where there was not a sufficient fly ash sample for both chemical and particle size analysis, lab personnel combined samples of various tests, leaving some analysis only an approximation.

Table 30.6 provides a comparison of measured and predicted collection efficiencies.

Table 30.6. Comparison of Actual and Predicted Collection Efficiencies.[a]

Test #	*Actual EFT*	*Deutsch-Anderson*	*Deutsch-Anderson Modified*	*CEGB*	*Selzler-Watson Original*	*Selzler-Watson Revised*	*Theodore*
A-1	95.8	84.57	81	97.79	65.48	87.92	97.44
A-2	95.6	85.24	81.5	98.12	67.14	88.49	97.21
A-3	98.0	87.58	82	98.2	67.17	88.18	98.49
A-4	94.4	82.8	78	96.14	48.43	84.63	97.50
A-5	96.6	85.36	83	96.92	57.59	85.68	96.23
A-6	95.7	85.36	82.5	96.44	53.17	84.14	95.25
A-7	94.9	79.67	77	96.24	49.33	86.41	95.19
A-8[b]	73.4	77.04	75	95.12	26.24	75.84	97.03
A-9[b]	83.3	77.25	75	96.14	37.22	80.99	97.80
A-10[b]	47.0	77.4	75	97.46	9.23	68.24	47.35
A-11[b]	68.7	75.28	73	96.87	29.66	80.40	90.54
A-12[b]	68.8	76.95	74	96.38	20.25	75.54	75.18
A-13[b]	68.9	74.93	72.5	95.93	19.77	74.88	72.85
A-14[b]	78.8	79.46	76	97.27	33.7	81.15	91.46
A-15[b]	81.8	77.54	74.5	97.34	25.16	78.39	87.47
A-16[b]	76.3	76.46	73	96.86	37.19	82.12	94.40
A-17[b]	85.8	75.43	73	96.31	33.96	81.3	91.75
A-18[b]	85.3	75.1	72.5	96.38	25.38	77.95	93.55
A-19	92.6	77.8	74.5	96.72	38.48	82.38	92.43
A-20[b]	77.8	79.09	76	96.88	26.7	78.36	92.75
A-21[b]	91.1	84.26	82	97.88	40.65	82.37	90.75
A-22	89.9	78.63	76	96.51	41.06	82.62	94.23
A-23[b]	71.3	78.63	75.5	96.56	25.61	77.77	88.48
A-24[b]	79.3	79.26	76	96.73	27.22	78.39	91.13
A-25	94.9	79.46	76	96.77	46.85	84.16	95.46
A-26[b]	88.3	79.46	76	96.57	27.88	78.44	85.02
A-27[b]	82.0	79.15	76.5	96.28	26.66	77.69	87.77
A-28[b]	81.3	77.54	75	96.13	31.66	79.86	94.65

Table 30.6 continued

Test #	Actual EFT	Deutsch-Anderson	Deutsch-Anderson Modified	CEGB	Selzler-Watson Original	Selzler-Watson Revised	Theodore
A-29[b]	74.0	76.34	75.5	96.23	25.36	77.46	86.45
A-30	85.6	89.59	84.5	96.43	38.36	81.81	95.36
A-31[b]	78.6	90.67	87.5	96.92	27.89	78.43	91.85
A-32	88.5	85.66	82	98.42	50.59	84.96	96.00
A-33[b]	88.0	86.28	84	98.04	42.96	82.36	92.22
A-34[b]	87.2	86.08	83.5	98.20	42.08	82.42	94.61
B-35	82.6	90.18	85	96.94	25.76	76.73	
B-36	63.7	79.26	74	94.05	16.36	73.45	
B-37	55.2	79.26	74.5	94.44	16.22	73.87	
B-38	72.6	84.4	78	95.35	19.7	75.43	
B-39	69.1	83.11	78.5	95.10	18.36	74.28	
B-40	59.8	72.63	—	93.03	15.46	73.31	
B-41	50.7	72.63	—	93.03	14.26	72.48	
B-42	85.2	84.92	78	96.14	31.68	81.03	
B-43	91.1	94.02	88	98.61	42.37	83.33	
B-44	92.8	93.87	87.5	98.81	42.94	84.21	
B-45	69.1	72.17	—	92.69	17.77	74.2	
B-46	71.1	72.63	70	93.24	16.94	74.44	
B-47	94.6	94.37	86.5	98.63	56.56	86.46	
B-48	94.8	94.95	89	98.79	56.68	86.24	
C-49	99.38	99.47	95	99.17	99.64	93.28	
D-50	99.19	99.13	93	99.11	99.53	93.08	
D-51	99.39	67.37	—	99.15	98.77	92.25	
C-52	99.17	75.29	—	98.96	99.52	93.18	
C-53	98.24	73.46	—	98.86	99.43	92.94	
C-54	97.42	91.69	79	99.34	99.55	93.32	
E-55	97.6	87.13	77	99.51	78.17	90.25	
F-56	99.2	92.95	88	99.59	83.05	91.12	

[b]Limestone injection.
[a]See Nomenclature for Tables, pp. 449-450.

NOMENCLATURE FOR TABLES

English

A = Area of the collecting surface (ft^2)
Buses = # of Buses per collector
Ac = Ash content of the coal (%)
AMV = sec EMV, apparent migration velocity
E_1 = Peak discharge electrode field strength KV/cm = 1.45–1.65 E_2
CnTime = contact time
CRT f1 = current in field #1
Currt = name plate current capacity of the trans.-rect. set
Ez = Average discharge electrode field strength KV/cm
Cx Area = IGCI definition of cross sectional area
EMV = Effective Migration Velocity, precipitation parameter

F = Specific collecting area (PA/V, sec/ft)
kw = Kilowatts, power supplied to discharge electrodes
g = specific surface area of inlet dust; (cm^2/g)
Gr Ld = grain loading corrected to 29.92 in Hg and 32°F
IKVA = installed kilovolt amps, design capacity of the transformer-rectifier system in the precipitator
Gr Ex = exit grain load corrected to 29.92 in Hg and 32°F
L ds wr = length of discharge wire per collector ft.
MMD = Mass median diameter
Q = gas flow rate actual ft^3/sec
P = Plate area per installed KVA, (ft^2/IKVA)
Trans = # of transformers per collector
PA = Plate area (ft^2)
Plt sp = spacing in cm
S = Sulfur content of the coal (%)
Spk rt = spark rate in the precipitator #/min
W = migration velocity
V = volumetric flow rate KACFM
Vh fl = average voltage drop in field #1

Greek

η = Efficiency
μ = gas viscosity (poise)
σ_g = standard geometric deviation

REFERENCES

1. White, H. J. "Electrical Precipitation," In *Chemical Engineering Progress*, Research-Cottrell, Inc., Bound Brook, New Jersey, (June, 1956), pp. 244-248.
2. Barrett, A. A. "An Analysis of Performance on Pulverised Fuel Ash, Part I," Central Electricity Generating Board, London, England, August, 1967.
3. Barrett, A. A. "An Analysis of Performance on Pulverised Fuel Ash, Part II," Central Electricity Generating Board, London, England, August, 1967.
4. Selzler, W. D. and W. D. Watson, Jr. " 'Hot' vs 'Enlarged' Electrostatic Precipitation of Fly Ash: A Cost Effectiveness Study," *J. Air Poll. Cont. Assoc.* **24**, (2), 115-121 (1974).
5. Reynolds, J. P., L. Theodore and J. Marino. "Calculating Collection Efficiences for Electrostatic Precipitators," *J. Air Poll. Cont. Assoc.* **25**, (6), 610-616 (1975).
6. Strauss, W. (ed.), *Air Pollution Control* (New York: Wiley-

Interscience), pp. 227-318.

7. Frisch, N. W. and D. W. Coy. "Commentary on 'Hot' versus 'Enlarged' Electrostatic Precipitation of Fly Ash: A Cost Effectiveness Study," *J. Air Poll. Cont. Assoc.* **24**, (9), 872-875 (1974).
8. Watson, W. D., Jr. "Author's Reply," *J. Air Poll. Cont. Assoc.* **24**, (9), 875-878 (1974).
9. Frisch, N. W. and D. W. Coy. "Specifying Electrostatic Precipitators for High Reliability," Presented at the Symposium on Electrostatic Precipitators for the Control of Fine Particles, Environmental Protection Technology series, January 1975, U.S. EPA Publication No. 650 12-75-016.
10. Results, Acceptance, and Performance Test Reports, Tennessee Valley Authority, Division of Power Production: Report No. 31, Kingston Steam Plant Unit 6, March 21-April 7, 1960; Report No. 34, Kingston Steam Plant Unit 6, December 5-21, 1960; Report No. 38, Widows Creek Steam Plant Unit 7, June 21-30, 1961; Report No. 43, Bull Run Steam Plant Unit 1, March 20-24, 1967; Report No. 50, Watts Bar Steam Plant Units A & B, July 22-29, 1969; Report No. 52, Tests after Electrical Field Modifications, Widows Creek Steam Plant Unit 8, Precipitator A, August 12-29, 1969; Report No. 53, Shawnee Steam Plant Unit 3, June 11-July 2, 1969; Report No. 54, Shawnee Steam Plant Unit 10, July 9-August 6, 1969; Report No. 55, Gallatin Steam Plant Unit 4, October 6-21, 1969; Report No. 56, Gallatin Steam Plant Unit 1, November 3-7, 1969; Report No. 59, Gallatin Steam Plant Unit 1, February 9-20, 1970; Report No. 62, Shawnee Steam Plant Unit 10 Precipitator A, June 9-July 15, 1970; Report No. 64, Paradise Steam Plant Unit 3, March-April, October, & November, 1970; Report No. 67, Kingston Steam Plant Units 1, 2, 5, 6 and 8, August-Sept., 1970; Report No. 69, Bull Run Steam Plant Unit 1, March 16-18, 1971; Report No. 72, Widows Creek Steam Plant Unit 7, June 26-27 & July 19-21, 1972; Report No. 73, Widows Creek Steam Plant Unit 8, Tests between 1967 and 1969; Report No. 74, Gallatin Steam Plant Units 2, 3, & 4, December 5, 1972-February 8, 1973; Report No. 75, Johnsonville Steam Plant Units 3 and 8, September 18 to November 22, 1972; Report No. 76, John Sevier Steam Plant Units 1 and 2, May-September, 1973; Report No. 78, Cumberland Steam Plant, Unit 1, July 12-26, 1973 and Unit 2, November 13-15, 1973; Report No. 79, Gallatin Steam Plant Unit 4, August 30, 1973; Report No. 80, Allen Steam Plant Units 1 and 2, March-April, 1973; Report No. 81, Johnsonville Steam Plant Unit 10, March 19-21, 1974; Report No. 82, Colbert Steam Plant Units 2 and 4, June-July, 1973; Report No. 83, Shawnee Steam Plant Units 4, 5, and 7, June 10-28, 1974.

APPENDIX

Method of Calculation

The comparisons between actual efficiencies and predicted efficiencies of various models necessitates a display of how the comparisons were made. A short description of the models used for the comparison and a hand calculation of each is given below using the values applicable to test A1.

A. Deutsch-Anderson Equation

$$\text{Efficiency } \eta = 1\text{-}e^{(-Aw/Q)}$$

where: A = Area of collecting surfaces, ft^2
Q = Gas flow rate, actual cubic feet per second
w = Feet per second, migration velocity, ft/sec

w is estimated by the following formula: $\dfrac{0.0058\ E_1\ E_2\ r}{\mu}$

where: E_1 = Peak discharge electrode, field strength, (KV/cm) $\simeq$ 1.45 to 1.63 times E_2 (14)
E_2 = Average discharge electrode field strength, (KV/cm)
r = Particle radius of the mass median diameter particle size
μ = Gas viscosity (poise)

Test A1

$$w - \frac{(0.058) \times (3.94) \times (6.12)\ (8.4 \times 10^{-4})/2 \text{ ft/sec}}{2.375 \times 10^{-4}} = 0.249 \text{ ft/sec}$$

$$\eta = 1 - \exp(-29{,}700 \times 0.249 \times 60/237{,}000) = 0.8464 \text{ or } 84.64\%$$

B. Deutsch-Anderson Equation Modified to accept Size Distribution Input. Given the Deutsch-Anderson exponent (Aw/Q) value, computing the natural log of the standard geometric deviation one can estimate efficiency.

Design Deutsch Exponent = 1.8734

$$\sigma_g = 1.99 \therefore 1\eta\sigma_g = 0.68813$$

$$\therefore \text{Efficiency} = 81\%$$

C. CEGB. The CEGB equation depends upon the Deutsch-Anderson equation for ultimate calculation of efficiency, however, the calculation of the migration velocity depends on the following equation:

$$\text{Emv} = 31.2 - 4.81\ \ln(F) - 8 \times 10^{-4}(P)$$

$$- 4 \times 10^{-8}(g)^2 - 8.1 \exp(-0.5S) - \frac{43.3}{Ac}$$

The "g" term is calculated from the Bahco Analysis in the following manner:

1 Micron Size	2 Percentage Range	3 Factor	4 2 x 3
<5		7500	
5 to 10		4000	
10 to 20		2200	
>20		950	
			TOTAL

Specific surface, $g = \frac{\text{total}}{100}$ (cm^2/gm)

where: F = Specific collecting area ($ft^2/ft^3/sec$)
P = Total Plate Area/installed power (ft^2/KVA)
g = Specific surface area of inlet dust (cm^2)
S = Total sulfur content of coal %
Ac = Total Ash content

$$EMV = 9.7038 - 2.475 \times 10^{-1} - 0.59028 - 1.5556 =$$
$$15.464 \text{ cm/sec} = 0.50735 \text{ ft/sec}$$
$$\eta = - \exp [A(EMV/Q)] = 97.796$$

D. Selzler-Watson Equation (original)

The Selzler-Watson Equation is:

$$\eta = 100 [1 - \exp (-.57 \times 203 \times (A/V)^{1.4} \times (Kw/V)^{.6} \times (S/Ac)^{.22})]$$

or

$$\eta = 100 [1 - \exp (-.57 \times 157.6 \times (A/V)^{1.4} \times (Kw/v)^{.6} \times (S/Ac)^{.22})]$$

where: A = Collecting plate area (1000s ft^2)
V = Flue gas volumetric flow rate (KA cfm)
kw = Power input to discharge electrode, (actual)
S/Ac = Sulfur to ash ratio by weight percent

$$\eta = 100 [1 - \exp(-.57 \times 157.6 \times (.125)^{1.4} \times (.126)^{0.6} \times (3.3/11.9)^{.22})] = 65.478$$

E. Revised Selzler-Watson Equation. The original Selzler-Watson was revised by Watson to yield the following:

$$\eta = 100 [1 - \exp (-0.57 \times 8.3 \times (A/V)^{0.32} \times (Kw/V)^{0.16} \times (S)^{.16})]$$

The constant 8.3 is for pulverized coal fired units.

$$\eta = 100 [1 - \exp(-.57 \times 8.3 \times (.125)^{.32} \times (.126)^{.16} \times (3.3)^{.16}] = 87.92 \text{ pct}$$

F. Theodore Model

$$\text{Efficiency} = 1 - c_i e^{-C_2 E_2}$$

where: $C_i = 0.77052$
$C_2 = 0.2044$
E = Average actual field strength (Kv/cm)

Therefore, for test #1 the predicted efficiency should be

$$\eta = (0\ 0.7052\ e^{-(0.2044 \times 4.027^2)})100$$
$$= 97.44$$

31

Study of Fiberglass Fabrics for the Collection of Fly Ash

K. E. Noll, W. T. Davis, and **G. W. Kelley**
The University of Tennessee, Department of Civil Engineering
Knoxville, Tennessee, and

H. E. Harper
Pollution Control Division, The Carborundum Company
Knoxville, Tennessee

INTRODUCTION

Fabric filtration of fly ash from industrial and power plant boilers is important because of the potential to remove fine particles at efficiencies greater than 99.9% and at a cost that is competitive with electrostatic precipitation. This is especially true when the fuel is coal with a low percentage of sulfur (typical of western coal) because this produces high-resistivity fly ash, which is difficult to collect with high efficiencies in an electrostatic precipitator. Although information concerned with the use of fabric filters for fly ash removal is limited, sufficient technical information has been reported in the literature to indicate both technical and economic feasibility on a number of oil- and coal-fired boilers. These fabric filter collectors generally used fiberglass filter media operating at air-to-cloth ratios between 2:1 and 7:1 and with maximum pressure drops between 2.5 and 6.0 inches of water. Collection efficiencies reported generally exceeded 99.5 percent.

Table 31.1 provides a summary of the performance data for a number of laboratory, pilot, and full-scale fabric filter collectors. In 1961, a pilot fabric filter study was conducted on an oil-fired steam plant of the Southern California Edison Company.[1,2] The pilot plant consisted of a single siliconized-Teflon® 10-oz/yd^2 12″ x 28′ fiberglass

Table 31.1. Collector Performance Summary.

Installation and Test Period	*Boiler Firing, Fuel and Effluent Treatment*	*Total Flow to Collector (acfm)*	*Temp. of Gas Entering Collector (°F)*	*Bag Life*	*Inlet Loading gr/SCF*	*Net Air-to-Cloth Ratio (FPM)*	*Collector Operating Pressure Drop (in H_2O)*	*Residual Filter Drag (in H_2O) / (FPM)*	*Reference*
So. Calif. Edison: El Segundo 9/61–3/63	Gas + Oil Dolomite Additives	a) 1431 b) 1508 c) 1480	a) 260 b) 258 c) 262	Unknown	Unknown	a) 7.0:1 b) 7.9:1 c) 7.6:1	a) 3.5 b) 2.6 c) 3.4	a) 0.5 b) 0.33 c) 0.43	1,2
So. Calif. Ed: Alamitos 2/65–3/68	Oil and/or Gas Dolomite Additives	820,000	258	>7 mo.	Unknown	6.5:1	5.7	0.88	3,4
Public Service Electric & Gas Co: Mercer Station, 7/65–1966	Coal Plus 20% Coke	21,600	270	Unknown	1.88–2.17	5:1	5.0	1.0	5,6
Pilot on Chem. Process Boiler, 1962	Coal	1,500	312	Unknown	3.7	3.5:1	4.0	1.4	7
Pilot on Australian Boiler, 1966	Coal	Unknown	<250	Unknown	Unknown	3.5:1	4.0	1.4	7

Pilot at Midwest Utility 1967–1969	Pulverized	2,880	250	>4000 hr	4–5	4:1	4.0	1.0	7
Pilot #4 1971–1972	Coal	Unknown	Unknown	a) <24 hr b) <24 hr c) >2000 hr	0.07	7.5:1	4.0	0.53	7
Colorado-UTE	Spreader Stoker Coal	80,000	311	Unknown	1.53	3.1:1	4.2	1.35	7
Penn. Pwr. & Light: Sunbury 1973	Anthr. + 15-35% Coke	125,000[a]	340	1 yr (approx)	Unknown	2.1:1	3.0	1.43	8
Kerr Ind.: Concord, NC	Spreader Stoker Coal	13,000	250	Significant Deterioration After 100 hr	Unknown (Nomex®)	7.0:1	6.0	0.86	9

[a]SCFM.

bag in each of two compartments with reverse air cleaning, an air-to-cloth ratio between 7:1 and 8:1, and a pressure drop of 2.5 to 3.5 inches of water (H_2O). The efficiency of particulate removal was greater than 99.9%. Testing also involved the addition of an alkaline material (dolomite) to lower the acid dewpoint from 285 to 120°F to extend bag life.

In 1964, based upon the preceding results, Southern California Edison constructed a full-scale prototype baghouse at their Alamitos Generating Station.[3,4] Several design and operational problems were encountered, and a number of modifications were made which resulted in operation at an air-to-cloth ratio of 6.5:1,. a pressure drop of 5.7 inches of water and a collection efficiency of 99+%. Effects of variations in the amount, size distribution and chemical composition of alkaline additions were studied along with variations in the fiberglass style with the objective of increasing bag life.

In 1965, a 36-bag 4-compartment pilot collector was installed on a coal-fired boiler at Mercer Generating Station by the Public Service Electric and Gas Company of New Jersey.[5,6] Feasibility was demonstrated for air-to-cloth ratios less than 6:1 with less than 5 inches H_2O pressure drop across the siliconized fiberglass filter media. Laboratory experiments performed as part of the study showed that redispersed fly ash in the laboratory gave results similar to those obtained on the field pilot unit with glass fabrics.

Adams[7] reported on four pilot studies with fabric filter collectors on coal-fired boilers and a 6-compartment modular collector on a Colorado-Ute Electric Association coal fired boiler. Shaker type collectors operated with pressure drops of 4 inches H_2O at air-to-cloth ratios of 3.5–4:1 and collection efficiencies in excess of 99%. One reverse pulse pilot collector operated at an air-to-cloth ratio of 7.5:1, pressure drop of 4 inches and an efficiency of 98.5%.

Pennsylvania Power and Light Company installed a 14-compartment structural baghouse on their Sunbury anthracite coal-fired boilers #1 and #2.[8] Using reverse air cleaning the Teflon®-coated fiberglass fabric, the operating differential pressure was 3 inches H_2O at an air-to-cloth ratio of 2:1.

McKenna[9] reported on the installation of a pilot plant reverse air collector on a 60,000 lb/hr spreader stoker. The pilot unit, capable of handling 11,000 acfm was operated at air-to-cloth ratios of 4-8:1 and utilized four 54-bag compartments. Filter media tested in the program were Nomex felt (14 oz/yd^2), Teflon woven (8 oz/yd^2), Teflon felt (24–26 oz/yd^2), and Gore-Tex® laminate on Nomex backing (4–5 oz/yd^2 + laminate). It was observed that the Gore-Tex fabric had significantly better release properties that the other media. The results of the Gore-Tex testing indicated that severe bag damage resulted from flexing and cage movement (spring filter cages). A change to rigid cages indicated no Gore-Tex wear or failure. Rigid cages were also found to be necessary for woven Teflon bags to prevent excessive bag wear. For the Nomex media the primary operating problem observed was fiber deterioration which was evident after only 100 hours of operation. This degradation was slowed down by precoating the Nomex bags with lime. It was concluded from the study that if bag wear and deterioration could be limited to 25% per year (by continuous injection of lime in the case of Nomex), that the fabric filter's annualized cost at an air/cloth ratio of 7.5:1 and a pressure drop of 5 inches H_2O would be less than for an electrostatic precipitator with similar efficiency (99.5%).

The general conclusion of the studies conducted on oil- and coal-fired boilers is that the technical feasibility of fly ash filtration has been demonstrated with one of the critical elements of design being fabric selection. One of the most prevalent problems is acid attack on the filter media. This has been reduced through the use of limestone or dolemite addition to the furnace and/or effluent and the selection of high-temperature fabrics such as fiberglass with abrasion- and acid-resistant protective finishes.

The objective of the study reported here was to compare the performance characteristics of different types of fiberglass fabrics when filtering fly ash. Laboratory test data were collected from both a bench-scale test unit and full-scale pilot plant by utilizing many different fiberglass fabrics and monitoring the change in pressure drop with filtration time at constant dust loading and air-to-cloth ratio. Outlet particle count data at various particle sizes as a function of cycle time were also obtained for each fabric/dust combinations. An evaluation of the data shows the importance of fabric selection with respect to its influence on maximum pressure drop, residual pressure drop, and the collection efficiency for fine particles.

THE PERFORMANCE CURVE CONCEPT

Historically, the development of the relationship predicting the rate of pressure drop increase of a fabric filtering a particle-laden medium was initiated by research in fluid flow through granular beds.[10,11] Williams, Hatch, and Greenburg[12] and Snyder and Pring[13] applied the basic filtration equations (as developed from Darcy's Law) to the filtration of particle-laden air flow through fabric filters.

For clean air and laminar flow the pressure loss across a filter is by Darcy's Law:

$$\Delta P_f = k_1 V \tag{1}$$

where ΔP_f is the pressure loss across the fabric and V is the face velocity (air-to-cloth ratio). The term k_1 is a function of gas viscosity, and filter characteristics such as thickness and porosity.

For the filtration head loss through a dust cake which is formed on a fabric filtering particle-laden gas, the following equation has been developed:

$$H = \left(\frac{k}{g}\frac{\mu}{\rho_g} \left(\frac{A}{V_a}\right)^2 \frac{(1-\epsilon)^2}{\epsilon^3} \frac{1}{\rho_p}\right) C_iV^2t \qquad (2)$$

where: H = increase in filter head loss due to dust load, feet of fluid-air (cgs units—cm of fluid—air)
k = constant $\simeq 5$ for many dusts
μ = gas viscosity
ρ_g = gas density—pounds/ft^3(g/cm^3)
g = acceleration of gravity
A = surface area of particles on filter—ft^2 (cm^2)
V_a = volume of particles on filter—ft^3 (cm^3)
ϵ = porosity of dust cake (unitless)
ρ_p = particle density—pounds/ft^3 (g/cm^3)
C_i = dust loading—pounds/ft^3 of dust (g/^{3})
V = face velocity of gas at filter—feet/minute (cm/sec)
t = filtration time—minutes (sec)

This equation can be re-expressed as

$$\Delta P_c = K_2C_iV^2t \qquad (3)$$

where ΔP_c is the pressure loss across the cake in inches of water gage (cm of H_2O). The dust resistivity, K_2, is dependent on the gas kinematic viscosity and the particle characteristics defined in Equation (2). In descriptive terms, K_2 can be expressed as follows in the English system:

$$K_2 = \frac{\Delta P_c}{C_iV^2t} = \frac{\text{inches of } H_2O \text{ (pressure loss)/fpm (face velocity)}}{\text{pounds (dust)/ft}^2\text{(cloth area)}}$$

The total pressure drop across the fabric and cake deposit is (from Equations 1 and 3)

$$\Delta P_{total} = \Delta P_f + \Delta P_{cake}$$

or

$$\Delta P_T = k_1V + K_2C_iV^2t \qquad (4)$$

As a result of the particle/fabric interactions which occur just after the filtration cycle begins, Equation (4) is not applicable to the filtration of industrial dusts.

For industrial filtration, the development of the performance curve concept[14] allows an evaluation of the pressure drop of fabric filters with time using graphic techniques. The performance curve is a graphic representation of the measured filter drag ($\Delta P/V$) across the fabric/dust layer as a function of the quantity of dust accumulated on the filter. A schematic representation of the performance curve is shown in Figure 31.1. The resistance to flow, *i.e.*, filter drag ($\Delta P/V$), is plotted versus the areal cake density, C_iVt (weight of dust/unit area of cloth). Typically, the curve is composed of two zones: a) the zone of cake repair, and b) the zone of homogeneous cake formation. The slope of the linear portion of the curve is the experimental dust resistivity, K_2, which is defined *theoretically* by Equation (2). Various research efforts have indicated that the slope of the operating curve, K_2, may be a function of other variables not included in Equation (2) such as fabric structure,[15] filtration velocity, and relative humidity.[16] These effects possibly account for the differences between theoretical and operating K_2 values.[17]

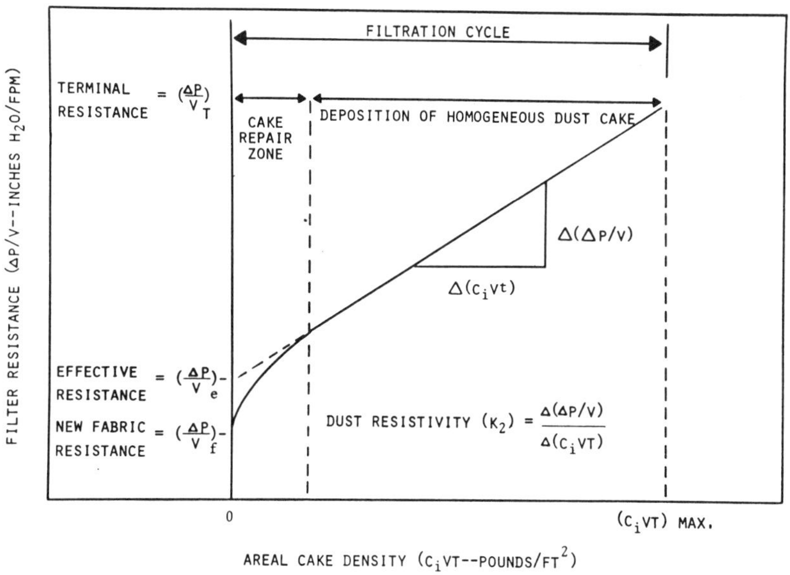

Figure 31.1. Basic performance curve of fabric filter.

Figure 31.1 represents the performance curve of a new fabric filtering dust. For fabrics which are repeatedly filtered and cleaned, an increase in the filter drag occurs due to the retention of a residual dust on the fabric. The difference between the effective resistance, $(\Delta P/V)_e$, and $(\Delta P/V)_f$ shown in Figure 31.1 is attributable to the fabric/dust interaction. For the case in which $(\Delta P/V)_e = (\Delta P/V)_f$, the performance curve is described by Equation (4), with the intercept being equal to k_1. This has been found to occur in the filtration of coarse materials.[13]

DESCRIPTION OF TEST FACILITIES

Test Procedures for the Generation of Performance Curves

Bench-Scale Unit

Two pilot plant test units were utilized to generate performance curves for fabric filters. The first test unit consisted of a bench-scale fabric filter test module with one square foot of filter cloth area. A schematic of the bench-scale unit is shown in Figure 31.2. The unit allowed testing of various fabric/dust combinations to determine the differences in pressure drop and particle collection efficiency with filtration time.

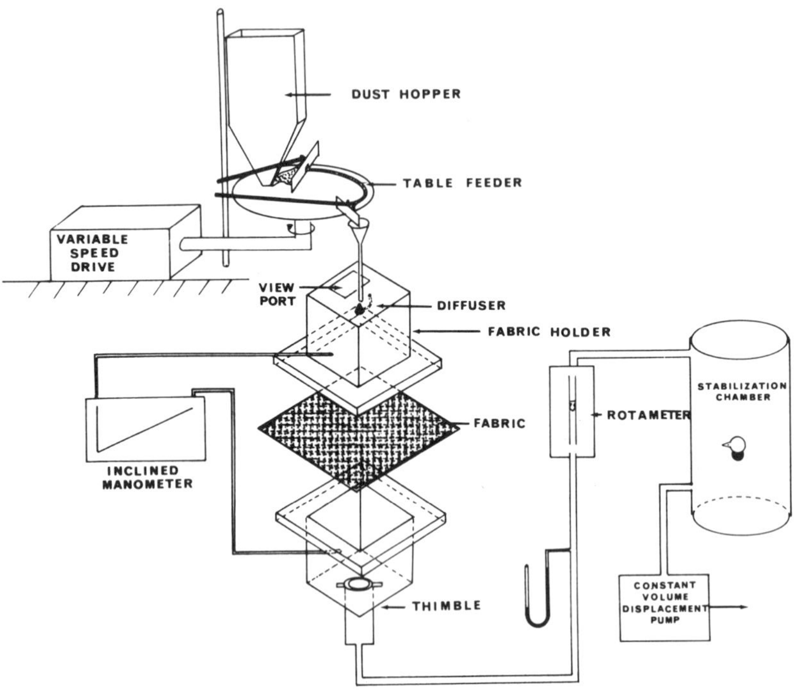

Figure 31.2. Schematic diagram of bench-scale test unit.

Test dust was metered from a rotating table feeder equipped with a variable speed drive to allow different dust concentrations to be filtered. To assist in the dispersion of the inlet dust, the collector module was equipped with a small Venturi throat and diffuser.

Flow through the module was produced by a constant volume displacement pump; flow was controlled by a regulator placed between the flow meter (rotameter) and the pump. A stabilization chamber was installed between the regulator and displacement pump with a bleed-in damper to reduce fluctuations of flow within the system.

The pressure drop across the fabric was monitored by an inclined manometer with ± .005-in. water-gage sensitivity. Efficiency information was obtained for each fabric/dust combination by monitoring isokinetically from the system into a Climet Model 250 particle counter which measured the number of particles/cf in selected size ranges between 0.5 and 10 μ. In this manner, inlet and outlet particle count and relative efficiencies at specified particle sizes were determined.

Full-Scale Unit

The second test unit was designed to simulate field operating conditions with intermittent cleaning of the filter fabric. The full-scale test program supported the bench scale tests by allowing the determination of stabilized (conditioned) residual pressure drop of fabrics after repeated dust loading and cleaning cycles, initial and stabilized performance curves (pressure drop versus time) for the fabrics, and initial and stabilized collection efficiencies for the fabrics.

The full-scale test unit consisted of two parallel-flow compartments with separate automatic damper flow control systems. (Figure 31.3.) Each basic compartment contained a tube sheet that held four 11-foot long by 4½-in. diameter filter bags. Individual compartment flow control was maintained by dampers which were activated by Dwyer Photohelic® pressure switches. The pressure differential was sensed across orifice plates located in the ductwork on the downstream side of the filter bags. By this method, the air-to-cloth ratio could be measured and maintained within approximately $\pm$ 2% of the desired constant value.

Flow was produced by two 15-in. vane axial fans operated in series (one on the upstream and one on the downstream side of the filter compartments). Flow entered the unit at the bottom of the compartments just above the dust hoppers, passed up through the inside of the filter tubes to the outside, then exited through the outlet located in the top of each compartment, passed through the orifice plate, and returned to the environmental control room for recirculation through six especially designed filter bags to eliminate any residual fines not collected by the test fabric.

Temperature and humidity control were maintained through the use of a 4000-watt thermostatically controlled heater, and an 8000 Btu humidistatically controlled air conditioning unit. The temperatures at critical points within the system were monitored by a strip chart recorder. The unit was operated at 100°F + 2°F and less than 30% relative humidity using the control system.

The grain loading was controlled through the use of Vibrascrew® dust feeder with a variable speed auger drive. Dust was fed from the Vibrascrew feeder directly into the inlet duct leading from the environmental control room to the unit and was conveyed at velocities of 2000–4000 fpm depending on the air/cloth ratio.

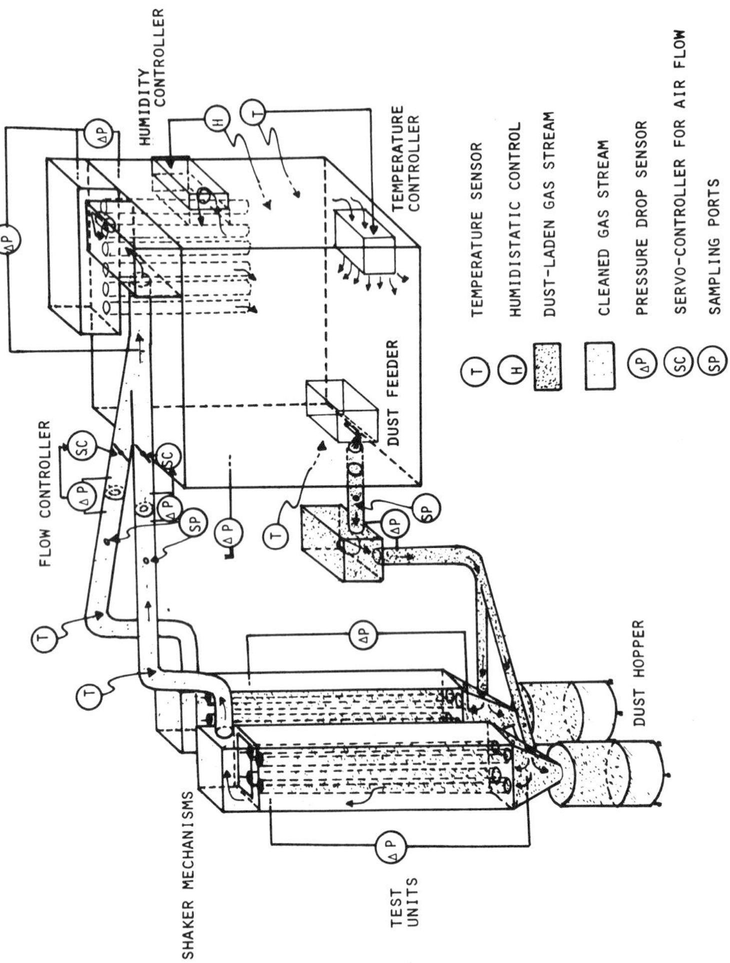

Figure 31.3. Schematic diagram of full-scale test unit.

The mode of cleaning the dust from the bags was by shaking the filter tubes with a shaking mechanism attached to the top of each filter bag compartment. The shaker mechanisms had variable amplitude and frequency of shake, as well as capabilities of vertical, horizontal, and rocking shaking modes.

An electronic control panel controlled the timing of the events throughout the cleaning cycle as follows:

1. Activation of fan motors
2. Delay time to allow unit to adjust to proper flow (~10 sec)
3. Activation of dust feeder
4. Deactivation of dust feeder
5. Delay time between dust feeder deactivation and fan deactivation
6. Delay time between fan deactivation and shaker activation
7. Activation of shakers
8. Settling period between shaker deactivation and fan reactivation to start the next cycle.

Pressure drop across the orifice plates in the system and across the bags in each compartment were monitored by Dwyer Magnehelic® gages and an inclined manometer in parallel for pressure drops between zero and 3 inches water gage; for pressure drops greater than 3 inches H_2O the inclined manometer was by-passed. All pressure taps are shown in Figure 31.3. The pressure drops and temperature readings were monitored from a central instrumentation panel.

The particle count as a function of particle size was determined for each fabric tested by withdrawing a slip stream from the outlet and monitoring the number of particles/cf using the Climet Model 250 particle counter. In this manner an indication of the change in efficiency with time was obtained.

Experimental Procedure

The objective of the fabric filtration test program was to obtain performance curves under controlled conditions and to determine their true value for documenting and interpreting the performance characteristics of baghouses operating under field conditions. The performance curve for each fabric was determined by monitoring the pressure drop across the bag as a function of time at constant loading concentrations and air-to-cloth ratios. Outlet particle counts at various diameters for each fabric were also taken both as a function of cycle time and as a function of the number of completed cycles.

The bench-scale test facility allowed many fabrics to be examined for their filtration characteristics by adding dust to each new fabric over a 90-minute period. The operating conditions were an air-to-cloth ratio of 3.1 ft/min, a temperature of 75°F, and a dust concentration of 6 g/ft^3 with fly ash as the test dust.

The full-scale test facility allowed each test fabric to be conditioned by repeated dust loading and cleaning until a stable residual pressure drop was achieved. Thus, performance curves and relative efficiency data were obtained for both new and conditioned fabrics.

The operating conditions for the full-scale test were a system flow of 175 cfm (an air-to-cloth ratio of 3.15), a temperature of 100°F and a dust concentration of 2.52 gr/ft^3 with a fly ash test dust. The shaker variables (amplitude, frequency, and time of shake) were set for maximum cleanability of the filter bags. The fiberglass fabrics and dust types tested on the two test facilities are described in Table 31.2. It should be noted that, in general, fiberglass fabrics would not be applied to shaker-type collectors. However, the basic objective of this report was not to determine fabric life, rather to obtain a comparison of the pressure loss and relative efficiency of the fabrics under similar cleaning accelerations. The experimental procedure allowed a comparison of:

1. Bench-scale and full-scale performance data for new fabrics
2. Conditioned fabric performance data (under identical cleaning conditions)
3. Full-scale laboratory data with data from baghouses operating under field conditions.

Table 31.2. Description of Fiberglass Test Fabrics and Dust Types.

Fabric	*Weight oz/yd^2*	*Fabric Description*	*Relative cost*
A	8.9	3 x 1 Twill Weave; all filament, light napped on one side, count 54 x 58; perm 45–55	1
B	14.5	Double-faced 4-harness satin; filament warp and fill/ 7% bulk in both warp and fill; count 60 x 28; perm 25–40	1.5
C	14.5	Double-faced 4 harness satin; 23% bulk fill, filament warp; count 60 x 28; perm 25–35	1.70
D	26.0	Filament warp/bulk fill	3.3
E	35.4	Special weave; nonstandard fabric (tricon)	4.3

Dust Type	*Mass-Mean Diameter (mmd)*	*Description*
Fly ash #1	9.6 μ[a]	Dust obtained from the electrostatic precipitator hoppers of a local power plant
Fly ash #2	20 μ[a]	Dust obtained from the electrostatic precipitator hoppers of a power plant in Wyoming
Nytal 400	5.4 μ[a]	A standard laboratory test dust (micronized talc)

[a]Dust was slightly agglomerated on the bench-scale test unit due to the feed mechanism producing a larger mmd than on the full-scale unit.

It is anticipated that these comparisons would also be representative of the fiberglass fabrics when utilized in other types of fabric filter collectors such as reverse air and pulse-type collectors.

SUMMARY OF TEST DATA

Bench-Scale Test Data

Data obtained from the bench-scale test unit were evaluated and performance curves were constructed by transposing pressure drop data (ΔP) to resistance data ($\Delta P/V$) and time data (t) into areal cake density data (C_iVt). Since the loading, C_i, and the velocity, V, were constants, the graph of $\Delta P/V$ versus C_iVt represented a plot of actual data (multiplied by a constant). Figures 31.4a and 31.4b are graphic summaries of the data obtained on the bench-scale unit showing the performance curves for the five fabrics listed in Table 31.2 for the Fly ash #1 test dust. Figure 31.4a illustrates the relationship between filter resistance and areal cake density (actual values of ΔP and time are also indicated on the axes). Figure 31.4b illustrates the relationship between the outlet particle count versus time for the five fabrics.

From the data obtained from the initial filtration performance tests performed on the square footer it can be seen that each pressure drop performance exhibits properties similar to the schematic performance curve presented in Figure 31.1:

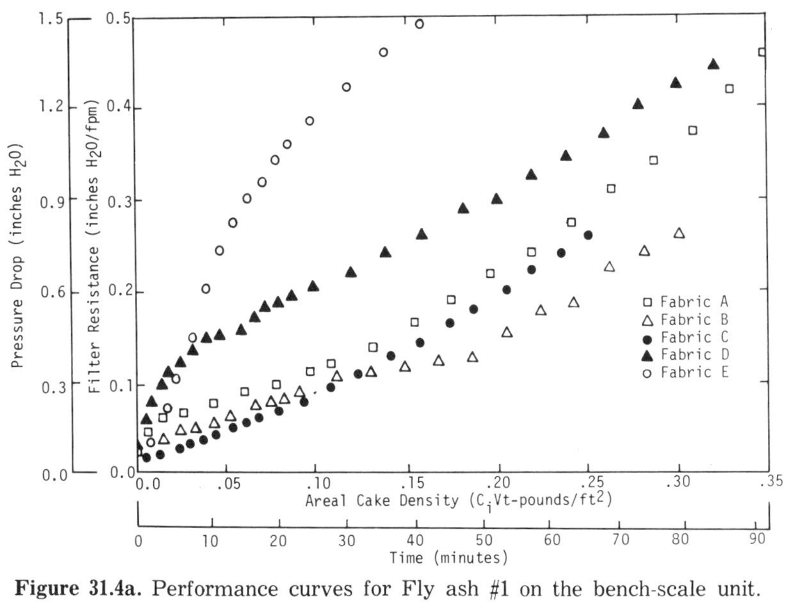

Figure 31.4a. Performance curves for Fly ash #1 on the bench-scale unit.

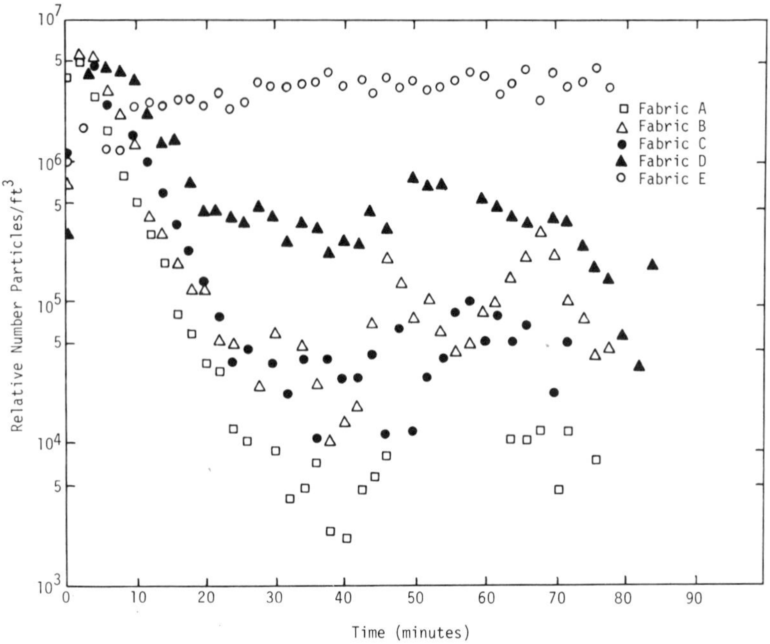

Figure 31.4b. Number of 0.5–1.0 μ particles versus time for Fly ash #1 on the bench-scale unit.

1. Each curve shows an initial pressure drop (resistance) increase which is characteristic of the fabric structure (for the specified dust characteristics). This initial nonlinear rise is an indicator of the manner in which the different fabric structures begin to form a dust cake.
2. Each curve eventually becomes linear indicating that a uniform cake has formed and that homogeneous cake filtration exists. The slopes, K_2, of these curves are relatively constant with a magnitude of 1.0–2.0 inches H_2O/fpm/lb dust/ft^2 cloth for $C_iVt \geqslant .10$ lb/ft^2.

The data in Figure 31.4b indicate that the outlet particle count is a function of fabric at the onset of filtration. It is also seen that the outlet particle count is influenced by the build-up of a dust cake. An analysis of the data presented in Figure 31.4a and 31.4b showed that fabrics D and E exhibited higher pressure losses and greater particle counts (less efficiency) than the other three fabrics (A, B, or C). Table 31.2 also indicated that the purchase costs of fabrics D and E were significantly greater than for fabrics A, B, or C. For these reasons, the emphasis of additional testing was to determine which of the fabrics A, B, or C had more optimum filtration characteristics.

From Figures 31.4a and b it can be seen that fabrics A, B, and C had similar pressure loss versus time relationships. The particle count relationships of these fabrics indicated that fabric A had less particle penetration than fabrics B or C, while fabrics B and C had similar values of particle penetration.

Fabrics B and C were also tested on the bench-scale test unit with two other test dusts: Fly ash #2 and Nytal 400®. The data obtained on these tests are shown in Figures 31.5a–5d. Again, fabrics B and C had similar pressure loss versus time relationships. Based on the data presented in Figures 31.4b, 31.5c and 31.5d it can be seen that fabric C had particle penetrations that were less than or equal to fabric B.

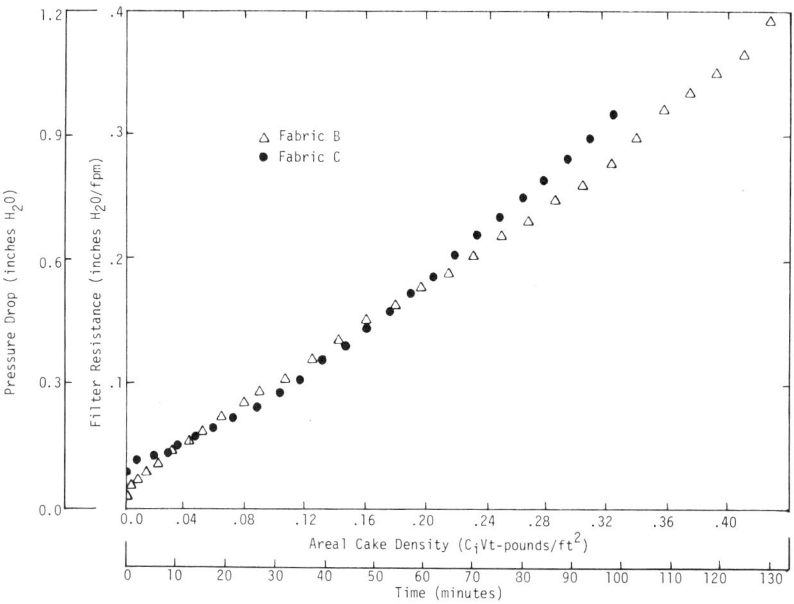

Figure 31.5a. Performance curves for Fly ash #2 on the bench-scale unit.

Full-Scale Test Unit

Tests were performed on fabrics A, B, and C on the full-scale test facility using Fly ash #1 to obtain a comparison of the filtration performance of the fabrics after having been conditioned. The performance curves for the initial 90-minute extended runs are shown in Figure 31.6 for comparison with the bench-scale data in Figure 31.4a. Again the performance curves varied depending on the fabric structure. The slopes of the curves in the straight line portion had a $K_2 \simeq$ 9-12 inches H_2O/fpm/lb dust/ft^2 cloth.

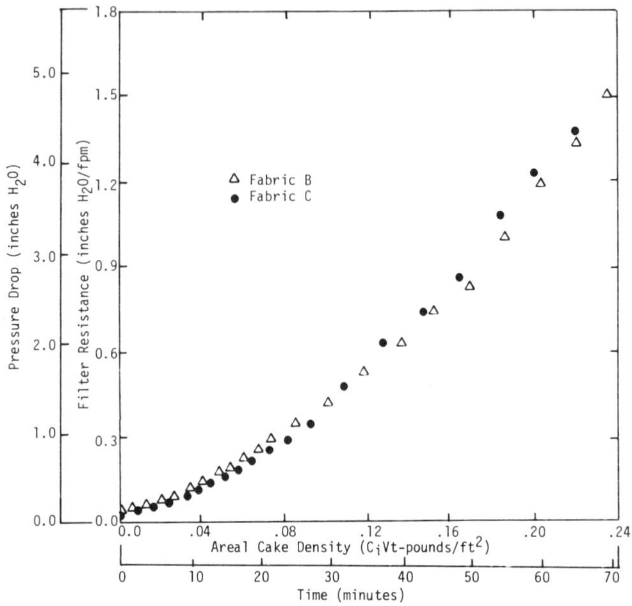

Figure 31.5b. Performance curves for Nytal 400 on the bench-scale unit.

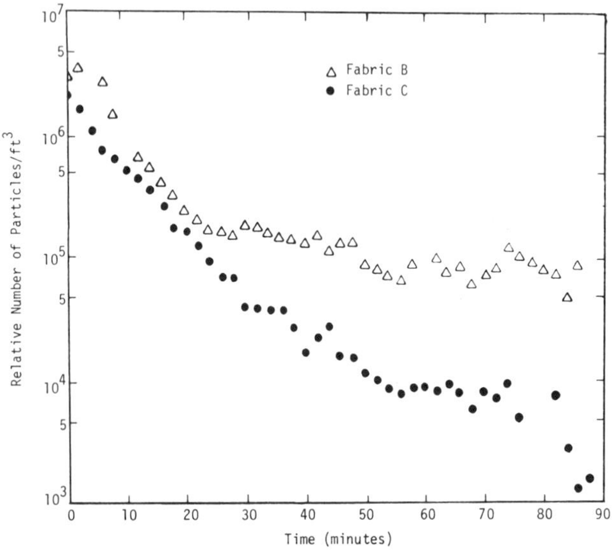

Figure 31.5c. Number of 0.5–1.0 μ particles versus time for Fly ash #2 on the bench-scale unit.

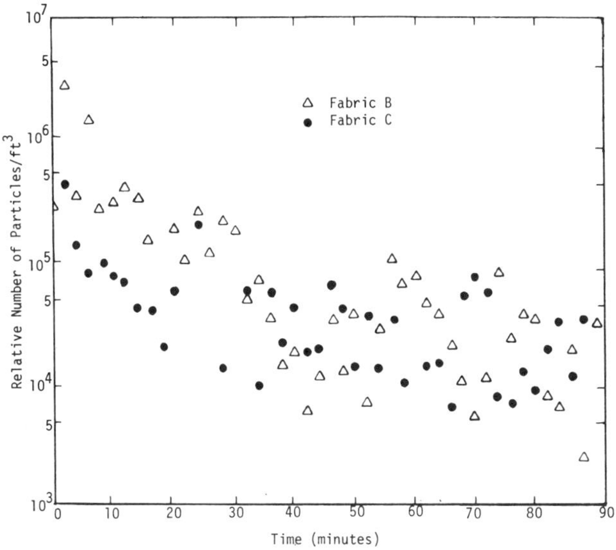

Figure 31.5d. Number of 0.5–1.0 μ particles versus time for Nytal 400 on the bench-scale unit.

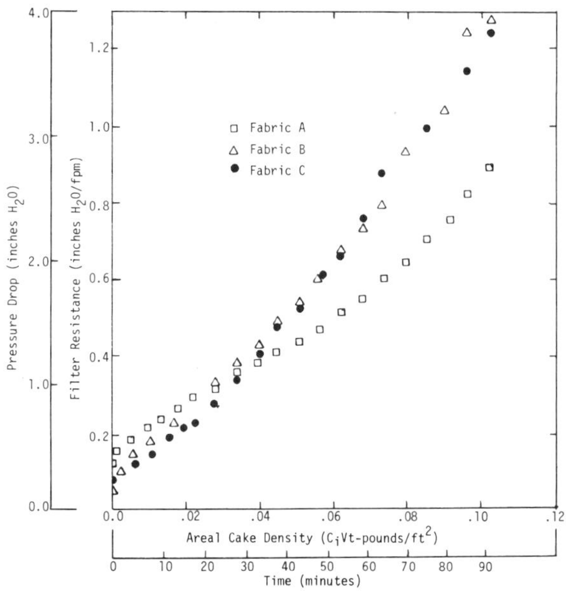

Figure 31.6. Initial performance curves for Fly ash #1 on the full-scale unit.

Full-scale tests allowed repeated dust loading/cleaning cycles until each fabric was conditioned. Figure 31.7 shows the performance curves for the three fabrics after having been conditioned. In this figure, C_iVt represents only the areal dust density loaded onto the fabric and does not include the residual areal density retained in each fabric after cleaning. A comparison of Figure 31.6 and 31.7 indicates that the conditioning of the fabrics resulted in higher residual resistances. The conditioned performance curves also exhibited linear portions with slopes between 4–13 inches H_2O/fpm/lb dust/ft^2. The pressure loss versus time curves for fabrics B and C are similar with the largest variation between the two being only 0.5 inches H_2O. The pressure loss across fabric A was less than that across fabric B and C for both the initial and conditioned cases.

Particle count data which were obtained for the initial and conditioned fabrics are presented in Figures 31.8 and 31.9 for the "0.5–1.0 μ" size interval. These data show the effect of fabric structure and cake formation on the relative particle penetration through the fabrics. The data indicate that fabric C had the least particle penetration of the three fabrics, while fabrics A and B had similar penetration values. A second important observation was that the penetration of 0.5–1.0 μ particles was relatively constant throughout the filtration cycle for the conditioned fabrics.

CONCLUSIONS

This study has provided a) a review of existing fabric filter installations on oil- and coal-fired boilers, and b) a review of experiments performed at the University of Tennessee test laboratory on the filtration of fly ash by fiberglass fabrics. The general conclusion of the historic review is that the technical feasibility of fly ash filtration has been demonstrated with one of the critical elements of design being fabric selection. The most prevalent problems found to exist in fabric filters applied to the control of boilers were acid attack and temperature damage to the filter media. These have been reduced through the use of limestone or dolemite addition to the furnace and/or effluent to reduce the acid dew point and the selection of high-temperature fabrics such as fiberglass with abrasion- and acid-resistant finishes. Existing fabric filter collectors on boilers have been found to have fabric pressure losses ranging from 2.6–6.0 inches of H_2O and filtration efficiencies of 99.0–99.9%.

The current study has provided a comparison of five potential fiberglass filter fabrics for use in the filtration of fly ash. Of these two (fabrics C and D) were found to have low potential use resulting from a combination of low filtration efficiency, high pressure losses, and high purchase cost. The other three fabrics (A, B, and C) were found to have more optimum performance characteristics: lower

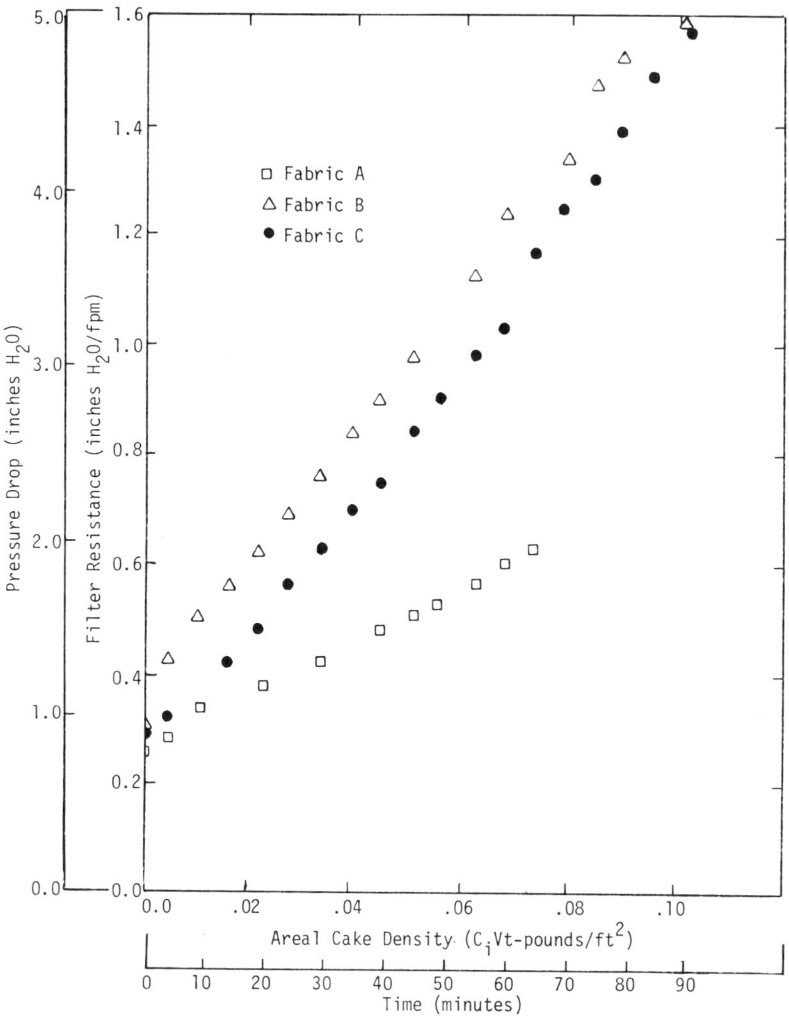

Figure 31.7. Conditioned performance curves for Fly ash #1 on the full-scale unit.

pressure drops, higher efficiencies (as inferred from the smaller number of particles penetrating the fabric), and lower purchase costs. Although fabrics A, B, and C were quite similar in performance, the experiments performed on the full-scale test unit indicated that fabric C had an efficiency greater than or equal to the other fabrics, while fabric A had pressure losses less than or equal to the other fabrics.

Based on the data obtained during the current study (Figure 31.7),

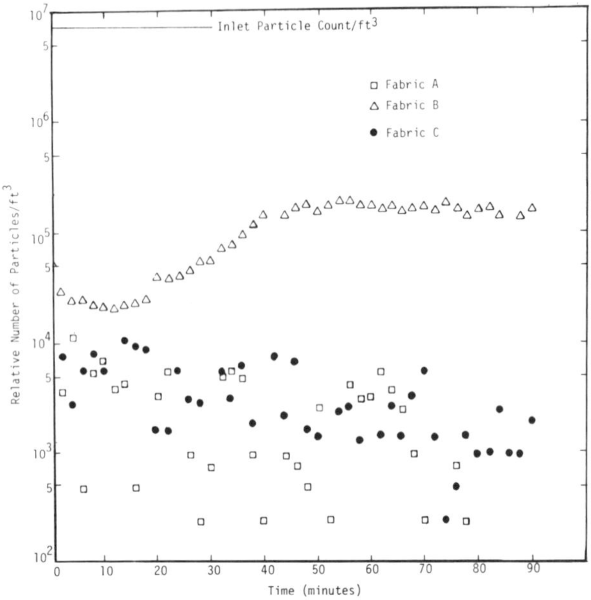

Figure 31.8. Number of 0.5–1.0 μ particles versus time for Fly ash #1 on the full-scale unit (new fabric).

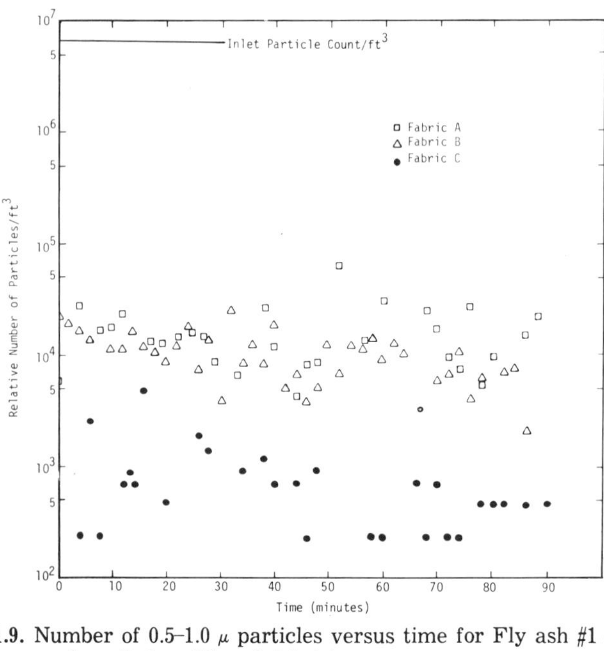

Figure 31.9. Number of 0.5–1.0 μ particles versus time for Fly ash #1 on the full-scale unit (conditioned fabric).

it was found that the pressure loss across the fabric filter at 50 minutes into the filtration cycle on the conditioned fabrics would range from 1.6 inches H_2O for fabric A to 3.1 inches H_2O for fabrics B and C for typical grain loadings of 2.5 gr/ft^3 and a filtration velocity of 3 ft/min (at 100°F). At temperatures typical of those encountered in industrial baghouses (~300°F), the pressure losses would range from 2.7 inches H_2O for fabric A to 5.3 inches H_2O for fabrics B and C (based on Equation 2). These values are similar to those reported above for operating field installations.

The results of this study have shown that the pilot plant fabric filter test collector can be used as a design tool and an aid in the prediction of the pressure drop of large-scale industrial fabric filter collectors as well as provide a means whereby the filtration performance can be evaluated for various fabrics leading to the choice of a fabric with the optimum operating characteristics of pressure loss and efficiency.

REFERENCES

1. Gosselin, A. E., Jr. "Pilot-Plant Investigation of the Bag Filterhouse for Control of Visible Stack Emissions from Oil-Fired Steam-Electric Generating Stations," *Proc. Amer. Power Conf.* **XXVI**, 128–137 (1964).
2. Gosselin, A. E., Jr. "The Bag Filterhouse for Oil-Fired Power Plants," *J. Air Poll. Control Assoc.* **15** (4), 179–180 (1965).
3. Bagwell, F. A., L. F. Cox and E. A. Pirsh. "Design and Operating Experience: A Filterhouse Installed on an Oil Fired Boiler," *J. Air Poll. Control Assoc.* **19** (No. 3), 149–154 (1969).
4. Felgar, D. N. and W. E. Ballard. "First Year's Experience with Full-Scale Filterhouse of Alamitos Generating Station," *Proc. Amer. Power Conf.* **XVIII,** 546–555 (1966).
5. Gosselin, A. E., Jr. and L. W. Lemon. "Bag Filterhouse Pilot Installation on a Coal-Fired Boiler — Preliminary Report and Objectives," *Proc. Amer Power Conf.* **XXVIII**, 534–545 (1966).
6. Borgwardt, R. H., R. E. Harrington and P. W. Spaite. "Filtration Characteristics of Fly Ash from a Pulverized Coal-Fired Power Plant," *J. Air Poll. Control Assoc.* **18** 387–390 (1968).
7. Adams, R. H. "Fabric Filters for Control of Power Plant Emissions," Paper No. 74-100, Presented at 67th Annual Meeting of the Air Pollution Control Association, Denver, Colorado, June, 1974.
8. Janoso, R. P. "Baghouse Dust Collectors on a Low Sulphur Coal-Fired Utility Boiler," Paper No. 74-101, Presented at the 67th Annual Meeting of the Air Pollution Control Association, Denver, Colorado, June, 1974.
9. McKenna, J. D. and R. Weisberg. "A Pilot Scale Investigation of Fabric Filtration as Applied to Coal-Fired Industrial Boilers," Presented at the Fourth Annual Industrial Air Pollution Control Conference, The University of Tennessee, Knoxville, March, 1974.
10. Carmen, P. C. "Fundamental Principles of Industrial Filtration (a Critical Review of Present Knowledge)," *Trans. Inst. Chem. Eng.* 168–188 (1938).
11. Fair, G. M. and L. P. Hatch. "Fundamental Factors Governing the Streamline Flow of Water Through Sand," *Trans. Am Water Wks. Assoc.* **25** 1551 (1933).

12. Williams, C. E., T. Hatch and Leonard Greenburg. "Determination of Cloth Area for Industrial Air Filters," *Heating, Piping, and Air Cond.* **12**, 259–263 (1940).
13. Snyder, C. A. and R. T. Pring. "Design Considerations in Filtration of Hot Gases," *Ind. Eng. Chem.* **47** (5) 960–966 (1955).
14. Spaite, P. W. and G. W. Walsh. "Effects of Fabric Structure on Filter Performance," *Amer. Ind. Hyg. Assoc. J.* **24**, 357–365 (1963).
15. Manual: Training Course 413: *Control of Particulate Emissions,* U.S. Environmental Protection Agency (June, 1973).
16. Durham, J. F. and R. E. Harrington. "Influence of Relative Humidity on Filtration Resistance and Efficiency," Presented at 63rd Annual Meeting of AICHE, December, 1970.
17. Billings, C. E. and J. Wilder. *Handbook of Fabric Filter Technology,* PB 200 648, Vol. I (Springhill, Virginia: National Technical Information Service).

32

The Techno-Economic Feasibility of Fabric Filtration as a Means to Control Fly Ash Emissions

John D. McKenna
Enviro-Systems & Research, Inc.
Roanoke, Virginia 24016

INTRODUCTION

This chapter reports on a techno-economic evaluation conducted from data obtained with a pilot plant installed on a slip stream of a coal-fired stoker boiler stack. Earlier studies indicated the critical parameters influencing pressure drop and outlet loadings. These early studies were mainly conducted on Nomex® felt.[1,2] A paper on the extension of the work on Nomex felt performance was also previously presented.[3] The data obtained on other filter media is presented in this chapter. These materials are Teflon® felt and Gore-Tex®, a PTFE laminate on a woven Nomex backing.

The fabric filter pilot plant unit employed is capable of handling 11,000 acfm of flue gas when operating at an air-to-cloth ratio of about six to one. This facility is installed on one of two stoker fed coal-fired boilers owned by Kerr Industries in Concord, North Carolina. The pilot plant and slip stream arrangement are shown in Figure 32.1. A schematic of the pilot plant is provided in Figure 32.2. The pilot unit contained a maximum of 2,480 ft^2 of cloth. It is divided into four cells, each containing a maximum of 54 bags. The bags are 5 in. in diameter and 8 ft 8 in. long. The bags are cleaned by closing off one cell at a time to the normal dust-laden gas flow and flushing the cell in the reverse flow direction with clean air.

Pilot plant tests were conducted on Teflon felt, Gore-Tex/Nomex, Dralon T felt and Nomex felt. Each set of bags were tested for par-

Figure 32.1. Fabric filter pilot plant installed at Kerr.

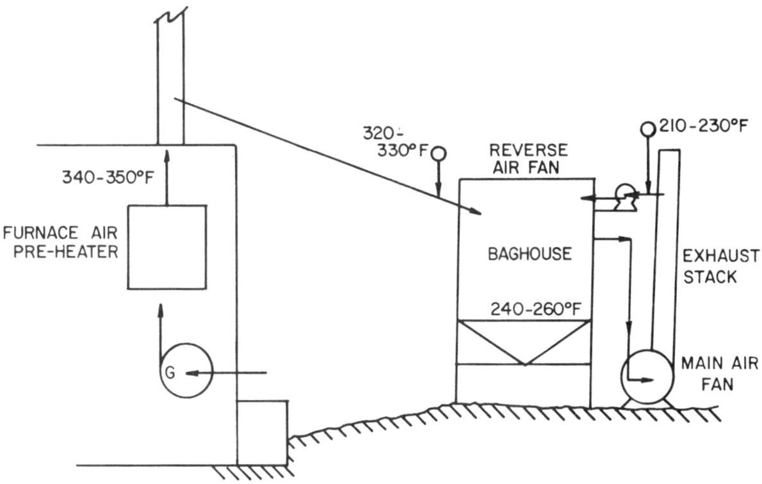

Figure 32.2. Pilot plant schematic diagram with temperature profile.

ticle size efficiency, pressure drop and clean-down profiles at three air-to-cloth levels. Outlet concentration data obtained at a number of air-to-cloth ratios for all of the filter cloths tested are given in Figure 32.3. Generally the Nomex felt provided the highest efficiency of removal. A comparison of operating pressure drop at various air-to-cloth ratios for the same four filter media is provided in Figure 32.4. Teflon felt has the lowest pressure drop curve while Nomex felt had the highest. Some of the more specific data obtained on Gore-Tex/Nomex and Teflon follow. Particle size analysis was obtained by the use of an Anderson *in situ* particle size analyzer (inertial impactor). Figure 32.5 is representation of the average inlet particle size distribution obtained via the Andersen sampler.

TEFLON FELT

Twenty Teflon felt bags, Style 2663, were tested in one cell. The Andersen sampler was utilized in obtaining *in situ* particle size data at air-to-cloth levels of 5.4 to 1, 8 to 1 and 14 to 1. Inlet flue gas volumes ranged between 1180 and 3260 acfm. Because of the limited amount of Teflon cloth on hand, difficulty was experienced in trying to operate at the lower air-to-cloth levels. To effect an air-to-cloth ratio of 5 to 1, the main fan was almost completely throttled. The reverse air volume was 3500 acfm; during outlet sampling the reverse air was not employed.

The particle size distribution for the individual runs was averaged

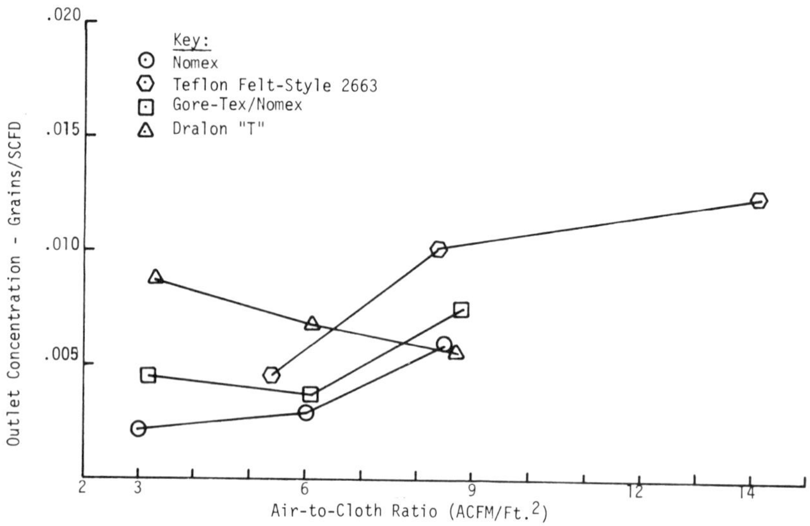

Figure 32.3. Outlet concentration vs. air-to-cloth ratio for various bag materials.

at each of the three levels of air-to-cloth. The comparison of the data is displayed graphically in Figure 32.6. The mass mean particle diameters were 1.4, 1.6 and 1.9 μ for air-to-cloth ratios of 5.4, 8 and 14 to 1 respectively.

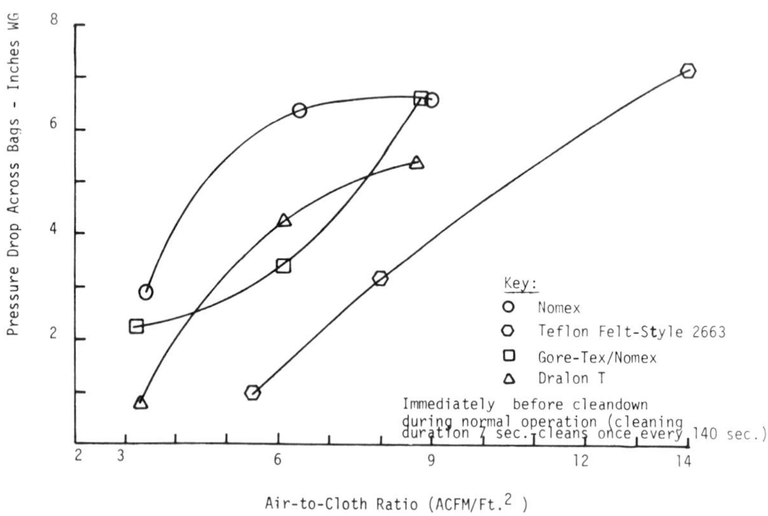

Figure 32.4. Comparison of operating pressures for various bag materials.

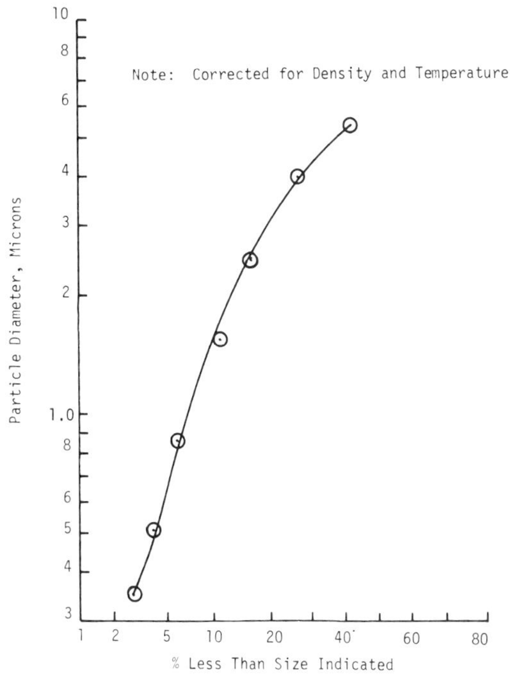

Figure 32.5. Inlet particle size distribution.

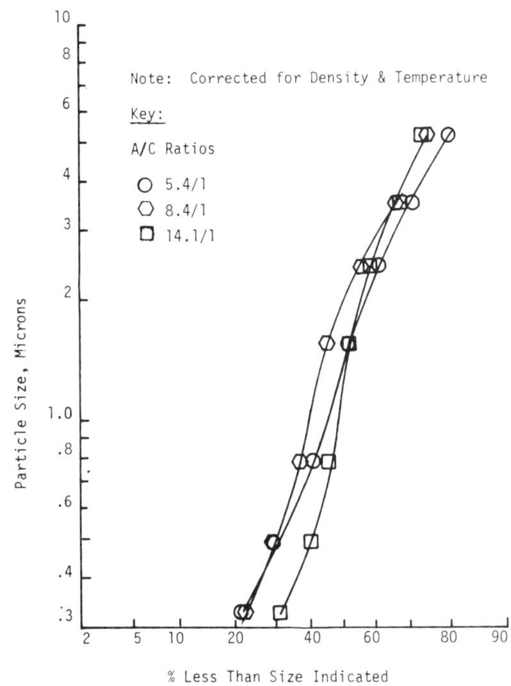

Figure 32.6. Outlet particle size distribution. Teflon felt-style 2663.

The size distribution curves indicate that about 56% was less than 3 μ at the lowest air-to-cloth, while 51% of the particulate, by weight, was less than 3 μ at 14 to 1.

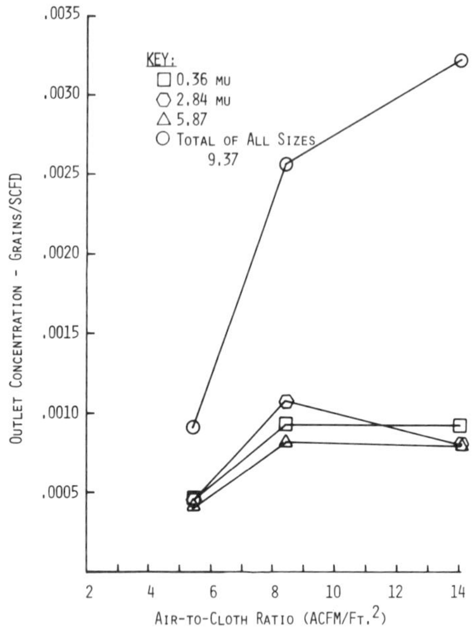

Figure 32.7. Outlet concentration by particle size vs. air-to-cloth ratio.

Average outlet concentrations are shown in Figure 32.7. It can be seen here that the largest particle size fraction is most sensitive to increases in the velocity or air-to-cloth ratio. Generally an increase in air-to-cloth ratio resulted in an increase in outloadings. For the smaller size fractions, the curve does appear to flatten above an air-to-cloth ratio of about 8. This is in general agreement with the data previously presented on Nomex felt.

Figure 32.8 shows penetration or 1-collection efficiency versus particle diameter. Three curves are shown, one for each of the three levels of air-to-cloth ratio. All three show the same general trend, with the curve sloping downward to the right indicating less penetration of the larger particles. Two of the curves indicate some leveling or decrease of penetration for the very small fractions. In the case of Nomex felt, this improved collection of the finest fractions was also present and more pronounced.

Pressure drop versus air-to-cloth data is presented in Figure 32.9 while examples of typical cleandown cycles are shown in Figure 32.10.

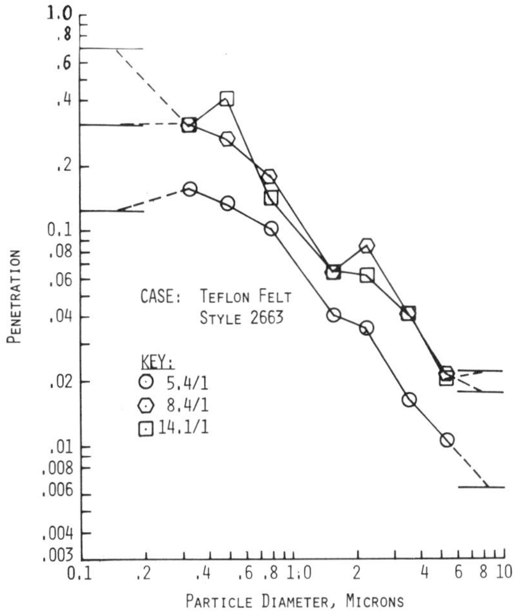

Figure 32.8. Penetration vs. particle diameter.

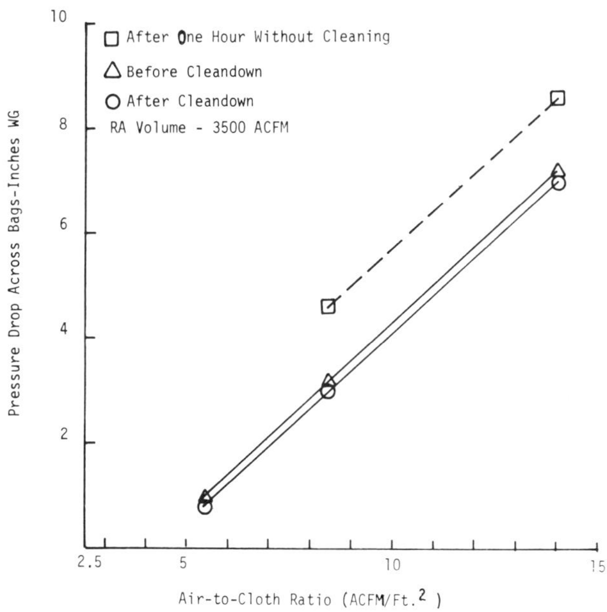

Figure 32.9. Pressure drop across bags vs. air-to-cloth ratio. Teflon felt—style 2663.

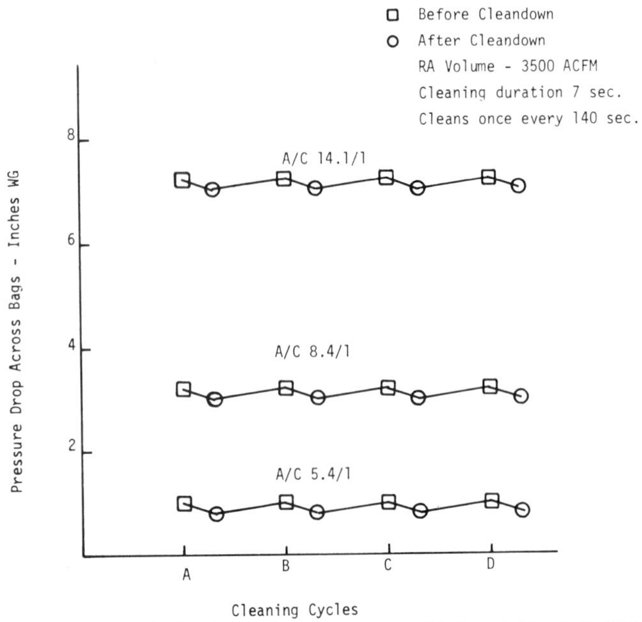

Figure 32.10. Typical cleaning cycles. Teflon felt—style 2663.

GORE-TEX/NOMEX

Thirty Gore-Tex/Nomex bags were placed in one cell. Inlet flue gas volumes ranged between 1000 and 3100 acfm and *in situ* particle size data was obtained at air-to-cloth levels of 3 to 1, 6 to 1 and 9 to 1. The reverse air fan was off during testing periods and no testing was carried out during boiler grate pulling operations.

The outlet particle size distribution for the individual runs was averaged at each of the three levels of air-to-cloth. The comparison of this data is shown in Figure 32.11 and indicates that about 46% of the particulate by weight was less than 2 μ at the lowest air-to-cloth, while the two other air-to-cloth levels showed about 37% of the particulate, by weight was less than 2 μ.

The mass mean particle diameters were 2.2, 3.0 and 2.8 μ for air-to-cloth ratios of 3, 6 and 9 to 1 respectively.

The performance of the Gore-Tex on submicron particles seems essentially the same at the three levels of air-to-cloth.

Outlet concentration versus air-to-cloth ratio for four particle size fractions are plotted in Figure 32.12. Like the Teflon felt case, the largest particle size fraction, *i.e.*, the total of all sizes greater than 9.55 μ, are most sensitive to increases in air-to-cloth ratio and an increase in air-to-cloth results in an increase in the outlet concentration.

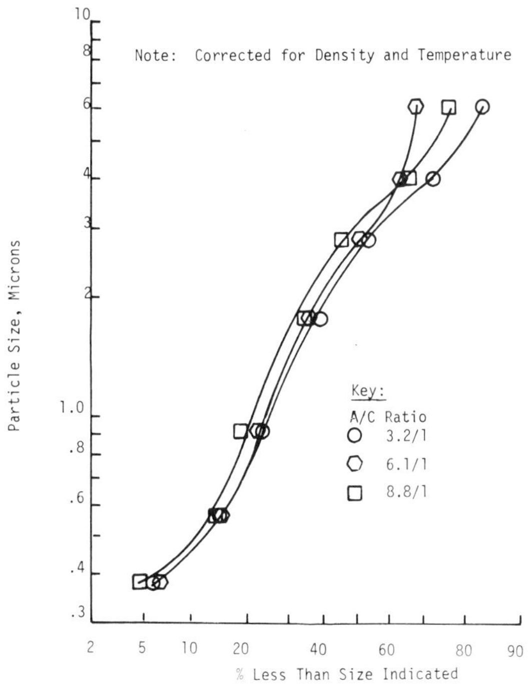

Figure 32.11 Outlet particle size distribution. Gore-Tex/Nomex.

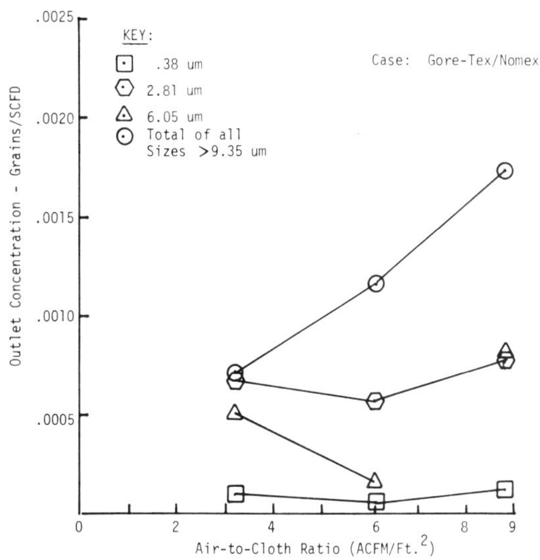

Figure 32.12. Outlet concentration by particle size vs. air-to-cloth ratio.

Figure 32.13 is penetration versus particle size. Three curves are shown, one for each of the three air-to-cloth ratios. These curves indicate an increase in penetration as the fractions become smaller from 10μ down to about ½ μ. Below ½ μ there is a sharp decrease in penetration for all three air-to-cloth ratios.

Pressure drop versus air-to-cloth ratio curves are provided in Figure 32.14. Typical cleaning cycles are shown in Figure 32.15.

ECONOMICS

The economics of applying fabric filters to industrial coal-fired boilers were evaluated, and comparative costs for an electrostatic precipitator (ESP) were developed. Installed costs, flange-to-flange hardware costs plus installation costs, were determined for a fabric filter dust collector sized for 70,000 acfm at 350° F. This was done for five fabric materials: Nomex felt, Gore-Tex on a Nomex woven backing, Gore-Tex on a Gore-Tex woven backing, Dralon T® and Teflon felt. Air-to-cloth (A/C) ratios considered for each fabric were 2.9, 5.8, 8.0 and 11.3. Operating and annualized (total cost of control) costs were also determined for each case.

Figure 32.16 is a plot of installed cost vs. air-to-cloth ratio for the case of Gore-Tex/Nomex. Between the ratios of 3 to 1 and 6 to 1, the installed cost came down rapidly due to the large influence of reduced amount of cloth required and the corresponding bag cost reduction. Above 6 to 1, the decrease in installed cost is not as rapid due to the greater importance of other relatively fixed cost such as installation. Similarly, Figure 32.17 shows the installed costs vs. air-to-cloth ratio for the case of Teflon felt bags.

A graphic comparison of the installed costs for the five bag materials is made in Figure 32.18. Teflon felt is seen to be the most expensive cost for all four air-to-cloth ratios investigated, and Dralon T is seen to be the least expensive. The curves draw closer together as the air-to-cloth ratio increases. This is due to a decreasing percentage of the total costs attributed to the bags as the size of the house decreases. A plot of installed cost versus dust removal efficiency for an electrostatic precipitator to handle the same case is given in Figure 32.19. A rapid increase in cost is seen in going from 90–99% efficiency.

Operating costs were determined for the five bag materials under study. These costs are based on 25% bag replacement per year and the actual pressure drops observed for each fabric at the different air-to-cloth ratios. These costs are represented in Figure 32.20. The most expensive case of $44,983/yr was Teflon felt at an air-to-cloth ratio of 2.9 to 1; while the least expensive case was Dralon T with an annual cost of $9,584 at 2.9 to 1.

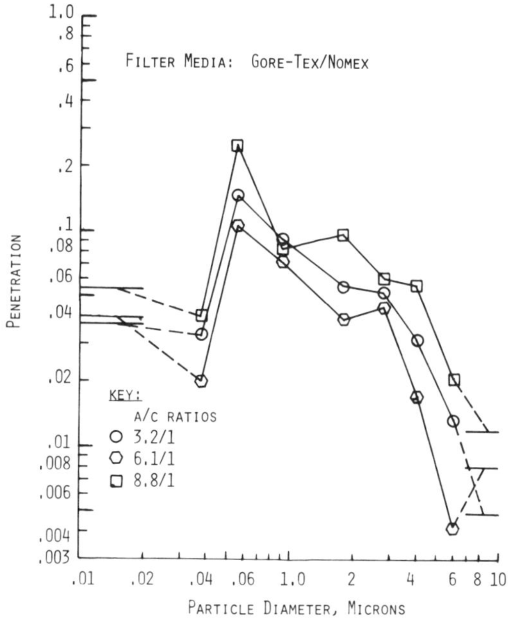

Figure 32.13. Penetration vs. particle diameter.

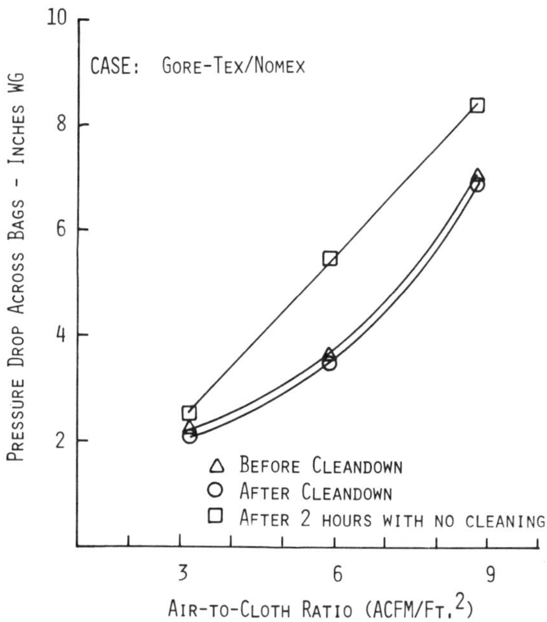

Figure 32.14. Pressure drop across bags vs. air-to-cloth ratio.

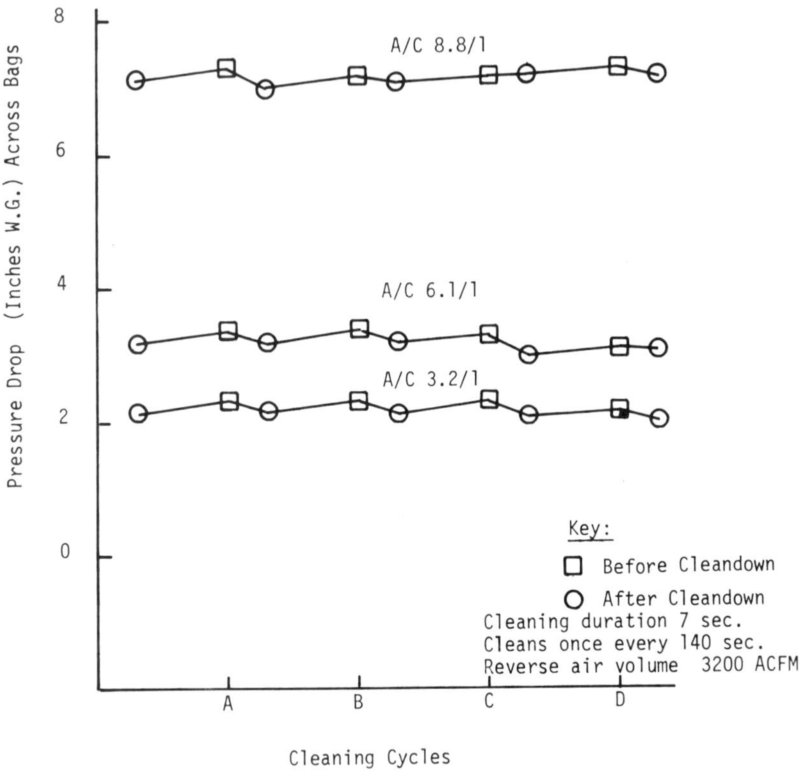

Figure 32.15. Typical cleaning cycles. Gore-Tex/Nomex.

Since no extended bag life data was obtained in this program, it was necessary to look at the operating costs for different bag life assumptions. Such a review for Teflon felt is displayed graphically in Figure 32.21. In this figure, annual operating costs are plotted versus air-to-cloth ratio with a family of curves which correspond to various bag life assumptions. Operating costs for the electrostatic precipitator are given in Figure 32.22 where a curve for costs versus efficiency is plotted.

The annualized costs or total costs of control were developed from the preceding installed and operating costs. These results, shown in Figure 32.23, were based on the following assumptions: First, hardware and installation costs are depreciated over 15 years. Second, the straight line method of depreciation (6 2/3% per year) is used. This method has the simplicity of a constant annual write-off. Third, other costs called capital charges, which include interest, taxes, insurance and other miscellaneous costs, are assumed equal to the amount of depreciation, or 6 2/3% of the initial installed cost. There-

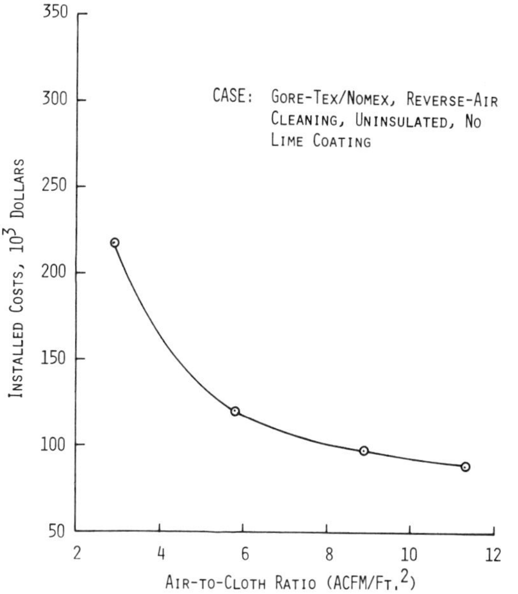

Figure 32.16. Installed costs vs. air-to-cloth ratio.

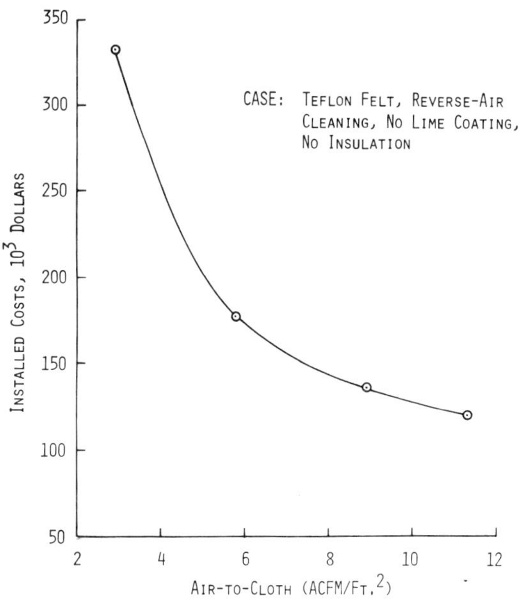

Figure 32.17. Installed costs vs. air-to-cloth ratio.

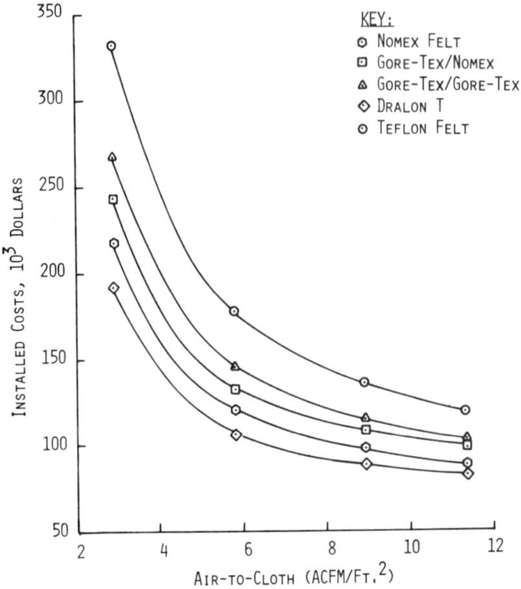

Figure 32.18. Installed costs vs. air-to-cloth ratio.

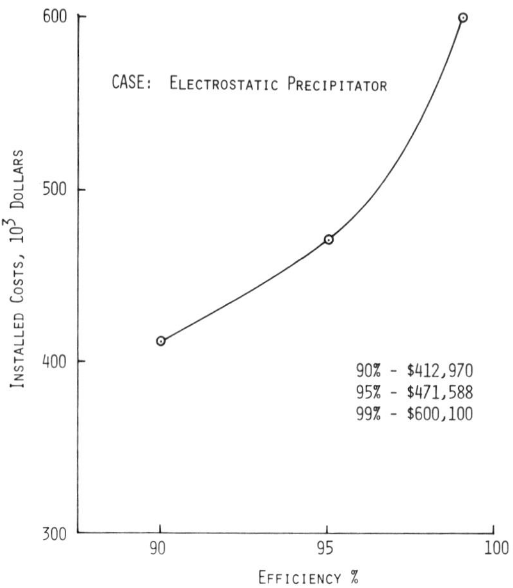

Figure 32.19. Installed cost vs. efficiency.

fore, depreciation plus these other annual charges amount to 13 1/3% of the installed costs.

The annualized costs for the electrostatic precipitation at three efficiency levels are provided in Figure 32.24. The annualized cost increased from $60,770 to $87,960 when the efficiency went from 90–99%.

Figure 32.25, provides a comparison of the annualized costs for the Dralon T and Teflon filter media cases with the three efficiency cases studied for the electrostatic precipitator. The two filter media chosen for this comparison were selected because they represent the least and most expensive filter media cases. Even the most expensive case, Teflon felt, compares very well with the electrostatic precipitator for all three efficiencies. It should be noted that extended bag life was assumed for this comparison.

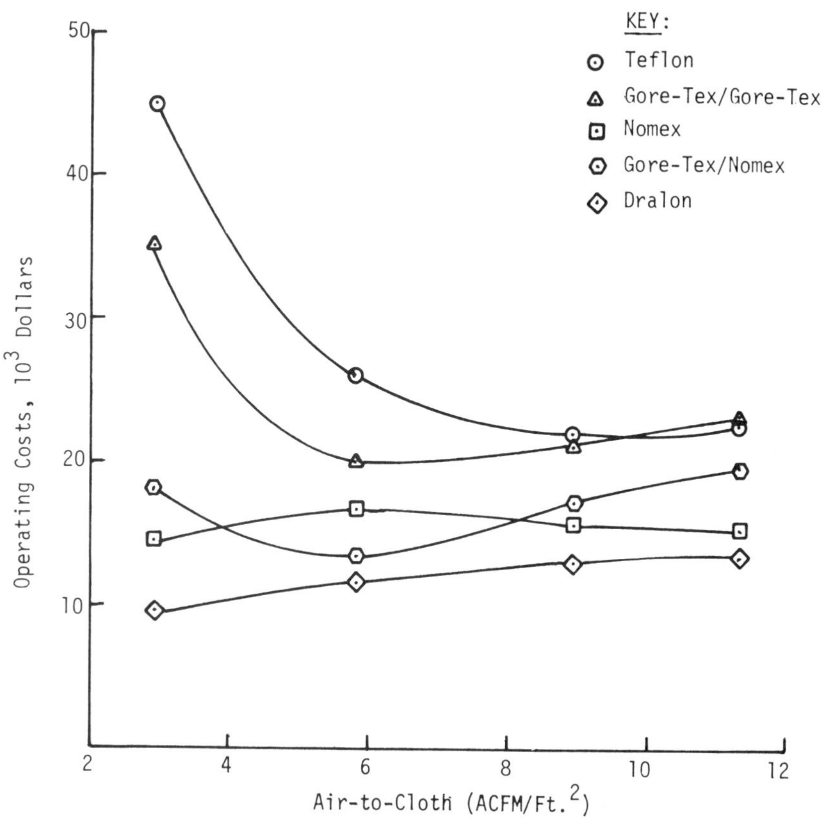

Figure 32.20. Operating costs vs. air-to-cloth ratio.

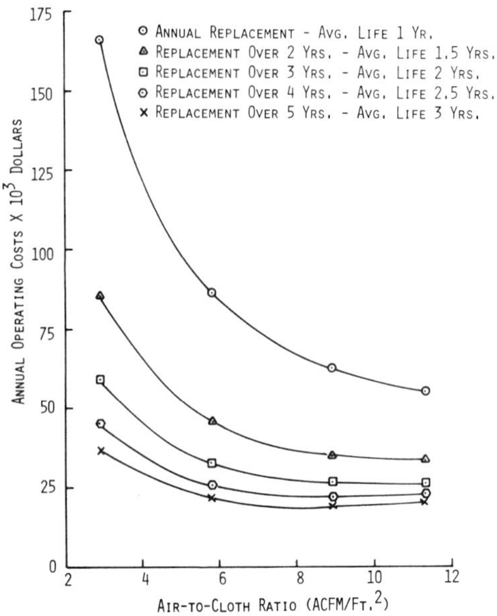

Figure 32.21. Teflon felt. Annual operating cost vs. air-to-cloth ratio for different bag life assumptions.

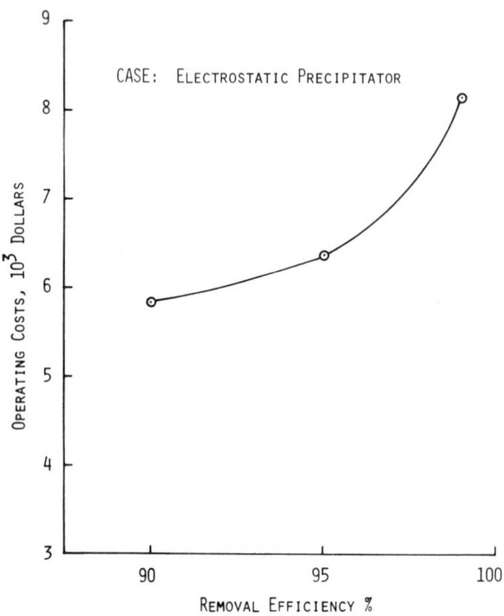

Figure 32.22. Operating costs vs. efficiency.

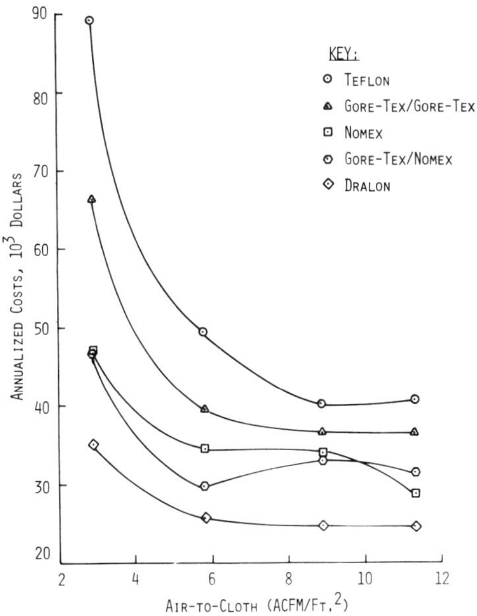

Figure 32.23. Annualized costs vs. air-cloth ratio.

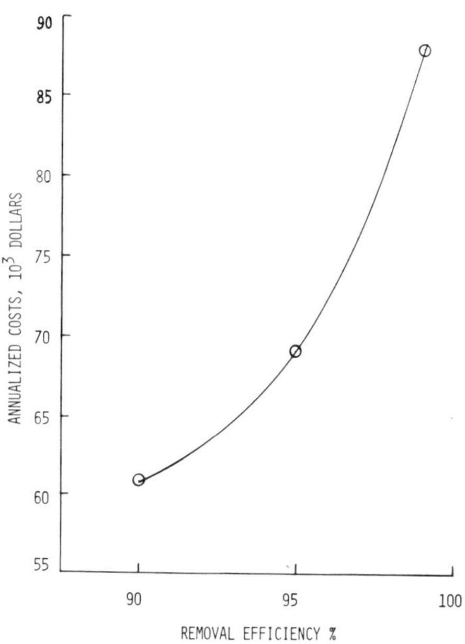

Figure 32.24. Annualized costs vs. efficiency.

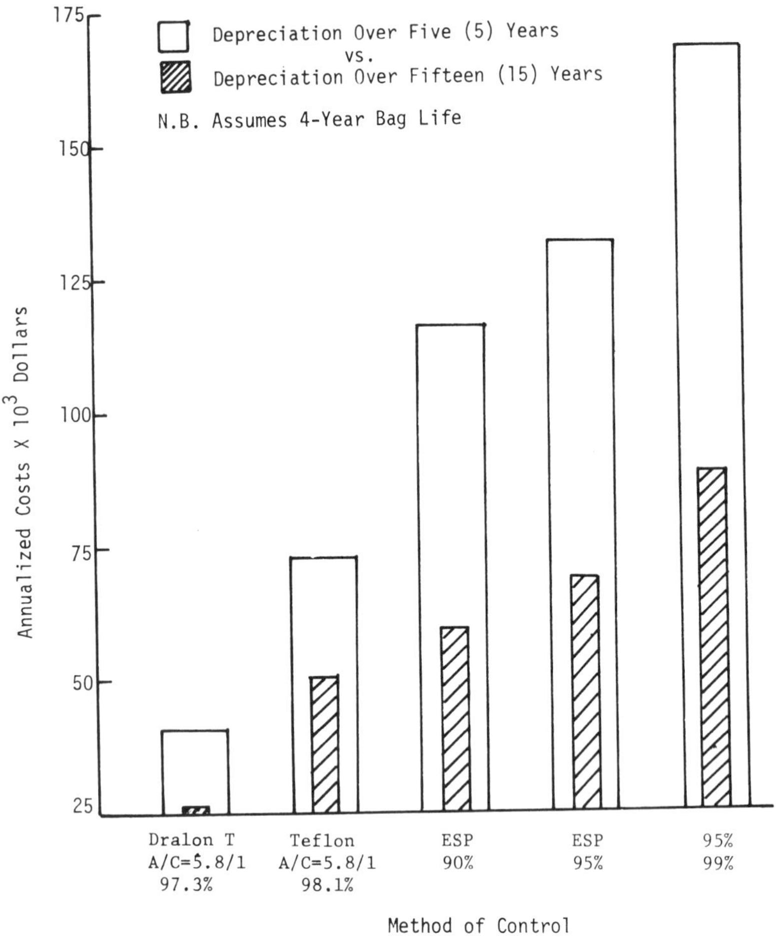

Figure 32.25. Annualized cost comparison.

CONCLUSIONS

Fabric filter dust collectors are suitable for control of fly ash from stoker-fed coal-fired industrial boilers in terms of both efficiency of dust removal and operating pressure drop. The main question yet unanswered is what bag life is achievable with continuous service. If two-year bag life can be achieved with Teflon felt bags, fabric filters appear economically more attractive for industrial coal-fired boilers than electrostatic precipitators.

Filter media selection influences both efficiency of removal and operating pressure drop. Test results indicate that a laminate on a woven backing can achieve performance and operating characteristics similar to those of a felt media.

The air-to-cloth ratio of superficial velocity is an important parameter influencing the outlet loading, for the type media tested. As the velocity is increased the overall outlet loading is increased. Different particle size fractions are affected differently by changes in velocity. Different filter media are influenced somewhat differently by velocity but those tested showed many similarities.

The question of bag life is the primary unanswered techno-economic question. It is anticipated that this question will be answered via a full-scale demonstration program.

ACKNOWLEDGMENTS

This program was jointly sponsored by The Federal Environmental Protection Agency, Kerr Industries and Enviro-Systems & Research, Inc. The author wishes to express his appreciation for the contributions of the foregoing as well as those of W. L. Gore & Associates and E. I. duPont.

This project has been funded at least in part with Federal funds from the Environmental Protection Agency under Contract Number 68-02-1093. The content of this publication does not necessarily reflect the views or policies of the U.S. Environmental Protection Agency, nor does mention of trade names, commercial products or organizations imply endorsement by the U.S. Government.

REFERENCES

1. McKenna, J. D. and R. Weisberg. "A Pilot Scale Investigation of Fabric Filtration as Applied to Coal Fired Industrial Boilers," *Proc. Fourth Annual Ind. Air Poll. Cont. Conf.*, University of Tennessee, Knoxville, March 28, 1974.
2. McKenna, J. D. "Applying Fabric Filtration to Coal-Fired Industrial Boilers," Environmental Protection Technology Series, July 1974, EPA-650/2-74-058.
3. McKenna, J. D., W. O. Lipscomb and J. C. Mycock. "A Pilot Study of Various Filter Media in the Control of Fly Ash Emissions," Presented at the Second National Conference on Energy and the Environment in Hueston Woods, Ohio, November 14, 1974.

33

Fabric Filter Design: Economics of Design, Operation and Maintenance

R. Filipczak
Pollution Control Division
Carborundum Environmental Systems, Inc., Knoxville, Tennessee

BACKGROUND

In the past, pollution control equipment utilizing textiles as filters was developed in an atmosphere that lacked the element of necessity. With few exceptions the purchase, use and maintenance of this equipment was not essential. During this period people watched ball games or birds instead of stacks, and an occasional dirty stack was more of a comfort than a catastrophe.

In general early fabric filter equipment controlled nuisance dust and was classified as nonproductive. This classification resulted in a traditional "least-first-cost" analysis for equipment purchase.

Basic research into product performance could not be justified, and developments were directed at proprietary changes that increased marketability instead of fabric filter technology. The least-first-cost concept reached into the engineering of the equipment without regard for functionality. For instance, the diameter of the textile filter tubes widely used in the industry was not established based on theoretical or experimental performance optimization; rather, this diameter was (and still is) used because it furnished 99% utilization of a textile loom width that was common and readily available.

Human engineering was also a victim of the least-first-cost tradition. While production equipment was being designed for higher efficiency, greater reliability and less maintenance, nonproductive fabric filter collectors were being forced into questionable reliability and almost impossible maintenance in the interest of the least-first-cost approach. Access to the filter equipment was designed for small, square midgets. Aisle widths were reduced for increased cloth area and, as a result, would accommodate only very thin men. Bag lengths are commonly over twice a man's reach and yet they still must be worked on at the top. Even when a maintenance man has a method of reaching the top of the bag, there is trouble. The clearance between the bag top and the roof has been reduced to a minimum. Since there is approximately 12 inches from a man's shoulder to the top of his head, it would be advantageous if men working on this equipment had heads that were countersunk about six inches. There are other areas in today's equipment that need a touch of human engineering to accommodate normal men.

EMPHASIS ON RELIABILITY

In December 1970, the Clean Air Act became law. It is now possible for enforcing agencies to issue "cease and desist" orders for noncompliance with air standards. These orders can shut down an operation or an entire plant. If equipment fails or must be shut down for maintenance the agency must be notified, and a temporary variance agreement must be issued. The Clean Air Act has supplied a very large dose of necessity to the design of air pollution control equipment.

Compliance with air standards is not a one-day, one-time, one-test condition. Air standards are continuous requirements that require continuous and effective operation of control equipment. Reliability is now an important factor. Ease of maintenance under normal operating conditions has thus become an essential feature. As such, when pollution control equipment is no longer classified as nuisance equipment but as part of a process scheme, its design must be reconsidered.

If production can be curtailed or stopped by failure of pollution control equipment, the reliability of the control equipment must be at least comparable to all other production equipment, and the reliability of the equipment becomes a factor of productivity. On the same basis, evaluation may indicate that pollution control equipment on secondary operations must be considered a necessary utility. In either case, it is a long stride forward from a "nonproduction" or "nuisance" classification. The air pollution control industry has the capability of designing and building better equipment right now. No major change in the state of the art would be required. All that is needed is to realize that the user of such equipment wants, and will pay for, reliable and efficient equipment.

DESIGN FEATURES FOR OPERATION AND MAINTENANCE

The step between wanting better equipment and purchasing better equipment requires a cost analysis. There are many ways to analyze the cost of owning and operating equipment. For this discussion, the total cost of operation is considered assuming a 15-year life for a fabric filter collector. The total cost of operation includes the installed cost, the cost of capital and the annual operating cost. Billings and Wilder[1] have published costs for typical fabric filter collectors, and these will be used. Because these numbers were developed in 1969, they have been converted to percentages which still should be proportionate to today's costs with the possible exception of electric power. The costs taken from the Billings and Wilder study are presented in Table 33.1 and are further detailed and listed in order of magnitude in Table 33.2. Using the results of Table 33.2, a brief trade-off study can be performed in which increasing the cost of one category by placing more attention and emphasis on it may reduce the total cost of operation while at the same time improve the reliability and ease of maintenance of the equipment. The following paragraphs will illustrate this trade-off using fabric filter replacement costs associated with three- or four-bag-reach collectors.

As seen in Table 33.2 the third largest contribution to the total cost of operation is the material and labor involved with the textile filter. Consider a fabric filter collector designed with four rows of 8-inch diameter filters on each side of a center aisle. A maintenance man without the special physical asset of extremely long arms cannot

Table 33.1. Fabric Filter Collector Distribution of Annualized Cost of Operation.[a]

Item		*Percent*
Installed equipment cost		
Fabric filter	5	
Fans and ducts	6	
Other	4	
	—	15
Annual Cost		
Power	11	
Labor	29	
Replacement fabric	10	
Plant overhead	24	
	—	74
Cost of capital		11
TOTAL		100

[a]Percentage of the cost $/CFM-year.

Table 33.2. Fabric Filter Collector Annualized Cost of Operation Arranged by Magnitude.[a]

Item		*Percent*
General maintenance and dust disposal labor (excluding labor associated with bag replacement)		27.8
Plant overhead (space, heat, insurance, etc.)		23.8
Fabric filter replacement		
Labor	10.8	
Filter material	9.5	20.3
Interest on unamortized part of installed cost at 10%		11.4
Electric power		11.4
FOB baghouse		5.2
Ducting		4.2
Foundation and erection		1.8
Fan and motor		1.6
Startup		0.6
Disposal equipment		0.6
Engineering-design-planning		0.6
Freight		0.3
Instrumentation		0.3
TOTAL		100

[a]Percentage, $/CFM-year.

reach to work on the fourth row. The terminology "reach to work" is worth repeating because maintenance requires the ability not only to touch but also to perform work motions. This design requires the removal of the first row of filters to change a failed filter in the fourth row. This is not an assumption for this discussion. Any maintenance man will be glad to explain the problems of reaching distant rows of filter units in almost any equipment.

The probability of failure of a fourth row filter (or for that matter any filter) is one in four if you consider any single failure as an independent event with equal chance of occurrence. For simplicity it is assumed that once a man can get to a failed filter, an equal number of work units is required for replacement. The three-bag-reach design, in which each filter can be worked on independently, would generate one work unit per filter failure. The four-bag-reach, which requires first row removal for access to the fourth row, requires five labor units to change four failed filters, or, on the average, 1.25 work units per filter replacement. This represents an increase of 25% in the labor required to change a failed filter in a four-bag-reach as opposed to a three-bag-reach design.

The Pollution Control Division of Carborundum Environmental Systems, Inc. and The University of Tennessee have developed computer programs that design and cost structural fabric filter collectors. Using these programs it has been determined that a collector with a three-bag-reach has an initial cost that is approximately 10% higher than one with a four-bag-reach. This figure is not exact for every design but it is typical.

If the 25% decrease in fabric filter replacement labor cost and the 10% increase in equipment cost are considered in the cost breakdown of Table 33.2, the results are enlightening. The labor to change bags is reduced to 8.1% from 10.8% but the 10% increase in the collector cost represents an increase in the total cost of operation of only about 0.5%. Thus, this trade-off of a more costly baghouse for a lower filter replacement labor cost results in a net decrease in the total cost of operation of 2.2%. This reduction is significant in itself, and if consideration is given to the increasing reluctance on the part of maintenance men to work in and on fabric filter equipment, it becomes more significant. Proper filter installation, which affects the incidence of premature failure and total filter life, while not quantifiable, is an important consideration. There is a higher probability that proper filter installation will result in a three-rather than four-bag-reach design; consequently, filter material costs (9.5% of the total cost of operation in Table 33.2) might also be reduced.

The above is an approximate cost trade-off analysis between initial collector cost and fabric filter replacement cost and was selected for the sake of simplicity. There are many other design factors that would benefit from an evaluation of annualized cost as opposed to a least-first-cost provided that ease of maintenance and reliability are accepted as evaluation factors. Access openings, aisles and the critical attachment of the fabric filter to metal equipment, each affect the labor involved in filter maintenance.

Instrumentation, which presently represents only 0.3% of the total cost of operation, should be increased so that failure of critical parts can be detected and/or anticipated. This would increase filter life and the overall reliability of the equipment. Again, the least-first-cost approach should be weighed against a total cost of operation evaluation.

CONCLUSION

To optimize the equipment now available, it is essential that reliability and ease of maintenance be recognized as contributors to initial equipment purchase cost. If the end user asks for and will pay for better equipment, it can be made available. Good cost analysis and wisely spent dollars can reinstate human engineering in the equipment to improve the ease of maintenance. Less emphasis should be placed on the original installed cost and more attention should be directed to the total cost of operation, a significant portion of which is maintenance and repair. Paying slightly more initially for reliability and maintenance features will prove to be more economical in the long run.

REFERENCE

1. Billings, C. E. and J. E. Wilder. *Handbook of Fabric Filter Technology,* GCA Corporation, prepared for HEW Contract CPA 22-69-38, Chapter 7, Table 7.3.

34

Fabric Filter Cost Optimization

Jerry W. Crowder
Knox County Department of Air Pollution Control
Knoxville, Tennessee

INTRODUCTION

While methods for the estimation of the cost of fabric filter systems have been available for a number of years, they have generally taken the form of a total system cost expressed as a dollars/cfm or dollars/ft^2 factor.[1,2] Only in more recent years[3-5] have models been presented where the costs of major subcomponents are related directly to the key factors of filter surface area, fabric type, pressure drop, flow rate and fabric filter performance. While such models represent significant improvements in the state-of-the-art of cost estimation, they do not generally provide a means whereby the optimum fabric filter system, in terms of cost and performance, can be selected. In this chapter we will present a proposed modeling scheme for optimizing the selection of a fabric filter system, and applications of this modeling scheme will be given based on the use of available cost estimation relationships.

VARIABLE DESCRIPTIONS

The variables affecting the cost of a fabric filter system may be categorized as either fixed, capital or operating expenses. Fixed expenses are considered to be those items whose costs are not directly related to classical operating parameters and include

1. Engineering studies (FA)—includes the engineering costs for a) analysis of the industrial process, b) determination of the best type of control device, c) sizing, costing and designing of the device, and d) supervision of construction.
2. Dust disposal and/or product recovery (FB)—a function of the decisions made during the engineering study and of the device

type and size chosen. Disposal itself is a function of the quantity of dust collected and the method of disposal chosen.

3. Investment return (FC)—a function of the efficiency of the optimization scheme employed and of the level (if any) of product recovery.
4. Taxes and insurance (FD)—a function of the capital cost, the process involved, the operational procedures, and the federal, state and local tax structures.

Capital expenses are a function of the total initial investment less the salvage value and include

1. Cost of device (CA)—the flange-to-flange cost FOB at the location where it was constructed. This cost can reflect either assembled units, bolt-together units or structural baghouse components.
2. Cost of appurtanences (CB)—includes gas conditioning equipment, fans, motors, ducts, conveyers, etc., FOB at the factory.
3. Cost of land (CC)—the cost of the land to be occupied by the filter system. This cost should be considered both from the standpoint of availability and value. It should also be recognized that the appreciation rate of the land may be different than the money rate over the depreciation period.
4. Cost of structures (CD)—includes cost of structures to support CA and CB. This may be only the duct supports or, in the case of the structural baghouse, may include the device housing.
5. Cost of shipping, installation and start-up (CE)—a function of CA and CB for shipping and installation and a function of the process for start-up. Labor costs for various specialized skills must be considered.
6. Salvage value (CF)—value of the device at the end of the depreciation period. Includes land value plus material value less removal cost.

Operating expenses include

1. Cost of power (OA)—a major operational cost that should be analyzed for the entire depreciation period. The cost of power should be considered to increase at a rate different from the value of the interest rate, since the current energy situation indicates that trend.
2. Cost of gas conditioning (OB)—a trade-off function between bag-life and maintenance cost. The conditioning cost versus any derived benefit must be evaluated.
3. Cost of bag replacement (OC)—a trade-off of benefit versus cost between gas-conditioning, filtering velocity and pressure drop.
4. Cost of operation and maintenance (OD)—related to the benefit versus cost analysis of OC and to the cost of labor at the required level of skill.

A summary of the functional dependency of each of the above variables is given in Table 34.1.

Table 34.1. Functional Dependency of Model Variables.

Variable	*Definition*	*Functional Dependency*
FA	Cost of engineering studies	Capital cost
FB	Cost of dust disposal and/or product recovery	Device type and size, dust characteristics
FC	Investment return	Device type and size, dust characteristics
FD	Cost of taxes and insurance	Capital cost, tax structure, process, operational procedures
CA	Cost of device	Flow rate, filtering velocity, pressure drop, dust and process characteristics, device type
CB	Cost of appurtanences	Flow rate, filtering velocity, pressure drop, dust and process characteristics, device type
CC	Cost of land	Flow rate, filtering velocity, pressure drop, device type
CD	Cost of structures	Flow rate, filtering velocity, pressure drop, device type
CE	Cost of shipping and installation	Flow rate, filtering velocity, pressure drop, device type
CF	Salvage value	Capital cost, land value, labor cost
OA	Cost of power	Flow rate, filtering velocity, pressure drop
OB	Cost of gas conditioning	Flow rate, filtering velocity, pressure drop, process
OC	Cost of bag replacement	Flow rate, filtering velocity, pressure drop, process
OD	Cost of operation and maintenance	Flow rate, filtering velocity, pressure drop, labor cost

MODELING SCHEME

If we assume that the engineering studies are capitalized, we may write the total annual cost (TAC) of the fabric filter system as

$$\begin{aligned} \text{TAC} = {} & [(\text{CA}+\text{CB}+\text{CC}+\text{CD}-\text{CF}+\text{FA})\ \text{ACF}] \\ & + [\text{FB}-\text{FC}+\text{FD}] + [\text{OA}+\text{OB}+\text{OC}+\text{OD}] \end{aligned} \tag{1}$$

where ACF is the annual cost factor and is a function of the value of money and the depreciation period and method used by the industry involved. The total annual cost includes all aspects of the control device cost for a year of operation at a given interest rate and over a given depreciation period.

Van Note and Weems[6] have discussed the numerical evaluation of filtration equipment costs in terms of the parameters given in Figure 34.1 and concluded that optimum cost is obtained when (1−A) +

(2–A) + (3–A) + (4–A) is at a minimum. Thus if the cost = $f^{1\text{-}4}$ minimum cost is obtained when

$$\frac{d\$}{d[f(1,2,3,4)]} = 0 \tag{2}$$

An anlysis of the variables involved in the fabric filter system annual cost shows that each of them is a function, either directly or indirectly, of both the filtering velocity (a function of the flow and the fabric surface area) and the pressure drop. It can be further generalized that the annual cost varies directly with the pressure drop and inversely with filtering velocity. Thus, if expressions can be developed relating cost to the pressure drop/filtering velocity ratio, minimum cost will be obtained when

$$\frac{d\$}{d(\Delta P/V)} = 0 \tag{3}$$

The pressure drop/filtering velocity ratio at this minimum cost sets the optimum operating conditions of the fabric filter system. A numerical evaluation scheme for determining the optimum operating conditions for the optimum type of fabric filter is given in Figure 34.2.

MODEL APPLICATION

The difficulty in applying the above model lies in the inability to obtain data of sufficient detail to allow development of the necessary

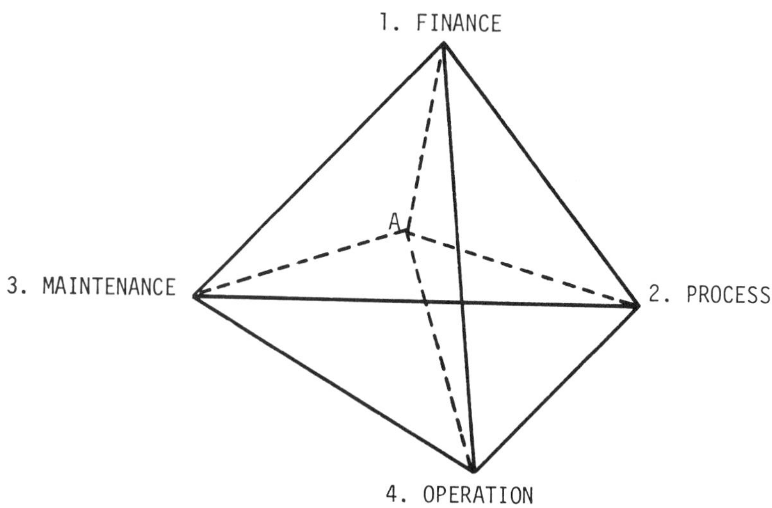

Figure 34.1. Filtration equipment cost parameters.

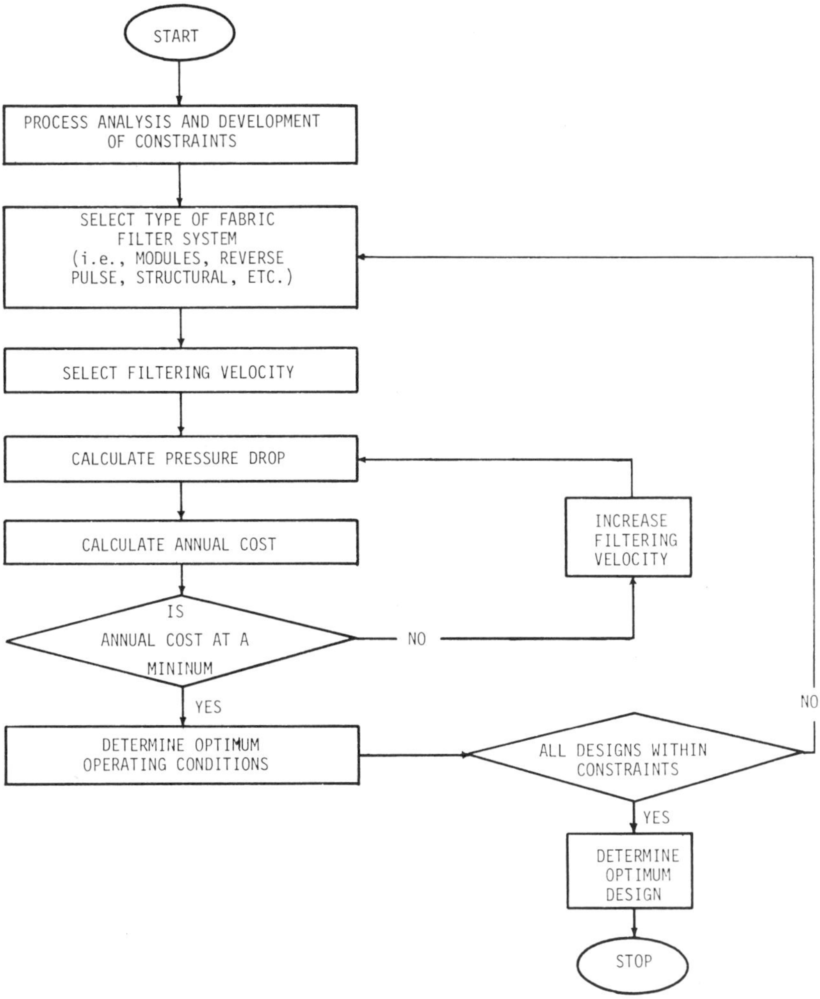

Figure 34.2. Modeling scheme.

relationships. In the absence of such information, available general equations may be used.

Following Fraser and Foley,[3] we may write the total annual cost as

$$TAC = [CA+CB+CC+CD)ACF] + [OA+OB+OC] \tag{4}$$

where the variables and their functions are given in Table 34.2. While the use of such relationships reflects a considerable simplification of the actual functional dependencies, its application to the proposed modeling scheme can provide useful information on cost trends.

Table 34.2. Definition of Model Variables.

Variables	*Definition*	*Function*[a]
CA	Cost of device and support less bags	$\$ = 0.802A^{1.074} - 0.4A$
CB	Cost of bags	$\$ = AC_B$
CC	Cost of appurtances	$\$ = 0.32Q^{1.018} + 6.39A^{0.911}$
CD	Cost of erection and installation	$\$ = 327.0A^{0.717}$
ACF	Annual cost factor	0.07
OA	Cost of power	$\$/yr = 0.0002125hQ\Delta PE$
OB	Cost of bag replacement	$\$/yr = AC_B/J$
OC	Cost of operational labor and maintenance	$/yr = 0.05 (Capital Cost)

[a]A = fabric area (ft^2), Q = flow rate (cfm), C = bag cost (\$/ft^2)
h = operating hr/yr, ΔP = total pressure drop (in H_2O),
E = electricity cost (\$/kwh), J = bag life (years)

Using the assumed systems characteristics given in Table 34.3 and a filter pressure drop relationship developed by Noll and Davis,[7] fabric filter system costs have been determined for 100,000, 200,000 and 300,000 cfm structural baghouse collectors. The capital and operating costs of these systems is presented in Figure 34.3 as a function of filtering velocity. It is interesting to note that, while capital

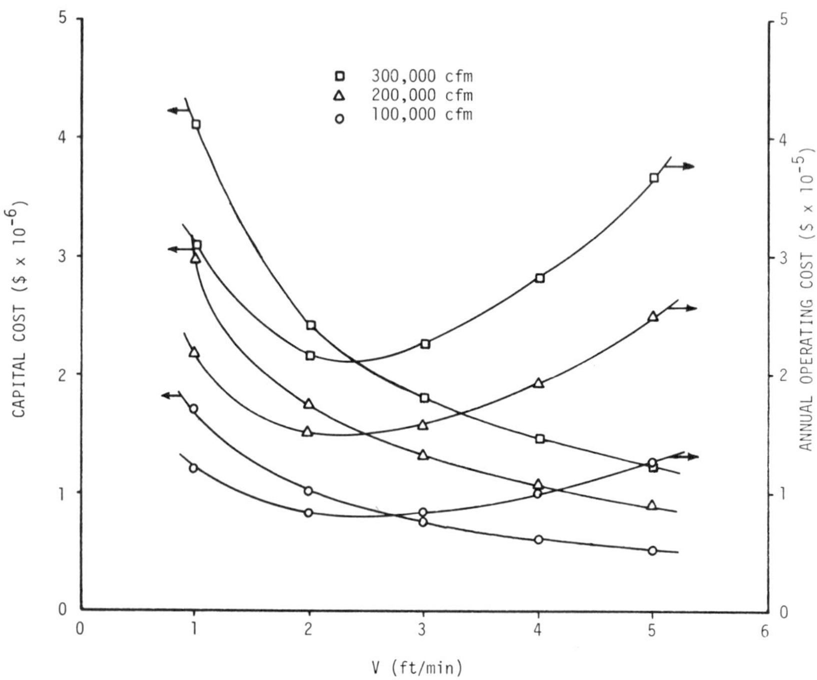

Figure 34.3. Capital and annual operating costs as a function of filtering velocity.

costs decrease with increasing filtering velocity, annual operating costs pass through a minimum. Such behavior occurs because of increasing power costs and decreasing bag replacement and maintenance costs and is a result of the modeling relationships used.

Total annual cost is presented in Figure 34.4 as a function of the pressure drop/filtering velocity ratio. In accordance with the previous discussion, the annual cost is found to be a minimum at $\Delta P/V = 4.3$ and, through the use of the filter pressure drop relationship, the optimum operating conditions for minimum annual cost can be selected.

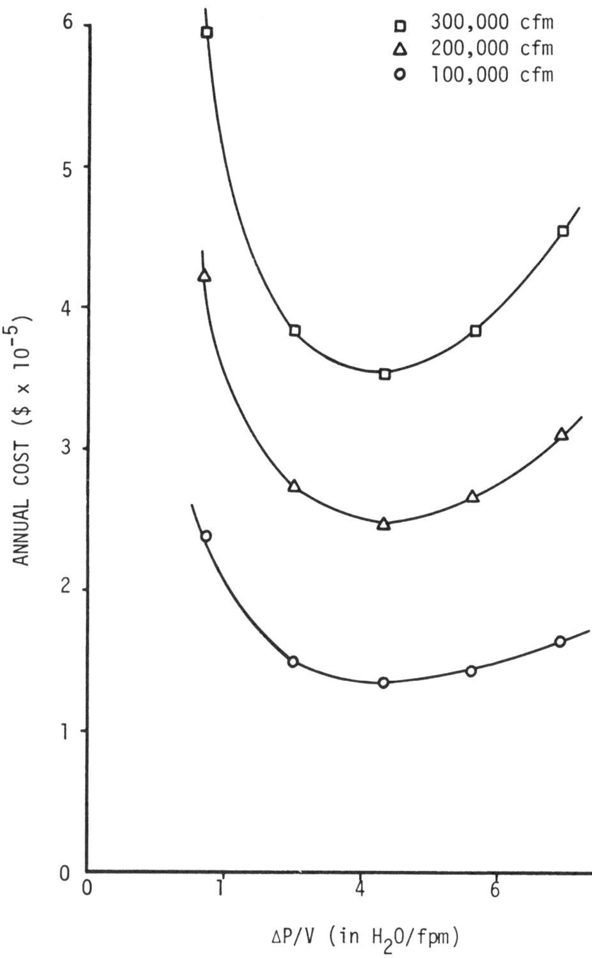

Figure 34.4. Total annual cost as a function of the pressure drop/filtering velocity ratio.

Table 34.3. Design Characteristics.

Filter cycle time	60 minutes
Average partial diameter	2.5 μm
Dust concentration	2.0 gr/ACF
Particle density	2.0 gm/cm^3
Temperature	70° F
Fabric type	Dacron
Bag life	1.5 yr
Bag cost	\$0.30/ft^2
Electricity cost	\$0.015/kwh
Operating hours	8760 hr/yr
External pressure drop (hoods, ducts, etc.)	10.0 in H_2O

SUMMARY

A modeling scheme for determining the minimum annual cost of a fabric filter system has been developed, and applications of this scheme using available cost relationships has been presented. It has been shown that with the use of appropriate performance relationships, optimum filter operating conditions for minimum annual cost can be selected.

REFERENCES

1. Alonso, J. R. F. "Estimating the Costs of Gas-Cleaning Plants," *Chem. Eng.* 86 (December 13, 1971).
2. *Industrial Ventilation*, 11th ed. (Lansing, Michigan: American Conference of Government Industrial Hygienists, 1970) pp. 11-22.
3. Fraser, M. D. and G. J. Foley. "Cost Models for Fabric Filter Systems," Presented at the 67th Annual Meeting of the Air Pollution Control Association, Denver, Colorado.
4. Fraser, M. D. and G. J. Foley. "A Predictive Performance Model for Fabric Filter Systems," Presented at the 67th Annual Meeting of the Air Pollution Control Association, Denver, Colorado.
5. Intertechnology Corporation. "Performance and Cost Models for Fabric Filter Systems," Draft Report, January 11, 1974.
6. Van Note, R. H. and F. T. Weems. "Filtration Equipment Applications," *I. & E.C.* 53, 546 (1961).
7. Noll, K. E. and W. T. Davis, private communications.

ACKNOWLEDGMENTS

The author wishes to express his thanks to Dr. K. E. Noll of the University of Tennessee and to Dr. S. P. Shelton of the United States Air Force for several useful discussions.

35

Computer Design of Fabric Filter Baghouses

K. S. Ramaswamy
Carborundum Environmental Systems, Inc.
Pollution Control Division, Knoxville, Tennessee

INTRODUCTION

Latest developments and successful application of fabric filter technology in controlling particulates virtually in every industrial pollution problem have substantially increased baghouse sales from $80 million in 1974 to a projected volume of $180 million annually by 1985. Because of the anticipated growth in this field, computer programs have been developed for predicting gas-to-cloth ratio versus pressure drop, for designing structural and mechanical components, and for cost evaluation. Optimum baghouse designs for a given set of conditions are achieved through the use of these programs.

TYPES OF BAGHOUSES

Depending upon the physical location of the baghouses in any pollution control system, baghouses are classified either as "suction" or "pressure" type. Suction baghouses (Figure 35.1a) are those in which the gases are pulled through the filter media by an induced draft fan located on the clean air side. Pressure baghouses (Figure 35.1b) are those in which the gases are pushed through the filter media by a forced draft fan located on the dirty air side.

Suction baghouse are of heavier airtight construction than are pressure baghouses. For systems above 100,000 cfm the filter media are usually 8-inch nominal diameter tubes 22 ft long or 12-inch nominal diameter tubes 30 ft. long. The fabrics can be cotton, polyester, Nobex® or fiberglass, depending upon process temperature and chemical properties of the gases and particles collected.

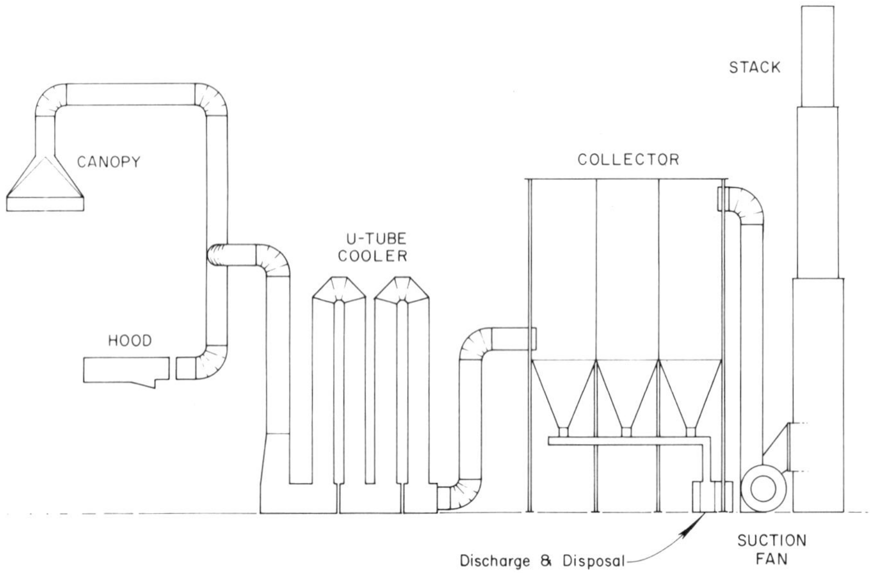

Figure 35.1a. Suction baghouse.

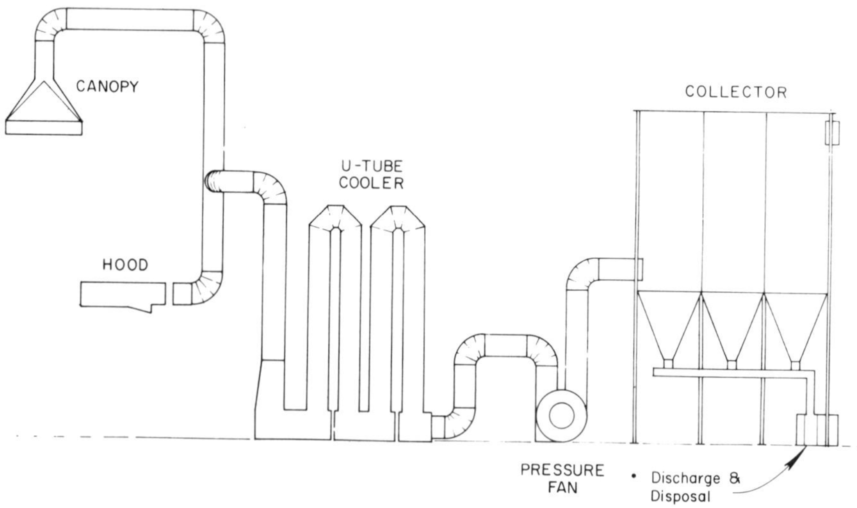

Figure 35.1b. Pressure baghouse.

VARIABLES AFFECTING DESIGN AND COST

The most significant variables that affect the design and cost of a baghousé are as follows:

1. Gas Volume—Normal range for structural baghouses are from 100,000–600,00 cfm with single jet plenum.
2. Gas-to-Cloth Ratio—These ratios vary from 1:1 through 4:1, depending on the nature of the dust, dust loading, size distribution, temperature, pressure drop, etc.
3. Temperature—Due to the availability of fabric, the maximum acceptable temperature is 550°F using a fiberglass fabric.
4. Bag Size and Spacing—8-inch nominal diameter 22 ft long bags are spaced on 9-inch centers (10-inch optional) and 12-inch nominal diameter 30-foot long bags are spaced on 14-inch centers.
5. Bag Reach and Arrangement—Several bag arrangements are possible with 3- and 4-bag reach for 8-inch diameter bags and 2- and 3-bag reach for 12-inch diameter bags. These are illustrated in Figure 35.2.
6. Bag Tension—Proper bag tensioning is vital for efficient cleaning of the bags. This varies from 15–75 lb tension depending on the application.
7. Number of Compartments—Proper consideration should be given for selecting the gross and net number of compartments without exceeding the allowable gas-to-cloth ratio while compartments are shut down for cleaning and for maintenance.
8. Bulk Density of Dust—in most applications dust density varies from 10 lb/ft^3 to 100 lb/ft^3 and this has a substantial effect on hopper and frame design as well as dust handling systems.

PARAMETRIC STUDY

Parametric studies were performed on baghouses using 8-inch diameter bags to investigate how the various input variables affect the final total weight (and hence the cost) of the baghouse. Due to the ever-changing prices of materials, all designs were compared on the basis of weight rather than cost since the weight of a baghouse is normally proportional to cost. The major objective was to arrive at the relationships of steel weight per cfm $\frac{(\text{lb-min})}{\text{ft}^3}$ versus compartment length (ft) and the steel weight per cloth area (lb/ft^2) versus cloth area per compartment.

Significant variables such as the gas volume, gas-to-cloth ratio, cleaning mechanism, wind and seismic loadings, and the bulk density of the dust were considered for these studies although 48 different input variables are considered in the computer program which could affect the design and cost of a baghouse.

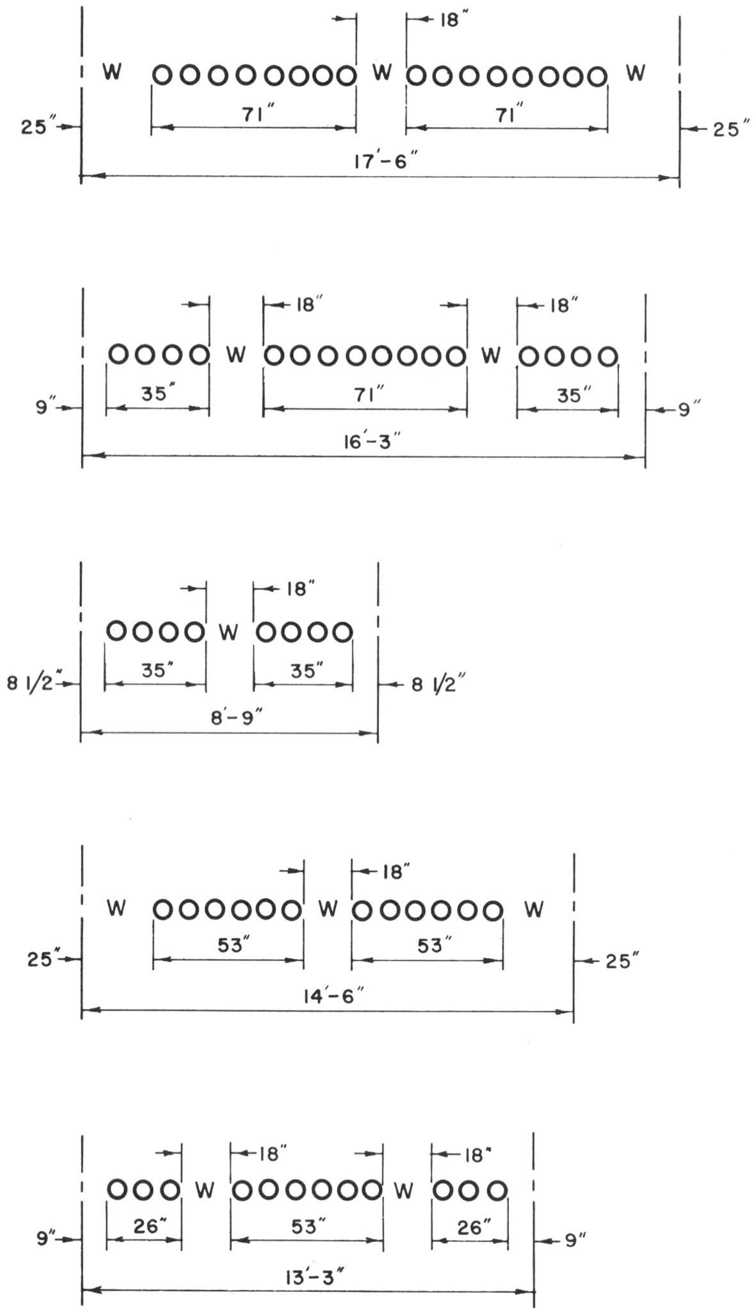

Figure 35.2. 8-inch diameter bag arrangement.

The following results are drawn from the parametric studies and are significant in the design of a baghouse:

1. Even though every input variable affects the design and cost, the optimum compartment length for all alternates is between 18 and 22 ft. This is shown in Figure 35.3.
2. As seen in Figure 35.4, economical design is achieved with approximately 19,000 ft^2 of cloth area for a 4-bag reach and 13,000 ft^2 of cloth for a 3-bag reach in each compartment.
3. Baghouses having a 3-bag reach are at least 10% heavier when compared to baghouses utilizing a 4-bag reach with 8-inch diameter bags.
4. The gas-to-cloth ratio has a significant effect on overall baghouse design. As seen in Figure 34.5, a baghouse is 45% heavier when the gas-to-cloth ratio is decreased from 3.8:1 to 2.5:1.
5. Baghouses with mechanical shakers are 4–5% heavier than baghouses with reverse air cleaning.
6. Changes in other design variables either decrease or increase the weight of the baghouse but the optimum compartment length range (18–22 ft) still remains the same.

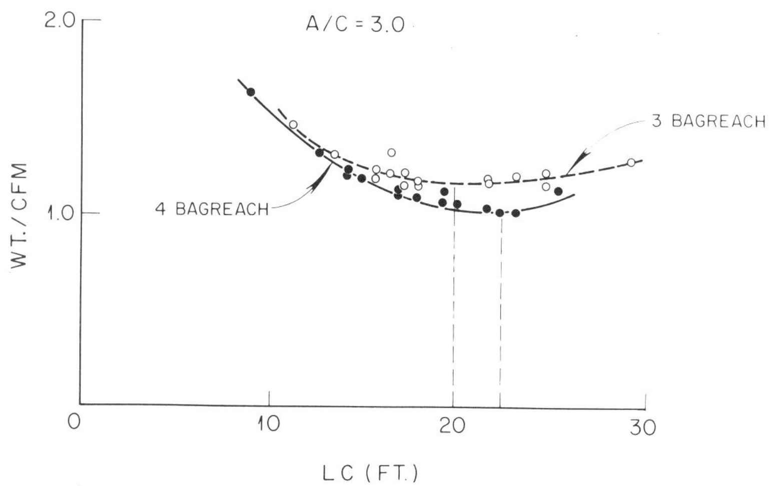

Figure 35.3. Weight/cfm vs. compartment length (8-inch diameter bags).

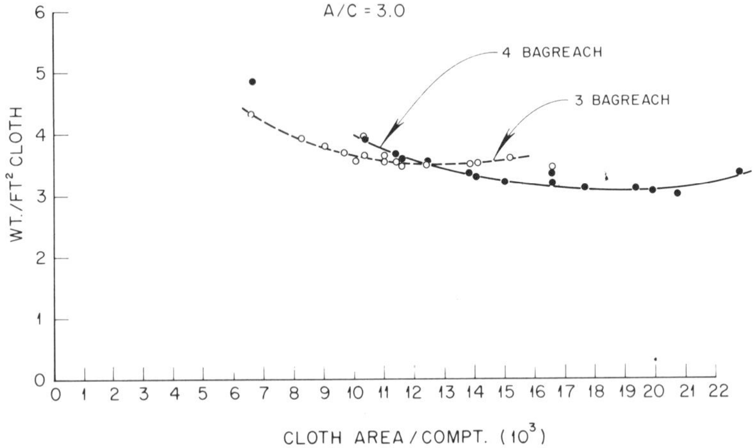

Figure 35.4. Wt/ft² cloth vs. cloth area/compartment x 10^3—(8-inch diameter bags).

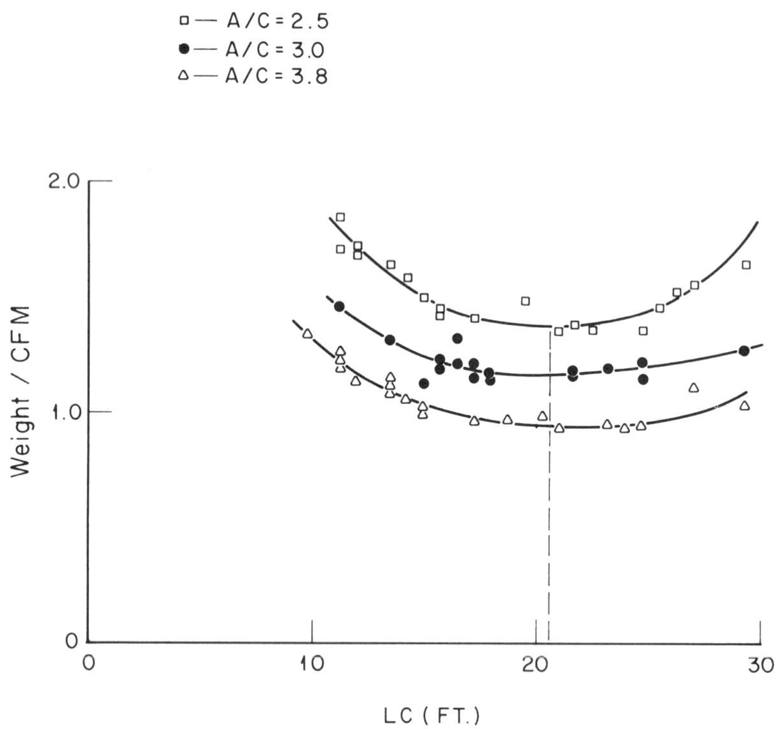

Figure 35.5. Weight/cfm vs. compartment length (8-inch nominal diameter bags).

36

Performance Characteristics of Seamless Filter Bags for Application to Industrial Gas Cleaning

Donald W. Wohlgemuth
Filter Media Division, Carborundum Environmental Systems, Inc.
Hagerstown, Maryland 21740

Wayne T. Davis and **Kenneth E. Noll**
Department of Civil Engineering, The University of Tennessee
Knoxville, Tennessee 37916

INTRODUCTION

The fabric filter has been widely applied in the collection of various types of atmospheric and industrial dust. The basic performance of fabric filter dust collectors can be measured in terms of the following characteristics: a) size and type collector (capital cost); b) pressure drop across the system (operating cost); c) bag life (capital and operating cost); and d) collection efficiency (government regulations).

The optimization of these parameters for different industrial applications has led to the development of filter fabrics constructed of various materials (*i.e.*, cotton, polyester, wool, acrylic, fiberglass Nomex®, Teflon®, etc.). Each filtration fabric, whether its form be woven, felted, or any other structure, has been used to satisfy specific requirements such as temperature resistance, cleanability, corrosion resistance, abrasion resistance, pressure drop, and collection efficiency. Historically, fabrics have been constructed from materials that have been commercially available and have been sewn into cylindrical or envelope type bags with one or more vertical seams.

The objectives of this chapter are to describe a new development in the fabric filter industry—the development of a seamless tube filter fabric designed specifically for application to dust control in industrial processes, and provide comparative results of the seamless tube fabric's performance characteristic with respect to other standard filter fabrics. The comparative evaluation of the new fabrics has been accomplished through extensive testing in three major phases: a) physical testing of the fabrics, b) pilot plant laboratory testing, and c) full-scale field testing.

The initial comparative evaluation of the seamless tube fabrics with other standard (seamed fabrics) has provided the following performance results:

1. The seamless-tube fabrics exhibited operating pressure drops less than or equal to (and efficiencies greater than or equal to) the standard fabric tested.
2. The seamless tube fabrics may be produced at a more competitive cost/bag than many of the standard fabrics due to their ability to be produced into cylindrical bags on a continuous basis.
3. The seamless tube fabrics' special weave allowed them to be installed in standard type shaker units taut in the maximum or minimum shaker tension modes without any observable effect on the residual pressure drop. This also reduces the possibility of damage either to the bags or to the shaker mechanism due to improper hanging.
4. The absence of the longitudinal seam as compared to standard filter bags eliminates the "banana" effect caused by sewn seams under tension. This also reduces the chance of bag wear caused by bag interference with collector walls and/or other bags. Additionally, high speed photography has shown that the seamless bags undergo a more uniform sinusoidal cleaning wave up and down the bags as opposed to the more random motion of seamed woven fabrics.
5. The absence of the longitudinal seam eliminates the needle holes which are generally present with a sewn seam. This provides less leakage potential and reduced seam failure potential. The seamless bag also provides approximately 2–3% additional usable filter area for a given size bag.
6. The absence of the longitudinal seam in the cuffs allows a better seal to the collector thimble or cell plate.
7. Greater resistance to inadvertent puncture exists due to the nature of the fabric (more resilient, heavier and thicker than most woven fabrics currently in use).

TESTING PROGRAMS

The seamless tube fabrics developed have been tested and evaluated in two specific modes: a) physical testing, and b) laboratory pilot plant testing. Full-scale field testing on an industrial process was being conducted at the time this chapter was being written.

Physical Testing of Seamless Tube Fabrics

Experimental Procedure

The physical test program consisted of conducting tests in accordance with nationally recognized standard test methods on the seamless fabrics and the "standard" woven filter fabrics. These tests were designed to yield the following information:

1. The weight of the fabric in ounces/yd^2
2. The thickness of the fabric in inches
3. The air permeability—flow through a one square foot area of cloth at 0.5-inch pressure drop across the fabric.
4. The Tabor abrasion number (the end point defined as the number of cycles required to produce yarn breaking or wear-through when observed with a fluorescent as background).

The Standard Method for each test is described in Table 36.1.

Table 36.1. Standard Test Methods Utilized in the Physical Testing Program.

Weight	Method 5401	Federal standard 191
Thickness	Method 5030	Federal standard 191
Air permeability[a]	Method 5450	Federal standard 191
Tabor abrasion[b]	Method 5306	Federal standard 191

[a]0.5 inches differential pressure.
[b]End point defined as the number of cycles to produce yarn breaking or wear-through when observed with a fluorescent as background.

The permeability test yields an indication of the new cloth pressure drop, which can be intuitively related to the pressure drop across operating field units. The Tabor abrasion number yields an indication of the potential bag life in that it is a measure of the number of cycles required to produce breaking of fibers. Table 36.2 lists the fabrics tested in the physical test program.

Results

Due to the large number of fabrics tested, the physical test data is broken down into two comparative tables; Table 36.3 includes those whose prime application is in shaker-type collectors, and Table 36.4 includes those fabrics whose prime application is in reverse air collectors. Fabric 0428 is listed in both tables since this fabric is in widespread use in both shaker and reverse air collectors.

The results of the physical test program show that the permeability weight and thickness of the seamless fabrics are within the range of the currently available woven fabrics. Secondly, the Tabor abrasion numbers of the seamless fabrics are within the range of the

Table 36.2. Description of Fabrics Tested in Physical Test Program.

Fabric Identification No.	*Description*	*Prime Application(s)*
A020	Seamless tube, polyester, heat set	Shaker
MA020	Seamless tube, polyester, heat set	Shaker
C020	Seamless tube, polyester, heat set	Shaker
A030	Seamless tube, polyester, heat set	Reverse Air
0335	3 x 1 Twill, heat set & scoured all spun	Shaker
0410	3 x 1 Twill, heat set & scoured siliconized, all filament	Reverse Air
0428	3 x 1 Twill, heat set & scoured combination	Shaker & Reverse Air
0431	2 x 2 Twill, heat set & scoured all spun	Shaker
0457	3 x 1 Twill, heat set & scoured siliconized, all filament	Reverse Air
0210	Cotton Sateen	Shaker
0216	Cotton Sateen, Flameproof finish	Shaker

Table 36.3. Results of Physical Testing of Fabrics Typically Used in Shaker Type Collectors.

	Fabric Identification No.							
Measured Parameter	*A020*	*MA020*	*C020*	*0335*	*0428*	*0431*	*0210*	*0216*
Weight (oz/yd^2)	9.7	8.9	10.0	7.5	6.4	11.8	9.6	8.3
Thickness (inches)	.025	.022	.025	.017	.013	.040	.025	.021
Air Permeability (ft^3/min/ft^2)	27.6	37.9	39.5	31.7	32.3	62.1	8.8	12.4
Tabor abrasion (cycles to failure)	1160	1013	755	430	716	920	1160	300

Table 36.4. Results of Physical Testing of Fabrics Typically Used in Reverse Air Type Collectors.

	Fabric Identification No.			
Measured Parameter	*A030*	*0410*	*0428*	*0457*
Weight (oz/yd^2)	12.2	5.3	6.4	5.0
Thickness (inches)	.027	.010	.013	.009
Air Permeability (ft^3/min/ft^2)	32.4	17.8	32.3	11.6
Tabor abrasion	>50,000	1070	716	1470

Standard fabrics or greater. The A030 seamless tube has a Tabor number significantly greater than most available fabrics indicating that it has potential for excellent durability in the field with respect to abrasion resistance.

Pilot Plant Testing of Seamless Tube Fabrics

Introduction

An extensive pilot plant testing program was conducted to evaluate the performance of three types of seamless tube filter bags (A020, MA020, and C020) against a standard sewn bag of a field-proven fabric (0335). (See Table 36.2 for fabric descriptions.) The pilot plant test unit was designed to simulate field operation, but with controlled conditions. Test results allowed a comparison of the control fabric and the three seamless tube fabrics for a) stabilized (conditioned) residual pressure drop, b) operating and maximum pressure drops, c) dust collection efficiencies, and d) bag stability.

Description of Experimental Set-up

The basic test unit (Figure 36.1) consisted of two parallel flow compartments with flow controls which allowed constant volume operation throughout the test cycle. (Only one compartment was operated during this experiment.) The compartment contained four bags with a total cloth area of 57 ft^2 (four 11 feet x 5 inch diameter bags).

A VibraScrew® dust feeder was employed to inject a specified quantity of dust into the air stream per unit time. The injected dust was dispersed and carried through the system by two 15-inch fans operated in series (one upstream and one downstream of the filter compartment).

The air-to-cloth ratio at the bags (or the flow through the system) was monitored through the use of a calibrated orifice plate located on the downstream side of the filter compartment. Constant flow conditions were maintained through the use of a Dwyer "photohelic"® pressure switch/butterfly value combination which were utilized to maintain a constant pressure drop across the orifice plate. Flow control was maintained within ±3% of the desired value.

A shaker cleaning mechanism with a rocking cleaning motion was employed to periodically clean the filter bags. It was similar to field-installed shaker mechanisms, and its frequency and amplitude were set to provide maximum cleaning (frequency-330 rpm, maximum amplitude — 0.4 inches peak-to-peak movement, shaker on-time—45 seconds).

The pressure drop across the filter bags was monitored with an inclined manometer in parallel with a Magnehelic® pressure gage. Pressure drops ranging from 0–20 inches were capable of being recorded during the filtration cycle.

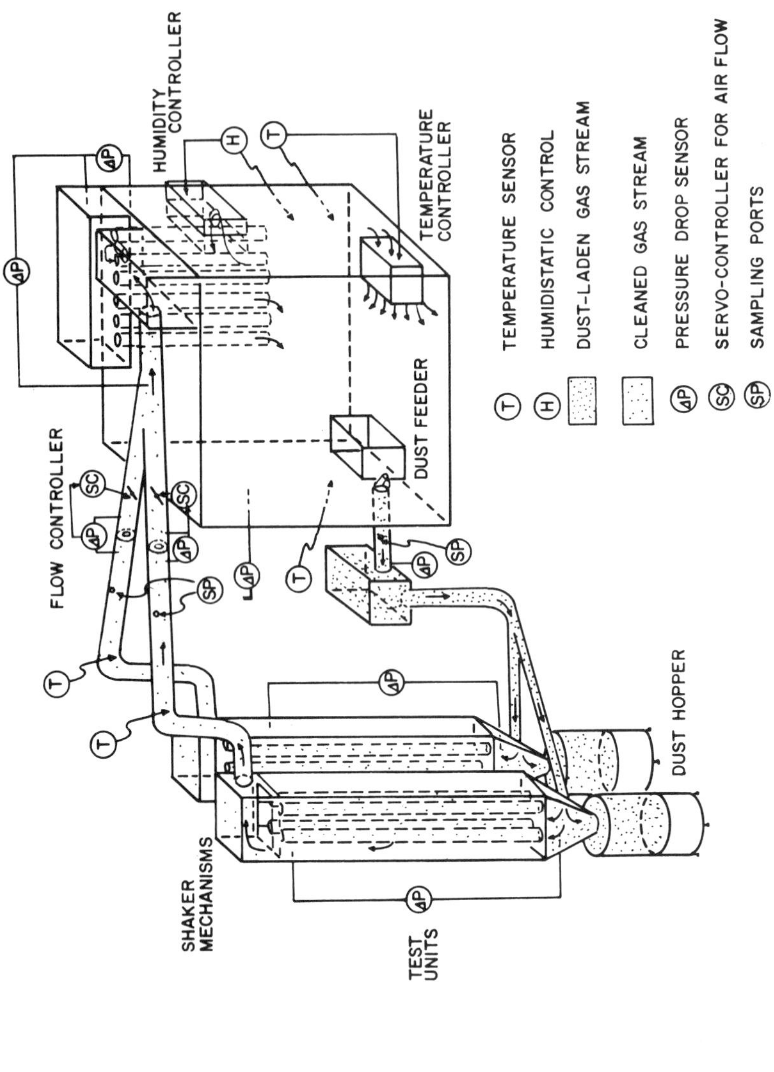

Figure 36.1. Schematic of the full-scale test unit.

The timing of each event within the cycle (VibraScrew on time, air flow on time, shaker on time, etc.) were controlled automatically through the use of an electric control panel with preset time intervals.

Experimental Procedure

The test program consisted of repeated dust loading of the seamless tube filter bags and the control filter bags under controlled conditions until the fabrics reached a conditioned state similar to that which would occur under field operating conditions. The number of repetitive runs of dust loading, followed by intermittent cleaning of the filter bags required to achieve a conditioned fabric was determined by a continuous updating of a graphical plot of residual pressure drop (just after cleaning) versus run number. Typically, 160 50-minute cycles were required. Measurements of pressure drop across the bags as a function of time were taken during each cleaning cycle with a constant grain loading of 4.24 gr/ACF, a gas temperature of 100° F, a filtration velocity of 3.15 ft/min, and a relative humidity of 20–30%.

The dust used was a standard test dust, Nytal 400®, with a mass mean diameter of 5.2 μ, a standard deviation of 1.2, and a particle specific gravity of 2.75. The conditions of air-to-cloth ratio (3.15 fpm) and grain loading (4.24 gr/ACF) were chosen to simulate extreme field operating conditions while the 50-minute cleaning cycle was typical of many field operating units.

After having conditioned each set of filter bags (at least 160 cycles), efficiency tests were performed on each set of bags. A minimum of one test (consisting of a 7–10 cycle average efficiency test) was performed on each set of bags before removing them from the unit. A second set of efficiencies was performed when the bags were reinstalled in the unit for an additional 10 runs and a 90-minute extended run. It was necessary to obtain an integrated sample over 7–10 consecutive runs in the outlet gas stream due to the small quantity of dust passing through the filter.

The sampling procedure consisted of continuously pulling an isokinetic slip stream from the exit gas stream. Sampling of the inlet was also performed as a check on data reliability. Inlet sampling data was in agreement with the gravimetrically measured feed rate of 5 gr/ACF (4.24 gr/ACF at the bags due to hopper fallout).

Results

The test data obtained from repetitive cycles of dust loading and removal allowed a comparison to be made between the seamless tube fabrics and the standard "control" fabric for: a) stabilized (conditioned)/residual pressure drop; b) operating and maximum pressure drops; c) dust collection efficiences, and d) bag stability during operation.

Figure 36.2 shows the final (90-minute) extended run for the three seamless tube fabrics and the control fabric. The data were collected after the four fabrics had been conditioned with test dust for at least 160 cycles of 50 minutes each and were stabilized with respect to residual pressure drop. The data show there was little difference in the performance of the three seamless tube fabrics while the control fabric showed a consistently higher residual and operating pressure drop. The residual resistances ($\Delta P/V$) for the seamless tube fabrics were near 0.3 inches H_2O/fpm (1.0 inches of H_2O), while that of the control fabric was near 0.6 inches H_2O/fpm (2.0 inches of H_2O). The maximum allowable operating pressure drop which would be typical of field operating conditions (6.0 inches pressure drop across a compartmental baghouse) was attained after about 40 minutes for the control fabric. At this same time in the cycle, there was about a 5.0-inch average pressure drop for the seamless tube fabrics. This represented a 15% difference in operating pressure drop and could represent a considerable economic savings for an operating baghouse.

The results of the dust collection efficiency tests performed on the stabilized fabrics are shown in Table 36.5.

Table 36.5. Efficiency Results Obtained on the Conditioned Test Fabrics.

Fabric	*Efficiency #1*[a] *(%)*	*Efficiency #2*[b] *(%)*	*Avg. Efficiency (%)*
0335	99.986, 99.9932[c]	99.9978	9.9973
A020	99.9966, 99.9960[c]	99.9986	99.9971
MA020	99.9984	99.9990	99.9987
C020	99.9989	99.9970	99.9979

[a]Efficiency test performed after conditioning of each set of bags.
[b]Efficiency test performed at end of all experiments in which each set of filter bags was reinstalled for a final 10 runs to obtain a final efficiency and extended 90-minute run.
[c]Duplicate tests performed to check reproducibility.

Full-Scale Field Testing

Introduction

Based on the results of the physical and pilot plant test programs, full-scale test programs are being undertaken to provide further credibility for using the seamless tube filter bags on industrial applications normally employing standard seamed woven filter bags.

Experimental Design and Procedure

The first field test of the MA020 seamless tube bags is being conducted in a three-compartment shaker-type bag house dust collector which had previously utilized a standard woven fabric, 0428 (pre-

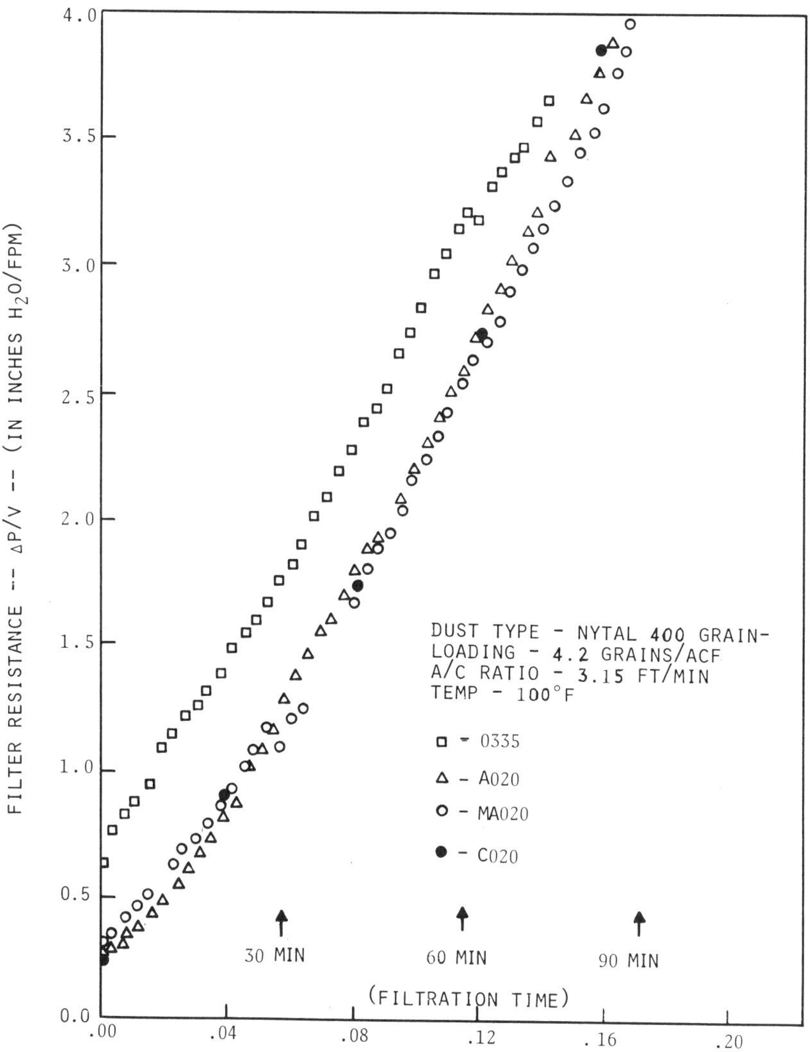

Figure 36.2. Weight of dust cake (lb/ft²). Resistance versus weight of dust cake for extended 90-minute run on the conditioned fabrics.

viously described). The collector description and operating parameters are described in Table 36.6.

The original 0428 filter bags were removed after nearly one year of successful operation. The new bags were installed and tests are being conducted currently to obtain performance data for compari-

Table 36.6 MA020 Field Test.

Collector Specifics
1. Carborundum CT-2 3-section Model 216 (welded modules)
2. Automatic continuous operation
3. Pressure operation
4. Original filter bags 0428 - (14-foot long) (648 total)
5. On-stream since January 1974
6. Location: Elyria, Ohio

Application Specifics
1. Specialty Foundry; electric arc furnaces, high alloy steel, collecting from one furnace on first shift and from two on second shift.
2. 3.4:1 A/C ratio
3. 5 or 6-day operation

son with the performance data of the original standard filter bags. Comparisons will be made of inlet/outlet particulate loadings, operating pressure drops, and power consumption of the two different filter materials.

CONCLUSIONS

The physical test program has shown that:

1. The "new" permeabilities, weights, and thicknesses of the seamless tube fabrics are within the range of currently available woven fabrics.
2. The Tabor abrasion numbers of the seamless tube fabrics are within the range of standard fabrics or greater. The A030 seamless tube has a Tabor number significantly greater than most available fabrics, indicating a potential for excellent durability with respect to abrasion resistance.
3. The seamless tube fabrics have potentially greater resistance to inadvertent puncture than typical woven fabrics due to their greater resilence, and different structure.

Further conclusions reached during the comparative pilot plant test program were as follows:

1. The seamless-tube fabrics exhibited operating pressure drops less than or equal to (and efficiencies greater than or equal to) the standard fabric tested.
2. The seamless tube fabrics' special weave allowed it to be installed in standard type shaker units taut in the maximum or minimum shaker tension modes without any observable effect on the residual pressure drop. This also reduces the possibility of damage either to the bags or to the shaker mechanism due to improper hanging.
3. The absence of the longitudinal seam as compared to standard

filter bags eliminates the "banana" effect caused by sewn seams under tension. This also reduces the chance of bag wear caused by bag interference with collector walls and/or other bags. Additionally, high-speed photography has shown that the seamless bags undergo a more uniform sinusoidal cleaning wave up and down the bags as opposed to the more random motion of seamed woven fabrics.

4. The absence of the longitudinal seam eliminates the needle holes which are generally present with a sewn seam. This provides less leakage potential and reduced seam failure potential. The seamless bag also provides approximately 2–3% additional usable filter area for a given size bag.
5. The absence of the longitudinal seam in the cuffs allows a better seal to the collector thimble or cell plate.

37

Air Pollution Control for Industrial Coal-Fired Boilers

Allen H. Jones
Western Precipitation Division,
Joy Manufacturing Company, Los Angeles, California 90039

This chapter is limited to the discussion of industrial coal-fired boilers, or the non-utility application. In the past coal-fired industrial boiler pollution control has been either by electrostatic precipitator or multicyclone. Today, all pollution control devices are being adapted to these boilers, including baghouses and wet scrubbers.

SULFUR PROBLEMS AND BENEFITS

Before selecting equipment for pollution control on coal-fired boilers, it is necessary to understand the role that sulfur plays in firing fossil fuels.

During combustion, sulfur forms with oxygen to make SO_2 and SO_3. About 95–97% of the sulfur combines to make SO_2—the remainder is SO_3. SO_2 and SO_3 combine with water to form sulfurous and sulfuric acid. These acids can form directly in the gas stream below the water dewpoint. These can cause extreme corrosion in a wet scrubber and, to a lesser degree, in the other types of units.

A more serious problem is the elevation of the dewpoint by SO_3. Most people involved with the burning of fossil fuels understand "acid dewpoint" which causes acid condensation at temperatures well above the "water dewpoint." However, the amount of SO_3 required to elevate the dewpoint is not alawys understood completely. To illustrate, we refer to Figure 37.1 showing the elevation in dewpoint due to the ppm of SO_3 in a gas stream. For example, 10 ppm in a flue gas can raise the dewpoint to 270° F. A 100 ppm can give a dewpoint of 310° F. To further put SO_3 in perspective, as a general rule of thumb whenever a 1% sulfur coal is burned under efficient conditions, approximately 800 ppm of SO_x will be generated. On the average, 3–5% of the SO_x becomes SO_3. Thus, with very little sulfur in the fuel, the SO_3 and the attendant elevation of dewpoint can be significant.

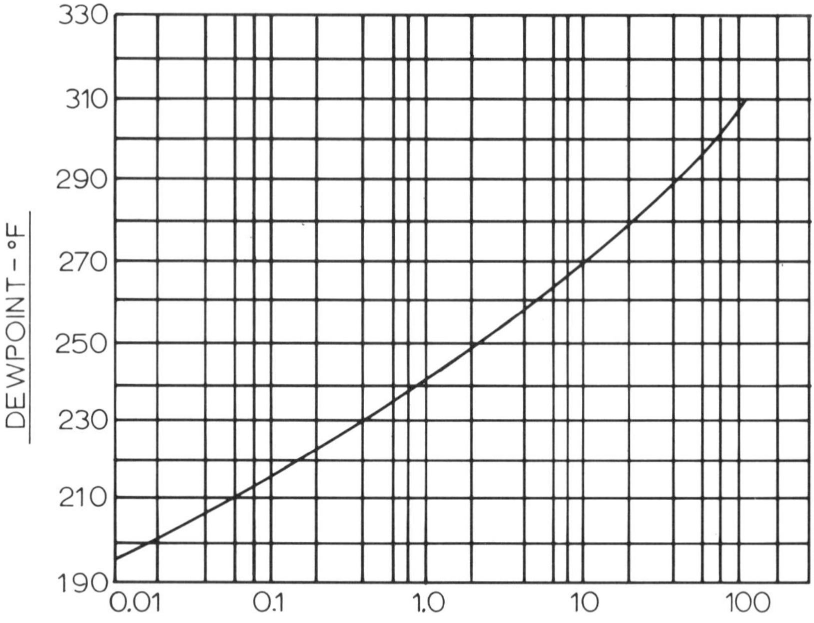

Figure 37.1. SO_3 (H_2SO_4) in flue gas, ppm.[1]

Paradoxically, the addition of SO_3 to the gas stream can improve the performance of an electrostatic precipitator. This is sometimes used to improve performance of coal-fired boilers using low-sulfur coal.

ELECTROSTATIC PRECIPITATOR

An electrostatic precipitator separates dust, fume, or mist from a gas stream by exposing the stream to a high voltage (20–80KV DC) electric field, which imparts an electric charge to solid particles in the stream. The particles then migrate to an oppositely charged collecting surface where they cling until removed.

The electric field is formed by radiation from electrodes, usually steel wires spaced equidistant from each other and the collecting surface. Particles are removed by periodically vibrating the collecting surface and the collected dust drops into hoppers.

The principal components of an electrostatic precipitator are:

1. Collecting System — This is a series of plates with fins to which the dust will be attracted and adhere. Plate rappers are provided to vibrate the plates to remove the collected dust and cause it to fall into the hopper(s). The spacing of these plates depends upon the voltage applied to the electrodes. Average voltages in industrial coal-fired boiler work average 45 KV with spacing between collecting surfaces of approximately 9 inches.
2. High Voltage System — These are usually wire electrodes by which the gas is ionized producing the electric field through which the gas must flow. Rappers are also provided to vibrate the wires at intervals to dislodge minor dust accumulation.
3. Power Supply — The high voltage required is produced by transformer-rectifiers, which increase commercial voltages to approximately 45 KV and convert alternating current to unidirectional current. Automatic voltage controls are required to maintain maximum voltages for maximum efficiencies as conditions vary.
4. Gas Distribution Device — It is imperative that the incoming gas flow be evenly distributed for nearly uniform velocity; this can be a simple perforated plate although certain inlet configurations may require special gas distribution components.

The sizing of an electrostatic precipitator depends upon these factors:

1. Collection efficiency (E) — Very high efficiencies in the 99+% range can be obtained since there is no theoretical lower limit to the size particle that can be collected. In theory, if a 90% efficiency could be obtained in one field, then each subsequent field would get 90% of the remainder from the preceding field leading to extremely high collection efficiency. This does not necessarily happen since some of those particles getting through each field are those that don't take an optimum charge, making the job of each succeeding field more difficult.
2. The volume (V) through a precipitator can be any number required. There are few restraints, large or small, on precipitator sizing except those provided by economics.
3. The apparent migration velocity (w) is the difficult factor to obtain except by experience. Dust conductivity (resistivity) is one factor having a great effect on migration velocity and precipitator performance. There is an optimum area of resistivity as shown in Figure 37.2. If the resistivity is too low, the particle gives up its charge to the collecting plate too quickly and may be reentrained into the gas stream. Carbon is an example of a dust with this characteristic.

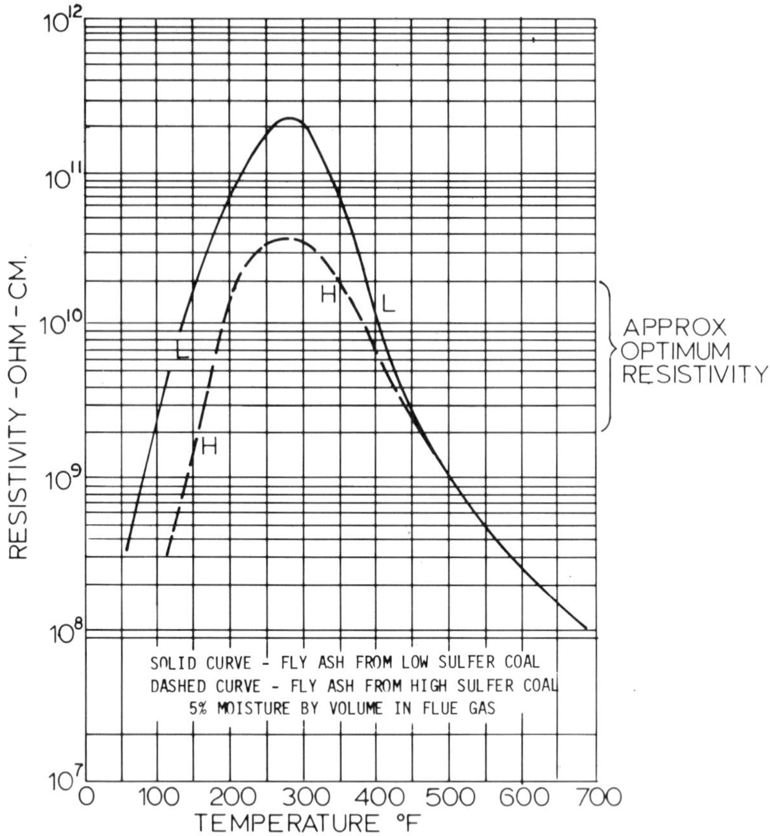

Figure 37.2. Typical plot of resistivity vs. temperature for fly ash.

If the resistivity is too high, the particle does not release its charge to the collecting plate, forming a layer that acts as an electrical insulator. This causes sparking, abnormal power consumption and loss of collection efficiency. The equation that expresses the relationship of the foregoing is

$$1 - E = e^{-w\left(\frac{A}{V}\right)}$$

Application of precipitators to industrial coal-fired boilers is not as clear-cut as it may seem. Looking at Figure 37.2, we can see that resistivity for fly ash increases with temperature to a maximum point and then decreases with further increases in temperature. In the typical curves the maximum resistivity is at about 300° F, which might be considered nearly normal for a boiler stack after the economizer. However, the resistivity is high as shown on the low-sulfur coal curve. To get to an area of better resistivity, it might be necessary to add SO_3 conditioning which reduces resistivity. The same result could be obtained by moving the precipitator ahead of the economizer, raising temperatures to perhaps 500° F and thus lowering the resistivity to a more appropriate level.

Figure 37.3 shows graphically where it may be best to use a precipitator, compared with a baghouse. Below 1% sulfur in the fuel and at temperatures below 500°, it is probably better to use some other device than a precipitator from an economic standpoint. The low amount of sulfur generates insufficient SO_3 to have the most beneficial effect. Also, the temperature effect on resistivity shown on Figure 37.2 mitigates against the precipitator at lower temperatures.

These negative factors do not absolutely preclude the use of a precipitator, but more fields, lower gas velocities, etc. will be required, making the units more costly. As temperatures increase over

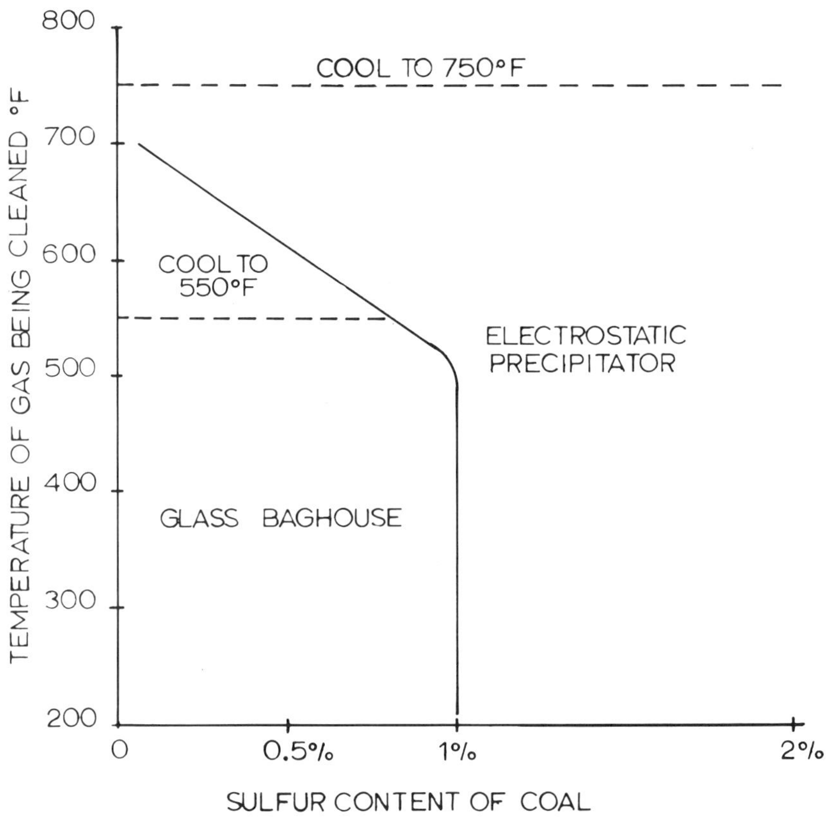

Figure 37.3. Precipitator or glass baghouse equipment selection for high efficiency fly ash collection.

550° F, resistivity of the fly ash increases, broadening the range in which the precipitator may be used, as shown in Figure 37.3. The carbon content in the fly ash may have a harmful effect on the collection efficiency of a precipitator on a coal-fired boiler. Carbon particles have a low resistivity and do not keep the charge making them difficult to collect. More than 15% carbon in the fly ash becomes troublesome. To overcome the harmful effects of excess carbon in the gas stream, cyclonic collectors are often used ahead of the precipitator. The cyclone may get an efficiency of 90% by itself, reducing to a great extent the unburned carbon to be handled by the precipitator. Often this results in getting by with a one-stage precipitator and a cost reduction.

CLOTH BAGHOUSES

The mechanism by which cloth filters dust is extremely complex, and as yet there is no agreement or complete understanding of all the factors involved. The cloth does not act as a screen which refuses to pass particles below a certain size. This is more easily understood when we consider that filter cloths are woven from fibers of approximately 20 μ in diameter and yet 0.5 μ and smaller particles are trapped and subsequently released. If screening was the mechanism of collection, it would be equivalent to trying to collect golf balls with a weave of tree trunks.

It has been demonstrated that the small particles in the air stream one micron and below attach to the fiber and to themselves to form chains which, in turn, become a loosely compacted mat which then acts as its own filter. It is theorized that this chain formation is a combination of inertial impaction, Brownian movement, and electrostatic charge depending on the size of the particle below one micron. Above one micron, gravitational settling and actual screening takes place. The capacity of cloth-to-filter air is stipulated by what is known as "ratio" or "air-to-cloth." This ratio is

Cubic feet air per minute/square feet of cloth

Cancelling out units shows that ratio also is the velocity through the cloth expressed in compatible units. For example

$$\text{Ratio} = 1000\ \text{cfm}/500\ \text{ft}^2 = 2\ \text{ft/min}$$

Typical ratios for handling fly ash from boilers range from 2/1 to 3/1 with woven glass cloth. The lower level is used with reverse air cleaning only, while the higher may be attained with a combination of shaking and reverse air. Higher ratios attained by shaking generally represent some loss of bag life.

The average pressure drop at which cloth will filter dust depends on ratio (velocity through the cloth) and the particle size of the dust. The volume of dust has little to do with pressure drop, but may increase the frequency of cleaning required as quantity increases.

Pressure drops vary from about 1.5 inches water column (w.c.) to 5 inches w.c. Pressure drops above this range generally mean the filter is under size, and premature bag wear will take place due to excessive cleaning.

There are two basic types of cloths used in filters on coal-fired boilers; namely, woven and needled felt.

Woven cloth is usually a simple sateen weave varying from about 9 oz/yd^2 to 18 oz/yd.^{2} High temperature woven cloths are commonly made from the following materials:

Material	*Temperature Limit*
Nomex®	350°–425° F
Glass	400°–500° F
Teflon®	400°–450° F

Ratios on woven cloth run between 1.5/1 and 3/1. This cloth is cleaned only by reverse low-pressure air or shaking.

Needled felt is made in a sandwich fashion with two felt batts laid on either side of a scrim (wide-open weave cheesecloth type). The felt batts are attached to the scrim by puncturing the sandwich with hundreds of strokes by needles with a small hook in the shank. This hook entangles the top fibers with the bottom fibers. The more needle punctures, the tighter the fabric and the greater resistance to air flow. The needling reduces the sandwich from a thickness of several inches to perhaps 0.25 inches. The scrim provides the cloth strength. Felts are made from synthetic fibers such as polyester, Nomex and Teflon. Needled felts cannot be made from cotton and other natural fibers or glass. Ratios on needled felts run from 3/1 to better than 10/1, depending on the type dust. These high ratios are only obtainable when using high-pressure compressed air cleaning.

Glass is the fabric most used on coal-fired industrial boilers. What has made this possible is the development of a coating, which prevents the glass fibers from literally cutting each other in two while in operation and during cleaning. Glass fibers are extremely strong but brittle and hard and without the lubrication between the fibers provided by the silicone coating, the fibers literally destroy each other as the bag is flexed through cleaning, etc. In subsequent years, coatings have been improved by the addition of graphite and, more recently, Teflon, so that bag life has been greatly increased. Glass is also relatively impervious to acid attack in the form of moisture, and does not react directly with the SO_3. Nomex, on the other hand, has a temperature limitation of about 350° with excursions to 400° and will not withstand sulfuric acid attack. In addition, recent experience may indicate that Nomex is attacked directly by SO_3. Teflon makes a fine high-temperature cloth but it is at least 10 times the cost of glass and, for this reason, is rarely used. Teflon also breaks down at temperatures in excess of 550°, giving off dangerous fluorine gas.

Reverse flow glass fabric filters are the most widely used on industrial coal-fired boilers. There are really two basic types of these reverse flow units, depending upon the location of the primary air moving fan. This fan may either blow the dirty air through the filter medium (pressure type) or can be used to pull the air through (suction type). The cleaning of the bags in both cases is reverse flow but slightly different methods are used.

To clean pressure types, the dampers to isolate a section are usually installed on the inlet side of the collector, and an auxiliary fan is used to suck air back through the section being cleaned. This air goes through a collapsing flue and into the suction fan which discharges it back into the inlet flue.

This reverse flow of air does two things. First, it collapses the bag and cracks the cake of dust loose from the inside of the bag. Secondly, the reverse air flow assists in further loosening the dust which falls into the hopper. The amount of reverse air used varies from approximately one-half to two-thirds of the main air stream quantity handled by one section. The cleaning cycle for one section takes about three minutes. The collapse valve is opened and closed several times during the cycle to allow dust to fall into the hopper. In recent years, bags with rings have come into use. These prevent the complete collapse of the bags. Complete collapse sometimes holds back the dust from dropping into the hopper.

To clean suction types, the dampers are usually located on the clean air side. Auxiliary fans blow into the section being cleaned, collapsing the bags and cleaning as described previously. The air-to-cloth ratio with reverse air flow glass fabric filters range around 2 to 1 to 2.4 to 1.

The suction type baghouse has generally replaced the pressure type, primarily since the dampers and fans are on the clean air side. Modular units on industrial boilers are generally finding more acceptance on coal-fired boilers due to lower costs. These modular units are shipped to the job assembled and insulated, cutting down on expensive field installation. A number of modules can be interconnected to make any size job. Installations of 150,000 CFM and more are not uncommon.

The filter bags in modular units come in various sizes, allowing for maximum flexibility when confronted with limited space. The bag sizes and cloth areas normally are

Diameter	*Height*	*Cloth Area*
5½″	10′6″	14.4 ft^2
8″	22′0″	46.0 ft^2
11½″	30′0″	94.0 ft^2

Shaker-type glass baghouses are being installed on some industrial boilers. They are similar in construction to the modular reverse air cleaning filters in size and bag configuration. The modules are isolated from the gas stream with dampers, as with the reverse air, while shaking takes place. A small amount of reverse air may also be utilized during shaking. Shaking allows for higher ratios up to 3 to 1. However, there have been some cases of accelerated bag wear on coal-fired boilers at these ratios.

Pulse jet-type filters are not common on coal-fired boilers because they must normally use either Nomex or Teflon felt for high-temperature work. Nomex is subject to acid attack and Teflon, while impervious to acid, is extremely expensive. However, there are examples of both of these in operation.

The tubular bags in a pulse jet filter are just the reverse of bags in the previously described glass baghouses. The bag is open at the top, and the dust is collected on the outside of the bag. The clean air passes to the inside and out the open end at the top. A spacer or cage is required to prevent the bag from collapsing in operation due to the gas flow from outside to inside.

Compressed air at 100 psig is used to clean the bags in a pulse jet unit. This air is injected into the bag at the top through a venturi section which entrains additional air. This air entering the top forms a shock wave which travels down the bag, dislodging the cake of dust on the outside of the bag. The compressed air shock wave stops the flow of air momentarily in the normal direction, further assisting in the dust cake removal. Some further assistance in cleaning is obtained by the actual passage of the cleaning air through the bag in the opposite direction.

Compressed air bursts of approximately 0.1 seconds are used for cleaning. From 8–12 bags are cleaned at any one time. Intervals between cleaning rows of bags run from 3 seconds to 20 seconds depending on the type of dust. A complete cycle of bag cleaning can range from 2–3 minutes to perhaps 20 minutes.

Due to the high-energy cleaning method and the utilization of felt cloth, a very high ratio of air-to-cloth can be used. In boiler work, these ratios run approximately 5 to 1. This makes the units very compact and they can be installed in relatively small areas.

A recent innovation in pulse jet collectors for coal-fired boilers is the utilization of a combination of gentle reverse air cleaning and pulse jet cleaning. It is claimed by infrequently using the jets and relying on the reverse air that a special woven glass fabric can be used which, heretofore, has not been possible. It is too soon to evaluate this new type collector.

The application of baghouses to coal-fired boilers is becoming increasingly necessary as tighter and tighter particulate regulations are applied. With low-sulfur (below 1%) coal-fired boilers and outlet temperatures below 500°, they represent a less expensive solution to the particulate problem from boilers. This is shown graphically in Figure 37.3.

Collection efficiency is seldom a real problem with a well-designed and maintained dust collector. Collection efficiency is independent of dust loading. Collection efficiencies of 99.9% are common. The units are more sensitive to condensation and require much more maintenance than do the other types of collectors.

The first baghouse on a coal-fired utility boiler and one of the first on any kind of boiler is the notable installation at Sunbury Station of Pennsylvania Power and Light at Shamokin, Pennsylvania. Beginning with a pilot unit in about 1970, the full-scale baghouses were started in February, 1973. Since that time, the units have operated continuously except for scheduled outages and two forced outages unrelated to baghouse performance. Of the 5040 bags, only 50 had been replaced after nearly two years of operation. Most of these were lost due to improper installation rather than bag wear. Within the next few months, a complete rebagging will take place, unit by unit, since it is felt that the bags are approaching failure. A few sections will be left with original bags to determine the maximum possible life.

The installation consists of four baghouses installed on two generating units, whose total capacity is 175 MW. The units are installed 140 ft in the air, sandwiched between the boilers. The fuel used is a combination of anthracite and petroleum coke. The anthracite has a sulfur content of 0.8% sulfur and the coke runs as high as 6%. The mixture runs between 15% and 35% coke. The pressure drop across the filter system does not exceed 5 inches, including the baghouse and the duct work system.

CYCLONIC COLLECTORS

Cyclonic collectors of the high-efficiency, multi-tubular type have been extensively applied to the collection of fly ash originating from industrial power boilers. However, under present day pollution control regulations, they are no longer acceptable as a primary collection device in most areas. Even in those parts of the country where a multi-tubular cyclone is acceptable, its application is very risky.

The basic principle of cyclonic separation of dust from an air stream is simple. The dirty gas is guided into cylindrical tubes at high velocity in a way that sets up a whirling action by means of fixed directional vanes located inside the tubes. The whirling, or tornado-like action generates centrifugal forces. These forces cause the dust particles, suspended in the gas, to be directed to the inside tube wall. The particles fall downward and out a discharge throat at the bottom of the collecting tube.

A reversed air flow condition is created within the collecting tube causing the cleaned gas to be directed upward to exit from the collector by a smaller diameter outlet gas tube which is partially located and centered within the collecting tube.

The centrifugal force developed is subject to the velocity of the gas and the radius of the tube in which it is whirled. Smaller tube diameters and higher inlet gas velocities create larger centrifugal forces. Particle size, weight and shape affect the centrifugal force acting on the particulate entering the collecting tube. Larger and denser particles are generally easier to collect than smaller or less dense particles.

The overall collection efficiency of a cyclonic collector is directly related to a) the quantity of particulate less than 10 μ in size, and b) the design pressure drop, or loss, across the unit, in direction of gas flow, measured in inches vertical water column.

The relationship of particle size to pressure drop is illustrated in this typical fractional efficiency tabulation:

Particle Size Range (μ)	*1-inch VWC Efficiency (%)*	*2-inch VWC Efficiency (%)*	*3-inch VWC Efficiency (%)*
0–10	62.5	67.0	70.0
10–20	95.8	97.8	98.0
20–44	98.6	99.1	99.3
+44	99.3	99.5	99.6

It will be evident from the foregoing that the smaller the quantity of less than 10 μ material, the higher is the resultant collection efficiency and vice versa. Efficiencies of multi-tubular cyclonic collectors on spreader stokers approach 95%. With pulverized fuel-fired boilers, the efficiency drops to 90%. Putting units in series only improves collection efficiency by a fraction.

The multi-tubular collectors are the cheapest of all collectors and the most trouble-free. Unfortunately, they are borderline in meeting efficiency requirements. In combination with a precipitator as described previously, they can materially reduce the cost of the entire pollution control package. In other applications, multi-cyclonic collectors are used as precleaners for scrubbers and baghouses to reduce the dust loading to the primary unit. The dust loadings from coal-fired boilers are not generally high enough to warrant precleaning prior to going into a baghouse or scrubber.

WET SCRUBBERS

If you have some way to dispose of the water from a wet scrubber, it makes an inexpensive collection device for effluent from coal-fired boilers.

The basic operating principles of scrubbers can be described most easily by the following relationship:

$$\text{eff} = \frac{DV}{d}$$

where d is the size of the water droplet, V is the relative velocity (between D & d), and D is the size of the particulate. The foregoing are the major influences on scrubber efficiency, but Brownian movement, agglomeration, etc., are other factors that have some small effect in scrubbing performance. Simply stated, the more energy put into a scrubber (V), the higher the efficiency. Also, the larger the particulate, the higher the efficiency. Within limits, the smaller the water droplet, the higher the efficiency. (If droplets get too small below 30 μ, efficiency will decrease to rapid acceleration of droplets to particulate velocity, thus reducing relative velocity.) Scrubbers must also be operated with enough contact to saturate the incoming gas stream, otherwise it will be a cooler and will not operate at maximum efficiency.

There are many types of scrubbers which embody the foregoing principles. In general, scrubbers are broken down into two categories of low energy 3–15 inches pressure drop) and high-pressure 15–100 inches pressure drop). Scrubbers for coal-fired boilers operate in the 8–15-inch pressure drop range. To gain these pressure drops there are numerous scrubber configurations at the lower energy levels. At the higher levels, venturi type scrubbers are the most common.

Scrubber application to a coal-fired boiler is not quite so easy as might be thought at first glance. Some of the things to consider are:

1. Scrubber efficiency is affected by particle size. Particle size information is essential if the scrubber is to be sized correctly.
2. Since a scrubber is essentially an adiabatic saturation machine, the outlet conditions depend on the inlet moisture condition which must be known.
3. The amount of water used depends on the amount of solids the particular unit in question can stand. Very few scrubbers for boilers can tolerate more than 10% solids, and enough water must go to drain to maintain this equilibrium.
4. Water circulation through coal-fired boiler scrubbers runs in the order 3 gpm/1000 cfm outlet volume. The amount to drain usually varies from less than .5 gpm to 1 gpm.
5. The concentration of acid that can be tolerated also has an effect on the amount of liquid to drain. This is, of course, tied to the materials of construction. Referring to Figure 37.4, it can be seen that at a temperature of 150° several stainless steels will tolerate sulfuric acid concentrations up to 40%. If scrubber adiabatic satura-

tion temperatures are higher, more liquid must be sent to drain to reduce acid concentration below corrosive levels.

Materials of construction for coal-fired boiler scrubbers varies with the concentration of sulfuric acid that can be maintained in the unit. As pointed out earlier, this depends upon the sulfur concentration in the fuel and the amount of liquid taken to drain.

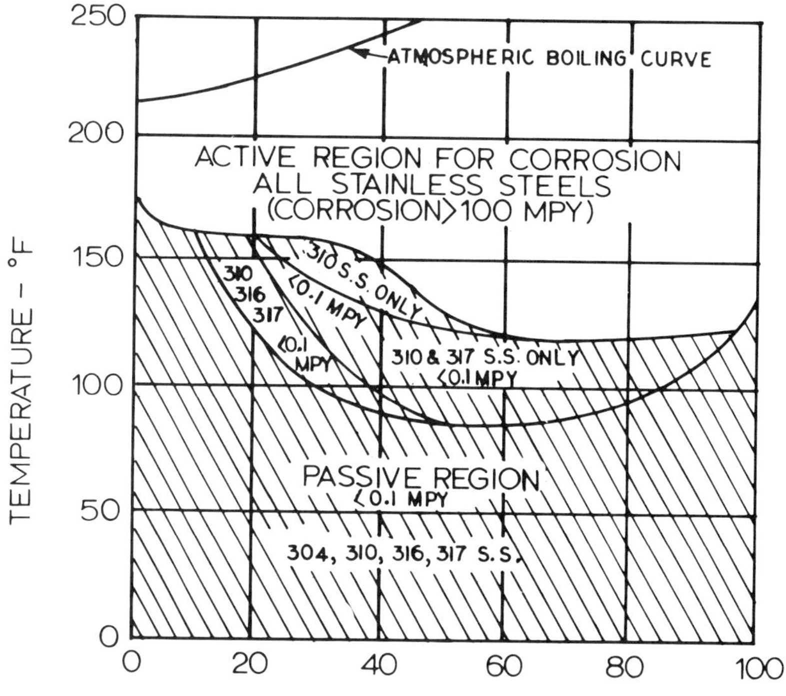

Figure 37.4. Sulfuric acid wt %.[2]

Stainless steels can be used on coal-fired boiler scrubbers, but many operators are unwilling to send the required amount of water to drain. These situations then require exotic metals such as titanium, hastelloy, etc.

FRP (fiberglass reinforced polyester) construction for scrubbers on coal-fired boilers is gaining increased acceptance as the resins are improved with regard to their ability to withstand high temperatures. Today's resins will withstand temperatures of 300° F with temporary excursions to much higher levels. Suitable quench chambers preceding the scrubbers, along with fail safe water spray provisions, protect against over temperature damage.

EQUIPMENT COSTS

The material cost relationship of the available particulate control devices is rarely easy to establish in such a manner as to be truly representative and meaningful. Conditions do vary from plant to plant; consequently, major design aspects, such as the desired collection efficiency, influence equipment selection parameters with a subsequent impact on material costs. It is possible, however, to give an order of magnitude comparison as follows:

Cyclonic Collectors (2-state):	\$ 20,000
Wet Scrubbers:	\$ 25,000
Fabric Filters:	\$100,000
Electrostatic Precipitators:	\$100,000

The cost disparity between the equipment selections is immediately evident. Obviously, if compliance with air pollution codes can be satisfied with an expenditure of \$20,000, rather than \$100,000, then this is the direction to pursue initially.

SULFUR DIOXIDE REMOVAL

At the present time, most SO_2 removal systems are confined to the large utility boilers. These are, for the most part, limestone or lime scrubbing.

Industrial boilers under 250,000,000 Btu/hr are not covered by federal regulations and are subject to state standards which vary considerably. For this reason, there are very few smaller non-utility boilers using SO_2 removal system. Those that are controlled for the most part are using the so-called "double alkali" system. The double alkali system involves scrubbing the flue gas with a soluble alkali such as sodium sulfite, which does not have the same scaling and plugging tendencies as do the calcium compounds of lime and limestone. The scrubbing liquor is regenerated with an insoluble alkali such as lime, producing a precipitate of calcium sulfite that is thrown away.

REFERENCES

1. Muller, Peter. *Chemie Ing. Techn.* 31 (5) (January 6, 1959).
2. "Twentieth Biennial C.E. Report," *Chem Eng.* (November 12, 1962).

38

Catalysis in Air Pollution Control

Erwin C. Betz
Catalytic Products International, Inc.

Catalysis is a technical process like flame combustion, filtration, decantation or evaporation. Although discovered as early as 1820, it was confined in its application to rather involved chemical, especially petrochemical, reactions.

Around the year 1950, catalytic oxidation was first applied as a production tool to convert hydrocarbons to harmless carbon dioxide and water vapor. This catalytic oxidation was developed and undertaken for controlling hydrocarbon emissions to the atmosphere, and for recovering otherwise wasted energy. It is the intent of this chapter to touch briefly on the principle of catalysis, and to describe catalytic systems, as they are practically applied to nuclear power plant effluents.

CATALYSIS

A catalyst, in the context of this discussion, is a substance which initiates the reaction of hydrocarbons with oxygen at 250°–650°F, and sustains this reaction to completion without being consumed itself. Depending on the physical state of the reactants, this oxidation can be carried out as a single or multi-phase reaction.

When catalysts are appraised, it is necessary to distinguish between the catalyst proper, and the carrier material, or substrate. Precious metals of the platinum group are generally used as catalysts. The carrier is either a metal or a ceramic material.

Material for the catalyst must have the following characteristics: a) resistance to oxidation, b) poor affinity for other chemicals, and c) low rates of recrystallization.

The carrier must have the following properties: a) high-temperature resistance, b) good mechanical and thermal shock resistance,

c) high thermal conductivity, and d) negligible liquid absorption rates.

In general use today are metal ribbons, wire mesh, metal wool, or shavings, and performed ceramic honeycomb structures, as well as spherical or cylindrical pellets. The selection of a catalyst must take into consideration the long-range reaction efficiency and the installed cost. Resuse of catalysts is also an important factor. Metallic catalysts can generally be recycled and reused in total, while ceramics permit recycling part of the precious metals only.

Performed catalysts of the metal and honeycomb type are generally built in modular form and contained in a mounting frame, while loose catalyst material is contained in a self-supporting cartridge. Whatever arrangement is selected, it should be easily dismountable.

APPLICATION—NUCLEAR INDUSTRY

The utilization of nuclear energy is generally accompanied by the release, or generation, of radioactive gases. Nuclear boiling water reactor systems release hydrogen and oxygen from the dissociation of water. Noble gases, which have a much longer half-life are also generated. These gases must lose their radioactivity before they are released to the atmosphere. To accomplish this decay, however, the easily flammable hydrogen and oxygen must be removed. This is another application where catalytic combustion, as a process tool, has proved most reliable.

Hydrogen and oxygen are recombined catalytically at temperatures as low as 150°F, and at concentrations near the explosive limit (Figure 38.1). The end purity of gases leaving the catalytic reactor must produce a bone-dry vent gas which contains less than 1 ppm of hydrogen. This application offers an excellent opportunity to compare the alternate method to the catalytic process; namely, thermal oxidation. Since the auto-ignition temperature is 1076°F, and the expected temperature rise from the oxidation of hydrogen is 600°F, the temperature of the thermal system can reach 1676°F. At this temperature, metal recrystallization jeopardizes the mechanical stability of a thermal recombiner system.

It is also important to point out the generation of halogens, especially iodine, during the nuclear process. Halogens will react with the catalyst and render the latter ineffective. They also weaken the crystalline structure of metals and increase the chance for failure on thermal recombiners. Therefore, halogens must be removed before attempting either catalytic or thermal destruction of hydrogen/oxygen. This is reliably accomplished with the use of scavengers.

These are suitable carriers, such as zeolite, charcoal, or Cr-Al ribbon coated with a halogen getter substance. These getters have

Figure 38.1. Nuclear recombiner catalyst assembly.

also proven useful in the removal of other catalyst poisons, such as lead, zinc, silicone, mercury, arsenic and phosphorus.

In summary, the fact of acceptance for the catalytic combustion principle to assure the environmental safety of the heavily debated nuclear power plants is most encouraging to those who seek out other applications for catalysis as an energy-saving production tool.

Index

C

I

L

M

N

O

P

R

X

Z